SOLAR GAMMA-, X-, AND EUV RADIATION

(1) J. Seibold
(2) J. M. Fontenla
(3) H. S. Ghielmetti
(4) S. Yousef
(5) H. Molnar
(6) G. M. Simnet
(7) C.-C. Cheng
(8) W. N. Glencross
(9) M. Machado

(10) G. Brueckner
(11) R. Ramaty
(12) T. Paneth
(13) J. T. Gosling
(14) R. P. Lin
(15) C. Y. Fan
(16) T. Takakura
(17) D. Datlowe
(18) C. Jordan

(19) J. Culhane
(20) J. Sahade
(21) G. Doschek
(22) R. Marabini
(23) L. de Feiter
(24) H. S. Hudson
(25) R. Hutcheon
(26) P. Pye
(27) J. H. Parkinson

(28) L. Golub
(29) S. Kahler
(30) A. Krieger
(31) Z. Švestka
(32) S. de Miceli
(33) D. Rust
(34) J. Underwood
(35) T. Gegerly
(36) J. Vorphal

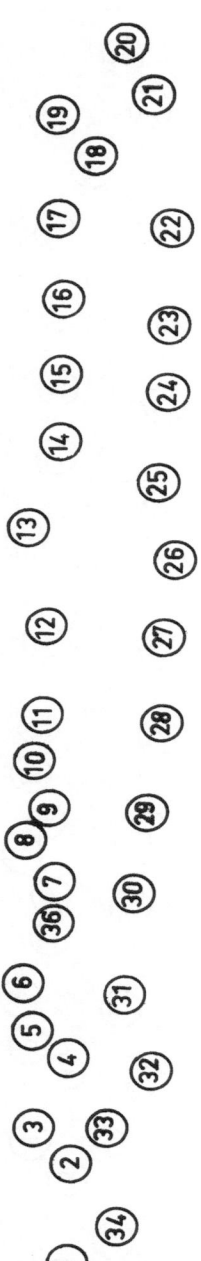

INTERNATIONAL ASTRONOMICAL UNION
UNION ASTRONOMIQUE INTERNATIONALE

SYMPOSIUM No. 68

HELD IN BUENOS AIRES, ARGENTINA, 11–14 JUNE 1974
ORGANIZED BY THE IAU IN COOPERATION WITH COSPAR

SOLAR GAMMA-, X-, AND EUV RADIATION

EDITED BY

SHARAD R. KANE

Space Sciences Laboratory, University of California, Berkeley, Calif., U.S.A.

D. REIDEL PUBLISHING COMPANY

DORDRECHT-HOLLAND / BOSTON-U.S.A.

1975

Library of Congress Cataloging in Publication Data

Main entry under title:

Solar gamma-, X-, and EUV radiation.

      (Symposium – International Astronomical Union ;
no. 68) Includes bibliographies.
      1.   Solar radiation — Congresses.   I.   Kane,
Sharad R. II. International Astronomical Union.
III. International Council of Scientific Unions. Com-
mittee on Space Research. IV. Series: International
Astronomical Union. Symposium ; no. 68
QB531.S57   523.7′2   75–6545
ISBN-13:978-90-277-0577-8   e-ISBN-13:978-94-010-1802-9
DOI: 10.1007/978-94-010-1802-9

Published on behalf of
the International Astronomical Union
by
D. Reidel Publishing Company, P.O. Box 17, Dordrecht, Holland

Sold and distributed in the U.S.A., Canada, and Mexico
by D. Reidel Publishing Company, Inc.
306 Dartmouth Street, Boston,
Mass. 02116, U.S.A.

TABLE OF CONTENTS

PREFACE   IX

LIST OF PARTICIPANTS   XI

PART 1:

GENERAL SOLAR ACTIVITY,
CORONAL HOLES AND BRIGHT POINTS

R. W. NOYES, P. V. FOUKAL, M. C. E. HUBER, E. M. REEVES, E. J. SCHMAHL,
J. G. TIMOTHY, J. E. VERNAZZA, and G. L. WITHBROE / EUV Observations
of the Active Sun from the Harvard Experiment on ATM   3
W. M. GLENCROSS / Holes in the Solar Corona   19
L. GOLUB, A. S. KRIEGER, J. K. SILK, A. F. TIMOTHY, and G. S. VAIANA / Time
Variations of Solar X-ray Bright Points   23
R. J. THOMAS / Solar Activity Observed in X-Rays and the EUV from OSO 7   25

PART 2:

ACTIVE REGIONS

J. H. PARKINSON / X-Ray Spectra of Solar Active Regions   45
J. P. PYE, R. J. HUTCHEON, J. H. PARKINSON, and K. A. POUNDS / X-Ray Spec-
troscopy of Solar Active Regions During the Third Skylab Mission   65
R. C. CATURA, L. W. ACTON, E. G. JOKI, C. G. RAPLEY, and J. L. CULHANE /
Spatially Resolved X-Ray Spectra of Coronal Active Regions   67
R. J. HUTCHEON / Classification of New Spectral Lines of Fe XVII Observed in
Solar Active Regions   69
S. YOUSEF / Statistical Methods in the Identification and Prediction of the Solar
X-Ray Spectral Lines   71
A. B. C. WALKER, JR. / Interpretation of the X-Ray Spectra of Solar Active
Regions   73
D. H. BRABBAN, E. B. DORLING, W. M. GLENCROSS, and J. R. H. HERRING /
Soft X-Radiation from Single Active Regions   101
A. S. KRIEGER, R. C. CHASE, M. GERASSIMENKO, S. W. KAHLER, A. F. TIMOTHY,
and G. S. VAIANA / Time Variations in Coronal Active Regions   103
G. E. BRUECKNER / Flare-Like Ultraviolet Spectra of Active Regions   105
C. JORDAN / The Structure of Solar Active Regions from EUV and Soft X-Ray
Observations   109

PART 3:

SOLAR FLARES

G. E. BRUECKNER / Ultraviolet Emission Line Profiles of Flares and Active Regions   135

K. G. WIDING / Fe XXIV Emission in Solar Flares Observed with the NRL/ATM XUV Slitless Spectrograph   153

G. A. DOSCHEK / X-Ray and EUV Spectra of Solar Flares and Laboratory Plasmas   165

P. R. SENGUPTA / Association of X-Ray Flares with Solar Coronal Active Regions   183

S. W. KAHLER, A. S. KRIEGER, J. K. SILK, R. W. SIMON, A. F. TIMOTHY, and G. S. VAIANA / Studies of the Dynamic Structure and Spectra of Solar X-Ray Flares   185

I. CRAIG / On the Thermal Structure of the Flare-Produced Plasma   187

D. W. DATLOWE / The Relationship between Hard and Soft X-Ray Bursts Observed by OSO 7   191

D. W. DATLOWE and H. S. HUDSON / Relationship between Hard and Soft Solar X-Ray Sources Observed by OSO 7   209

S. W. KAHLER / Thermal and Nonthermal Interpretations of Flare X-Ray Bursts   211

P. HOYNG, J. C. BROWN, G. STEVENS, and H. F. VAN BEEK / High Time Resolution Analysis of Solar Flares Observed on the ESRO TD-1A Satellite   233

J. A. VORPAHL and T. TAKAKURA / Rise Time of Hard X-Ray Bursts   237

J. C. BROWN and H. F. VAN BEEK / Determination of the Height of Hard X-Ray Sources in the Solar Atmosphere by Measurement of Photospheric Albedo Photons   239

D. M. RUST / Inference of the Hard X-Ray Source Dimensions in the 1972, August 7, White Light Flare   243

J. C. BROWN / The Interpretation of Spectra, Polarization, and Directivity of Solar Hard X-Rays   245

L. D. DE FEITER / Solar Flare X-Ray Measurements and Their Relation to Microwave Bursts   283

T. TAKAKURA / Relation of Microwave Emission to X-Ray Emission from Solar Flares   299

R. TALON, G. VEDRENNE, A. S. MELIORANSKY, N. F. PISSARENKO, V. M. SHAMOLIN, and O. B. LIKIN / X- and γ-Ray Measurements during the 1972, August 2 and 7 Large Solar Flares   315

E. L. CHUPP, D. J. FORREST, and A. N. SURI / High Energy Gamma-Ray Radiation above 300 keV Associated with Solar Activity   341

R. KOGA, G. M. SIMNETT, and R. S. WHITE / Measurements of a Gamma-Ray Burst above 1 MeV   361

R. RAMATY and R. E. LINGENFELTER / Gamma-Ray Lines from Solar Flares   363

R. P. LIN / Fast Electrons in Small Solar Flares   385

C. Y. FAN, G. GLOECKLER, and D. HOVESTADT / Nuclei of Heavy Elements
from Solar Flares   411

C. C. CHENG and D. S. SPICER / Implications of NRL/ATM Solar Flare Observations on Flare Theories   423

H. S. HUDSON, T. W. JONES, and R. P. LIN / Nonthermal Processes in Large
Solar Flares   425

Z. ŠVESTKA / On the Acceleration Processes in Solar Flares   427

PREFACE

The symposium on 'Solar Gamma-, X- and EUV Radiation' was held at Buenos Aires, Argentina, from 11 June to 14 June 1974. It was sponsored jointly by the International Astronomical Union (IAU) and the Committee on Space Research (COSPAR). The Organization Committee responsible for the program consisted of Drs K. A. Anderson (Chairman), J. L. Culhane, G. Elwert, B. B. Fossi, S. L. Mandel-s'tam, W. M. Neupert, V. K. Prokofiev, and J. Sahade and representatives of COSPAR, Drs H. Friedman and Z. Švestka. During the symposium Dr Švestka kindly represented the chairman of the Organizing Committee who was unable to attend the symposium.

The local arrangements in Buenos Aires were made by Drs J. Sahade (Chairman), H. S. Ghielmetti, M. J. Gulich, H. Molnar, J. J. Tasso and N. Martinez Riva de Tropper.

This symposium brought together the observational and theoretical aspects of the Solar Gamma-, X-, and EUV Radiation and other related solar emissions such as radio and energetic particles. There were three specific topics for the symposium, viz. X-ray and EUV emissions from solar active regions, EUV, X- and Gamma-emissions from solar flares, and mechanisms of hard photon emissions. The large improvement in our understanding of the physical processes in the active regions and flares, made possible by the various spacecraft and ground-based observations during the past few years could be clearly seen from the papers presented during the Symposium. Although only a fraction of the Skylab observations were analyzed at that time, several Skylab experimenters discussed their measurements related to the active regions and flares as well as the newly discovered solar features such as *coronal holes* and *bright points*.

The order in which the papers are included in these proceedings is somewhat different from the order in which they were presented during the symposium. Most of the papers submitted for publication were specifically related to either active regions or flares. These are included in Parts 2 and 3 respectively of these proceedings. The few papers which dealt with several aspects of solar activity or discussed coronal holes or bright points are included in Part 1. It is hoped that this organization of the proceedings will help to bring into focus the physical process occurring in the two major forms of solar activity, viz. the active regions and flares.

The excellent work done by the Organizing Committee in preparing a good program for the meeting can be clearly seen from the large number of new results presented during the symposium. About sixty scientists from twelve different countries participated in the Symposium. A total of forty-five papers were presented indicating the large amount of research currently being done in the various aspects of the Solar Gamma-, X-, and EUV Radiations.

An important part of the proceedings of a symposium is the edited transcripts of the discussions following each paper. Dr Sahade made a great effort to record on magnetic tapes the discussions that followed the presentations of the various papers. However, due to technical difficulties, it was not possible to make transcripts which could be reproduced in these proceedings.

Publication of proceedings, such as these, is possible only with the cooperation and help from many people. I am particularly thankful to Dr K. A. Anderson, Chairman of the Organizing Committee and Dr Edith A. Müller, Assistant General Secretary of the IAU, for keeping me informed about the various aspects of the program and proceedings of the symposium. Material related to the list of participants, a photograph of the participants and tapes and transcripts of the discussions were kindly provided by Dr J. Sahade. Ms Carol Legge helped with a large part of the clerical work needed in the organization of the program and the preparation of the manuscript for publication. Finally, thanks are due to Mr J. F. Hattink of D. Reidel Publishing Company for his help in bringing the proceedings to the present published form.

Space Sciences Laboratory   SHARAD R. KANE
University of California
Berkeley, Calif., U.S.A.

LIST OF PARTICIPANTS

Albano, J., Observatorio Astronomico, La Plata, Argentina
Anzer, U., Max Planck Institute, Munich, West Germany
Azcarate, I., Instituto de Astronomia y Fisica del Espacio, La Plata, Argentina
Brown, J. C., Space Research Laboratory, Utrecht, The Netherlands
Brueckner, G., Naval Research Laboratory, Washington, D. C., U.S.A.
Castore de Sistero, Observatoria Astronomico, Cordoba, Argentina
Cheng, C. C., Naval Research Laboratory, Washington, D. C., U.S.A.
Culhane, J., Lockheed Palo Alto Research Laboratory, Palo Alto, Calif., U.S.A.
Datlowe, D., University of California, San Diego, Calif., U.S.A.
de Feiter, L., Space Research Laboratory, Utrecht, The Netherlands
Doschek, G., Naval Research Laboratory, Washington, D.C., U.S.A.
Fan, C. Y., University of Arizona, Tucson, Ariz., U.S.A.
Fontenla, J. M., Observatorio Nacional de Fisica Cosmica, San Miguel, Argentina
Fortis, A. M., Observatorio Nacional de Fisica Cosmica, San Miguel, Argentina
Garriott, O., Johnson Space Center, Houston, Tex., U.S.A.
Ghielmetti, H. S., Instituto de Astronomia y Fisica del Espacio, La Plata, Argentina
Glencross, W. M., University College, London, United Kingdom
Gosling, J. T., High Altitude Observatory, Boulder, Colo., U.S.A.
Golub, L., American Science and Engineering, Cambridge, Mass., U.S.A.
Hoyng, P., Laboratorium voor Ruimteonderzoek, Utrecht, The Netherlands
Hudson, H., University of California, San Diego, Calif., U.S.A.
Hutcheon, R. J., Leicester University, Leicester, United Kingdom
Jordan, C., Culham Laboratory, Abingdon, Berkshire, United Kingdom
Kahler, S., American Science & Engineering, Cambridge, Mass., U.S.A.
Konig, P., P. U. for C.H.E., Potchefstroom, South Africa
Krieger, A., American Science & Engineering, Cambridge, Mass., U.S.A.
Lawlor, S. McKenna, Bunboyne, Co. Meath, Ireland
Lin, R. P., University of California, Berkeley, Calif., U.S.A.
Machado, M., Observatorio Nacional de Fisica Cosmica, San Miguel, Argentina
Marabini, R., Observatorio Astronomico, La Plata, Argentina
deMiceli, M. C., Observatorio Nacional de Fisica Cosmica, San Miguel, Argentina
Molnar, H., Observatorio Nacional de Fisica, San Miguel, Argentina
Monteagudo, V., Instituto de Astronomica y Fisica del Espacio, La Plata, Argentina
Mugherli, V. J., Instituto de Astronomia y Fisica del Espacio, La Plata, Argentina
Noyes, R. W., Center for Astrophysics, Cambridge, Mass., U.S.A.
Paneth, T., Observatorio Nacional de Fisica Cosmica, San Miguel, Argentina
Parkinson, J., University of Leicester, Leicester, United Kingdom

Peretti-Molemaert, A., Observatorio Nacional de Fisica Cosmica, San Miguel, Argentina

Pye, J. P., University of Leicester, Leicester, United Kingdom

Ramaty, T., Goddard Space Flight Center, Greenbelt, Md., U.S.A.

Rust, D., Sacramento Peak Observatory, Sunspot, N.M., U.S.A.

Sahade, J., Instituto de Astronomia y Fisica del Especio, La Plata, Argentina

Selzer, E., Institute de Physique du Globe, Paris, France

Seibold, J., Observatorio Nacional de Fisica Cosmica, San Miguel, Argentina

Sengupta, P. R., Institute of Applied Manpower Research, New Delhi, India

Simnett, G. S., University of California, Riverside, Calif., U.S.A.

Sistero, R., Observatorio Astronomico, Cordoba, Argentina

Švestka, Z., American Science and Engineering, Cambridge, Mass., U.S.A.

Takakura, T., Tokyo Astronomical Observatory, Tokyo, Japan

Thomas, R., Goddard Space Flight Center, Greenbelt, Md, U.S.A.

Underwood, J. H., Aerospace Corporation, Los Angeles, Calif., U.S.A.

Valcinek, B., Astronomical Institute of Czechoslovak Academy, Ondřejov, Czechoslovakia

Vedrenne, G., Centre d'Etude Spatiale des Rayonnements, Toulouse, France

Vorpahl, J., Aerospace Corporation. Los Angeles, Calif., U.S.A.

Widing, K., Naval Research Laboratory, Washington, D. C., U.S.A.

Yousef, S., Cairo University, Cairo, Egypt

Vernazza, J., Center for Astrophysics, Cambridge, Mass., U.S.A.

Walker, A. B. C., Stanford University, Stanford, Calif., U.S.A.

PART 1

GENERAL SOLAR ACTIVITY, CORONAL HOLES AND BRIGHT POINTS

EUV OBSERVATIONS OF THE ACTIVE SUN FROM THE HARVARD EXPERIMENT ON ATM

R. W. NOYES, P. V. FOUKAL, M. C. E. HUBER*, E. M. REEVES,
E. J. SCHMAHL, J. G. TIMOTHY, J. E. VERNAZZA, and G. L. WITHBROE

*Center for Astrophysics, Harvard College Observatory and
Smithsonian Astrophysical Observatory, Cambridge, Mass. 02138, U.S.A.*

Abstract. In this paper we review some preliminary results from the Harvard College Observatory Extreme Ultraviolet Spectroheliometer on ATM that pertain to solar activity. The results reviewed here are described in more detail in other papers referred to in the text. In the following paragraphs we first describe the instrument and its capabilities, and then turn to results on active regions, sunspots, flares, EUV bright points, coronal holes, and prominences.

I. The Instrument

The HCO scanning spectroheliometer consists of an 18-cm f/12.5 off-axis paraboloid feeding a normal-incidence concave grating spectrometer through a 5″-square entrance aperture. Seven detectors, located in the spectrometer focal plane, simultaneously record emission from a variety of wavelengths in the range 300 Å to 1335 Å. In the 'polychromatic' position of the grating, the seven wavelengths correspond to strong emission lines or continua ranging in temperature of formation from 10^4 K to 1.6×10^6 K, and other emission features can be selected by repositioning the grating. After the grating position is selected, the imaging mirror can perform a two-dimensional raster in angle, building up a set of simultaneous spectroheliograms each covering a 5′-square area with 5″ resolution. The time required to perform such a raster is 5.5 min. Alternatively, a one-dimensional scan, covering a region 5′ long but only 5″ wide, may be performed in 5.5 s, thus allowing data with high time resolution to be obtained for flares and other transient phenomena. A third observational mode is to fix the imaging mirror and position the ATM so that a solar feature of interest lies on the entrance slit, and rotate the grating continuously to build up a spectrum of the feature over the region 300–1335 Å with 1.5 Å spectral resolution and 3.3 min temporal resolution. For more details on the experiment and its capabilities see Reeves *et al.* (1974).

II. Active Regions

The HCO EUV observations of active regions are valuable because they can be used to study simultaneously the chromospheric and coronal layers where many solar transient phenomena occur. Figure 1 shows spectroheliograms constructed from digital data using a computer-driven cathode-ray tube display. Two active regions near the east limb are visible in these pictures, which illustrate how dramatically the

* On leave from ETH, Zürich, Switzerland.

Sharad R. Kane (ed.), Solar Gamma-, X-, and EUV Radiation, 3–17. All Rights Reserved.

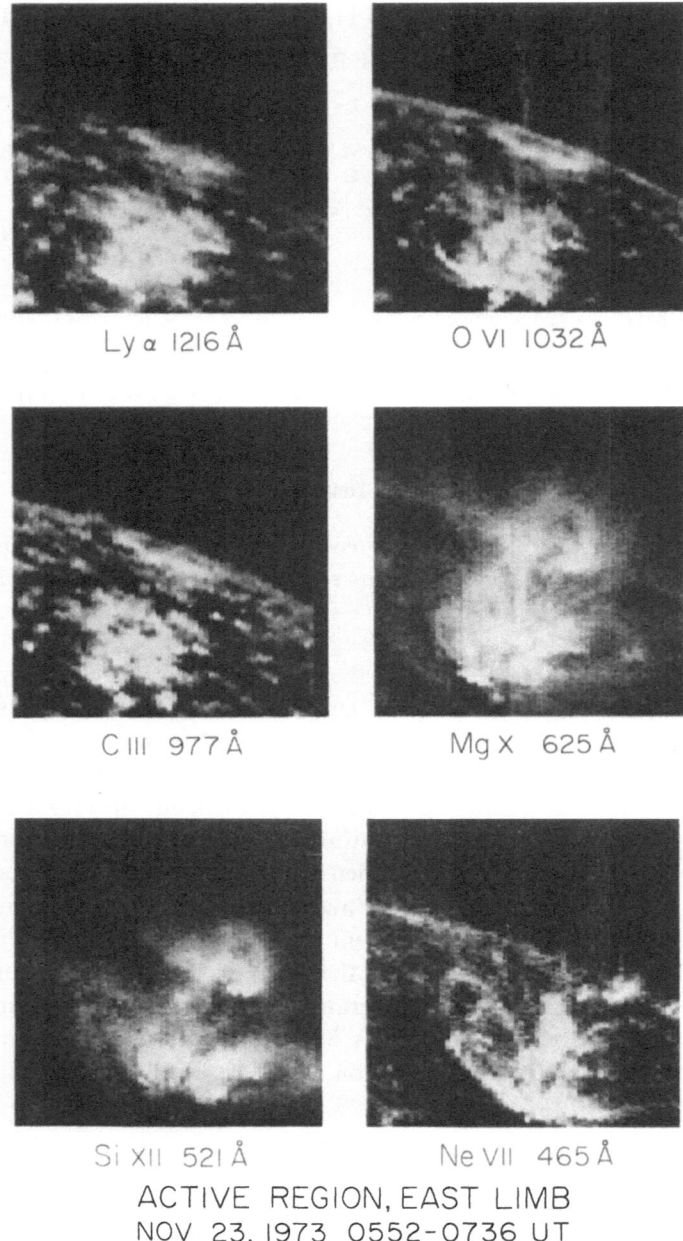

ACTIVE REGION, EAST LIMB
NOV 23, 1973 0552-0736 UT

Fig. 1.   Active region near the limb, viewed by the HCO instrument in its 'polychromatic' grating position. Images cover 5′ square, with 5″ spatial resolution, and are gray-level equivalents of the original digital data.

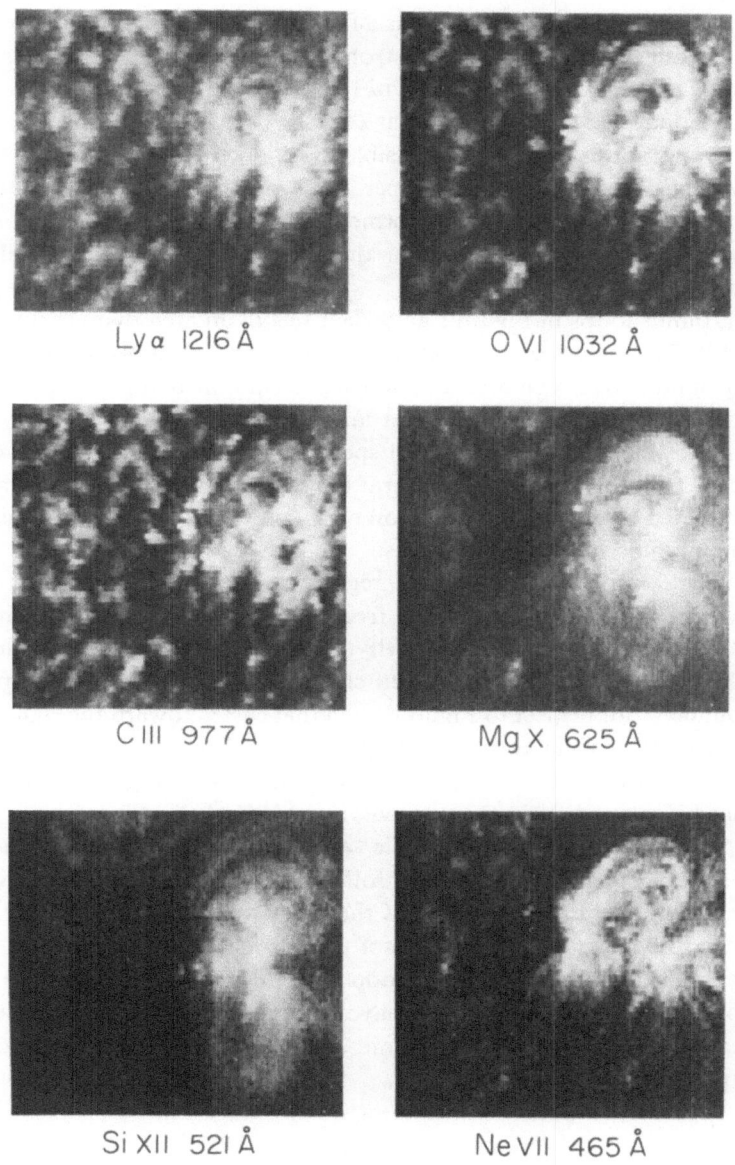

ACTIVE REGION LOOPS
NOV 24, 1973 0208-0235 UT

Fig. 2.   Same active region viewed 18 h later, showing the development of loop structures.

appearance of an active region changes as a function of temperature. The temperatures at which the different lines are most strongly emitted are 2×10^4 K for Lα $\lambda 1216$, 8×10^4 K for C III $\lambda 977$, 3×10^5 K for O VI $\lambda 1032$, 6×10^5 K for Ne VII $\lambda 465$, 1.5×10^6 K for Mg X $\lambda 625$, and 2.5×10^6 K for Si XII $\lambda 521$.

Several coronal loops or arches are visible in Figure 1. Generally these features are most easily seen in lines formed at temperatures of a few hundred thousand degrees and higher. Loops observed in lines formed between 10^5 K and 10^6 K are usually sharper and have higher contrast than the loops seen in lines formed at higher temperatures.

The foregrounds loops in Figure 1 have their feet in an area that is extremely bright in Ne VII $\lambda 465$. That area lies over and adjacent to a large sunspot, the leading sunspot in the foreground active region. The following sunspot also appears to be marked by bright Ne VII emission, in this case by a large column or plume of Ne VII emission visible in the lower center of the Ne VII spectroheliogram. This column of material remained over the active region for several hours and must have a temperature close to 6×10^5 K, the temperature of formation of Ne VII, since the feature is nearly absent in spectroheliograms made in other lines.

Coronal loops associated with active regions often exhibit significant changes in EUV brightness on time scales ranging from minutes to hours. These changes may or may not be associated with flare activity. Figure 2 illustrates how the loops in the region in the foreground of Figure 1 had changed in about 18 h. These pictures are rotated about 60° with respect to Figure 1 such that east is toward the right in Figure 2 and toward the upper right corner of the pictures in Figure 1. The series of loops on the southern side of the active region (toward the left in Figure 1 and the top of Figure 2) has brightened up in O VI, Ne VII, and Mg X. As in Figure 1 the feet of the loops are located in an area bright in Ne VII that lies over and adjacent to the large leading sunspot in the active region. The following sunspot is located below the small dark area near the right (eastern) side of the plage observed in C III $\lambda 977$. This spot also appears to lie near the feet of some of the coronal loops.

The increased EUV brightness of the loop configuration on the southern side of the active region as seen in Figure 2 may have been influenced by flares that occurred near the leader sunspot between the times of Figures 1 and 2, as described in a separate paper (Withbroe, 1975).

III. Flares

The HCO spectroheliometer has the capability to make observations of emission at several characteristic temperatures simultaneously, with high temporal resolution (5.5 s), and with high spatial resolution (5″) along a one-dimensional scan of 5′ length. By revealing the comparative behavior of the emission at different temperatures, such observations can restrict the range of possible mechanisms operative in the flare.

Preliminary analysis of such data (Noyes et al., 1975) indicates that the impulsive rise in EUV emission occurs essentially simultaneously at all levels from the transition zone to the corona, at least to within the 5.5 s resolution of the instrument. In many cases the simultaneous rise extends down to the Lα chromosphere, although in others (Figure 3) the chromospheric emission is delayed relative to the transition zone and coronal emissions and is not impulsive. However, two-dimensional rasters of flares (Withbroe, 1975) reveal that chromospheric flares tend to occur at the footpoints of

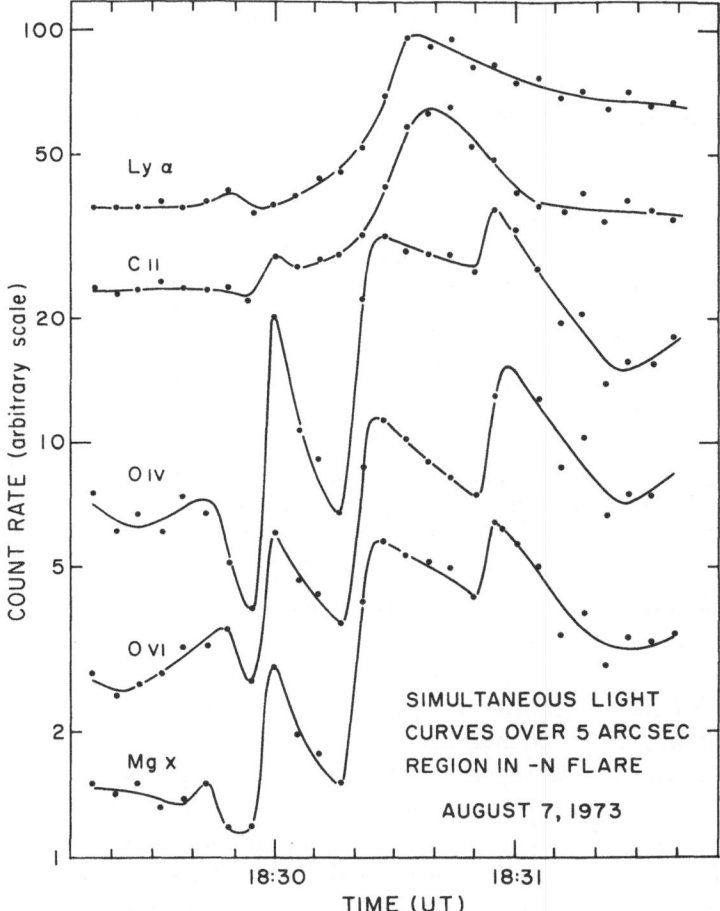

Fig. 3.   Simultaneous light curves of a 5″ region in a small flare, showing impulsive brightening simultaneously in transition zone and corona, but time lags in the underlying chromosphere.

higher-temperature loops, which may be inclined to the vertical; therefore it seems possible that the lack of simultaneity in some of the high-time-resolution one-dimensional scans results from the scan slightly missing the chromospheric footpoint.

The rise time of impulsive emission in flares at all levels has been observed to be quite rapid, often with *e*-folding time shorter than the 5.5 s time resolution of the instrument.

As expected, the maximum EUV intensity of flares, when averaged over the 5″ entrance slit of the ATM instrument, is higher than previous measurements using instruments with poorer spatial resolution that were less able to isolate the peak flare emission. In fact, for many flares the ATM instrument either saturated or tripped out when observing strong emission lines. From a spectrum near the peak of a small flare we obtain preliminary intensities (subject to slight modification when the final instrument calibration is available) as shown in Table I.

TABLE I

Intensities near maximum of 1 N flare, 1973 Sept. 2, 0236 UT, averaged over 5″ scan aperture

Line	T_{max}[a]	I_{flare}[b] (erg cm^{-2} s^{-1} sterad^{-1})	I_{quiet}[c] (erg cm^{-2} s^{-1} sterad^{-1})	I_{flare}/I_{quiet}
Lα	2×10^4	$>7.5 \times 10^5$	64000	>12
Lβ	2×10^4	2×10^5	730	270
LC, integrated	8×10^3	1×10^6	4000	250
N III 991	1×10^5	1.44×10^5	56	2500
O VI 1032	3×10^5	$>3.3 \times 10^5$	330	>1000
O V 760	2×10^5	4.4×10^4	44	1000
O V 630	2×10^5	6.7×10^4	525	125
Mg X 625	1.6×10^6	6.5×10^3	30	200
Fe XVI 335	3×10^6	1.0×10^5	<600	>160

[a] Approximate temperature of maximum emission in non-flaring Sun.
[b] Spectral scan, 1973 Sept. 2, 0235 UT, using preflight intensity calibration.
[c] Spectral scan, disk center, 1973 Aug. 31, 1313 UT, using preflight intensity calibration.

The data in Table I are only illustrative of flare intensities seen within a restricted 5″ aperture, since their relative values vary considerably from flare to flare (Noyes *et al.*, 1975).

The two O V lines have been included in Table I because their ratio is density-sensitive, and the greater enhancement of O V 760 over O V 630 in the flare is indicative of a large increase of density at a temperature of 200 000 K in the flare plasma. Calculations of Munro (1973) place the density well in excess of 10^{11} cm^{-3} although uncertainties in atomic data as well as the assumption of a statistically steady population of the O V energy levels prevent the assignment of a precise value.

The higher chromospheric densities in the flare plasma have an interesting effect on the hydrogen Lyman continuum emission spectrum (Noyes *et al.*, 1975). The slope and intensity of the continuum show it to be emitted nearly under LTE conditions (i.e., the high density drives the ground-state departure coefficient b_1 nearly to unity), although the electron temperature at optical depth unity in the Lyman continuum remains nearly unchanged from the quiet Sun value.

Localized flare brightenings, although seen simultaneously in lines with characteristic temperatures from chromospheric to coronal, are often followed by brightening in chromospheric and transition zone lines that move upward along pre-existing active region loops. This is suggestive of the ejection of chromospheric material into the loop structures (Withbroe, 1975). It is clear from the ATM data that flares are intimately connected with overlying coronal loop structures, and that a full empirical understanding of the interaction requires simultaneous data in several lines over a two-dimensional field with high temporal and spatial resolution.

IV. Sunspots

HCO observations of sunspots (Foukal *et al.*, 1974) reveal a very intense emission in transition zone lines (2×10^5 K $< T_{max} < 5 \times 10^5$ K, where T_{max} is the temperature of maximum relative abundance of the ion in question). The enhancement may reach an order of magnitude above that of the surrounding plage region, although it is generally unremarkable for chromospheric or coronal lines. Figure 4 shows a set of EUV

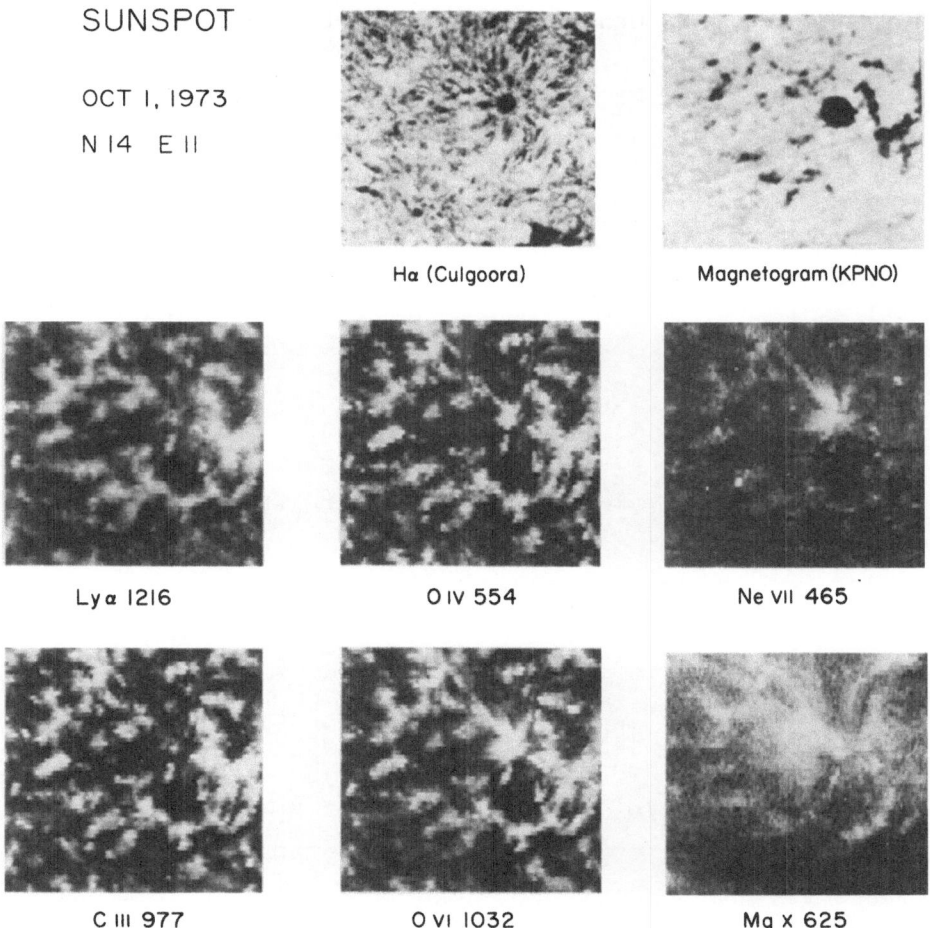

SUNSPOT

OCT 1, 1973

N 14   E 11

Hα (Culgoora)   Magnetogram (KPNO)

Lyα 1216   O IV 554   Ne VII 465

C III 977   O VI 1032   Mg X 625

Fig. 4.   Sunspot observed near disk center, showing the very large enhancement of lines formed between 2×10^5 K and 8×10^5 K.

rasters over an isolated sunspot near disk center, in which very strong emission from transition zone lines appears directly over the sunspot umbra. In order of increasing temperature of formation, the remarkably strong emission first appears in O IV and other lines formed around 2×10^5 K; the emission is very localized, to a region centered on, and smaller than, the photospheric umbra. The halfwidth of the emission is seen in Figure 3 to increase regularly with height throughout the transition region, probably reflecting the fact that the magnetic field is diverging with height above the umbra.

The EUV spectrum of sunspots (Noyes, 1975) is remarkable, showing great enhancements of lines of N IV, V, O IV, V, VI, Ne IV, V, VI, VII, Mg VI, VII, VIII, and S IV, VI. In addition, intersystem lines of O V, Ne VI, and Ne VII are well-observed (Figure 5).

The very large transition zone emission above spots could be due to a localized increase of density in the transition zone, were it not for the fact that the intensity

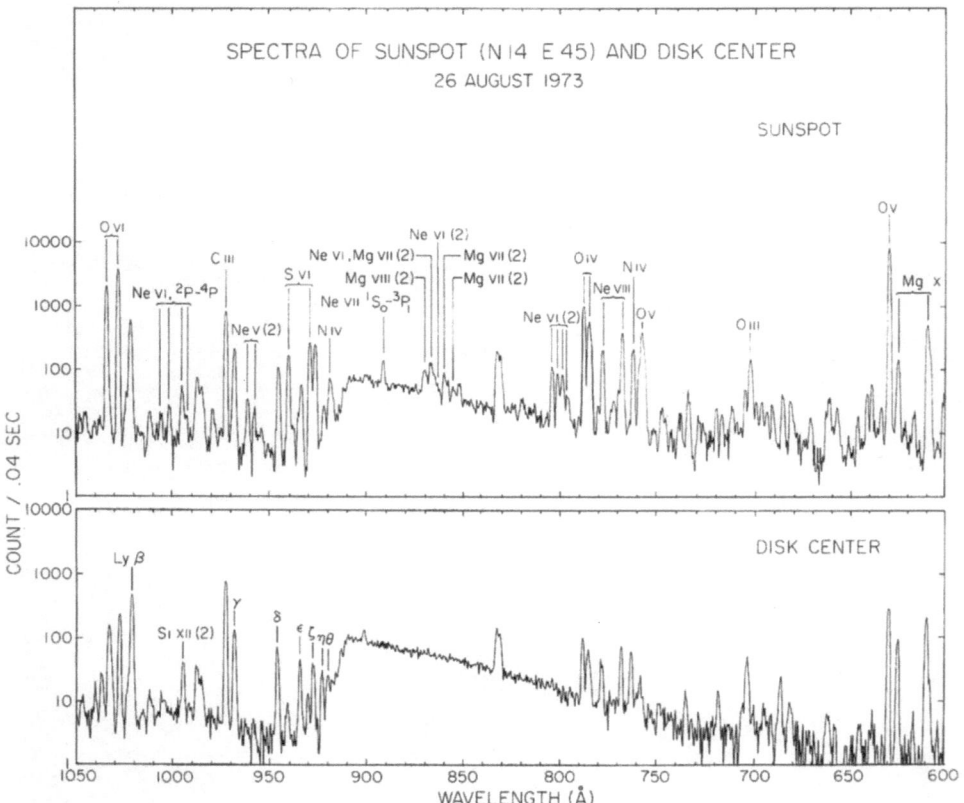

Fig. 5.   Spectrum of 5″ region over sunspot umbra, showing very large enhancement of lines with characteristic temperatures between 200 000 K and 800 000 K. The Ne VI intersystem transitions near 1000 Å are seen for the first time in this spectrum.

ratios of density-sensitive Be-like lines indicate if anything a *decrease* of density over that of the surrounding plage. A more reasonable interpretation is that the temperature gradient in the region 2×10^5 K $< T < 5 \times 10^5$ K above the umbra is decreased by up to an order of magnitude over that in the plage, thus increasing by a like factor the volume able to emit at a given temperature.

Although the intensity is enhanced by a factor of 10 in the transition zone over the umbra, the total emission is still negligible compared to the 'missing flux' from the photosphere in the spot umbra.

V. EUV Bright Points

Bright point sources of EUV radiation (Timothy *et al.*, 1975) have been observed in all locations on the disk. The sources can be most easily identified in Mg x 625.3 Å. Figure 6 shows a matrix of Mg x rasters covering the entire Sun, in which bright points appear to be uniformly distributed over the disk. Several hundred bright points, having dimensions of 30″ or less, can be seen in this image; many appear to be associated with bipolar magnetic-flux regions, based on an initial analysis of Kitt Peak National Observatory magnetograms taken on the same day (courtesy J. Harvey).

Bright points are most clearly observed in coronal holes (see Figure 7). Observations at the limb do not reveal any great differences in the apparent height of the peak of emission for Mg x 625.3 Å compared with that of H Lα 1215.7 Å. Consequently it is clear that the bright points do not extend to great distances in the corona and have a maximum altitude of the order of 1500 km or less. Some bright points are observed in transition region lines such as Ne vii 465.2 Å (see Figure 7), and yet are not visible in the coronal line Mg x 625.3 Å. Bright points which are clearly evident in transition region and chromospheric lines but invisible in coronal lines have also been observed outside of coronal holes. These points could represent a séparate class of features or an earlier stage in the development of the bright points observed in coronal lines. A detailed analysis of the complete life history of a coronal bright point will be necessary to resolve this question.

No systematic differences have been observed in the emission from bright points inside and outside coronal holes. Many points appear as double features at transition region and chromospheric temperatures and as a laterally displaced single feature at coronal temperatures. This strongly suggests that a bright point has a basic loop structure.

Spectra of bright points for the wavelength range 1336 Å to 360 Å show an increase in the emission at all wavelengths of at least a factor of five compared with the average emission from a quiet region on the disk. Qualitatively, bright points appear to have a density about a factor of four greater than that of the surrounding medium, but show only a very small increase in temperature compared to that of the quiet corona.

A significant fraction of bright points exhibit flaring phenomena (Figure 8). Here the emission increases by an order of magnitude for a total lifetime of the order of that of a single raster scan (330 s). In at least one example the point has been observed

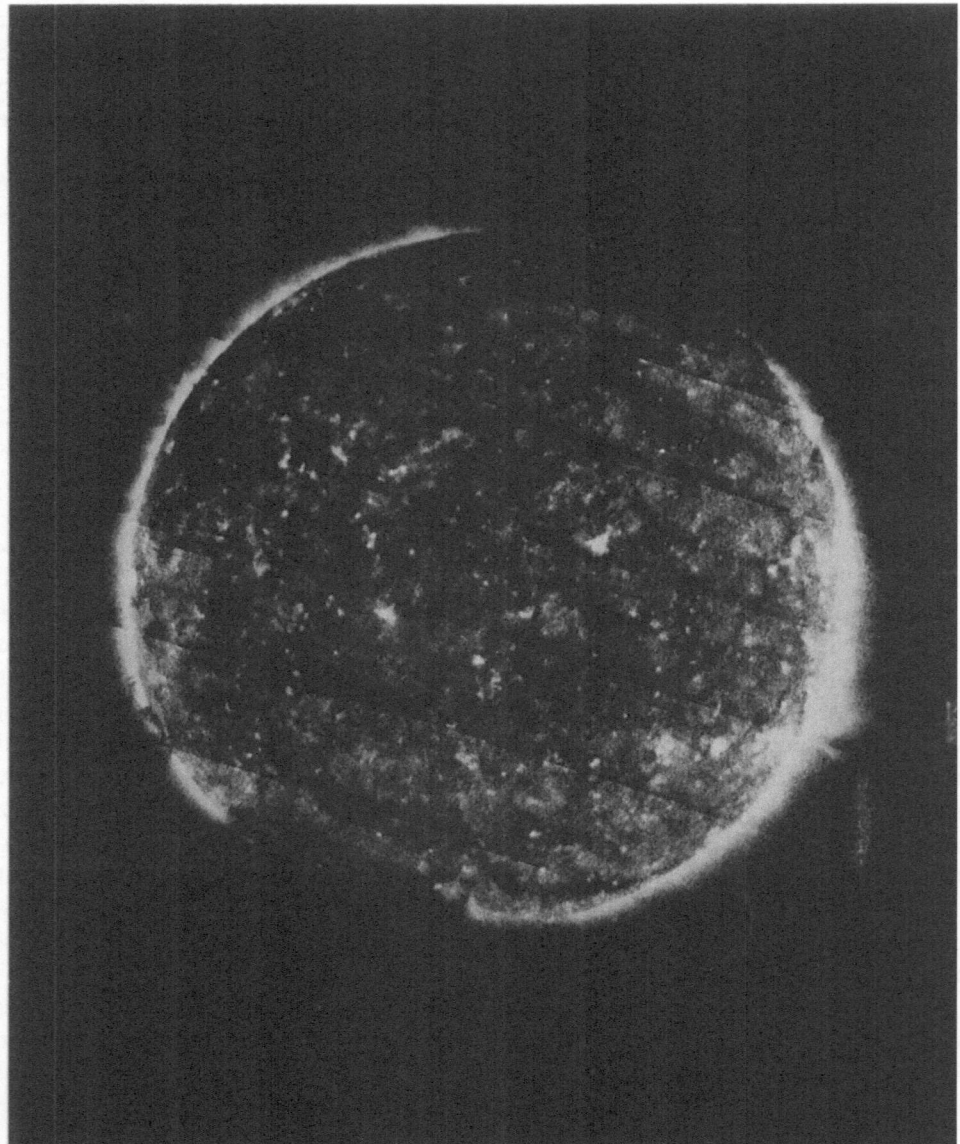

Fig. 6.   EUV bright points seen in Mg x 625, in a matrix of 5′ rasters covering the entire solar disk, observed 28 January 1974. The banded structure is due to a defect in the plotting device.

to flare initially at transition region wavelengths. Furthermore, for a large fraction of flaring events a bright point several thousand kilometers away has been seen to flare simultaneously. No connecting loop structures between these regions have so far been observed.

POLAR SCAN   13 DECEMBER 1973

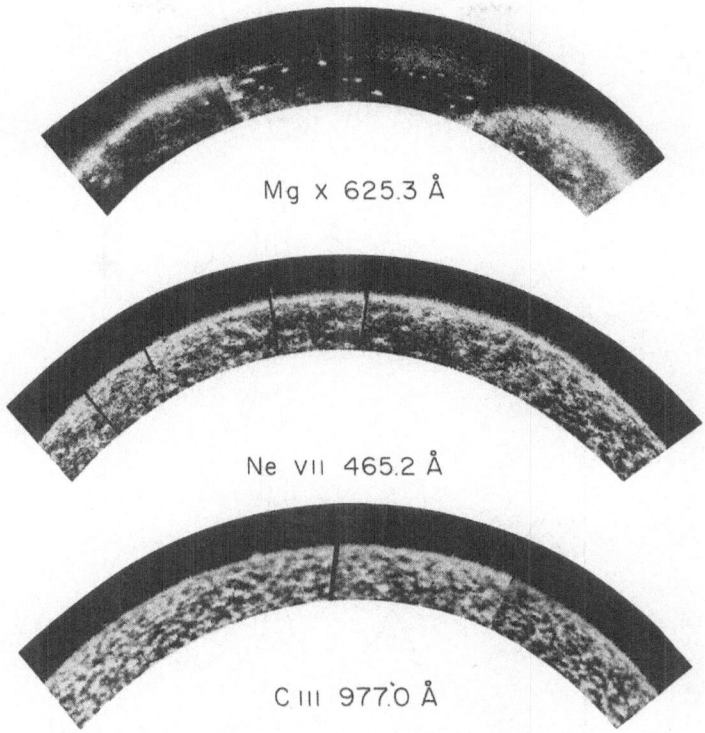

Fig. 7. Coronal hole at the south solar pole, showing EUV bright points in Mg x and a greater number in the transition zone line Ne vii.

We are now undertaking a detailed analysis of the structure of a series of bright points for which data have been recorded over the complete lifetime of several hours. These results will be correlated with X-ray and magnetic observations in order to produce a complete picture of the structure and energetics of these features in both the flare and the non-flare modes.

VI. Coronal Holes

Although coronal holes represent in one sense the antithesis of solar activity, their relevance for the overall structure of the general solar magnetic field and their apparent relation to the solar wind (Krieger *et al.*, 1973) are potentially important to several aspects of solar activity, so we include in this paper a short discussion of their properties as revealed by the HCO EUV data.

Coronal hole boundaries are obvious in coronal lines (Mg x 625 Å, for example) and are to some extent visible in lines formed above 7×10^5 K (e.g., Ne viii, Ne vii). The ATM observations revealed that coronal holes actually manifest themselves in

14   R. W. NOYES ET AL.

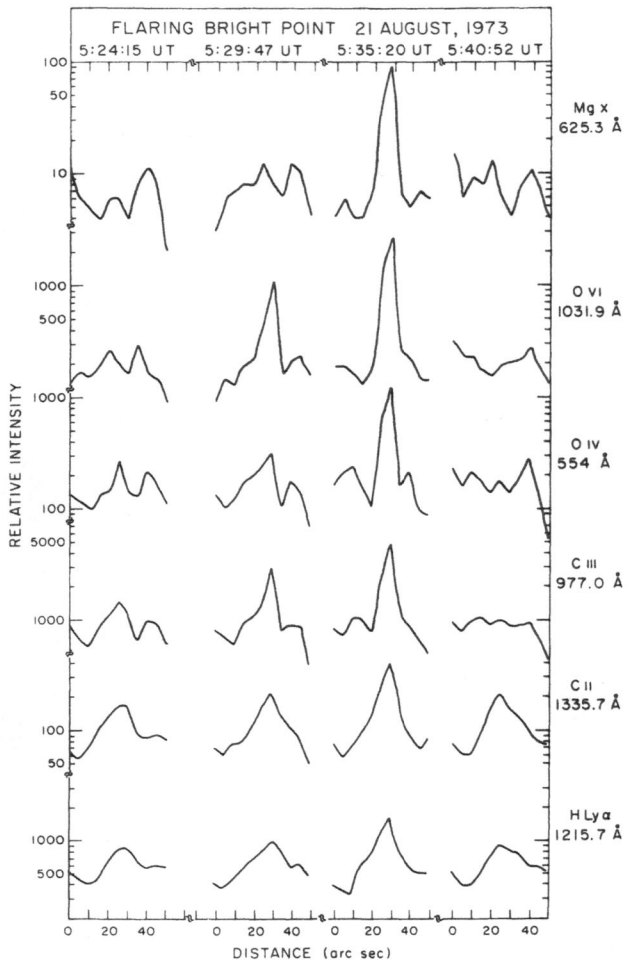

Fig. 8.   Traces across flaring EUV bright point in four successive rasters in the polychromatic
position. Note the early brightening in the transition zone.

all layers of the upper solar atmosphere (Huber *et al.*, 1974). In particular, the contrast
between supergranulation boundaries and cells as seen in spectral features emitted
by the chromosphere and transition zone is reduced inside coronal holes. The indications of lower density in the transition zone and chromosphere underlying coronal
holes, as derived from the density-sensitive intensity ratio of the 1176 and 977 Å lines
of C III (Munro and Withbroe, 1972) and from the color temperature of the Lyman
continuum (Vernazza and Noyes, 1972), respectively, have been confirmed.

    Furthermore, direct evidence for the resulting model having a thicker transition
zone at the lower density inside holes (Munro and Withbroe, 1972) was obtained from
limb observations of a polar coronal hole. Owing to the improved spatial resolution

available on the ATM instrument, we can directly observe the increased thickness of the transition zone as increased height of emission of transition zone lines. Figure 9 (Huber *et al.*, 1974) shows measured displacements of the limb above the polar coronal hole of Figure 7, relative to the nearby quiet atmosphere outside the hole. From this result we infer a reduction of downward thermal conduction in the hole by a factor of about six relative to the quiet Sun, in agreement with the earlier findings of Munro and Withbroe (1972).

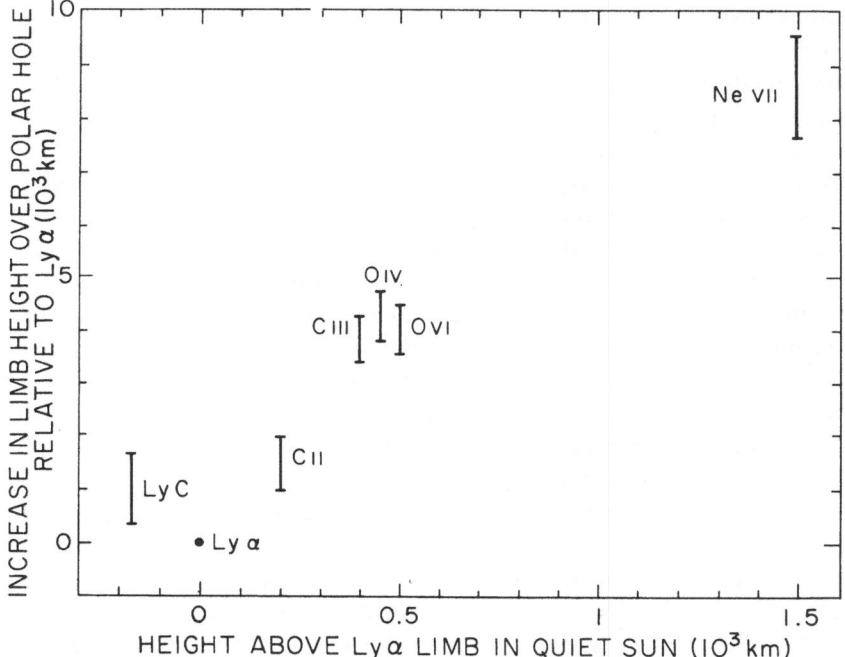

Fig. 9.   Increase in limb height of various lines over coronal hole of Figure 7, compared with nearby quiet limb.

VII. Prominences

HCO observations of quiescent prominences (Schmahl *et al.*, 1974) include spectroheliograms in the strong lines and continua of hydrogen and helium, by means of which the cool threads of prominences and filaments absorb and radiate most of their energy. In addition the data include spectroheliograms of many important lines radiated by the transition sheath between the threads and the surrounding corona. Prominences appear to have a cold central core ($\sim$ 6000–7000 K) surrounded by a thin transition zone which merges with the hot corona (Noyes *et al.*, 1972) analogous to the temperature profile of the normal chromosphere, transition zone, and corona. This similarity makes it possible to derive properties of prominences by comparisons of line intensities in prominences and in the quiet solar atmosphere.

When prominences are observed near the limb in Lα and the Lyman continuum, the limb is entirely obscured by the prominence. This implies that prominences are optically thick at these wavelengths, in agreement with theoretical expectations (Hirayama, 1964). Furthermore, the 5″ slit is filled by the emitting material. Simultaneous spectroheliograms at two wavelengths in the Lyman continuum determine a color temperature of the emitting hydrogen. This temperature lies between 5500 and 8000 K, with the lower values tending to occur near the middle and the higher values near the edges and top of prominences. This result confirms earlier, lower resolution EUV data (Noyes et al., 1972).

From the color temperature and the intensity of the Lyman continuum, we can determine the departure coefficient b_1 for the ground state of hydrogen. We expect b_1 to be of the order of unity since the kinetic temperatures are comparable to the radiation temperature, and this indeed is the case.

All of the lines (exclusive of helium) emitted by the hotter portions of prominences are optically thin, for prominences appear transparent in these lines when they are observed against the solar disk. The optical depth of prominences can be determined by the relative brightnesses of the prominences against the limb, just above the limb, and the limb itself. (The filling of the slit can be guaranteed by the opaque appearance of the prominence in Lα or the Lyman continuum.) For C III $\lambda 977$, the value of the optical depth ($\tau \approx 0.3$) is about half that found by Withbroe (1970) for the quiet transition zone. This implies that the column density of the C III-emitting prominence sheath (i.e., within the temperature range 5×10^4 K $\lesssim T \lesssim 1.2 \times 10^5$ K) is about half that of the quiet transition zone.

The thickness of the prominence transition sheath can be estimated from the ratio $\tau^2/I \propto (N_e \, dh)^2/N_e^2 \, dh$. This result yields a thickness of about 10 km for the C III transition sheath on either side of the threads. In addition, by the slightly different technique of comparing the O VI emission in the prominence with that in the quiet corona off the limb, we can deduce a thickness of about 40 km on either side of the threads for the temperature range 2×10^5 K $\lesssim T \lesssim 5 \times 10^5$ K. Both of these determinations lead to a prominence transition zone thickness about a factor of two thinner than that for the quiet solar atmosphere (Dupree, 1972).

The thickness inferred above, when combined with the observed intensities, implies an electron density about half that at the same temperature in the quiet solar atmosphere. This is a surprising result, which is not clearly understood. It is worth noting that the density as inferred independently from the density-sensitive line ratio $I(\text{C III } \lambda 1176)/I(\text{C III } \lambda 977)$ is also about half that of the quiet Sun at the same temperature.

The intensity of the optically thin lines formed at temperatures $> 2 \times 10^4$ K increases to a maximum near the top of prominences. This result suggests that the sheath thickness increases with height, or that the temperature gradient between the hydrogen threads and the corona decreases with height. A possible explanation is that the conductive flux inward from the corona is balanced locally by the radiative flux outward. Then, since the emissivity of the threads (as seen in the Lyman lines and the Lyman continuum) is a decreasing function of height, the conductive flux inward must also decrease with height, and hence the sheath increases in thickness to accomplish this.

Much of the analysis for quiescent prominences may be extended to active loops and filaments, surges, and eruptive prominences. Work is progressing along these lines in the analysis of ATM data.

Acknowledgments

The results reported here reflect the long labors of very many individuals, without which there would be no results to report. The authors wish to express their gratitude for the wisdom, foresight, dedication, and courage of the NASA administrators, scientists, engineers, and astronauts who made Skylab the success it was. Portions of this work were supported through NASA contract NAS 5-3949.

References

Dupree, A. K.: 1972, *Astrophys. J.* **178**, 527.
Foukal, P. V., Huber, M. C. E., Noyes, R. W., Reeves, E. M., Schmahl, E. J., Timothy, J. G., Vernazza, J. E., and Withbroe, G. L.: 1974, *Astrophys. J. Letters* **193**, L 143.
Hirayama, J.: 1964, *Publ. Astron. Soc. Japan* **16**, 104.
Huber, M. C. E., Foukal, P. V., Noyes, R. W., Reeves, E. M., Schmahl, E. J., Timothy, J. G., Vernazza, J. E., and Withbroe, G. L.: 1974, *Astrophys. J. Letters* **194**, L 115.
Krieger, A. S., Timothy, A. F., and Roelof, E. C.: 1973, *Solar Phys.* **29**, 505.
Munro, R. H.: 1973, Thesis, Harvard University.
Munro, R. H. and Withbroe, G. L.: 1972, *Astrophys. J.* **176**, 511.
Noyes, R. W.: 1975, to be submitted to *Astrophys. J.*
Noyes, R. W., Dupree, A. K., Huber, M. C. E., Parkinson, W. H., Reeves, E. M., and Withbroe, G. L.: 1972, *Astrophys. J.* **178**, 515.
Noyes, R. W., Foukal, P. V., Huber, M. C. E., Reeves, E. M., Schmahl, E. J., Timothy, J. G., Vernazza, J. E., and Withbroe, G. L.: 1975, to be submitted to *Solar Phys.*
Reeves, E. M., Timothy, J. G., and Huber, M. C. E.: 1974, *Proc. S.P.I.E.* **44**, 159.
Schmahl, E. J., Foukal, P. V., Huber, M. C. E., Noyes, R. W., Reeves, E. M., Timothy, J. G., Vernazza, J. E., and Withbroe, G. L.: 1974, *Solar Phys.* **39**, 337.
Timothy, J. G., Foukal, P. V., Huber, M. C. E., Noyes, R. W., Reeves, E. M., Schmahl, E. J., Vernazza, J. E., and Withbroe, G. L.: 1975, to be submitted to *Solar Phys.*
Vernazza, J. E. and Noyes, R. W.: 1972, *Solar Phys.* **26**, 325.
Withbroe, G. L.: 1970, *Solar Phys.* **11**, 208.
Withbroe, G. L.: 1975, in preparation.

HOLES IN THE SOLAR CORONA

W. M. GLENCROSS

Mullard Space Science Laboratory, Dept. of Physics and Astronomy,
University College London, WC1E 6BT, England

Summary. Babcock (1961) outlined the sequence of events which takes place in the Sun's atmosphere during a solar cycle. Magnetic field loops, having preferred directions, emerge from the solar surface and thereafter merge with neighbouring loops to produce more extended structures. Although flux tubes emerge with a strong E-W field component, having the field direction reversed from one side of the equator to the other, there is a tendency for the longer loops produced by merging to have a significant N-S alignment (Hansen *et al.*, 1972).

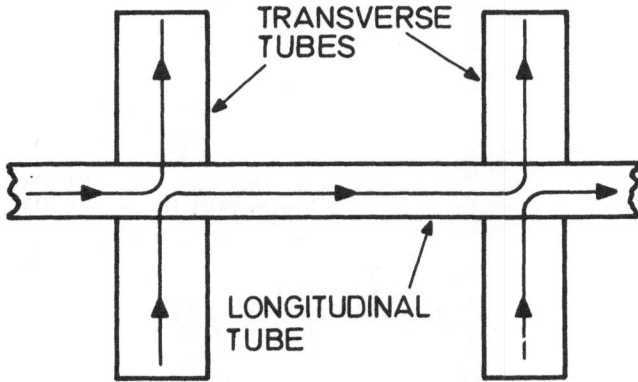

Fig. 1.   When each of a pair of parallel flux tubes makes contact with a perpendicular tube, it is possible for field lines (represented as curves with arrow heads) to cross-connect so as to pass from one member of the pair to the other.

Consideration is given here to how the Sun's field might develop once some long flux tubes have been formed with a predominantly N-S alignment. (Such tubes might be part of the general dipole field structure of the Sun). New magnetic arches, having a significant E-W component, emerge from the solar surface and come into contact with the older N-S field so that merging occurs between the approximately perpendicular ropes. Figure 1 shows a plan view along a short length of a N-S aligned flux rope (denoted as the 'longitudinal tube') when two transverse tubes develop beneath it. These latter structures will tend to have parallel fields if they are on the same side of the solar equator. Merging of fields could allow some flux lines to pass between one transverse tube and the other in the manner illustrated.

Figure 2a shows the situation when the N-S flux tube crosses the solar equator. Because transverse structures on either side of this boundary are anti-parallel, any

Sharad R. Kane (ed.), Solar Gamma-, X-, and EUV Radiation, 19–21. All Rights Reserved.
Copyright © 1975 by the IAU.

field line linking them has both foot-points on the same side of the longitudinal flux tube. Figure 2b shows a perspective view of the extreme case when the two anti-parallel flux tubes have merged completely via the longitudinal rope. In practice there would be cross-connection between numerous transverse structures on one side of the

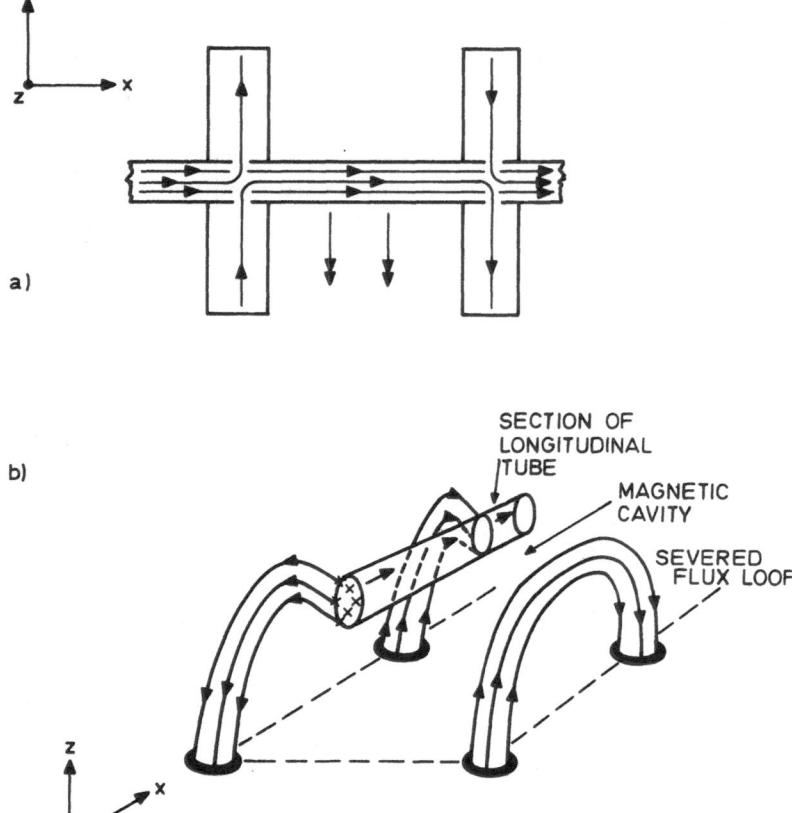

Fig. 2. Diagram (a) is similar to Figure 1, except the transverse arches are now anti-parallel. Diagram (b) is a perspective view of the extreme situation when all field lines from one transverse arch pass into the other.

equator and a similar number on the other side, with the resultant production of a long magnetic cavity.

   Conditions shown in Figure 2 can also be expected to form at intermediate latitudes in each hemisphere as a new solar cycle begins to develop. Flux tubes produced here will have polarities which are the reverse of those in the surrounding regions which are still associated with the earlier cycle.

   Apart from the coronal holes, it is known that some filaments also have well de-

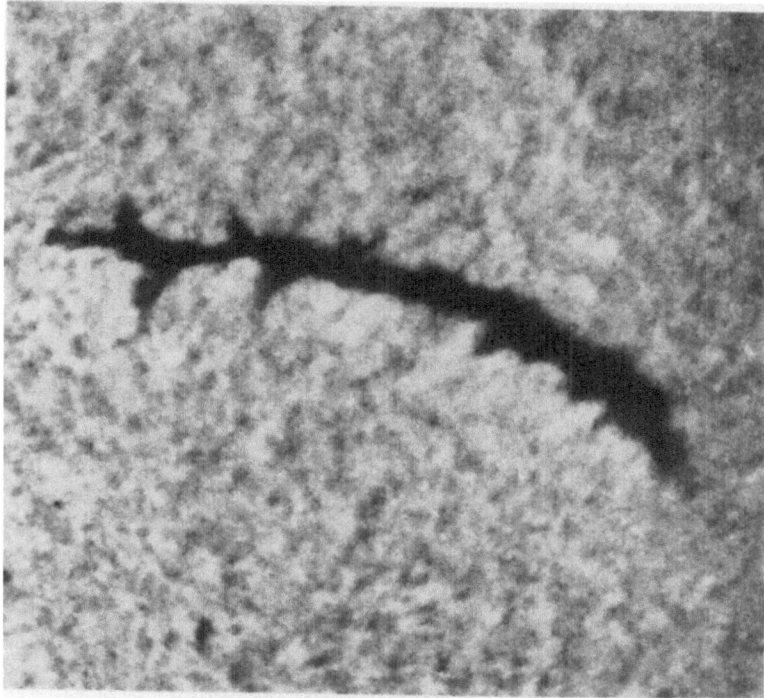

Fig. 3. Observations suggest that supporting magnetic field lines lie perpendicular to the axis of a quiescent prominence, although they are deflected to pass along the filament material itself. Where the filament shown here is viewed from above it seems possible to recognise some flux tubes lying transverse to the axis. If this interpretation is correct, the magnetic structure could be similar to that shown in Figure 1. Prominences formed in favourable locations might be expected to develop the geometry shown in Figure 2, in which case filament cavities might be formed by a mechanism similar to that of coronal holes.

veloped dark cavities associated with them when they are observed in soft X-radiation (Vaiana *et al.*, 1973). There is in fact observational evidence (Ioshpa, 1968) that some prominence structures have magnetic configurations composed of a longitudinal magnetic field, containing condensed material, which is supported by transverse fields along its length. If merging between fields develops, the geometries shown in Figures 1 and 2a might form. A short length of the filament in Figure 3 appears to show transverse flux ropes, containing condensed material, lying approximately perpendicular to the main line of the prominence.

References

Babcock, H. W.: 1961, *Astrophys. J.* **133**, 572.
Hansen, S. F., Hansen, R. T., and Garcia, C. J.: 1972, *Solar Phys.* **26**, 202.
Ioshpa, B. A.: 1968, in K. O. Kiepenheuer (ed.), 'Structure and Development of Solar Active Regions', *IAU Symp.* **35**, 261.
Vaiana, G. S., Kreiger, A. S., and Timothy, A. F.: 1973, *Solar Phys.* **32**, 81.

TIME VARIATIONS OF SOLAR X-RAY BRIGHT POINTS

L. GOLUB, A. S. KRIEGER, J. K. SILK, A. F. TIMOTHY

American Science & Engineering, Cambridge, Mass., U.S.A.

and

G. S. VAIANA

Center for Astrophysics, Cambridge, Mass., U.S.A.

Summary. An example of the overall view of the X-ray corona (nominal filter passband 2–32 Å and 44–54 Å) showing a coronal hole, filament activity, bright points and the large scale-scale loop structures, is shown in Figure 1. This is one of the 32 000 X-ray images obtained with the AS & E X-ray telescope on Skylab. A comprehensive review describing the characteristics of the various features and their implications regarding the high velocity solar streams, evolution of magnetic fields

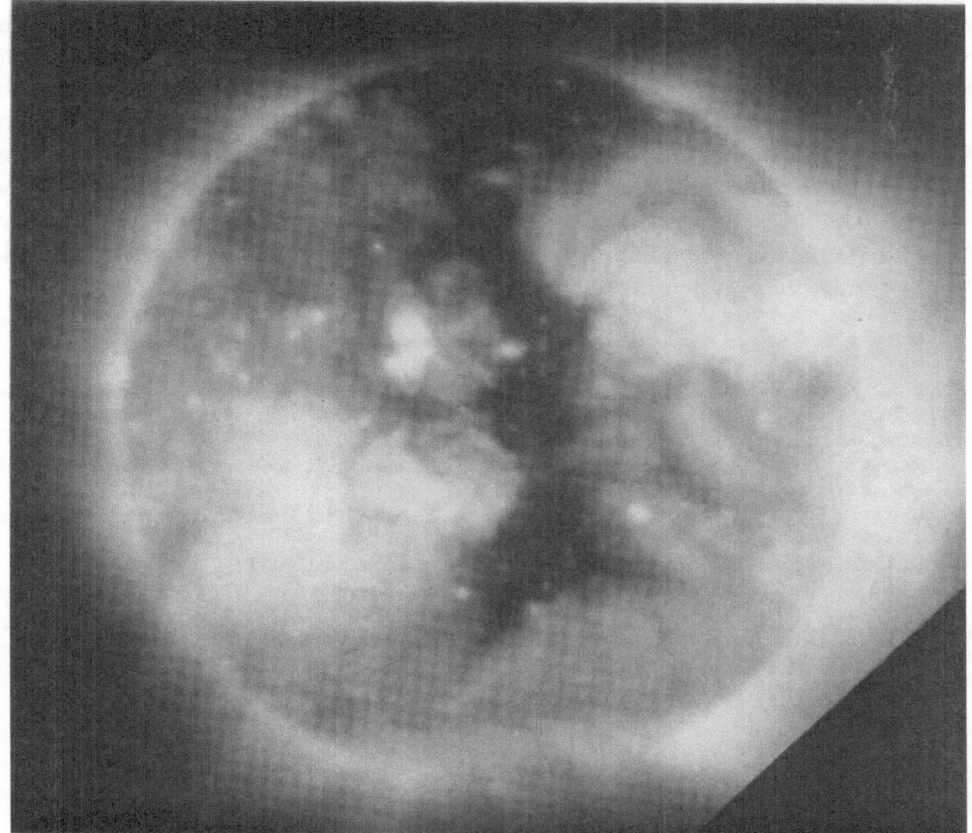

Fig. 1. Overall view of the X-ray corona on June 1, 1973 (nominal filter passband 2–32 Å and 44–45 Å), showing a coronal hole, filament cavity, bright points and the large-scale loop structures.

Sharad R. Kane (ed.), Solar Gamma-, X-, and EUV Radiation, 23–24. All Rights Reserved.
Copyright © 1975 by the IAU.

in active regions, and sources of soft X-ray emission has been given by Vaiana *et al.*
(1975). In the present summary we will only be concerned with the bright points.
Studies of solar X-ray bright points, show that these features represent a distinct
class of solar activity. Bright points appear first as a diffuse cloud of soft X-ray
emission typically growing to 30″ in diameter, with growth rates of ~ 1 km s^{-1}.
Several hours after the point first becomes visible a bright compact core forms,
growing to 10″. The lifetime distribution of bright points follows a Poisson distribution with a mean of eight hours (see references). The points are distributed uniformly
over the entire solar surface, with approximately 500 on the Sun at any time. Their
occurrrence appears to be independent of major active regions, except for a visibility
factor near high loop structures or a possible decrease in number in active region
latitudes.

Bright points show a one-to-one correspondence with bipolar magnetic features
when comparison is made with simultaneous magnetograms having arcsec resolution
and high sensitivity (< 10 G). Total flux values per point are 10^{19}–10^{20} Mx. There
is also a strong correlation between X-ray observations and those made in transition
region and in chromospheric lines, such as the EUV lines or Ca K. However, the
chromospheric observations show both bipolar and unipolar magnetic enhancements,
so that X-ray photos are needed to identify the locations of bright points.

Bright points are also seen to undergo flare activity at all latitudes from the equator
to the poles. Approximately 10% of all points flare during their lifetimes. A preliminary
study of these flares indicates variability approaching that found in major active
regions. Risetimes from < 10 s to > 10 min and decay times of < 1 to > 30 min have
been observed. Corresponding brightenings can be seen in chromospheric lines and
in high resolution Hα filtergrams.

References

Golub, L., Krieger, A. S., Silk, J. K., Timothy, A. F., and Vaiana, G. S.: 1974, *Astrophys. J. Letters*
    189, L93.
Harvey, K. L. and Martin, S. F.: 1973, *Solar Phys.* **32**, 389.
Krieger, A. S., Viana, G. S., and Van Speybroeck, L. P.: 1971, in R. Howard (ed.), 'Solar Magnetic
    Fields', *IAU Symp.* **43**, 397.
Vaiana, G. S., Davis, J. M., Giacconi, R., Krieger, A. S., Silk, J. K., Timothy, A. F., and Zombeck,
    M.: 1973, *Astrophys. J. Letters* **185**, L47.
Vaiana, G. S., Krieger, A. S., and Timothy, A. F.: 1973, *Solar Phys.* **32**, 81.
Vaiana, G. S., Krieger, A. S., Van Speybroeck, L. P., and Zehnpfenning, T.: 1970, *Bull. Am. Phys.
    Soc.* **15**, 611.
Vaiana, G. S., Krieger, A. S., Timothy, A. F., and Zombeck, M.: *Proc. IAU Colloquium*, No. 27,
    in press.

SOLAR ACTIVITY OBSERVED IN X-RAYS
AND THE EUV FROM OSO 7

ROGER J. THOMAS

Laboratory for Solar Physics and Astrophysics, NASA-Goddard Space Flight Center,
Greenbelt, Md., U.S.A.

Abstract. Since 1971 the Goddard X-ray and EUV spectroheliograph aboard OSO 7 has been measuring the spatial distribution and time-variations of localized temperature and density features in solar active centers and flares. In some cases the sizes, shapes, orientations and locations of emitting plasmas at temperatures ranging from 10^4 K (Hα) to as high as 2×10^7 K (Fe xxv) have now been measured simultaneously. Our observations of active regions are consistent with the coronal structure being made up of nested systems of arches with footpoints in areas of opposite magnetic polarity. Temperatures seem to increase for arches nearer the center and also towards the top of any given magnetic arch, the innermost loops having the highest temperature gradients. There is also some evidence for electric current flow along such loops. Radiative cooling is significant for the region's hot central core which therefore must be maintained by a more or less continuous injection of energy.

A nested arch structure is also indicated for XUV flares of the two-component type, which likewise may require continuous energy input since conduction cooling should otherwise be very rapid. Multiple spikes during the impulsive phase seem to represent the consecutive triggering of different sources within the region and may occur outside of any detectable pre-existing coronal feature. Comparison of spatial distributions at several wavelengths during various stages of flare events provides information on interactions between the wide range of atmospheric levels involved. We have evidence for polarization of about 20% in a number of X-ray bursts, continuing throughout the decay phase. At least for some flares, our measurements seem to contradict the model of electron beams being radially injected into the chromosphere.

I. Introduction

The Goddard X-Ray and EUV spectroheliograph launched aboard OSO 7 on 29 September 1971 was designed to investigate a wide range of solar phenomena at wavelengths emitted by the corona and upper chromosphere. I will limit myself here to a discussion of just some of the results we have recently obtained on active regions and flares, using principally the spectroheliograms made simultaneously at several individual wavelengths between 1.7 and 400 Å with a spatial resolution of $20'' \times 20''$ or better. In addition, I will report on the initial analysis of our X-ray polarimeter measurements and the bearing they have on current flare theories. The instrument itself has already been described by Underwood and Neupert (1974).

II. Active Regions

Prior to the advent of observations from space, total solar eclipses and artificial eclipses achieved with ground-based coronagraphs provided the only opportunities for recording the form and structure of the high temperature corona. These observations, made with the Sun's disk occulted, provided valuable information on characteristic structures such as streamers and coronal condensations. However, the fundamental limitation that the underlying solar atmosphere could not be simultaneously recorded

Sharad R. Kane (ed.), Solar Gamma-, X-, and EUV Radiation, 25–42. All Rights Reserved.
Copyright © 1975 by the IAU.

except at the limb made it difficult to associate coronal forms with more commonly observed photospheric and chromospheric features.

Now, results from X-ray and EUV telescopes carried above the Earth's atmosphere on rockets, satellites, and ATM have shown that much of the structure of the inner corona is in the form of loops, occurring either as connections between areas of opposite magnetic polarity within a given bipolar region or as higher arches connecting seemingly independent active regions (e.g., Vaiana *et al.*, 1968). It has also become possible to measure directly the physical characteristics of these coronal condensations on the disk, thus providing a solid basis for an attack on the critical problem of determining exactly how they are formed and maintained. For example, using the Harvard spectroheliograph on OSO 4, Noyes *et al.* (1970) found that active regions are characterized on the average by greater electron pressure (by a factor of 5), higher coronal temperature (2.5×10^6 K), and greater conductive flux back to the chromosphere (by a factor of 5) over that observed in the undisturbed corona. However, the obvious importance of fine-scale structure recently emphasized by the magnificent ATM observations presents a clear challenge to existing, highly simplified models of these coronal enhancements.

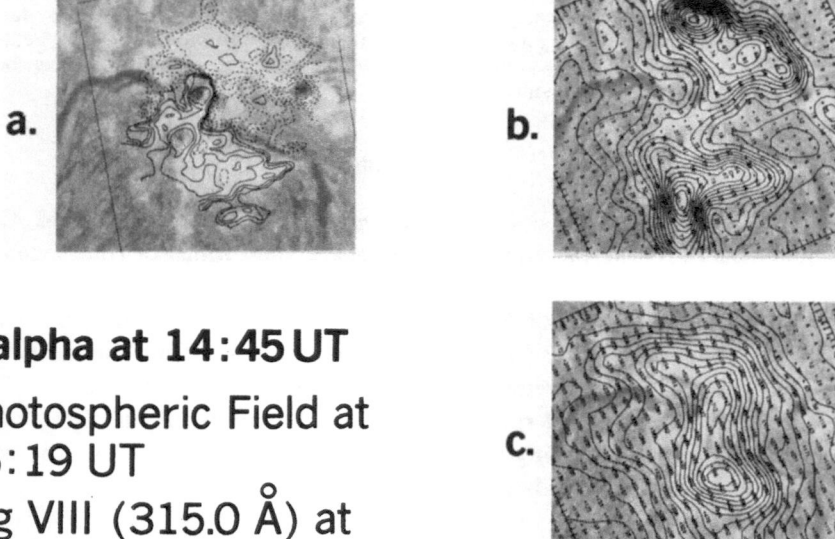

H-alpha at 14:45 UT

a. Photospheric Field at 16:19 UT
b. Mg VIII (315.0 Å) at 15:56 - 16:15 UT
c. Fe XVI (335.4 Å) at 15:55 - 16:16 UT

Fig. 1.   Comparison of the photospheric magnetic field measured by Sacramento Peak Observatory with the EUV emission observed by OSO 7 for an active region at 15°S, 13°E on 1972, January 19. Each map is overlayed on an Hα photograph of the region and covers an area 5′ on a side.

During the $2\frac{1}{2}$ yr lifetime of OSO 7, we have observed a vast number of active regions in all stages of development. I have selected one to describe in some detail, a dynamic flaring region that crossed the disk in January 1972, because it typifies many of the characteristics we have seen in other regions of this type.

Figure 1 shows the EUV emission of Mg VIII and Fe XVI from this region in the form of isophote contours overlayed on an Hα photograph taken at the Sacramento Peak Observatory. The photospheric magnetic field as measured by Rust (1974), is also indicated, with the neutral line running approximately east-west bisecting the region. The principal EUV structure is made up of a pair of sources, one on each side of the neutral line and therefore overlying areas of opposite magnetic polarity. In the transition zone and lower corona, however, numerous secondary enhancements appear, not all of which are associated with bright Hα plage, as can be seen in the Mg VIII map of Figure 1.

22:37 - 22:51 UT

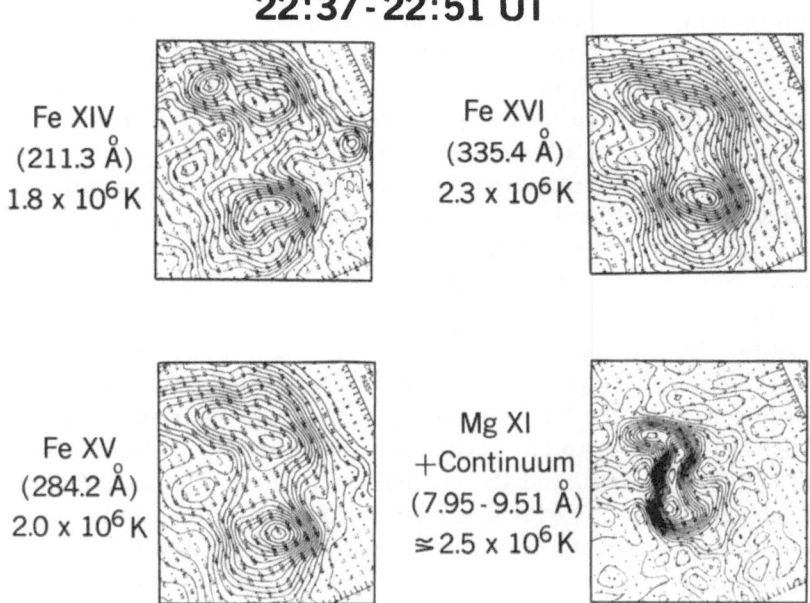

Fe XIV (211.3 Å) 1.8 x 10⁶ K

Fe XVI (335.4 Å) 2.3 x 10⁶ K

Fe XV (284.2 Å) 2.0 x 10⁶ K

Mg XI +Continuum (7.95 - 9.51 Å) ≲ 2.5 x 10⁶ K

Fig. 2. Spatial distribution of EUV and X-ray emission arranged in a temperature sequence for the active region at 15° S, 9° E on 1972, January 19. The temperature given is that at which the contribution function for that line has its maximum value. Maps are roughly $3.7 \times 4.2'$ in area. North is at the top in this and the preceding figure.

In Figure 2, several EUV maps of the region are arranged in a sequence of increasing temperature of emission, from Fe XIV at 1.8×10^6 K to Mg XI at 2.5×10^6 K. It should be noted that the temperature given is not that at which the ion population is greatest, but rather the temperature at which the contribution function has its maximum, a more appropriate value as pointed out by Jordan (1975). At elevated coronal temperatures, the structure is seen to simplify, becoming much less contorted, and often takes on a linear form with bright Hα plage appearing near the extremities of the coronal emission. At the highest temperatures, it becomes very compact, giving the appearance of a hot central core reported by earlier observers such as Blake *et al.* (1964) and Pounds and Russell (1966).

The bridge of EUV emission between the two principal sources also seems to change its orientation with increasing electron temperature, for example rotating from a northwest-southeast direction to a directly north-south orientation in the Mg VIII and Fe XVI maps of Figure 1. This may be evidence for the existence of non-potential magnetic fields, and therefore electric currents, in the coronal enhancement overlying the plage. Nakagawa (1974, private communication) has calculated possible coronal field configurations for the region satisfying force-free field conditions and using a variety of assumptions about the direction and magnitude of the electric current flow. Although there are difficulties with the interpretation of such calculations, preliminary comparisons with our EUV observations suggest that various regimes of temperature in the corona may correspond to different strengths of the electric current, and that the flow may even reverse direction at certain levels. However, this tentative conclusion will require further analysis for confirmation.

Another area of research presently in progress involves an attempt to determine the spatial distribution of temperatures within active regions. Using ratios of the Fe XV to Fe XVI intensities at various positions in maps such as those of Figure 2, it is possible to obtain at least a qualitative measure of how the coronal temperature varies across the region. We find that the highest temperature in an active region does not seem to occur at either of the Fe XV or Fe XVI bright points, but rather may occur at the location of the relatively compact X-ray source.

Now, within the limits of our spatial resolution, it is a straightforward matter to estimate the physical characteristics of just this small, highest-temperature EUV feature averaged along the line of sight. For the active region of January 1972, the electron density turns out to be 3×10^9 cm^{-3}, based on a thickness of 14000 km as measured at the west limb, and the thermal energy density is therefore 3 erg cm^{-3} if a peak temperature of 2.7×10^6 K is used. For such a feature, it can be shown that radiative losses are not at all negligible compared to the conductive loss rates, contrary to what has sometimes been suggested to be the case for non-flaring active regions (Tucker, 1973). This is especially true if, as we believe, the high temperature feature is at the top of a magnetic arch where conduction will be somewhat inhibited. In fact, based on the loss rates calculated by Cox and Tucker (1969), the entire kinetic energy of this source should be radiated away in about 6000 s, or less than two hours. Since the enhancement actually persists for many days, energy must be more or less continuously supplied in some way at the rate of roughly 2×10^{-3} erg cm^{-3} s^{-1} throughout its lifetime. A similar though much less stringent conclusion was reached by Reidy *et al.* (1968) who found that the radiation cooling time of an active region could be as short as a half-day.

Our observations seem consistent with the following general picture. The coronal enhancement overlying an active region appears to be made up of a nested set of arches or loops, with footpoints in regions of opposite magnetic polarity. The outermost arches are coolest, while those nearer the center are hotter. But even along a given field line there is a temperature gradient such that the highest temperature occurs at the highest point of the magnetic arch. This temperature gradient is smallest for the cool, outer loops, and becomes significantly greater for the loops near the center. Such a picture appears to be quite compatible with the Skylab and rocket results already presented at this symposium.

III. Flares

Turning now to the question of flares, I think it is necessary to emphasize that this term may actually cover a number of fundamentally different phenomena, each presumably having its own energy conversion mechanism (see, for example, Tandberg-Hanssen, 1973). Thus, when using results from different flares, one must be careful to consider only events of the same basic type. A similar difficulty arises in regard to the XUV bursts that accompany flares. De Jager (1965) was the first to suggest that X-ray bursts be divided into two classes, which are presently referred to as impulsive and gradual bursts. Often, a single event will contain both of these components. It is not yet clear to what extent such two-component bursts represent an additional unique class of events, or merely a combination of the two previously mentioned types.

I will restrict myself here to just the type which exhibits both components, a typical example of which is shown in Figure 3 (taken from Frost, 1969). The impulsive phase of such a burst consists of one or more extremely rapid spikes and normally lasts for only about two minutes. These spikes invariably occur during the rise of the associated gradual component, which itself may have the appearance of high-frequency fluctuations superimposed on a much more gradual rise and fall. Kane (1969) and others have shown that the impulsive component usually dominates at higher energies because of its harder spectrum. For that reason, and because of the rapid time variation, this component is almost certain to be non-thermal in character.

The situation is not so clear for the gradual component. Although it shows many thermal properties, there are several indications that non-thermal processes are still taking place during this phase. One of the most striking is the discovery from the New Hampshire experiment on OSO 7 that nuclear reactions are occurring in a flare for more than ten minutes (Chupp *et al.*, 1973), that is, well after the impulsive burst. Another is the hardening of the spectrum above 12 keV that Kahler *et al.* (1970) and others have seen throughout the gradual phase of some flares.

The close correspondence between the hard X-ray and microwave bursts is quite evident from Figure 3. In addition, Donnelly (1967, 1969) has demonstrated that variations in broad-band EUV emission are also very similar to those in the hard X-ray burst, at least during the impulsive phase. Using the spectral resolution of the Harvard OSO experiments, Donnelly *et al.* (1973) later concluded that the EUV emission formed at temperatures up to a million degrees exhibited this impulsive behavior.

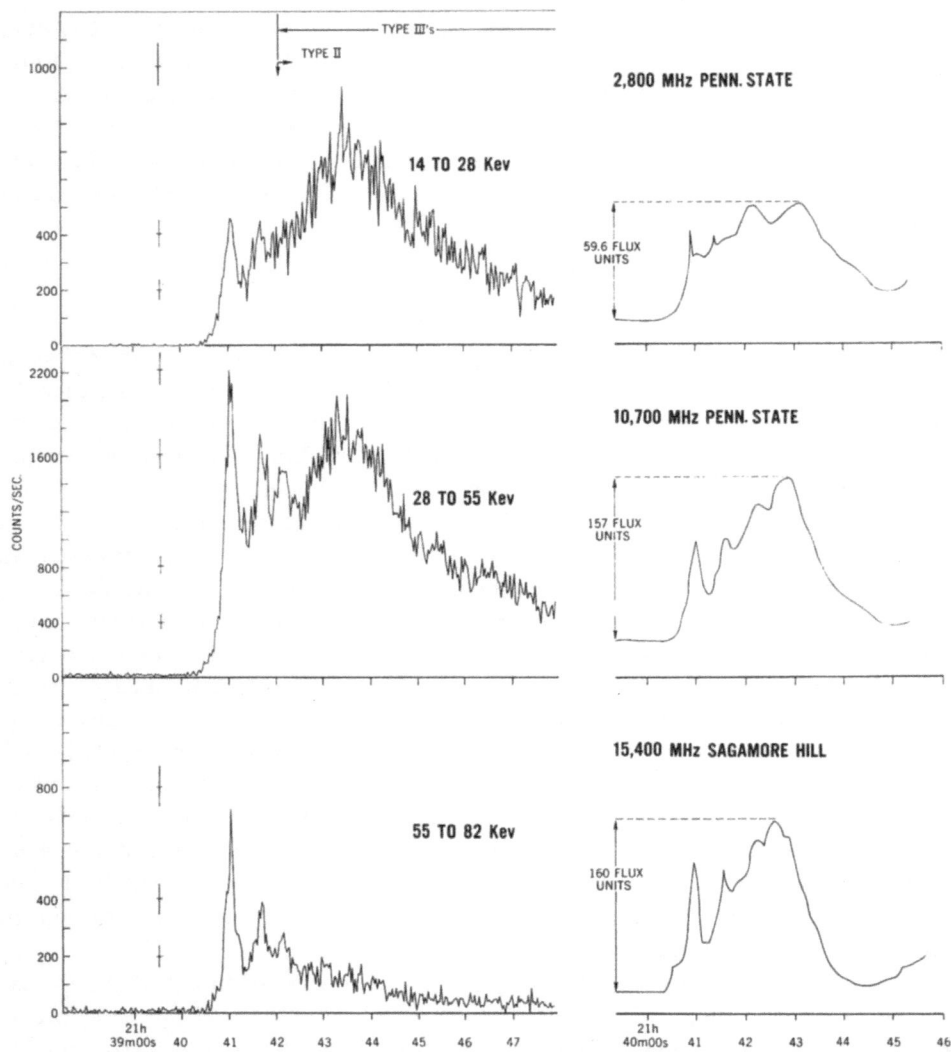

Fig. 3. A two-component solar burst on 1969, March 1 observed in hard X-rays by OSO 5 and at several radio frequencies by Penn. State and Sagamore Hill. The three leading impulsive spikes are followed by a component with a more gradual envelope and a distinctly different spectral shape. Taken from Frost (1969).

On the basis of X-ray measurements from OGO-5, Kane and Donnelly (1971) suggested that the EUV emission is produced thermally in a region heated by collisional losses of the non-thermal electrons responsible for the impulsive X-ray and microwave bursts. They believe this heating must occur at densities above 10^{12} cm^{-3}, that is, in the chromosphere. Such a result fits in very nicely with the discovery by Vorpahl and Zirin (1970) that impulsive bursts are associated with the chromospheric brightening of small kernels of Hα emission.

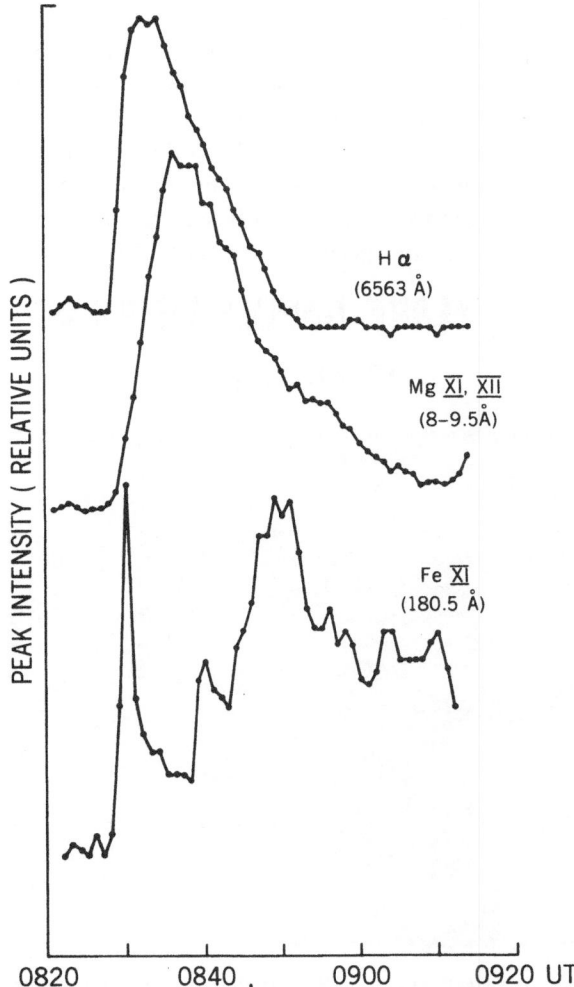

Fig. 4.   Time history of the peak intensity of a flare on 1972, February 13 observed by OSO 7 at three wavelengths in the visible (Hα), soft X-rays (MgXI, XII), and EUV (FeXI).

The exact relation between the impulsive and gradual phases of this type of event is very much unknown, at least to me. From the University of California, San Diego experiment on OSO 7, Datlowe and Hudson (1975) have been able to show that the total energy contained in the electron beam responsible for the impulsive burst is adequate to account for the energy in the X-ray emitting plasma during the gradual phase, under certain assumptions. But whether or not the electron beam really is the only source remains an open question.

Now let me discuss a few of our Goddard OSO 7 flare observations that bear on some of these points. Figure 4 shows the time variations of a flare's peak intensity at three different wavelengths: Hα, soft X-rays, and an EUV line. The coronal X-ray

emission of Mg XI and XII which is formed at about 9×10^6 K and the chromospheric
Hα radiation both have just a simple rise and decay. On the other hand, the Fe XI
emission at 1.2×10^6 K shows a great deal of structure, with an initial impulsive phase
and then a rather complex gradual phase. At the instant of the impulsive burst,
there is an abrupt rise in both Hα and X-rays. With the one-minute time resolution of
our observations, the impulsive EUV burst seems to be a single spike. But higher
resolution data, such as from the New Hampshire X-ray monitor on OSO 7, indicate
there were at least two spikes and perhaps even more during this interval.

FLARE IMPULSIVE PHASE

13 FEB 1972 OSO-7

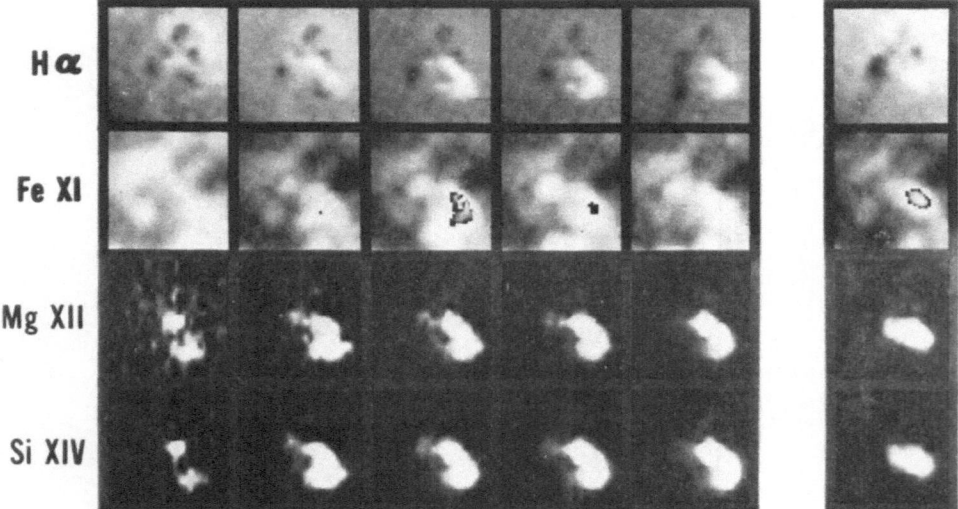

Fig. 5.—Time sequence from left to right of OSO 7 spatial maps in a grey-scale representation for the
flare shown in Figure 4. From top to bottom are maps 5′ square in Hα, Fe XI, Mg XI–XII, and
Si XIII–XIV. The sequence on the left covers the impulsive phase at roughly one-minute intervals. The
set on the right were all made at the peak of the gradual component. Dark markings in the center of
some Fe XI maps are due to computer overflow and actually represent the very brightest locations.

Each curve in Figure 4 gives the intensity of the flare's brightest point as a function
of time, regardless of where its spatial location might be. But our EUV maps during
the gradual phase show that there were actually two different locations that brightened
and faded in an apparently independent manner, one peaking around 08 40 UT and
the other at 08 50 UT. This independent behavior of different points seems to be typical
of the gradual component formed in the transition zone and lower corona. It unfortunately makes the interpretation of spectra taken without spatial resolution very
difficult, since the radiation may well be a combination from two distinct sources,
one brightening while the other fades.

We also found for this particular event that the impulsive EUV emission came from at least two different locations. Figure 5 is a time sequence from left to right of our spatial maps in a grey-scale representation. The top set are Hα maps, the second row are Fe xi, and the bottom two rows are maps in two soft X-ray bands. Incidently, these Hα observations were made from OSO 7, in fact through the same entrance aperture as the EUV, so there is no uncertainty whatsoever in the spatial correspondence with the other EUV maps. The sequence on the left covers the impulsive phase at roughly one-minute intervals, while the set on the right were all made at 08 50 UT, the peak of the gradual component. Our computer overflowed for the very brightest Fe xi points, and these appear as odd dark markings in the center of some of the maps.

The dark speck in the second Fe xi map is the site of the initial impulsive burst. One minute later, the peak EUV emission appears in a different location, some 30000 km away. Apparently, multiple spikes do not represent the re-energizing of the same feature, but rather the consecutive triggering of different sources within the region.

Note that the Hα flare starts in a well-defined kernel seen in the second map of the top row of Figure 5. It is obvious, however, that this Hα kernel is not at all in the same location as the simultaneous EUV spike. The only way I can see to reconcile this observation with the otherwise attractive hypothesis of Kane and Donnelly (1971) is if the non-thermal electron beam entered the chromosphere at a very large slant-angle. But this would seriously conflict with the radial-beam model suggested by the work of De Jager and Kundu (1963) that many believe is correct.

Another puzzling aspect is the fact that the impulsive emission seen in the third Fe xi map of Figure 5 is coming from a much larger region than the gradual source shown on the far right. If the total flux we measure can be used to indicate relative size (which seems reasonable since the peak intensities were almost identical), the volume of the impulsive source is three times that of the gradual source. Perhaps the impulsive phase involves a large number of separate electron beams scattered over a wide area producing a 'shot-gun' effect, whereas the heating during the gradual phase occurs in a more compact region.

As can be seen in the bottom two rows of Figure 5, the coronal X-ray enhancement, which was a double source before the flare, appears as a single elongated feature whose main axis seems to rotate during the event. I would assume that this rotation corresponds to a re-alignment of the coronal magnetic field configuration during the flare. If the flaring plasma cools by expansion, we see no evidence of it in our maps. In fact, the XUV sources may actually appear to contract as they fade.

Figure 6 shows the intensity variations at two EUV wavelengths of the brightest point in another flare, an event recently analyzed by Neupert et al. (1974). Although it does have a gradual phase, the burst is clearly dominated by the impulsive component in this case. The radiation at 295 Å in the lower curve is almost entirely continuum and scattered light, principally from H i and He ii, and so is characteristic of chromospheric temperatures. The top curve, however, is mainly Fe xiv, which radiates at about 1.8×10^6 K. This extends the impulsive behavior to much higher temperatures than those reported by Donnelly et al. (1973).

34 ROGER J. THOMAS

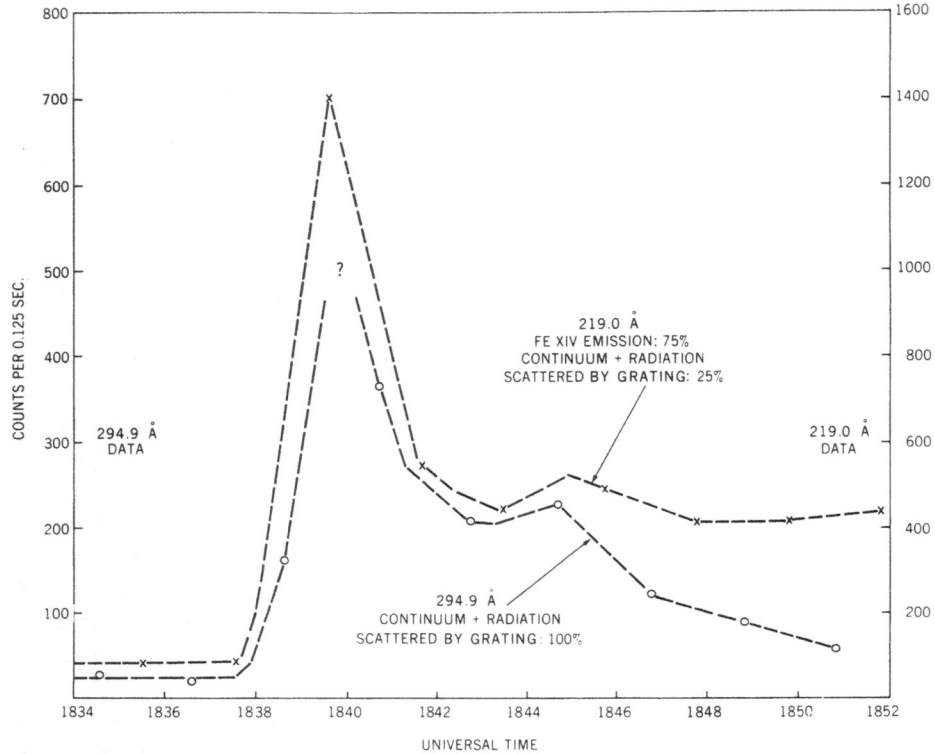

Fig. 6. The intensity of the brightest point in a flare on 1972, August 2 at two EUV wavelengths. The impulsive phase between 1838 and 1842 UT clearly dominates the following gradual component in this event. Figures 6, 7, and 8 are taken from Neupert *et al.* (1974).

Figure 7 shows the spatial distribution at these two wavelengths, as well as that of Fe XVII which radiates at about 3×10^6 K. The maps on the right are during the impulsive burst, those on the left were made just prior to the event. The XUV enhancements before the flare are associated with areas of the active region that were bright in Hα. But the flare emission is from a distinctly different location, some 50 000 km away from the main pre-flare enhancement. Clearly, the flare does not originate in any pre-existing coronal feature, at least none that we can detect. In fact, one can show that simply heating or compressing the pre-existing coronal material at this location cannot account for the strength of the emission measured during the flare itself. We believe this means that the XUV-emitting flare plasma originated as a result of the heating of chromospheric, or perhaps even photospheric, material. For other flares, the situation is not always as clear as it was in this case. But I have not yet seen any flare in our OSO 7 data where the impulsive burst occurred at the brightest point of a pre-existing coronal feature.

Three spectroheliograms made during the gradual phase of the previous flare are shown in Figure 8, each superimposed on the same off-band Hα photograph from the Lockheed Solar Observatory. In addition to the chromospheric radiation at 295 Å

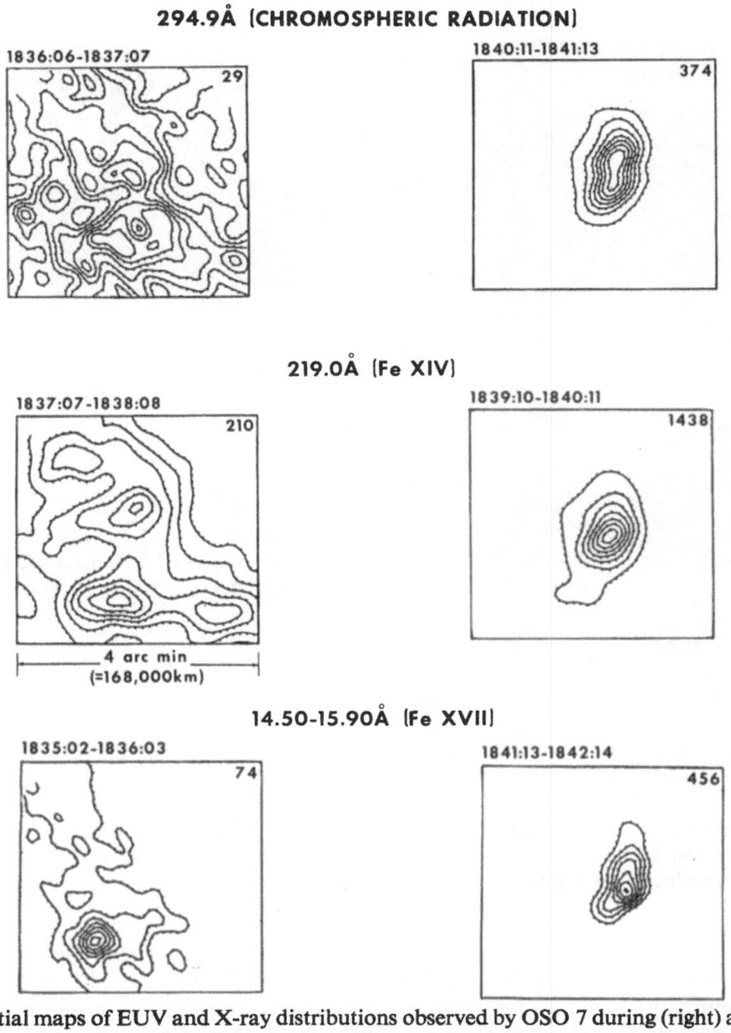

294.9Å (CHROMOSPHERIC RADIATION)

219.0Å (Fe XIV)

14.50-15.90Å (Fe XVII)

Fig. 7.   Spatial maps of EUV and X-ray distributions observed by OSO 7 during (right) and just prior to (left) the impulsive phase of the flare shown in Figure 6. Isophote contours are scaled to one-eighth of the observed peak count which is indicated in the upper right corner of each map. Note that only weak EUV and no measurable soft X-ray emission emanated from the site of the flare before the event began.

(map a), we have isophotes of X-ray emission formed at about 9×10^6 K (map b) and at around 30×10^6 K (map c). All three sources are quite elongated, as is the Hα flare. However, only the features at 10×10^6 K or less are exactly coincident with the Hα, and they lie directly over the magnetic neutral line that separates the two flare ribbons. The 30×10^6 K feature is distinctly displaced, which we interpret to mean that it is some 35000 km above the other sources, since our line-of-sight is 30° from the normal to the Sun's surface. Although measured in X-rays, this is very similar to the result for the Fe xxiv EUV source that Widing (1975) described earlier in this symposium.

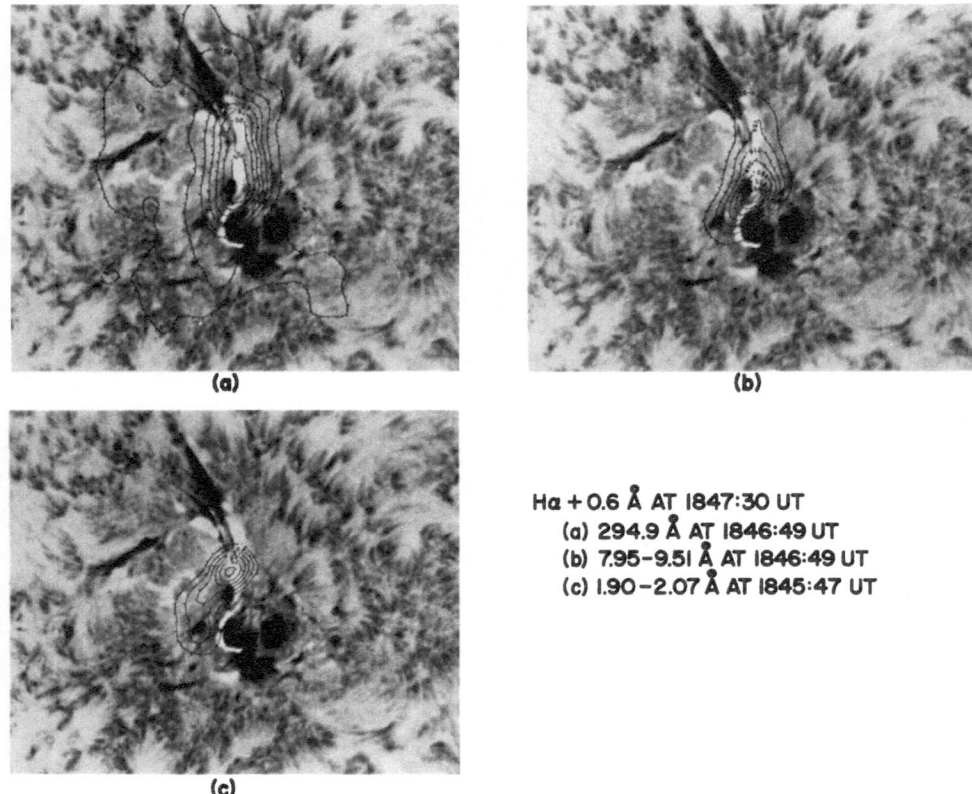

(a)

(b)

Hα + 0.6 Å AT 1847:30 UT
   (a) 294.9 Å AT 1846:49 UT
   (b) 7.95-9.51 Å AT 1846:49 UT
   (c) 1.90-2.07 Å AT 1845:47 UT

(c)

Fig. 8.   Comparison of XUV spatial distribution during the gradual phase of a flare on 1972, August 2 with an off-band Hα photograph made at the Lockheed Solar Observatory. The XUV emissions are: (a) chromospheric, (b) Mg xi–xii formed at about 9×10^6 K, and (c) X-ray continuum characteristic of around 30×10^6 K.

As in the active region studies, our flare observations are consistent with the coronal structure being a nested set of arches or loops connecting regions of opposite magnetic polarity. But in the flare case, the greatest temperature is found at the top of the highest arch system, not immediately above the chromosphere where the fast electrons are thought to be heating the material.

The 30×10^6 K Fe xxv feature was observed to cool to 10×10^6 K in about 10 minute's time, but it is easy to show that radiation cooling was negligible during that period (unlike the case for a non-flaring active region discussed above). In fact, based on a simple model suggested by Culhane *et al.* (1970), conduction cooling along a magnetic arch is such an efficient process in this type of flare that the feature should have cooled about four times faster than it actually did. This means either that the hot plasma was trapped in a magnetic bottle which effectively isolated it from the chromosphere, or else that energy was being continuously supplied to the region throughout its decay.

I have been describing some rather indirect evidence for coronal loop structures from observations made on the disk. Much more obvious are our EUV and X-ray observations of post-flare loops on the limb, shown in Figure 9. In a sense, this combines results that had been made from several different experiments on ATM.

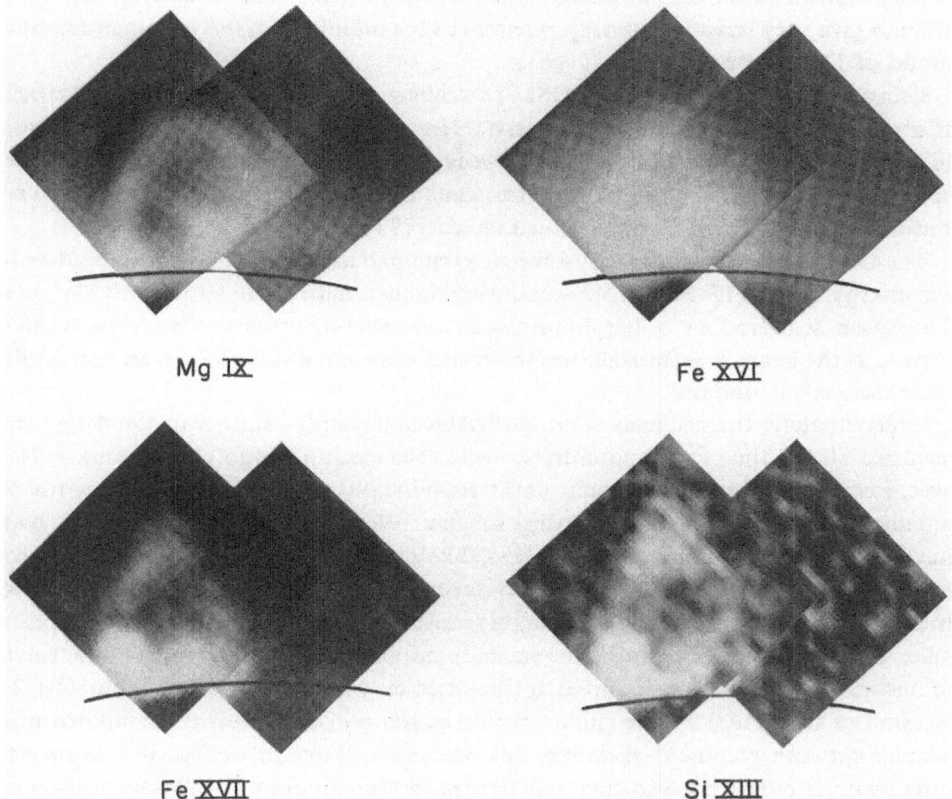

Mg IX   Fe XVI

Fe XVII   Si XIII

Fig. 9.   EUV and X-ray observations from OSO 7 of post-flare loops on the south-east limb at around 0200 UT on 1972, February 9. Each image is a composite of two maps with $5 \times 5'$ fields of view. The approximate position of the limb is indicated.

The appearance in Mg IX at 0.9×10^6 K is very similar to the classical Hα loop system, while in Fe XVI at 2.3×10^6 K the EUV emission is diffuse and almost featureless. In X-rays, at even higher temperatures, the loop structure reappears but with a distinctly different shape. Furthermore, the top of the X-ray loop is particularly enhanced, as is at least one footpoint. These X-ray emitting postflare loops are clearly not isothermal.

IV. X-Ray Polarization

It has been recognized for some time that X-ray polarization measurements could provide direct evidence for non-thermal processes in flares, as pointed out by Korchak

(1967) and Elwert (1968). Recently, Tindo *et al.* (1970, 1972a) have reported positive detection of polarization in several X-ray bursts. By normalizing their measurements to an assumed zero polarization in the decay phase, they conclude that the impulsive component has a polarization of 10–30%, although their latest paper (Tindo *et al.*, 1972b) indicates that this level may also persist into the flare's decay. Their results seem to give very strong support in general to the radial-beam theory suggested by the model of De Jager and Kundu (1963).

Using a small polarimeter on OSO 7, we likewise have evidence for polarization of about 20% in a number of X-ray bursts. However, we have found that polarization often seems to continue throughout the decay phase, sometimes even appearing to be strongest then. Furthermore, at least for some flares, our measurements seem to contradict the radial-beam theory (Nakada *et al.*, 1974).

The OSO 7 polarimeter is very similar to the one used by Tindo's group, although our energy range (15–30 keV) is somewhat higher than theirs. Incident X-rays are Thompson scattered by a beryllium block into the surrounding three pairs of detectors. If the beam is polarized, the scattering is asymmetric, and the response of the three channels is unequal.

Unfortunately, the channels were not balanced exactly, and so unequal responses occurred all the time (as apparently was also the case in Tindo's experiment). However, instead of normalizing each event individually to attempt to correct for this imbalance, we adopted the following scheme. We assumed that the X-rays had a power-law spectrum throughout the 15–30 keV range of our instrument. Numerous measurements of solar bursts indicate this is reasonable (e.g., Frost, 1969; Kane, 1969). In that case, the ratios of the channel responses will vary depending on the spectral index of the X-ray beam. But if the beam is unpolarized, those ratios will be related to one another in a manner similar to the solid curve labeled $P=0$ in Figure 10. The position of the actual P-zero curve on the graph will depend on the unknown imbalance between channels. However, any deviation from this curve will occur only if the beam is polarized, and then will depend on the amount of the polarization and its position angle, as shown by the three closed curves in Figure 10. The dashed lines show the amount of deviation from the P-zero curve that would indicate polarization of at least 20%.

The results using this method for eight different flares in June 1972 are displayed in Figure 11, which gives the observed ratios of channel count rates averaged over one or two minute intervals and with background subtracted. The flares occurred close enough in time that any possible drift in the imbalance between channels could not affect the results significantly. Error bars are given unless they were smaller than 4%. We assumed that the P-zero line could be estimated by a least-squares fit through the plotted points. Any other line would only serve to increase the total amount of polarization inferred. As before, the dashed lines on the graph indicate positions that would imply polarization of at least 20%.

To the lower right of Figure 11, the location of each event on the solar disk is shown, with north at the top. The cross-hatched areas are the locations on the disk

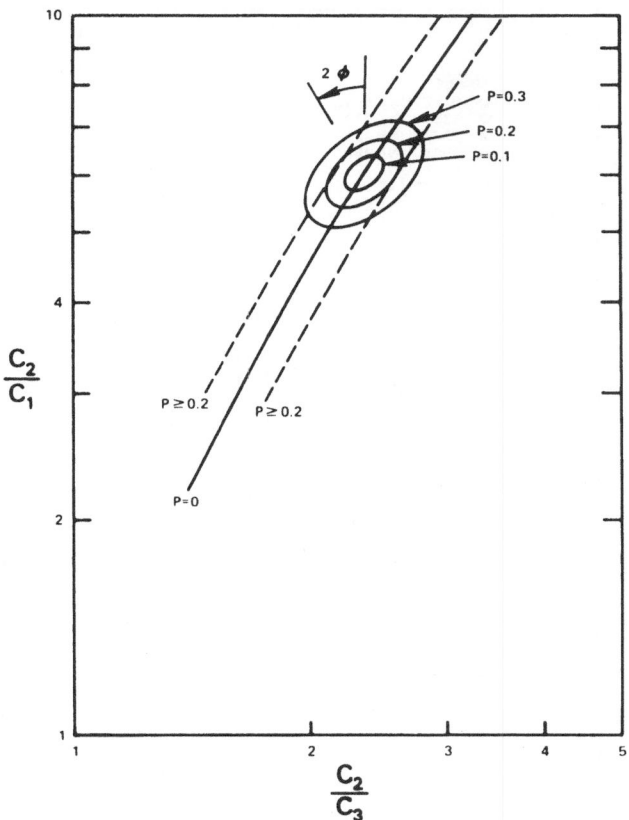

Fig. 10. Typical curves relating the ratios of counting rates for the three channels of the OSO 7 polarimeter if the incident X-rays have power-law spectra. For zero polarization, the observed ratios should fall somewhere on the line labeled $P=0$ depending on the steepness of the power law. The closed curves show deviations from a given point on the P-zero curve for various strengths of polarization and values of the position angle.

where the radial-beam theory predicts polarization position angles such that deviations would be along the P-zero line. Thus, if this theory were correct, flares such as the one on June 17 should have ratios that lie almost directly on the P-zero line regardless of their level of intrinsic polarization. Yet this particular flare actually exhibited the largest displacement we have yet measured. On the other hand, a limb flare outside the hatched area, such as the June 11 event, should have a very large displacement according to the radial-beam theory, whereas in fact we find it to have almost none. For these two flares at least, the belief by Tindo *et al.* (1972b) that polarization measurements are consistent with the radial-beam theory does not appear to be supported.

Our results for the series of great flares in August 1972 are similar in that we again measure displacements during several flares that imply a polarization of about 20%.

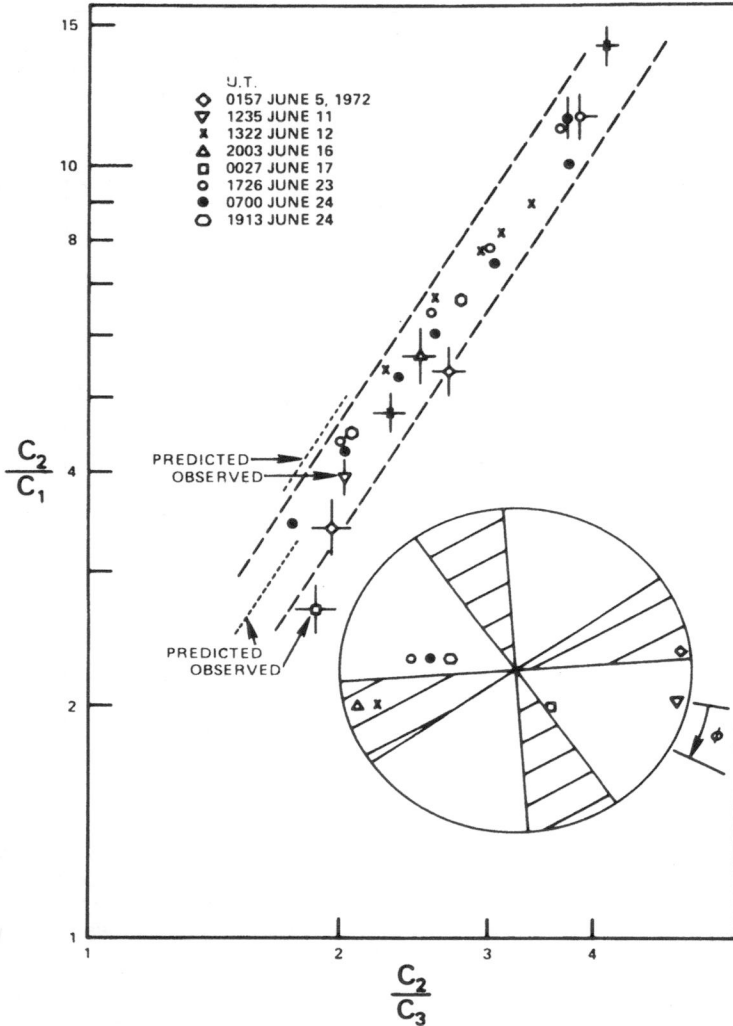

Fig. 11. Observed values of ratios of counting rates for eight X-ray events during June 1972. Standard deviations less than 4% of ratios are not shown. The dashed curves correspond to inferred polarizations of at least 20%. In the lower right, the locations of the events on the solar disk are indicated, with north at the top. Figures 10 and 11 are taken from Nakada *et al.* (1974).

We had a particularly complete coverage of the August 7 flare, which occurred near disk center but outside of a hatched area. Figure 12 gives the time history of this X-ray burst, along with the inferred polarization limits throughout the event. Surprisingly, the polarization appears smallest during an impulsive burst at 0350 UT, and becomes steadily stronger in the gradual phase and as the flare emission declines.

These results are exactly opposite to what I would have expected. Notice that if we had normalized this flare to an assumed zero polarization at its decay phase in the

OSO-7 POLARIMETER MEASUREMENTS

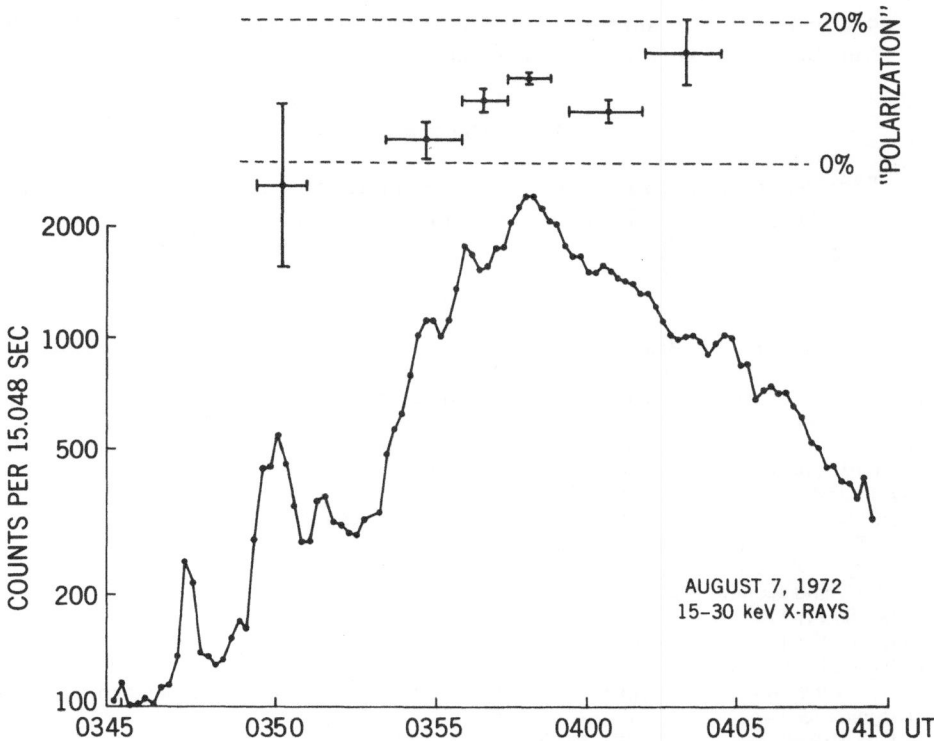

Fig. 12.   Time history of the inferred polarization limits for an event on 1972, August 7 along with the X-ray flux profile for the burst. Note that the polarization may be smallest during an impulsive spike at 03 50 UT, but seems definitely greater than 10% well into the decay of the gradual phase.

way that Tindo *et al.* (1970, 1972a) did for their measurements, the strongest polarization would then appear in the impulsive burst, just as the Tindo group reported. But if that were true, it would require that other flares in August 1972 had polarizations exceeding 40% in their declining phases.

It is of course possible that the impulsive burst did have strong polarization, but with a position angle that made our technique unable to detect it. In any case, if our reduction scheme is correct, one is forced to conclude that X-ray polarization of about 20% must exist during the decay phase of at least some flares. This might indicate that non-thermal acceleration processes are going on more or less continuously throughout the gradual component of these bursts. On the other hand, Angel (1969) has pointed out that even a thermal source can provide polarized emission if it is highly asymmetric as our observations certainly show the flare sources to be. Another possibility is that radiation backscattered from the photosphere could give rise to X-ray polarization as noted by Brown *et al.* (1974), although Beigman (1974) claims that this effect can only account for a few percent of the measured value. Perhaps the solution is actually some combination of these factors.

However it turns out, I am sure there will be many other surprises before we finally understand these strange animals called solar flares.

Acknowledgments

Parts of the research described above were originally carried out by W. M. Neupert, M. P. Nakada, and R. D. Chapman. I very much appreciate their suggestions during the preparation of this paper.

References

Angel, J. R. P.: 1969, *Astrophys. J.* **158**, 219.
Beigman, I. L.: 1974, Preprint No. 68, Lebedev Physical Institute, Moscow.
Blake, R. L., Chubb, T. A., Friedman, H., and Unzicker, A. E.: 1964, *Space Res.* **IV**, 785.
Brown, J. C., McClymont, A. N., and McLean, I. S.: 1974, *Nature* **247**, 448.
Chupp, E. L., Forrest, D. J., and Suri, A. N.: 1973, in R. Ramaty and R. G. Stone (eds.), *High Energy Phenomena on the Sun*, NASA SP-342, p. 285.
Cox, D. P. and Tucker, W. H.: 1969, *Astrophys. J.* **157**, 1157.
Culhane, J. L., Vesecky, J. F., and Phillips, K. J. H.: 1970, *Solar Phys.* **15**, 394.
Datlowe, D. W. and Hudson, H. S.: 1975, This volume, p. 209.
De Jager, C.: 1965, *Ann. Astrophys.* **28**, 125.
De Jager, C. and Kundu, M. R.: 1963, *Space Res.* **III**, 836.
Donnelly, R. F.: 1967, *J. Geophys. Res.* **72**, 5247.
Donnelly, R. F.: 1969, *Astrophys. J.* **158**, L165.
Donnelly, R. F., Wood, A. T., and Noyes, R. W.: 1973, *Solar Phys.* **29**, 107.
Elwert, G.: 1968, in K. O. Kiepenheuer (ed.), 'Structure and Development of Solar Active Regions', *IAU Symp.* **35**, 444.
Frost, K. J.: 1969, *Astrophys. J.* **158**, L159.
Jordan, C.: 1975, This volume, p. 109.
Kahler, S. W., Meekins, J. F., Kreplin, R. W., and Bowyer, C. S.: 1970, *Astrophys. J.* **162**, 293.
Kane, S. R.: 1969, *Astrophys. J.* **157**, L139.
Kane, S. R. and Donnelly, R. F.: 1971, *Astrophys. J.* **164**, 151.
Korchak, A. A.: 1967, *Soviet Phys. – Dokl.* **12**, 192.
Nakada, M. P., Neupert, W. M., and Thomas, R. J.: 1974, *Solar Phys.* **37**, 429.
Nakagawa, Y.: 1974, Private Communication.
Neupert, W. M., Thomas, R. J., and Chapman, R. D.: 1974, *Solar Phys.* **34**, 349.
Noyes, R. W., Withbroe, G. L., and Kirshner, R. P.: 1970, *Solar Phys.* **11**, 388.
Pounds, K. A. and Russell, P. C.: 1966, *Space Res.* **VII**, 38.
Reidy, W. P., Vaiana, G. S., Zehnpfennig, T., and Giacconi, R.: 1968, *Astrophys. J.* **151**, 333.
Rust, D. M.: 1974, Private communication.
Tandberg-Hanssen, E.: 1973, *Earth Extraterrestrial Sci.* **2**, 89.
Tindo, I. P., Ivanov, V. D., Mandelstam, S. L., and Shuryghin, A. I.: 1970, *Solar Phys.* **14**, 204.
Tindo, I. P., Ivanov, V. D., Mandelstam, S. L., and Shuryghin, A. I.: 1972a, *Solar Phys.* **24**, 429.
Tindo, I. P., Ivanov, V. D., Valnicek, B., and Livshits, M. A.: 1972b, *Solar Phys.* **27**, 426.
Tucker, W. H.: 1973, *Astrophys. J.* **186**, 285.
Underwood, J. H. and Neupert, W. M.: 1974, *Solar Phys.* **35**, 241.
Vaiana, G. S., Reidy, W. P., Zehnpfennig, T., Van Speybroeck, L., and Giacconi, R.: 1968, *Science* **161**, 564.
Vorpahl, J. and Zirin, H.: 1970, *Solar Phys.* **11**, 285.
Widing, K. G.: 1975, This volume, p. 153.

PART 2

ACTIVE REGIONS

X-RAY SPECTRA OF SOLAR ACTIVE REGIONS

JOHN H. PARKINSON

Mullard Space Science Laboratory, Dept. of Physics and Astronomy,
University College London, Holmbury St. Mary, Dorking, Surrey, England

Abstract. The last few years have seen great progress in our understanding of X-ray spectra of solar active regions. This paper demonstrates both the usefulness and the limitations of the techniques, both scientific and instrumental, that have recently become available. Improvements in spectral resolution led to the discovery of weak satellite lines to helium-like ions; the quantitative theory for these lines is also discussed. The observed intensities of the Fe XVII lines are also investigated and found to be in agreement with calculations that allow for cascading processes.

I. Introduction

For many years now it has been recognised that the solar corona has a temperature in excess of one million degrees and emits most of its energy in the X-ray region. However it is only in the last ten years or so that it has been possible to obtain X-ray spectra of sufficient resolution that the dominance of line emission was recognised. Further, it is only in the last three years that spatially resolved high resolution spectra have been obtained. The excellent review by Walker (1972) covers much of this work. Although this Symposium takes place at a time when the Sun is in the 'quiet' part of its cycle, the last few years, as we will see, have been anything but 'quiet' for those working in the X-ray spectral region!

The corona is an object of great interest for the solar astronomer and the atomic physicist alike. By observing the spectrum below approximately 25 Å the solar astronomer can obtain information on the physical state of the plasma with temperatures $\gtrsim 10^6$ K, he can also directly observe the coronal condensations above active regions as they rotate across the solar disc. The atomic physicist is able to observe a plasma which has physical conditions very different to those he is able to create in the laboratory. The spectrum contains lines due to a wide variety of transitions; in addition to the normal allowed electric dipole transitions both magnetic dipole and quadrupole transitions are observed, mainly from H-, He- and Ne- like ions.

This paper does not aim to be a comprehensive review but rather to highlight the techniques, both scientific and instrumental, that are available and to demonstrate both their usefulness and limitations. In a following paper Dr Walker (1975) will discuss more of the implications of some of the observations discussed here. We start by discussing the experimental techniques and some of the improvements that have been made recently in increasing spectral resolution. We will see how this increased resolution has allowed us to use new methods for the diagnosis of the temperature and density structure of coronal features. Finally we investigate the atomic physics that can be derived from X-ray spectral observations. The spectrum emitted by an active region between 2 and 5×10^6 K is very different to that emitted by a flare. We will always restrict ourselves to observations of non-flaring active regions.

Sharad R. Kane (ed.), Solar Gamma-, X-, and EUV Radiation, 45–64. All Rights Reserved.
Copyright © 1975 by the IAU.

II. Experimental Techniques

The main dispersive instrument that has been used below 25 Å is the scanning Bragg crystal spectrometer, the diffracted photons being detected by proportional counters. A wide variety of crystals is available with various lattice spacings and rocking curves, so that the choice of crystal can be optimised for the particular problem being studied. Much ingenuity has been shown in building smooth, linear drive systems for scanning the crystals and detectors, and many Bragg angle readout systems have been flown in rocket and satellite experiments.

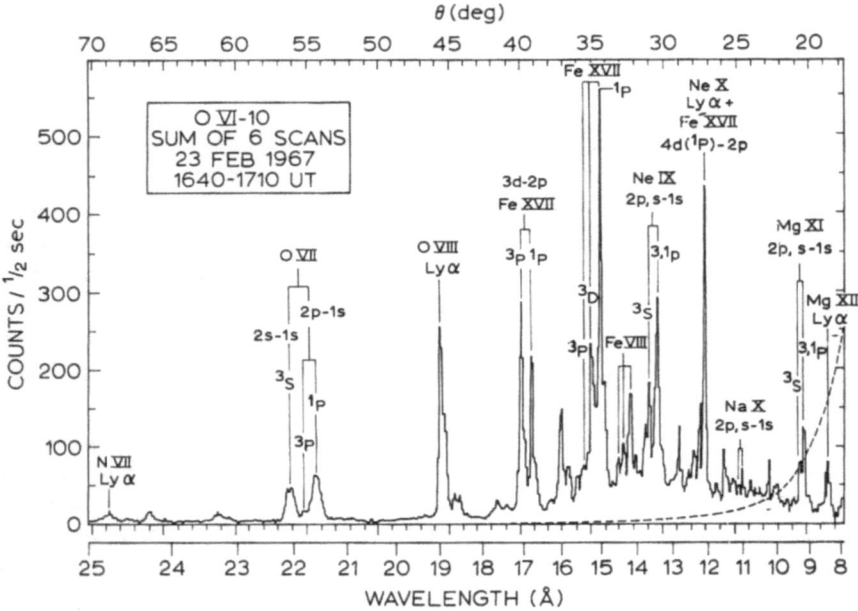

Fig. 1.   Non-flare spectrum obtained with a KAP spectrometer on the OVI-10 satellite on 1967, February 23 when the solar X-ray emission was dominated by a single active region (Rugge and Walker, 1968).

The early observations of Blake *et al.* (1965), Fritz *et al.* (1967), Evans and Pounds (1968), Rugge and Walker (1968) and Neupert (1971) were from rocket and satellite instruments which viewed the whole Sun and integrated the emission from the corona and all of the active regions present. Examples of these observations are shown in Figures 1 and 2 and it can be clearly seen that the spectrum is dominated by lines of Fe XVII together with lines from H- and He- like Mg, Ne and O. The spectral resolution is quite low and many lines are blended together.

Figure 3a shows a scan through the Mg XI lines around 9 Å made with a rocket spectrometer that had better spectral resolution than previous instruments. At the time of the rocket flight there were five active regions on the Sun and each one pre-

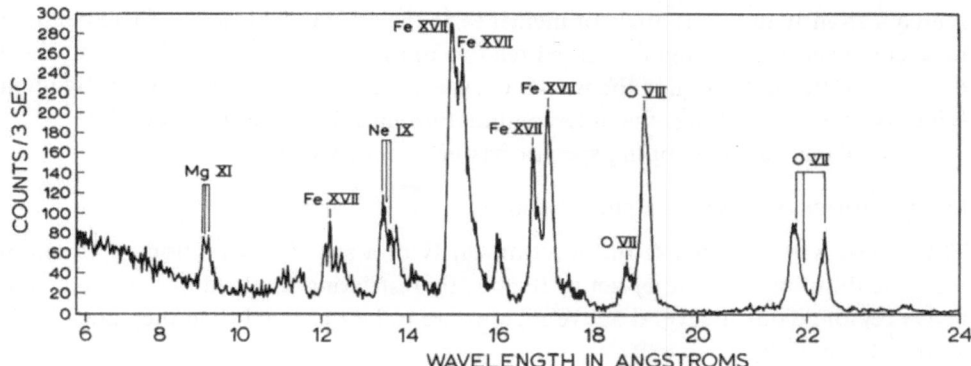

Fig. 2. Non-flare spectrum obtained on 1969, January 27 with a KAP spectrometer on the OSO 5 satellite (Neupert, 1971).

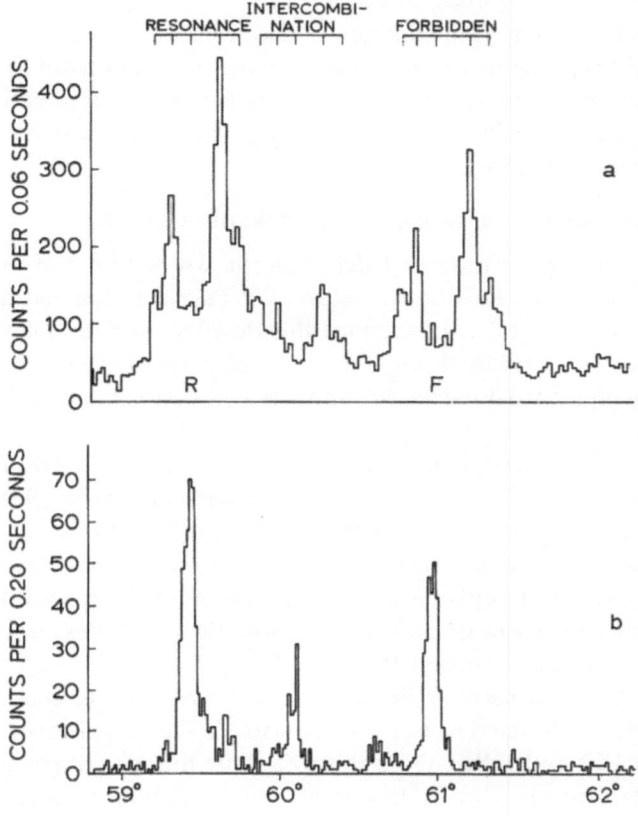

Fig. 3. Scans through Mg XI (a) with an uncollimated spectrometer showing spectra for five active
regions (Batstone, 1970). (b) with a spectrometer collimated to 4′ FWHM, showing several satellite
lines (Parkinson, 1971a).

sented a slightly different angle of incidence to the crystal. This led to there being five different spectra, each slightly shifted relative to the next. Batstone (1970) was able to deconvolve this data to give the individual spectrum for each active region, by identifying five resonance lines, five intercombination lines and five forbidden lines.

The problem of overlapping spectra has been approached in two ways.

(a) INSTRUMENTS VIEWING A SINGLE REGION

The easiest way to improve the spectral clarity of a set of observations is to remove the contributions from other active regions by restricting the field of view to a single active region. Thus if several active regions are to be observed then they must be observed sequentially.

Figure 3b shows the results obtained by using a spectrometer of the same design as that used to obtain the results in Figure 3a, but with the field of view restricted to 4' in order to isolate a single active region (Parkinson, 1971a). Here we can see the main lines resolved, there are no overlapping spectra, and many weak satellite lines are observed; these will be discussed in Section IV.

The most useful design of collimator is one where grids are aligned and spaced in a geometrical progression which depends on the hole: bar ratio of the grids. This type of construction has an advantage over the Soller type of collimator as reflections off the slats at grazing incidence are removed. The technique is described in more detail by Parkinson (1971b).

(b) INSTRUMENTS CAPABLE OF PRODUCING SPECTROHELIOGRAMS

Several workers have used devices which scan the whole Sun and remove the need to conduct observations sequentially. Acton *et al.* (1972) collimated their instrument in only one direction, to 1.7', and scanned this slowly across the solar disc while two spectrometers scanned rapidly through pre-selected parts of the spectrum. The rocket was rolled through approximately 60° and the collimator scanned back across the solar disc. Both of the spectrometers used KAP crystals, one was set to scan through the lines of O VII between 21.38 and 22.27 Å and the other scanned through the lines of Ne IX between 13.34 and 13.78 Å. The crystals scanned their pre-selected wavelength ranges every 1 s and the rocket moved through 1.5'. in 5 s In this way it was possible to derive spectra for each active region present on the solar disc. Figure 4 shows such spectra for a single active region and also the sum of all the spectra from the active regions. The form of the spectra and other information from this rocket flight will be discussed in more detail in Section IV.

A similar method was used by Bonnelle *et al.* (1973) who also used two spectrometers which were collimated in one direction to 3'. This experiment was pointed at the centre of the Sun and rolled about this axis while the spectrometers scanned their pre-set wavelength ranges. Two ADP crystals were used, one covering the range 8.39 to 8.53 Å which contains the Mg XII Lα line, the other the Mg XI lines in the range 9.15 to 9.34 Å. A wavelength scan took 17.3 s and the payload rolled at a rate of 20.2° s^{-1}. ADP has a narrow rocking curve so this was exploited to give both spectral

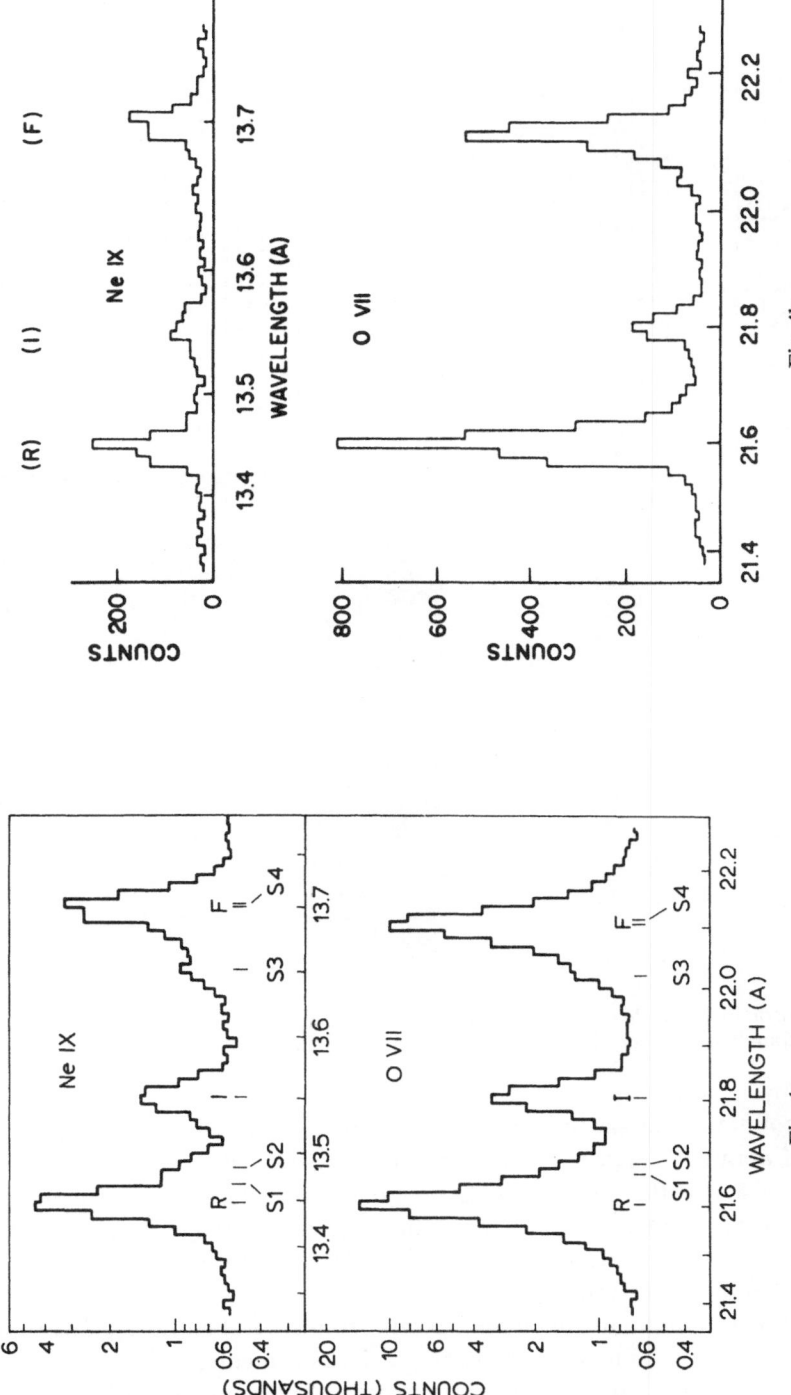

Fig. 4b.

Fig. 4a.

Fig. 4a–b.   Scan through Ne IX and O VII. (Acton *et al.*, 1972). (a) for the whole Sun. (b) for a single active region.

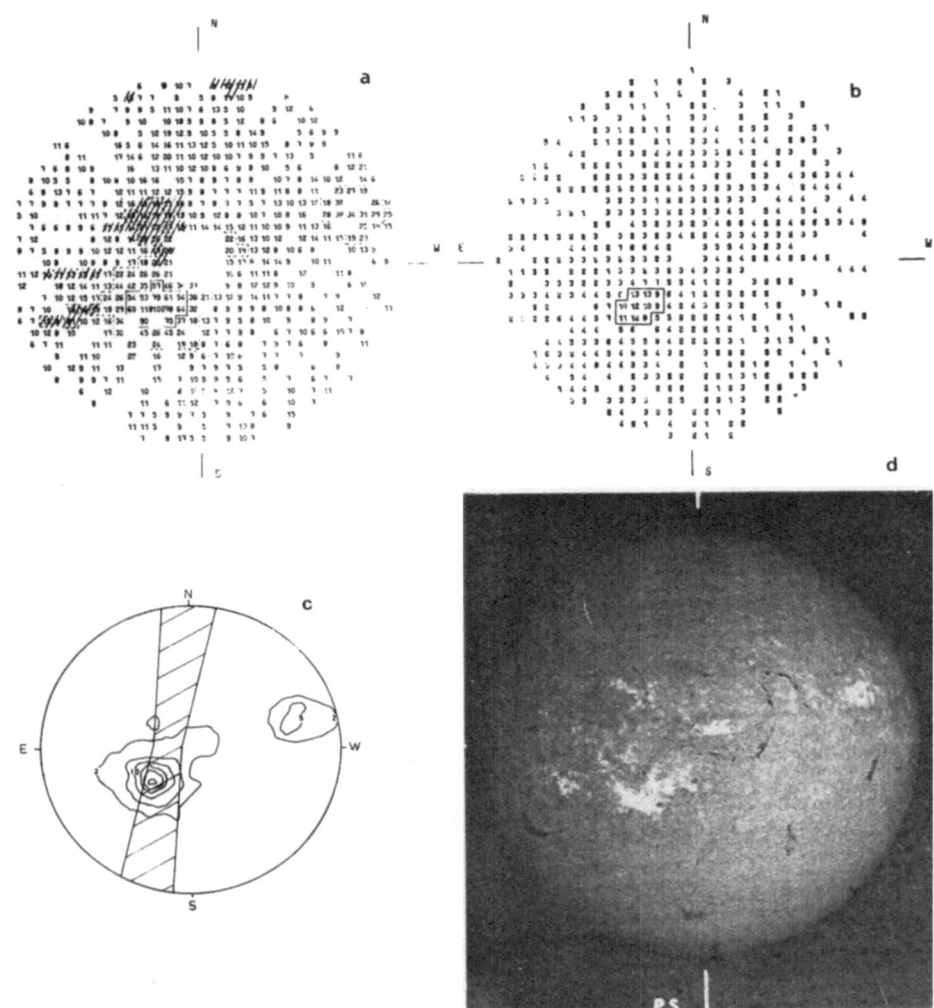

Fig. 5.   Comparison by Bonnelle *et al.* (1973) of (a) Spectroheliogram in the resonance line of Mg xi. (b) Spectroheliogram in the Lα line of Mg xii. (c) 8.4–9.6 Å map from OSO 5. (d) Hα spectroheliogram.

and spatial resolution, and spectroheliograms were constructed in a single line with a resolution of about 1′. Figure 5 shows two such spectroheliograms, one is in the resonance line of Mg xi the other in the Lα line of Mg xii, these are compared with an Hα image and an OSO-5 8.4–9.6 Å map, (this wavelength range is dominated by lines of Mg xi and xii, Parkinson and Pounds (1971)) and there is good agreement between these two quite different instruments.

A very different technique for obtaining spectra from several active regions simultaneously has been proposed by Brabban *et al.* (1971). This involves placing a rotation modulation collimator in front of a crystal spectrometer and reconstructing an image of the Sun in the radiation emitted in each spectral line. The first results (Brabban, 1973) showed that this technique was successful when used in a rocket experiment.

III. The Analysis of an Active Region X-Ray Spectrum

In order to make use of the intensity of a line it must be expressed in energy units by taking into account the instrumental parameters.

The line flux is given by

$$E = \frac{2 \times 10^{-8} N\omega}{A R_c \lambda P T_1 T_2} \ \text{erg cm}^{-2} \text{s}^{-1},$$ (1)

where N is the number of counts in the line; ω is the crystal scan rate in radians s^{-1}; A is the effective crystal area; R_c is the crystal integrated reflection coefficient in radians; λ is the wavelength of the line; P is the photon detection efficiency at wavelength λ; T_1 is the atmospheric transmission at wavelength λ; and T_2 is the transmission of the collimator.

Some of these parameters are comparitively easy to measure or calculate, others are more difficult. For example the reflectivity of the crystals is an important parameter and Leigh and Evans at the University of Leicester and Blake at the University of Chicago have spent many hours carefully measuring the variation of this parameter with wavelength for many crystals.

In order to analyse the spectrum of an active region the lines must first be identified correctly, in general this requires both good spectral resolution and accurate theoretical predictions. At the present time it has only been possible to measure the intensities of lines and we now consider what we can learn from such measurements.

Following Pottasch (1964) we can write the intensity of a line as

$$E = 7.75 \times 10^{-43} g f A_z \int G(T_e) N_e^2 \, dV,$$ (2)

where

$$G(T_e) = T_e^{-0.5} 10^{-\left(5040 \frac{E_0}{T_e}\right)} A_{zi}$$ (3)

g is a Gaunt factor; f is the oscillator strength of the transition being considered; A_z is the abundance of the element relative to hydrogen; T_e is the electron temperature; and A_{zi} is the fraction of element z in ionisation state i at temperature T_e.

If an active region were isothermal it would be a trivial problem to solve Equation (2), however, in general, an active region contains material at a variety of temperatures. A single ion is observed to radiate over a range of temperatures and this range can be much less than the range of temperatures in an active region. Thus it is important to observe a number of lines which cover a wide range of temperatures rather than just one or two lines. The resonance and Lα lines of He- and H-like ions of O, Ne, Mg, Si have generally been used as these are strong lines and the atomic physics of such simple ions is relatively well understood. Figure 6 illustrates the function $G(T_e)$ (Jordan, 1969) for several He-like resonance lines and clearly shows the importance of observing several lines from an active region.

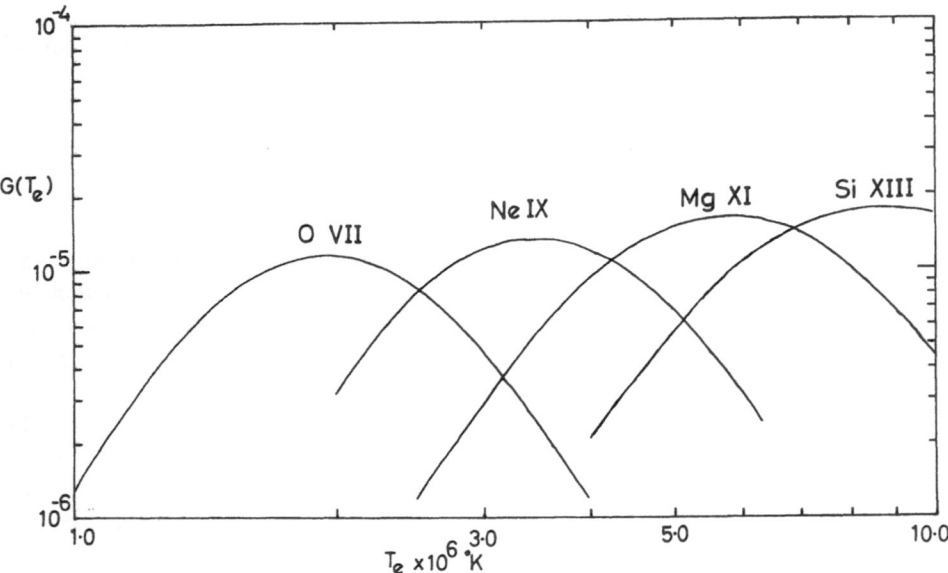

Fig. 6.   Curves of $G(T_e)$ calculated using Equation (3) and the ionisation balance of Jordan (1969).

The method of analysis developed by Batstone et al. (1970) replaces the integral in Equation (2) by summation over several temperature intervals of width ΔT_e, thus

$$E = 7.75 \times 10^{-43} g f A_z \sum_{\Delta T_e} G(T_e)(N_e^2 V)_{T_e}. \tag{4}$$

A number of lines covering a wide range of temperatures were used and a set of over-determined simultaneous equations was solved to give an emission measure, $N_e^2 V$, variation with temperature, T_e.

Chambe (1971) has since suggested a method which assumes a differential emission measure of the form

$$\frac{d}{dT_e}\int N_e^2 \, dV = C \times 10^{-T_e/T_0}, \tag{5}$$

where C and T_0 are constants which are determined from the analysis. This method is useful when only a few lines have been observed.

After the temperature structure of a plasma has been diagnosed by using the intensities of several lines it is possible to use the model to estimate oscillator strengths and collision strengths for ions that were not used in constructing the model.

It seems worthwhile at this point to illustrate this model fitting procedure with a spectrum I obtained with a spectrometer launched at 05 29 UT on 30 November 1971 from Woomera, South Australia. The spectrometer contained three (ADP, Gypsum and KAP) crystals each with an effective area of 50 cm^2 and collimated to 3' FWHM. The whole payload was pointed at N15 W35 (McMath Region 11621), however a small error in the pointing control system meant the centre of the active region was not viewed with maximum collimator efficiency. All three systems gave good spectra, and absolute intensities for 98 lines between 9 and 22.5 Å were measured using the reflectivities of Leigh and Evans (1972). Where the same strong line was observed on two or more crystals the intensities always agreed to better than 25%.

TABLE I

Intensities, oscillator strengths and abundances for the prime lines of
O VII, O VIII, Ne IX, Ne X and Mg X

Ion	Line intensity, E 10^{-6} erg cm^{-2} s^{-1}	Oscillator strength f	Abundance $A_z \times 10^{-5}$	$\Sigma_\Delta T_e (N_e^2 V)_{T_e}$ $G(T_e) \times 10^{41}$
O VII	91.0	0.69	50	17.0
O VIII	120.0	0.42	50	36.8
Ne IX	24.4	0.73	7	30.8
Ne X	4.7	0.42	7	10.3
Mg XI	8.2	0.74	6	11.9

TABLE II

A model of an active region based on
Equation (4) and Table I

Temperature interval $T_e \times 10^6$ K	Emission measure $N_e^2 V \times 10^{47}$ cm^{-3}
1.5–2.5	0.8
2.5–3.5	1.5
3.5–4.5	0.6
4.5–5.5	0.1

Table I shows the intensities of the prime lines of O VII, O VIII, Ne IX, Ne X and Mg XI together with the oscillator strength and abundances used (Pottasch, 1967). A model was constructed from this data using Equation (4) and is shown in Table II. We will refer back to this model and illustrate its usefulness later in the paper.

IV. Helium-like Ions

(a) THE 2^3S FORBIDDEN LINE

As the spectral resolution of the early observations increased it became apparent that on the long wavelength side of the helium-like resonance and intercombination lines there was a further line almost as strong as the resonance line. This was interpreted by

Gabriel and Jordan (1969a) as due to the magnetic dipole transition $1s^2 . {}^1S_0 - 1s2s . {}^3S_1$ which had previously been assumed to decay by a two-photon process. Gabriel and Jordan (1969b) investigated interchange between the 2^3S and 2^3P levels and found that as the density increased so the collision rate $2^3S \rightarrow 2^3P$ increased and competed with the spontaneous decay of the 2^3S level. In this way the intensity ratio, R, of the forbidden to the intercombination line depends on the electron density, N_e. Solving the statistical equilibrium equations for the ground state and the six $n=2$ levels gives

$$R = \frac{A(2^3S \rightarrow 1^1S)}{N_e C(2^3S \rightarrow 2^3P)(1+F) + A(2^3S \rightarrow 1^1S)} \left(\frac{1+F}{B} - 1\right) \tag{6}$$

(neglecting photo-excitation from 2^3S to 2^3P), where

$$F = \frac{C(1^1S \rightarrow 2^3S)}{C(1^1S \rightarrow 2^3P)} \tag{7}$$

and is usually taken as 0.35.

The effective branching ratio

$$B = \frac{1}{3} \frac{A(2^3P_1 \rightarrow 1^1S)}{A(2^3P_1 \rightarrow 1^1S) + A(2^3P \rightarrow 2^3S)} +$$
$$+ \frac{5}{9} \frac{A(2^3P_2 \rightarrow 1^1S)}{A(2^3P_2 \rightarrow 1^1S) + A(2^3P \rightarrow 2^3S)},$$

where the A's are spontaneous decay rates, and the C's are collision rates.

For low densities the collision rate $2^3S \rightarrow 2^3P$ will be small compared to the spontaneous decay rate $2^3S \rightarrow 1^1S$ and R will have a maximum value of

$$R_0 = \left(\frac{1+F}{B} - 1\right). \tag{9}$$

It is only when the density has increased sufficiently that the value of R will deviate from the value R_0. The density at which this happens is called the critical density and using the best available data Gabriel and Jordan (1972) give values of 7.3×10^9 cm^{-3} for O VII, 1.5×10^{11} cm^{-3} for Ne IX, 1.8×10^{12} cm^{-3} for Mg XI and 1.3×10^{13} cm^{-3} for Si XIII. Apart from the value for O VII these densities are much higher than are currently believed to exist in active regions.

The observations of Acton et al. (1972) referred to earlier have allowed them to directly measure the ratio R for O VII and Ne IX in several active regions during their rocket flight. Chambe's exponentially decreasing emission measure model was used and a characteristic temperature for the formation of each ion in each active region was deduced. The temperature dependence of the ratio R was calculated by Blumenthal et al. (1972) who found a strong dependence on temperature. Gabriel and Jordan (1973) disputed their calculations on the grounds that they had treated the contribution from dielectronic recombination incorrectly so finding a strong dependence on temperature. Figure 7 shows the calculations of Blumenthal *et al.* and Gabriel and Jordan together with the observations of Acton *et al.* for O VII and Ne IX. We conclude that there has been no observed variation of R with temperature, but under certain particular circumstances variations from the low density limit may be found.

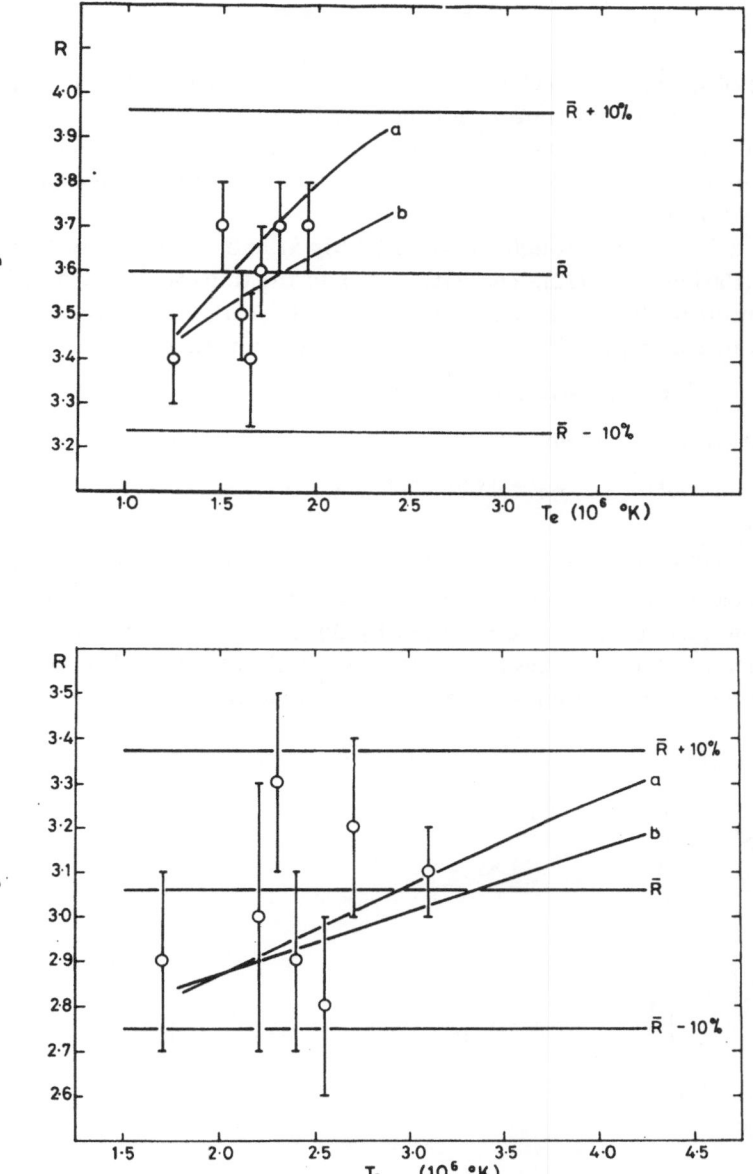

Fig. 7. The variation of R with T_e. (a) O VII. (b) Ne IX. The observations are by Acton *et al.* (1972), curve (a) is calculated by Blumenthal *et al.* (1972) and curve (b) by Gabriel and Jordan (1973).

56   JOHN H. PARKINSON

(b) SATELLITE LINES

Greater spectral clarity and increased instrument sensitivity also led to the discovery
of satellite lines on the long wavelength side of He-like resonance lines. The satellite
lines had been observed in laboratory plasmas for a number of years (e.g. Edlen and
Tyren, 1939) and they were classified as being transitions of the type

$$1s^2 . n1 - 1s2p . n1 .$$

It was the observation of such lines in the solar X-ray spectrum that prompted theoretical studies of the configurations giving these lines.

   We have already seen that the observations of Acton *et al.* (1972) in Figure 4 show
satellite lines in He-like Ne and O. Similar data has been obtained by Walker and
Rugge (1971) for silicon.

   Figure 8 is a scan through the lines of Mg XI and shows many satellites clearly
resolved (Parkinson, 1972). The letters refer to the transitions listed in Table III. A
similar spectrum has been reported by Bonnelle *et al.* (1973) from the experiment
discussed in Section 2. The lines marked R_3 are from transitions

$$1s^2 2s . ^2S_{1/2} - (1s2p \, ^1P) \, 2s . ^2P_{3/2,1/2}$$

those marked R_4 are from

$$1s^2 2p . ^2P_{1/2, 3/2} - 1s2p^2 . ^2D_{3/2, 5/2}$$

R_1 and R_2 are

   $1s^2 . n1 - 1s2p . n1$ with $n=4$ and 3 respectively (Summers, 1973). Figure 8 demonstrates clearly that the dominant method of formation of these weak lines is by
dielectronic recombination rather than by direct excitation of inner shell electrons.
This is because all of the lines from transitions of the type $1s^2 . 21 - 1s2p . 21$, for which
the upper levels are auto-ionising in the LS approximation with the $1s^2 + e$ continuum,

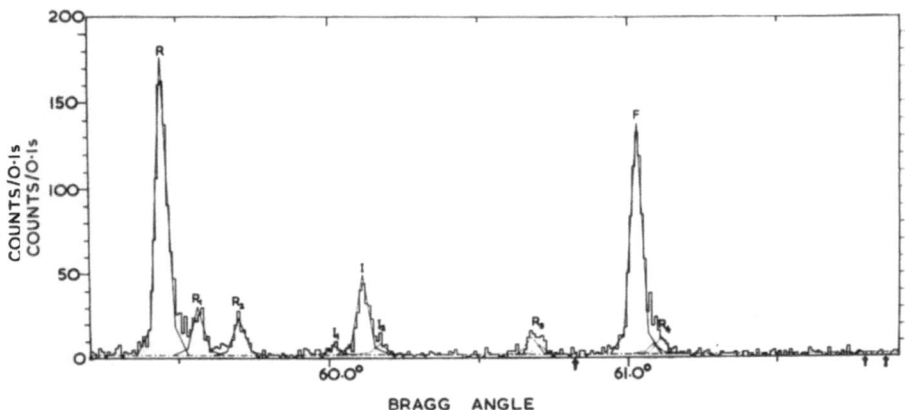

Fig. 8.   The lines of Mg XI observed with a collimated ADP spectrometer (Parkinson, 1972). The
letters refer to the transitions listed in Table III. The arrows indicate the positions of lines that would
be produced by direct excitation of inner shell electrons.

TABLE III
Observed and calculated wavelengths and relative intensities

	Transition Lower level	Upper level	Wavelength (Å) This work	Calculated	Relative Counts	Relative intensity Observed	Calculated
R	$1s^2$ $^1S_0$	$1s2p$ $^1P_1$	9.169	9.168	885	100	100
R_1	$1s^2 4l$	$1s2p\,4l$	9.180		146	16.5±1.5	
R_2	$1s^2 3l$	$1s2p\,3l$	9.193		115	12.9±1.2	
I_1	$1s^2 2p$ $^2P_{1/2}$ $^2P_{3/2}$	$1s2p^2$ $^2S_{1/2}$	9.223	9.221 9.224	22	2.5±0.7	0.5 1.0
I	$1s^2$ $^1S_0$	$1s2p$ $^3P_2$ $^3P_1$	9.232	9.227 9.230	235	26.1±1.8	
I_2	$1s^2 2s$ $^2S_{1/2}$	$(1s2p\,^3P)\,2s$ $^2P_{3/2}$ $^2P_{1/2}$	9.237	9.234 9.235	30	3.3±0.7	0.7 0.5
R_3	$1s^2 2s$ $^2S_{1/2}$	$(1s2p\,^1P)\,2s$ $^2P_{3/2}$ $^2P_{1/2}$	9.284 9.286	9.283 9.285	56 29	6.2±0.9 3.2±0.7	6.3 3.1
F	$1s^2$ $^1S_0$	$1s2s$ $^3S_1$	9.315	9.314	717	78.2±3.6	
R_4	$1s^2 2p$ $^2P_{1/2}$ $^2P_{3/2}$ $^2P_{3/2}$	$1s2p^2$ $^2D_{3/2}$ $^2D_{3/2}$ $^2D_{5/2}$	9.319 9.323	9.320 9.323 9.324	33 52	3.6±0.7 5.7±0.8	3.4 1.0 5.5

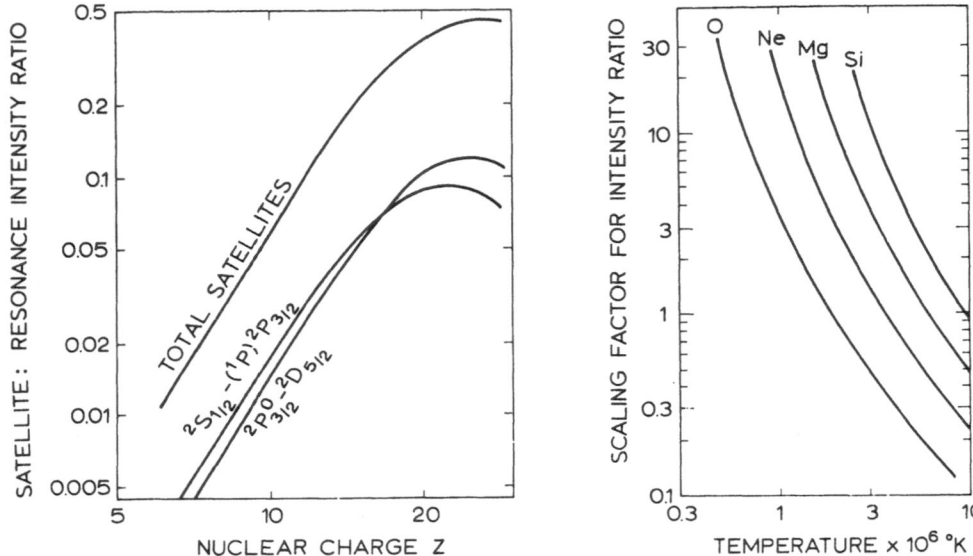

Fig. 9. Satellite line intensities (Gabriel, 1972). (a) Variations of satellite: resonance intensity ratio with nuclear charge, Z, at the characteristic temperature T_m. (b) Scaling factors for intensity ratios at other temperatures.

are observed and none of the lines from non-auto-ionising upper levels (e.g. $1s^2 2p \cdot {}^2P - 1s2p^2 \cdot {}^2P$ at 9.302 Å) are observed.

Note that the absolute wavelength accuracy of the strong lines here is ± 0.001 Å and the resolving power of this spectrometer, using an ADP crystal is close to 5000.

The intensity of these satellite lines has been investigated in detail by Gabriel (1972). He found that the ratio of the intensity of the lines marked R_3 in Figure 8 to the intensity of the resonance line, R, is strongly temperature dependent. Figure 9 summarises these calculations and shows how the satellite line intensity varies with nuclear charge, Z, and with temperature. Thus the observations of the satellite lines is another important diagnostic tool since the variation of the satellite line intensity with temperature is different from the variation of the resonance line intensity with temperature.

TABLE IV

Relative intensities of the O, Ne and Mg lines
emitted at different temperatures

Ion	Temperature interval $\times 10^6$ K			
	1.5–2.5	2.5–3.5	3.5–4.5	4.5–5.5
O VII	0.56	0.40	0.04	–
O VIII	0.17	0.63	0.19	0.01
Ne IX	0.09	0.63	0.25	0.03
Ne X	–	0.25	0.61	0.14
Mg XI	0.02	0.40	0.46	0.12

We are now in a position to compare the accuracy of Gabriel's calculated variation of satellite resonance intensity with the observations in Figure 8 and the model in Table II. Table IV lists the fraction of each line emitted in each temperature bin so by reference to Figure 9 we can predict a satellite: resonance ratio for the whole model. This gives

Ratio satellite, R_3 : resonance, R

	Observed	Predicted
Ne IX	0.017	0.039
Mg XI	0.093	0.130

The line R_3 is somewhat weaker than predicted, this may be due to slight errors in the rate coefficients used. For Mg XI we can compare the ratio of $R_3 : R_4$ and find good agreement between the observed ratio of 9.4 : 9.3 and the predicted ratio of 4.6 : 4.5.

V. Hydrogen-Like Ions

The ratio of the intensities of the Lα and Lβ lines for a hydrogen-like ion is sensitive to temperature. Hutcheon and McWhirter (1973) have investigated this temperature variation for various density regimes. For the hydrogen-like ions observed in the solar corona and discussed in the present paper the variation is very slow, a factor two change in temperature giving only a 10% change in the Lyman ratio. Hutcheon and McWhirter point out that the laboratory theta pinch observations give measured Lyman ratios for C VI and N VII of 60–80% of their theoretically predicted values and explain the discrepancy as due to self absorption in the plasma. As recent solar observations of O VIII and Ne X give similarly reduced ratios this explanation is open to question.

Satellite lines to Lα lines have been observed in the solar spectrum of Mg XII by Walker and Rugge (1971). These lines are very weak and so may not prove to be useful for plasma diagnostic work.

VI. Neon-Like Ions

(a) Fe XVII

We have seen from Figures 1 and 2 how lines of Fe XVII dominate the solar spectrum below 25 Å. It is only recently that we have been able to account accurately for their intensities.

The strong lines correspond to transitions from the $2p^5 3s$ and $2p^5 3d$ levels to the ground state $2p^6 . {}^1S_0$ and occur around 17 Å and 15 Å respectively. It has long been recognised (Pottasch, 1966) that each excited level cannot be accounted for by the simple coronal equilibrium conditions, where excited levels are populated solely by collisional excitation. The early theoretical studies Garstang (1966) and Froese (1967) did not allow for the population of certain levels by cascades from higher levels. An investigation by Beigman and Urnov (1969) showed the true complexity of the problem in that the $2p^53s$ levels are populated by the cascade chain.

$$2s2p^63d \rightarrow 2s^22p^53d \rightarrow 2s^22p^53p \rightarrow 2s^22p^53s$$

Unfortunately they were only able to treat the complete multiplets and were not able to deal with individual transitions. The problem has recently been solved by Loulergue and Nussbaumer (1973) who considered excitations to and cascades between 36 levels, and calculated the relative intensities of the 8 observed lines for transitions from $2p^53s$, $2p^53d$ and $2s2p^63p$ to the ground state. These relative intensities were found to be in agreement with the recent observations of Parkinson (1973) shown in Figure 10.

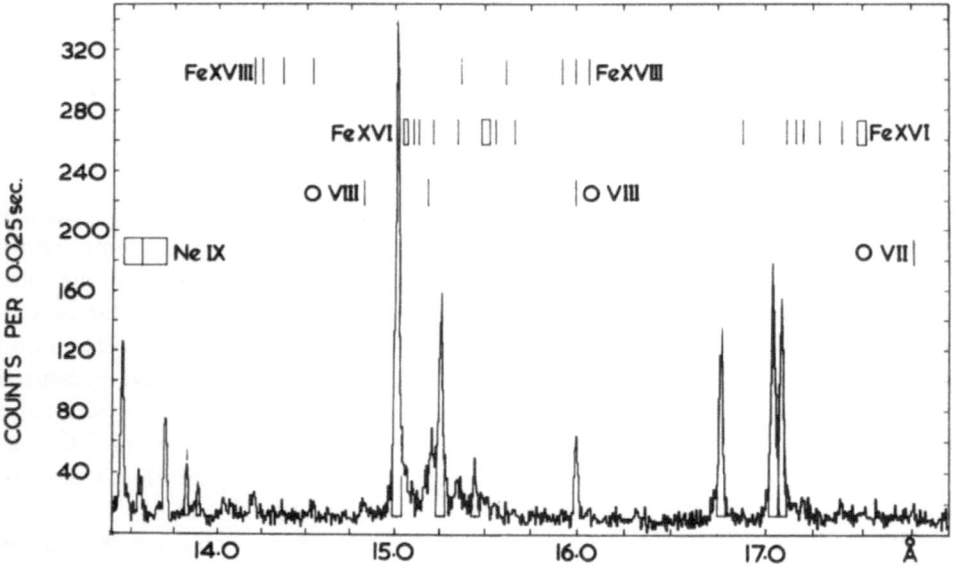

Fig. 10. Part of a collimated spectrum showing the dominance of Fe xvii between 13 and 18 Å. (Parkinson, 1973). The intensities of these lines are given in Table V.

In the light of the more recent observations it is now clear that the line ratios derived from older observations are considerably in error, mainly because the Fe xvii lines had not been clearly resolved. Figure 11 shows the $2p^53s$ lines around 17 Å. This section of spectrum is particularly interesting as the $^3P_1$ and $^3P_2$ lines at 17.041 and 17.086 Å are clearly resolved. The $^3P_2$ transition is particularly interesting as this is a magnetic quadrupole transition, i.e. $\Delta J = 2$, the $^3P_2$ level being the lowest level of the first excited state (Garstang, 1969).

Many of the energy levels giving higher transitions in Fe xvii are not known accurately and extrapolations along series' are difficult. The $2p^54d$ lines at 12 Å are well

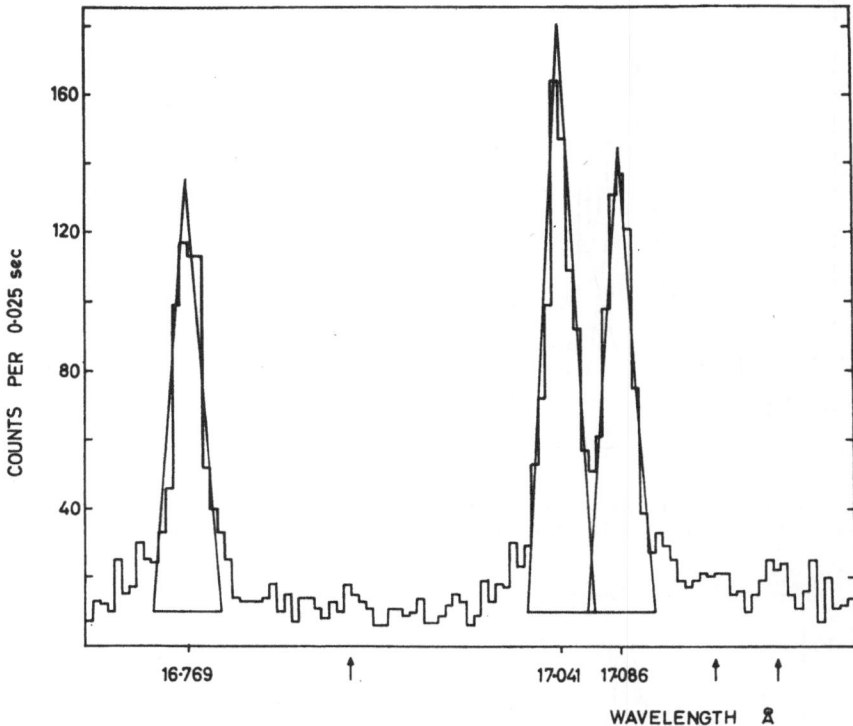

Fig. 11.   Enlargement of part of Figure 10 showing the 3s Fe XVII lines at 17 Å. The $^3P_1$ and $^3P_2$ lines at 17.041 and 17.086 Å are clearly resolved. The arrows denote the positions of possible satellite lines.

known but the $4d.\,^1P_1$ resonance line at 12.121 Å is generally blended with the Ne x Lα line at 12.132 Å. Figure 12 shows these lines clearly resolved for the first time together with the $4d\,^3D_1$ line at 12.263 Å for comparison. The observed ratio of $^1P_1/^3D_1$ is 1.2 and this compares very favourably with the value of 1.1 calculated by Froese. The $5d\,^1P_1$ and $^3D_1$ lines are identified at 11.130 and 11.251 Å and the $6d\,^1P_1$ and $^3D_1$ lines at 10.660 and 10.771 Å. The ns series is more difficult, the 4s lines are expected to lie between 12.50 and 12.80 Å but this region of the spectrum is very crowded as can be seen in Figure 13 which shows the same section of spectrum observed simultaneously on a KAP and a Gypsum crystal.

(b) RELATIVE INTENSITIES OF Fe XVII LINES

Loulergue and Nussbaumer referred all of their intensities to the $3d\,^1P_1$ line at 15.013 Å as the population of this upper level by cascades was very small. If a reliable oscillator strength for this line can be calculated then by using the model in Table II the abundance of Fe can be calculated. To a first approximation I have used the oscillator strength of Froese of 2.22 and this gives an Fe abundance of 6.4×10^{-5} in good agreement with the value obtained by Pottasch (1967) of 6.5×10^{-5}.

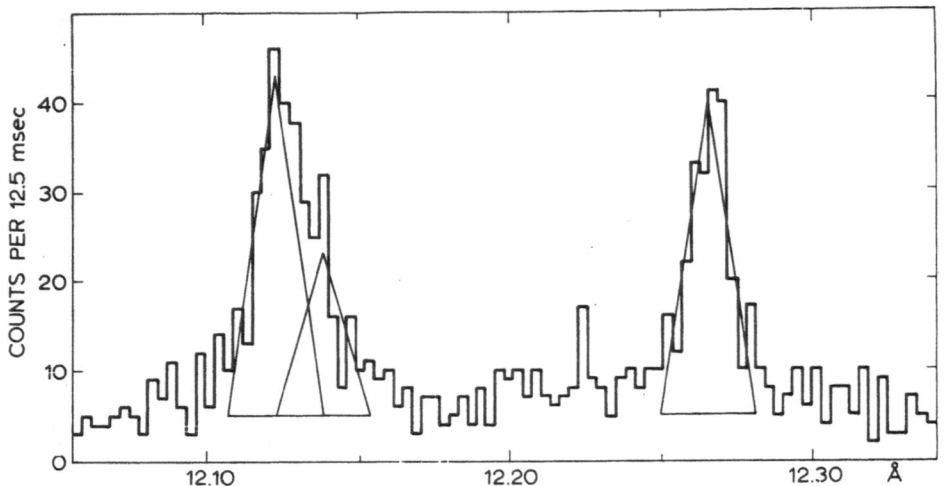

Fig. 12.   A continuation of the spectrum in Figure 10 showing the $4d$ Fe XVII lines at 12 Å. The $4d\,^1P_1$ Fe XVII line at 12.121 Å and the Ne X Lα line at 12.132 Å are clearly resolved.

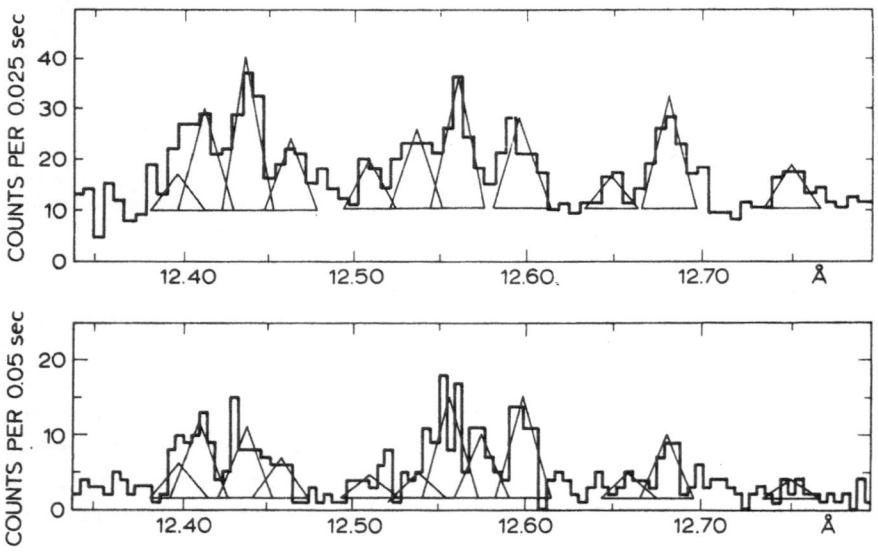

Fig. 13.   The complex spectrum between 12.3 and 12.8 Å observed simultaneously with collimated KAP (upper) and Gypsum (lower) crystals. The Ni XIX $3d$ lines are at 12.436 and 12.659 Å; the Fe XVII $4d\,^3P_1$ and $4s$ lines also occur in this part of the spectrum.

In Table V I have listed all of the other Fe XVII lines that it has been possible to identify together with their intensities relative to the 15.013 Å line. Using the abundance just derived I have calculated oscillator strengths and collision strengths. These are also given in Table V.

TABLE V

Observed intensities of Fe xvii lines relative to the intensity of 15.013 Å line

Transition		Observed			
Lower level	Upper level	Wavelength λ, Å	Intensity E, $\times 10^{-6}$ erg cm^{-2} s^{-1}	Oscillator strength f	Effective collision strength
$1s^22s^22p^6.^1S_0$	$1s^22s^22p^53s.^1P_1$	16.769	71.6	0.94	0.050
$1s^22s^22p^6.^1S_0$	$1s^22s^22p^53s.^3P_1$	17.041	103.7	1.35	0.074
$1s^22s^22p^6.^1S_0$	$1s^22s^22p^53s.^3P_2$	17.086	81.8	1.07	0.058
$1s^22s^22p^6.^1S_0$	$1s^22s^22p^53d.^1P_1$	15.013	111.3	2.22	0.107
$1s^22s^22p^6.^1S_0$	$1s^22s^22p^53d.^3D_1$	15.259	53.4	1.06	0.052
$1s^22s^22p^6.^1S_0$	$1s^22s^22p^53d.^3P_1$	15.449	12.8	0.26	0.013
$1s^22s^22p^6.^1S_0$	$1s^22s\ 2p^63p.^1P_1$	13.824	7.9	0.21	0.009
$1s^22s^22p^6.^1S_0$	$1s^22s\ 2p^63p.^3P_1$	13.889	4.4	0.12	0.005
$1s^22s^22p^6.^1S_0$	$1s^22s^22p^54s.^1P_1$	⎰ 12.509	1.0	0.04	0.001
		⎱ 12.599	2.8	0.10	0.004
$1s^22s^22p^6.^1S_0$	$1s^22s^22p^54s.^3P_1$	⎰ 12.681	2.6	0.09	0.004
		⎱ 12.751	0.9	0.03	0.001
$1s^22s^22p^6.^1S_0$	$1s^22s^22p^54d.^1P_1$	12.121	10.5	0.37	0.014
$1s^22s^22p^6.^1S_0$	$1s^22s^22p^54d.^3D_1$	12.263	8.8	0.31	0.012
$1s^22s^22p^6.^1S_0$	$1s^22s^22p^54d.^3P_1$	⎰ 12.398	1.0	0.04	0.001
		⎱ 12.409	2.6	0.09	0.004
$1s^22s^22p^6.^1S_0$	$1s^22s^22p^55d.^1P_1$	11.130	2.9	0.13	0.005
$1s^22s^22p^6.^1S_0$	$1s^22s^22p^55d.^3D_1$	11.251	3.2	0.14	0.005
$1s^22s^22p^6.^1S_0$	$1s^22s^22p^56d.^1P_1$	10.660	1.3	0.06	0.002
$1s^22s^22p^6.^1S_0$	$1s^22s^22p^56d.^3D_1$	10.771	2.0	0.09	0.003

(c) Ni xix

Ni xix is also a Ne-like ion but makes a much weaker contribution to the solar spectrum than Fe xvii due to its low abundance and the higher temperatures needed to excite the ion. The Ni xix $3s\ ^1P_1$, $^3P_1$ and $^3P_2$ lines are observed at 13.776, 14.037 and 14.081 Å respectively together with the $3d\ ^1P_1$, $^3D_1$ lines at 12.436 and 12.659 Å. A similar procedure to that used for Fe xvii above gives an Ni abundance of 3.6×10^{-6}, after assuming an oscillator strength of 2.2 for the $3d.^1P_1$ transition.

References

Acton, L. W., Catura, R. C., Meyerott, A. J., Wolfson, C. J., and Culhane, J. L.: 1972, *Solar Phys.* **26**, 183.
Batstone, R. M.: 1970, Private communication.
Batstone, R. M., Evans, K., Parkinson, J. H., and Pounds, K. A.: 1970, *Solar Phys.* **13**, 389.
Beigman, I. L. and Urnov, A. M.: 1969, *Optika Spectrosk.* **27**, 380.
Blake, R. L., Chubb, T. A., Friedman, H., and Unzicker, A. E.: 1965, *Astrophys. J.* **142**, 1.
Blumenthal, G. R., Drake, G. W. F., and Tucker, W. H.: 1972, *Astrophys. J.* **172**, 205.
Bonnelle, C., Senemaud, C., Senemaud, G., Chambe, G., Guionnet, M., Henoux, J. C., and Michard, R.: 1973, *Solar Phys.* **29**, 341.
Brabban, D. H.: 1973, Thesis, University of London.
Brabban, D. H., Glencross, W. M., and Herring, J. R. H.: 1971, *New Techniques in Space Astronomy*, p. 135.

Chambe, G.: 1971, *Astron. Astrophys.* **12**, 210.

Edlen, B. and Tyren, F.: 1939, *Nature* **143**, 940.

Evans, K. and Pounds, K. A.: 1968, *Astrophys. J.* **152**, 319.

Fritz, G., Kreplin, R. W., Meekins, J. F., Unzicker, A. E., and Friedman, H.: 1967, *Astrophys. J.* **148**, L133.

Froese, C.: 1967, *Bull. Astron. Inst. Neth.* **19**, 86.

Gabriel, A. H.: 1972, *Monthly Notices Roy. Astron. Soc.* **160**, 99.

Gabriel, A. H. and Jordan, C.: 1969a, *Nature* **221**, 947.

Gabriel, A. H. and Jordan, C.: 1969b, *Monthly Notices Roy. Astron. Soc.* **145**, 241.

Gabriel, A. H. and Jordan, C.: 1972, *Case Studies in Atomic Collision Physics* **II**, North Holland, Amsterdam, p. 109.

Gabriel, A. H. and Jordan, C.: 1973, *Astrophys. J.* **186**, 327.

Garstang, R. H.: 1966, *Publ. Astron. Soc. Pacific* **78**, 399.

Garstang, R. H.: 1969, *Publ. Astron. Soc. Pacific* **81**, 488.

Hutcheon, R. J. and McWhirter, R. W. P.: 1973, *J. Phys. B.* **6**, 2668.

Jordan, C.: 1969, *Monthly Notices Roy. Astron. Soc.* **142**, 501.

Leigh, B. and Evans, K.: 1972, Private communication.

Loulergue, M. and Nussbaumer, H.: 1973, *Astron. Astrophys.* **24**, 209.

Neupert, W. M.: 1971, *Physics of the Solar Corona*, Reidel, Holland, p. 237.

Parkinson, J. H.: 1971a, *Nature Phys. Sci.* **233**, 44.

Parkinson, J. H.: 1971b, Thesis, University of Leicester.

Parkinson, J. H.: 1972, *Nature Phys. Sci.* **236**, 68.

Parkinson, J. H.: 1973, *Astron. Astrophys.* **24**, 215.

Parkinson, J. H. and Pounds, K. A.: 1971, *Solar Phys.* **17**, 146.

Pottasch, S. R.: 1964, *Space Sci. Rev.* **3**, 816.

Pottasch, S. R.: 1966, *Bull. Astron. Soc. Neth.* **18**, 443.

Pottasch, S. R.: 1967, *Bull. Astron. Soc. Neth.* **19**, 113.

Rugge, H. R. and Walker, A. B. C.: 1968, *Space Res.* **VIII**, North-Holland, Amsterdam, p. 439.

Summers, H. P.: 1973, *Astrophys. J.* **179**, L45.

Walker, A. B. C.: 1972, *Space Sci. Rev.* **13**, 672.

Walker, A. B. C.: 1975, This volume, p. 73.

Walker, A. B. C. and Rugge, H. R.: 1971, *Astrophys. J.* **164**, 181.

X-RAY SPECTROSCOPY OF SOLAR ACTIVE REGIONS
DURING THE THIRD SKYLAB MISSION

J. P. PYE, R. J. HUTCHEON, J. H. PARKINSON*, and K. A. POUNDS

Dept. of Physics, University of Leicester, England

Summary. This paper describes the analysis of soft X-ray spectra of solar active regions observed on a Skylark sounding rocket flight. The experiment was launched from Woomera, South Australia, at 0535 UT on 26th November 1973. The payload consisted of 3 plane scanning Bragg crystal spectrometers, covering the wavelength range 4 to 23 Å, and collimated to 3 arc min FWHM. The instrument was the same as that used by Parkinson (1972, 1973). The launch was part of a collaborative observing program with American Science & Engineering who obtained X-ray photographs of the Sun with the S0-54 ATM X-ray telescope, simultaneously with the Leicester observations. We report preliminary results for McMath region 12624 (S10 W28). This region was observed about 60 min after the peak of an importance −N, class CO flare.

In He-like ions the intensity ratio of the resonance line to dielectronic recombination satellites is sensitive to temperature. This temperature dependence has been calculated by Gabriel (1972). We have measured the ratios, and, using the stronger satellites in He-like O, Ne, Mg, Si, have derived electron temperatures for each ion (Table I).

TABLE I

Temperatures derived from satellite to resonance
line intensity ratios in He-like ions

Ion	Satellite to resonance line photon intensity ratio, $\dfrac{I(1s^2 2s\,^2S_{1/2} - (1s2p\,^1P)\,2s\,^2P_{1/2,\,3/2}}{I(1s^2\,^1S_0 - 1s2p\,^1P_1)}$	$T_e(10^6\ \mathrm{K})$
O VII	$\leqslant 0.04^{+0.08}_{-0.04}$	$\geqslant 1.1_{-0.4}$
Ne IX	0.011 ± 0.004	$6.3^{+2.7}_{-1.2}$
Mg XI	0.042 ± 0.009	$6.4^{+1.1}_{-0.7}$
Si XIII	$\leqslant 0.14 \pm 0.14$	$\geqslant 5.5_{-0.7}$

Analysis of the observed intensities of several of the resonance lines of the H- and He-like ions by the method of Batstone *et al.* (1970) indicates a total emission measure, in the temperature interval 1.5 to 6.5×10^6 K, of about 2.2×10^{47} cm^{-3}.

---

* Present address: Mullard Space Science Laboratory, University College London, England.

Sharad R. Kane (ed.), Solar Gamma-, X-, and EUV Radiation, 65–66. All Rights Reserved.

Measurements of the wavelengths and relative intensities of Ne IX lines and the dielectronic satellite lines are in good agreement with the calculations of Gabriel (1972). Three unclassified lines which may be due to satellites with the outer electron having $n \geqslant 3$ are observed at 13.463, 13.472 and 13.490 Å. The last 2 lines have been previously observed by Parkinson (1971). Several other lines which occur in the same wavelength range (13.26–13.83 Å) have been tentatively identified as transitions in Fe XVIII, XIX, XX.

We have obtained relative intensities of $n = 3 \rightarrow 2$ and $4 \rightarrow 2$ lines of Ne-like Fe and Ni and compared them with the calculations of Loulergue and Nussbaumer (1973, 1974). For the observed transitions there is good agreement between the measured and predicted intensity ratios. In cases where a statistically significant signal ($\gtrsim 4\sigma$) is not observed, the upper limits obtained for the intensities are consistent with the calculations. For the Fe XVII $n = 3 \rightarrow 2$ transitions the present observations are in good agreement with the observed intensity ratios of Parkinson (1973). The unpublished calculations of Loulergue and Nussbaumer are preliminary figures only.

Acknowledgments

The authors wish to express their thanks to Drs Loulergue and Nussbaumer for permission to quote their work in advance of publication. R.H. acknowledges receipt of an SRC Fellowship and J.P.P. acknowledges receipt of an SRC research studentship.

References

Batstone, R. M., Evans, K., Parkinson, J. H., and Pounds, K. A.: 1970, *Solar Phys.* **13**, 389.
Gabriel, A. H.: 1972, *Monthly Notices Roy. Astron. Soc.* **160**, 99.
Loulergue, M. and Nussbaumer, H.: 1973, *Astron. Astrophys.* **24**, 209.
Loulergue, M. and Nussbaumer, H.: 1974, Private communication.
Parkinson, J. H.: 1971, *Nature Phys. Sci.* **233**, 44.
Parkinson, J. H.: 1972, *Nature Phys. Sci.* **236**, 68.
Parkinson, J. H.: 1973, *Astron. Astrophys.* **24**, 215.

SPATIALLY RESOLVED X-RAY SPECTRA OF CORONAL ACTIVE REGIONS

R. C. CATURA, L. W. ACTON, and E. G. JOKI

Lockheed Palo Alto Laboratories, U.S.A.

and

C. G. RAPLEY and J. L. CULHANE

Mullard Space Science Laboratory, University College London, England

Summary. X-ray spectra from a number of coronal active regions were obtained during ATM support rocket flights carried out by the Lockheed group on June 11 and December 19, 1973. Multi-grid collimators were used to provide fields of view of 40″ diameter and 90″ diameter for a number of scanning crystal spectrometers and a bent crystal spectrometer which employed a position sensitive proportional counter to register the diffracted spectrum. A solar image was produced on film and on a TV camera on board the rocket with the aid of a 1 Å Hα filter. A small part of the X-ray collimator was used to generate a multiple spot diffraction pattern which was superimposed on the Hα image and the composite picture was transmitted to the ground. Pre-launch calibrations allowed the spot corresponding to the X-ray collimator axis to be identified and so the collimator pointing direction on the solar disc was controlled from the ground by means of commands sent to the rocket.

In the June 11 flight, spectra of both active region material and of the quiet corona were obtained in a number of different wavelength ranges with the scanning spectrometers and included emission features of Ne IX, Fe XVII, O VIII and O VII. The bent crystal spectrometer obtained a spectrum in the range 10–18 Å and so established the usefullness of this new technique (Catura *et al.*, 1974). For the flight of December 19, all of the spectrometers were collimated to 40″. The six wavelength ranges scanned by the plane spectrometers included emission features of Mg XI, Ne IX, Ne X, Fe XVII, Fe XVIII, O VIII and O VII. The bent crystal spectrometer was used in the range 16–17.2 Å and demonstrated its ability to resolve the closely spaced $3p_1$ and $3p_2$ Fe XVII lines at 17.041 Å and 17.086 Å.

Comparison of the collimated spectra with X-ray images obtained with the S0–56 telescope on ATM, kindly supplied by Dr J. H. Underwood, established that the higher temperature material is concentrated in compact features surrounded by more extended cooler material. The corona outside of the active regions was only detected in the lines of O VII and O VIII. Further analysis of these data will include a search for density effects in the He-like line ratios, the estimation of temperature from satellite line intensities and the construction of models of the relation between emission measure and temperature for different parts of the active regions.

The Lockheed work was supported by NASA-Ames Research Center.

References

Catura, R. C., Joki, E. G., Bakka, J. C., Rapley, C. G., and Culhane, J. L.: 1974, *Monthly Notices Roy. Astron. Soc.* **168**, 217.

CLASSIFICATION OF NEW SPECTRAL LINES OF Fe XVII OBSERVED IN SOLAR ACTIVE REGIONS

R. J. HUTCHEON

Dept. of Physics, University of Leicester, England

Summary. This summary reports new emission lines of Fe XVII observed in two coronal active regions (McMath regions 12624, 12628). The lines were recorded by a Bragg plane crystal spectrometer collimated to 3′ FWHM, with a gypsum crystal scanning the wavelength range 8–14 Å. This spectrometer, part of a three spectrometer package previously used by Parkinson (1972, 1973), was mounted aboard a Skylark rocket (SL 1206) launched from Woomera at 0535 UT on 26th November 1973. One of the active regions studied (12624) was observed about 60 minutes after the peak of an importance $-N$ class CO flare.

TABLE I

Measured and predicted wavelengths for newly identified Fe XVII transitions

Classification		Wavelength	
Pair coupling notation	LS coupling notation (upper level)	Predicted	Observed
$2s^2\,2p^6\,{}^1S_0 - 2s^22p^5({}^2P_{3/2})\,4s\,[3/2]_2$	$2p^54s\;{}^3P_2$	12.696	12.698
$2p^5({}^2P_{3/2})\,5s\,[3/2]_2$	$2p^55s\;{}^3P_2$	11.422 ⎞	11.419 B
$2p^5({}^2P_{3/2})\,5s\,[3/2]_1$	$2p^55s\;{}^3P_1$	11.415 ⎠	
$2p^5({}^2P_{1/2})\,5s\,[1/2]_1$	$2p^55s\;{}^1P_1$	11.282	11.285 B
$2p^5({}^2P_{3/2})\,7s\,[3/2]_2$	$2p^57s\;{}^3P_2$	10.545 ⎞	10.544
$2p^5({}^2P_{3/2})\,7s\,[1/2]_1$	$2p^57s\;{}^3P_1$	10.543 ⎠	
$2p^5({}^2P_{1/2})\,7s\,[1/2]_1$	$2p^57s\;{}^1P_1$	10.429	10.431
$2p^5({}^2P_{1/2})\,7d\,[3/2]_1$	$2p^57d\,{}^1P_1$	10.382	10.385
$2p^5({}^2P_{1/2})\,8d\,[3/2]_1$	$2p^58d\,{}^1P_1$	10.217	10.217

B: Blend (see text)

The new lines are listed in Table I with their measured and predicted wavelengths. The proposed classifications are given in the pair coupling notation as this is the most appropriate. (The statement by Kastner *et al.* (1967) that the LS coupling scheme best describes the $2p^53d$ terms in Fe XVII does not apply to the other $2p^5$ ns, nd terms). However since the LS designations of many Fe XVII terms appear in the literature, the corresponding LS classifications are also given to aid reference.

Each wavelength was measured twice, once in each active region, except the 10.431 Å and 10.544 Å lines which were only observed in the decaying flare region. The measured wavelengths given for the other lines are the average of two values which generally differed by less than 0.005 Å. The final uncertainty is probably within 0.003 Å.

Sharad R. Kane (ed.), Solar Gamma-, X-, and EUV Radiation, 69–70. All Rights Reserved.

The predicted wavelengths were extrapolated from previously known Fe XVII terms (Swartz *et al.*, 1971; Tyren, 1938) by Ritz's Law (Edlen, 1964). To satisfy one requirement of this procedure the base and predicted terms were limited to those described by the same vector coupling scheme (pair coupling). The predicted wavelengths should be accurate to within 0.005 Å, the main source of uncertainty being that in the base wavelengths, and therefore agree satisfactorily with experiment. These identifications are supported further by the smooth decrease of intensities towards the series limit.

The 11.285 Å and 11.419 Å features coincide in wavelength with laboratory lines classified as Fe XVIII $2p^5-2p^4 4d$ transitions by Swartz *et al.* (1971). The intensities of other Fe XVII and Fe XVIII lines observed in our results suggest that these features are blends with significant contributions from the Fe XVIII transitions mentioned above and the Fe XVII transitions shown in Table I. The lines between 10.2 Å and 10.6 Å may possibly be blended with currently unknown Fe XVIII $2p^5-2p^4 5d$ transitions which are expected to lie in this region. It is intended to discuss this further in a future publication.

The lines at 10.217 Å and 11.419 Å in our scans may account for lines at 10.23 Å and 11.41 Å reported, but not identified, by Neupert *et al.* (1973). However, their 10.23 Å feature may also be due to Ne X $L\beta$ (at 10.239 Å). These authors also report a line at 10.86 Å. This may be due to the Fe XVII transition $2p^6\ ^1S_0-2p^5\ (^2P_{3/2})$ $6s[\frac{1}{2}]_1$, for which we predict the wavelength 10.849 Å.

References

Edlen, B.: 1964, 'Atomic Spectra', in *Handbuch der Physik* **27**, Ed. by S. Flügge, Springer-Verlag, Berlin, p. 80.
Kastner, S. O., Omidvar, K., and Underwood, J. H.: 1967, *Astrophys. J.* **148**, 269.
Neupert, W. M., Swartz, M., and Kastner, S. O.: 1973, *Solar Phys.* **31**, 171.
Parkinson, J. H.: 1972, *Nature (Phys. Sci.)* **236**, 68.
Parkinson, J. H.: 1973, *Astron. Astrophys.* **24**, 215.
Swartz, M., Kastner, S. O., Rothe, E., and Neupert, W. M.: 1971, *J. Phys. B.* **4**, 1747.
Tyren, F.: 1938, *Z. Phys.* **111**, 314.

STATISTICAL METHODS IN THE IDENTIFICATION AND PREDICTION OF THE SOLAR X-RAY SPECTRAL LINES

SHAHINAZ YOUSEF

Astronomy Department, Faculty of Science, Cairo University, Cairo, Egypt

Summary. With the accumulation of numerous identifications of the solar X and XUV spectral lines, the need is felt for systemizing such identification. The results are presented below.

The degree of ionization distribution of spectral lines at different wavelength bands is shown in Figure 1, where each stroke represents a spectral line that has been identified in the solar spectrum. A preliminary list has been used in preparation of this figure.

It is obvious that the X-ray spectral lines with wavelength below 60 Å form two main sequences. The first one contains lines of light elements from carbon to sulphur. It starts from the XVI degree of ionization at $\lambda \geqslant 2$ Å and proceeds to lower stages of ionizations at longer wavelengths where it ends at the vth stage for λ about 40 Å.

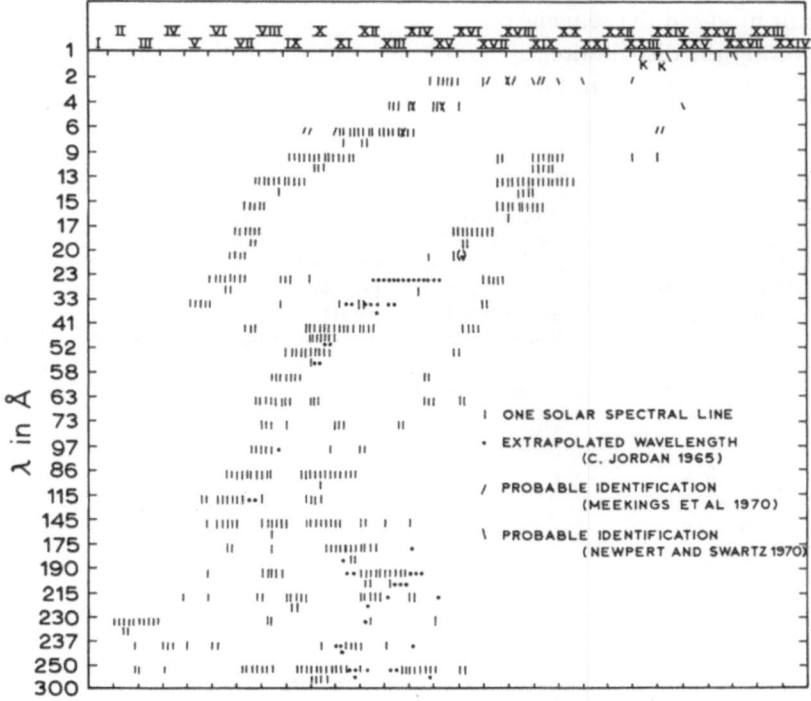

Fig. 1. The degree of ionization distribution of spectral lines at different wavelength bands. Each stroke represents a spectral line that has been identified in the solar spectrum.

Sharad R. Kane (ed.), Solar Gamma-, X-, and EUV Radiation, 71–72. All Rights Reserved.
Copyright © 1975 by the IAU.

The second main sequence contains lines of heavier ions and is expected by extrapolation to start from the XXIII or XXIV degree of ionization for $4 < \lambda < 2$ Å and extends to the IX degree of ionization at $\lambda \leqslant 63$ Å.

There is a gap of lines in the observed solar spectrum in this sequence between 20–40 Å. However, a number of lines that would fit nicely in this gap was theoretically predicted by Jordan in 1965. It is thus suggested that more lines are expected to be observed in the solar spectrum in order to fill this gap. This expectation has been confirmed by Freeman and Jones (1970), who identified one line of S XII at 36.39 Å and reported three other unidentified lines in the vicinity of 35 Å. Forecast is also made for the identification of more spectral lines in the XIX to XXV stages of ionization and wavelength between 2–9 Å is made. This was partially confirmed by the identification of two Fe XXIII lines at 8.56 Å and 8.8 Å and a Fe XXII line at 8.98 Å (Doschek *et al.*, 1973).

In addition to these two main sequences, there might be an iron lines subsequent between 23 and 60 Å.

It is also noticed that some other lines lie off the sequence; these, in my opinion, are probably wrong and need re-identification. For instance, the lines at ~ 11.0 Å and 12 Å identified by Chubb *et al.* (1964) as due to transitions in Fe XXIV and XXIII respectively are wrong. The first one was later re-identified by Connerade *et al.* (1970) in laboratory spectrum as due to Fe XIX. This identification would then make the line fit nicely in the sequence.

On plotting the distribution of ionization potential with wavelength for the same lines shown in Figure 1, all lines form one sequence only.

References

Chubb, T. H., Friedman, H., Kreplin, R. W.: 1964, *Space Res.* **IV**, 759.
Connerade, J. P., Peacock, N. J., and Speer, R. J.: 1970, *Solar Phys.* **14**, 159.
Doschek, G. A., Meekins, J. F., and Cowan, Robert D.: 1973, *Solar Phys.* **29**, 125.
Freeman, F. F. and Jones, F. B.: 1970, *Solar Phys.* **15**, 288.
Jordan, Carole: 1965, Ph.D. Thesis, University of London.
Meekins, J. F., Dorschek, G. A., Friedman, H., Chubb, T. A., and Kreplin, R. W.: 1970, *Solar Phys.* **13**, 198,
Neupert, W. and Swartz, M.: 1970, *Astrophys. J.* **160**, L189.

INTERPRETATION OF THE X-RAY SPECTRA OF
SOLAR ACTIVE REGIONS*

ARTHUR B. C. WALKER, JR.

*Institute for Plasma Research and Department of Applied Physics,
Stanford University, Stanford, Calif., U.S.A.*

Abstract. This paper presents a review of recent analytical studies of the coronal X-ray spectrum below 25 Å. The techniques used to compute the theoretical coronal spectrum, and the currently available atomic rate constant data are reviewed first. Spectroscopic techniques which have been proposed for the determination of coronal temperature and density structure, and the results derived from their application to coronal spectra are also reviewed.

A number of coronal models based on X-ray observations have been developed recently, and the coronal temperature structure and composition predicted by these models is discussed, and compared with models of the corona and transition region derived from studies of the solar EUV spectrum.

I. Introduction

In this review the interpretation of the soft X-ray spectrum of the solar corona, and the development of models of coronal composition and density and temperature structure will be discussed.

Comprehensive reviews of spectroscopic observations of the corona in the grazing incidence region of the spectrum ($\lambda \lesssim 300$ Å), and of the techniques used to interpret these observations, were presented at the Third Symposium on Ultraviolet and X-ray Spectroscopy of Astrophysical and Laboratory Plasmas by Doschek (1972) for flare observations, and by Walker (1972) for the non-flaring Sun. Culhane and Acton (1974) have reviewed more recent work.

In the preceding paper Dr Parkinson (1975) has reviewed the observational techniques used, and the results of recent spectroscopic studies of the coronal spectrum below 25 Å. In the present paper, the analytical techniques used to interpret the coronal X-ray spectrum, and the coronal models derived with these techniques will be emphasized.

II. Techniques for the Construction of Coronal Models from Spectroscopic Observations

Walker (1972) and Gabriel and Jordan (1972) have reviewed the calculation of the coronal spectrum, and the spectroscopic techniques which have been used to construct models of the corona. Consequently, I shall only briefly summarize the status of calculations of the coronal spectrum, with emphasis on the most recent results. The spectroscopic techniques used for the development of coronal models have also been adequately treated in the above references, and a brief summary will suffice here.

The development of coronal models from coronal permitted line observations was first discussed in a general way by Pottasch (1964, 1966, 1967). Pottasch's analysis

* This work was supported by NASA Grant, NGR 05-020-695.

was confined to transition region lines whose excitation functions have a sharply peaked dependence on temperature, and he was able to replace the explicit temperature dependence of the emission function of each line by an average emission function evaluated at the temperature of most efficient emission for that line. A number of authors (Batstone *et al.* (1970), Chambe (1971) and Walker (1972)) have revised Pottasch's analysis for coronal lines, by including the temperature variation of each line explicitly. Walker *et al.* (1974a) have presented a systematic formulation of this approach. Since the spectroscopically resolved observations of the corona which have been available prior to the launch of the Apollo Telescope Mount (ATM) solar observatory on Skylab have not had sufficient spatial resolution to allow the physical structure of coronal X-ray emitting regions to be studied, analyses of these observations have ignored the geometry of the emitting region. The observation of X-ray emitting arch filament systems by Skylab (Vaiana *et al.*, 1973; Vaiana, 1974; Underwood, 1974) has emphasized the importance of the magnetic structure of coronal active regions. Accordingly, I will conclude this section with a discussion of the modification in the analytical approach developed by Pottasch required to take account of the geometrical constraints imposed by the coronal magnetic structure.

(a) CALCULATION OF THE CORONAL LINE SPECTRUM

The intensity observed in an emission line (with upper level i and lower level j) in ionization stage z of element Z is given by

$$I_{Zzij} = \int \int \int E_{Zzij} [T_e(x_v), n_e(x_v)] \, dx_1 \, dx_2 \, dx_3, \qquad v = 1, 2, 3. \qquad (1)$$

The integral is over the coronal volume unresolved by the spectroscopic observation.
  The emission function for a particular line may be written

$$E_{Zzij} = \frac{hc}{\lambda_{ij}} A_{ij}^r \left[n_{zi}(T_e, n_e) + \sum_{n'l' > n_i + 1} n_{(z-1)in'l'}(T_e, n_e) \right], \qquad (2)$$

where A_{ij}^r is the transition probability for transition ij, n_{zi} is the population of the excited level i in the ion whose charge is z, and $n_{(z-1), in'l'}$ is the population of the doubly excited level in the ion whose charge is $(z-1)$, with the same configuration as the level i, but with an additional electron in the level $n'l'$. As a number of authors have pointed out (see, for example, Walker (1972) and Gabriel and Jordan (1972)), if $n'l'$ is sufficiently large (in practice $> n_i + 1$) the transitions $in'l' \to jn'l'$ cannot be resolved from the transition $i \to j$, and therefore effectively increase the flux in the line $i \to j$. For $n' < n_i + 1$ the transition $in'l' \to jn'l'$ can be resolved as a satellite line from $i \to j$, and in that case the emission is given by

$$E_{Zzij} = \frac{hc}{\lambda_{ij}} A_{ijn'l'}^r [n_{(z-1)in'l'}(T_e, n_e)], \qquad (3)$$

where we have no longer made the assumption $A_{ij}^r \simeq A_{ijn'l'}^r$.

Gabriel and Jordan (1972) have discussed the equation of statistical equilibrium which must be solved in order to calculate the state population functions. For most cases of interest, we may considerably simplify the equation of statistical equilibrium, and consequently, the excitation function can be expressed in simple form. For singly excited states, direct collisional excitation is the principal mechanism of population, and we may express the upper state population as

$$n_{zi} = \frac{n_e n_{zg}(T_e)\, \alpha_{gi}^{ex}(T_e)}{\Gamma_i^r} \tag{4}$$

where Γ_i^r is the total radiative width of the state i [i.e. $\Gamma_i^r = \sum_k A_{ik}^r$] and $\alpha_{gi}^{ex}(T_e)$ is the collisional excitation rate from the ground state.

We may express the ground state density, n_{zg}, in terms of the relative population of the ionization stage of interest, $a_z{}^*$, the relative abundance of the element, A_Z, and the number of hydrogen atoms per electron, a_H. This latter quantity is approximately constant since the hydrogen and helium in the corona are completely ionized.

$$n_{zg} = (n_{zg}/n_Z)\,(n_Z/n_H)\,(n_H/n_e)\,n_e = a_z A_Z a_H n_e.^{**}$$

We shall also have occasion to use the fractional population of an excited state

$$n_{zi} = (n_{zi}/n_Z)\,(n_Z/n_H)\,(n_H/n_e)\,n_e = f_{zi} A_Z a_H n_e.$$

The emission function can be written as

$$E_{Zzij}^{ex} = \frac{hc}{\lambda_{ij}}\frac{A_{ij}^r n_e^2 a_H A_Z a_z(T_e)\,\alpha_{gi}^{ex}(T_e)}{\Gamma_i^r}. \tag{5}$$

This equation is called the coronal excitation equation. We shall now discuss the modifications to Equation (5) which are necessary in order to discuss the line emission functions of ions for the hydrogen-like through neon-like isoelectronic sequences which are responsible for the coronal lines below 25 Å. We shall refer to the ions which have no metastable excited levels, and no levels with strong excitation rates from the ground level which have forbidden decays, as belonging to Class I. The coronal excitation equation will provide a good approximation in calculating the spectrum of these ions. Hydrogenic ions and lithium-like ions satisfy these criteria.

If an ion has a metastable excited level, this level will generally be strongly coupled to adjacent levels by radiative and collisional processes. It is then possible to solve the equation of statistical equilibrium for this group of strongly coupled levels separately. The population of other levels in the ion will still be governed by the coronal excitation equation. An example of this situation is provided by the metastable $1s2s\,^3S$ state in

---

* Jordan (1969, 1970) and Landini and Fossi (1972) have calculated a_z for the abundant elements.
** This equation is not strictly valid if the ion contains metastable states which may have an appreciable steady state population.

helium-like ions, which is strongly coupled with the $1s2p\ ^3P_{2,1,0}$ states. The relative intensities of the lines from these ions becomes density dependent at sufficiently high densities. The density dependent emission functions of the triplet states of helium-like ions has been discussed extensively in the literature (Gabriel and Jordan, 1972, 1973). We shall refer to ions with metastable excited levels as belonging to Class II.

In deriving Equation (5), we neglected population processes such as radiative recombination, and cascade from higher levels. The assumption that these processes are of minor importance is generally valid as long as the decay of the higher levels to the ground level is not forbidden. If such decays are forbidden, these levels will contribute to the population of the lowest lying excited levels by cascade processes, and Equation (5) will no longer describe the level populations. If there are no metastable states, we must solve a set on N simultaneous equations, which involve population of each level by collisional excitation from the ground term, and by radiative decay from higher levels. The level population will then be expressed as a sum of terms such as appear in Equation (5), with coefficients which depend on the radiative decay rates from the levels responsible for the cascade. Walker et al. (1974c) show that the emission function may be written

$$E_{ij}^{ex}(T_e, n_e) = \frac{hc}{\lambda_{ij}} A_{ij}^r n_e n_{z,g} \frac{\sum_{k=i}^{N} \alpha_{gk}^{ex} |\Gamma_m^r \delta_{mp} - A_{mp}^r|}{\prod_{k=i}^{N} \Gamma_k^r}, \tag{6}$$

where $m = i+1, \ldots N$; $p = i, \ldots k-1, k+1, \ldots, N$ (i.e., $p \neq k$) and $A_{mp}^r = 0$ for $p \geq m$. We have defined the total radiative width of the state m, $\Gamma_m^r = \sum_{k=1} A_{mk}^r$. The symbol δ_{mp} refers to the Kronecker delta function $[\delta_{mp} = 1, m = p, \delta_{mp} = 0\ m \neq p]$.

The neon-like ion Fe XVII provides an example of an ion of this type. Figure 1 (taken from Walker et al., 1974c) shows the level structure of the excited levels of Fe XVII with principal quantum number 3, and the principal decay paths of each level. We will refer to ions of this type as Class III.

A number of ions have metastable low lying excited states within the ground term, or within a low lying excited term. At sufficiently high temperatures and densities, the population in these low lying excited states can become substantial, and may exceed the population of the ground level. The coronal excitation equation must quite obviously be modified for these ions. Generally, the populations of the ground term, and of the low-lying excited states may be computed by regarding these levels as strongly coupled, and solving the statistical equilibrium equation ignoring all higher levels. Once the populations of these metastable levels have been determined, we can calculate the population of excited states by applying Equation (5) to each highly populated level. If $f_g, f_{g'}, f_{g''}, \ldots$ are the fractional populations of these levels then we may write the emission function of a line as

$$E_{Zzij}^{ex} = \frac{hc}{\lambda_{ij}} \frac{A_{ij}^r n_e^2 a_H A_z a_z(T_e) \sum f_{zg} \alpha_{gi}^{ex}(T_e)}{\Gamma_i^r}. \tag{7}$$

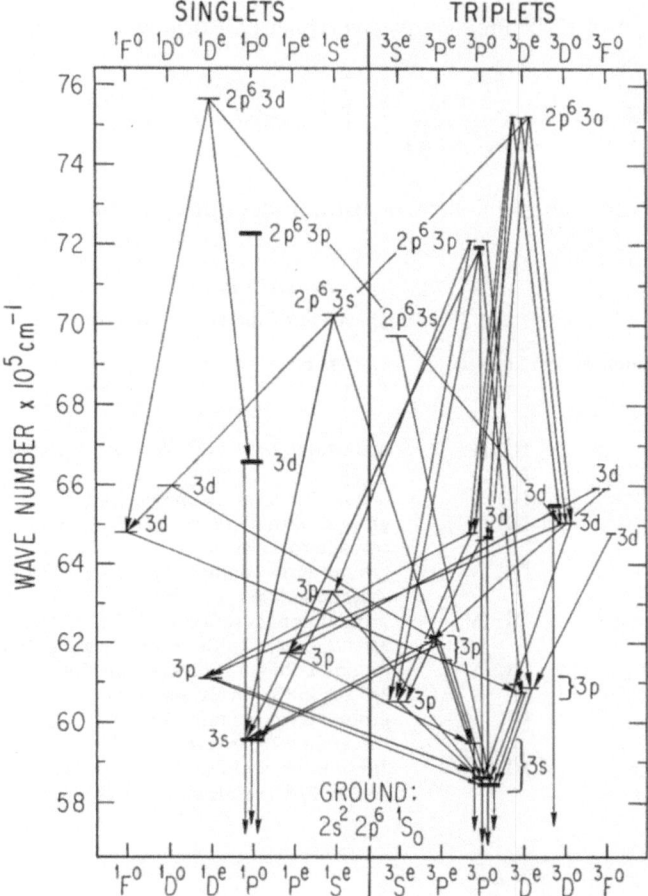

Fig. 1. Level diagram for Fe XVII, showing the principal decay paths for the first 36 excited levels. Those levels with transitions to the ground level are shown in bold type.

Examples of ions with metastable low lying excited states are those of the beryllium-like and boron-like isoelectronic sequences. The density dependent populations of the $2S^2\,{}^1S$ ground state and $2s2p\,{}^3P_{2,1,0}$ excited states of beryllium-like ions have been extensively discussed by Jordan (1971) and Munro *et al.* (1971). We will refer to ions with metastable low lying excited states as Class IV. In Table I, we have classified the ions of the ten simplest isoelectronic sequences.

The collisional excitation function which appears in Equation (5) is generally expressed in terms of a collision strength, Ω_{gi}, which is defined as

$$\sigma = \frac{\pi a_0^2 \Omega_{gi} I_{\mathrm{H}}}{(2J_g+1)\,E},$$

where σ is the excitation cross section, a_0 the Bohr radius, I_{H} the rydberg, J_g the spin of the initial state, and E the incident electron energy. For a Maxwellian distribution, the excitation rate is

$$\alpha_{gi}^{ex} = \frac{2\sqrt{2}}{(2J_g + 1)} \frac{\pi a_0^2}{\sqrt{m\pi kT}} I_H \int\limits_{E_i}^{\infty} \Omega_{gi} \exp\left(- E/kT\right) \frac{dE}{kT}. \tag{8}$$

E_i is the threshold energy of the excitation under consideration.

TABLE I

Classification of the H- Ne-like Ions

Sequence	Class	Comments
H I	I	
He I	II, III	metastable level $1s2s\ ^3S$; $1sn\ ^3L$ cascade to $1s2\ ^3L$
Li I	I	
Be I	IV	metastable levels $2s2p\ ^3P_{2,1,0}$
B I	III, IV	ground term contains $2p\ ^2P_{3/2,\ 1/2}\ 3p\ ^2P$ levels decay by cascade to $3s\ ^2S$
C I	III, IV	ground term contains $2p^2\ ^3P_{2,1,0}$, $2p3p$ levels decay by cascade to $2p3s$
N I	III, IV	ground configuration contains $2p^3\ ^4S_{3/2}$ and $2p^3\ ^2D_{5/2,\ 3/2}$; $2p^23p$ levels decay by cascade to $2p^23s$
O I	III, IV	ground configuration contains $2p^4\ ^3P_{2,1,0}$, $2p^4\ ^1D$ and $2p^4\ ^1S$; $2p^33p$ levels decay by cascade to $2p^33s$
F I	III, IV	ground term contains $2p^5\ ^2P_{1/2,\ 3/2}$, $2p^43p$ levels decay by cascade to $2p^43s$ levels
Ne I	III	$2p^53p$ levels and $2s2p^63l$ levels decay by cascade to $2p^53s$ and $2s^22p^53l$ levels respectively.

The collision strength can be expressed in terms of the oscillator strength, f, and the gaunt factor g, using the modified Bethe approximation of Van Regemorter (1962).

$$\Omega_{gi} = \frac{8\pi}{\sqrt{3}} \frac{I_H}{E} f_i g_i. \tag{9}$$

For two excitations which are similar, we may presume that the gaunt factors are equal, and derive the useful relation

$$\Omega_{gi}(E_i) = \frac{f_i}{f_{i'}} \frac{E_{i'}}{E_i} \Omega_{gi'}\left(\frac{E_{i'}}{E_i} E\right). \tag{10}$$

In the succeeding subsections, we will briefly discuss the determination of the atomic rate parameter required to evaluate Equations (5), (7), (8), and (10). Emphasis will be placed on the calculation of collisional excitation cross sections. References for oscillator strengths and transition probabilities are given by Gabriel and Jordan (1972) and Walker (1972).

(i) *Hydrogenic Ions*

Burgess *et al.* (1970) have calculated the collisional excitation cross sections for the $1s-2s$ and $1s-2ps$ excitations in hydrogenic ions, using the Coulomb Born II approximation.

Walker (1972) has tabulated the temperature averaged collision strength $z^2\Omega$ for these transitions

$$\overline{z^2\Omega} = \frac{1}{2J_g + 1} \int\limits_1^\infty d\varepsilon \{z^2\Omega \exp[-(\varepsilon - 1)/\theta]\}/\theta$$

(where $\varepsilon = E/E_i$ and $\theta = kT/E_i$) using the collision strengths given by Burgess *et al.* We may obtain the collision strength for higher excitations ($1s-ns$ and $1s-np$) by using Equation (10). Recently, Walker (1974) has calculated the cross section for $1s-2s$ and $1s-2p$ excitation using relativistic wave functions. He finds substantial corrections to the results of Burgess *et al.*, for large Z, however, for values of Z of importance for coronal studies, ($Z \leqslant 28$) the corrections are small.

(ii) *Helium-like Ions*

Until quite recently, there were few calculations available for the collision strengths of excitations in helium-like ions. Collision strengths were derived from the hydrogenic calculations of Burgess *et al.* for the singlet (^1S and ^1P) excitations using Equation (10) by Walker (1972), and by Gabriel and Jordan (1972) using the R matrices tabulated by Burgess *et al.* Gabriel and Jordan were able to derive collision strengths for excitation of both the singlet (^1P and ^1S) and triplet (^3P and ^3S) lines. Tully (1974) has recently calculated the excitation cross section for the $1sns$ $^1S(n=2,\ldots 6)$ and $1\ snp$ $^1P(n=2,\ldots 6)$ levels in the Coulomb Born Approximation. The cross section obtained by this more accurate technique are equal, within a few percent, to the cross section derived by Walker (1972), using Equation (10), from the hydrogenic cross sections, as shown by Table II.

TABLE II

Comparison of O VII collision strengths

Energy (in units of threshold energy)		1	2	3
$[(f\lambda)_{\rm OVII1s-2p}/(f\lambda)_{\rm OVIIIL\alpha}]$	$z^2\Omega^{\rm CBO}(1\ ^2S-2\ ^2P)$[a]	2.765	4.171	5.444
	$z^2\Omega^{\rm CB}\ (1\ ^1S-2\ ^1P)$[b]	2.825	4.675	5.975
$[(f\lambda)_{\rm OVII1s-2p}/(f\lambda)_{\rm OVIIIL\alpha}]$	$z^2\Omega^{\rm CBO}(1\ ^2S-2\ ^2S)$[a]	0.676	0.718	0.741
	$z^2\Omega^{\rm CB}\ (1\ ^1S-2\ ^1P)$[b]	0.803	0.868	0.880
$[(f\lambda)_{\rm OVII1s-3p}/(f\lambda)_{\rm OVIIIL\alpha}]$	$z^2\Omega^{\rm CBO}(1\ ^2S-2\ ^2P)$[a]	0.508	0.767	1.000
	$z^2\Omega^{\rm CB}\ (1\ ^1S-3\ ^1P)$[b]	0.615	0.996	1.256

[a] Tully (1974).
[b] Burgess *et al.* (1970).

However, it is recommended that the more accurate cross section of Tully be adopted for future work. For the triplet levels, the cross sections given by Gabriel and Jordan are the most reliable available.

(iii) *Lithium-like Ions*

Gabriel and Jordan (1972) and Sampson and Parks (1974) have given extensive accounts of theoretical and experimental studies of the cross sections of the lithium-like ions.

Bely (1966a, 1966b) and Bely and Petrini (1970) and Flower and Launay (1972) have calculated threshold collision strengths for the transitions $2s - nl$ and $2p - nl$ in lithium-like ions.

The calculations of Bely and Bely and Petrini appear to result in good agreement with observation, except for the $2s - 3p$ excitation, where the cross sections of Burke *et al.* (1966) appear to result in closer agreement. Heroux and Cohen (1971), Heroux *et al.* (1972), and Malenovsky and Heroux (1973) have analyzed the relative intensities of the lines of the lithium-like ions O VI, Ne VIII, and Mg X. They find that their observed intensities are consistent with the theoretical cross sections.

(iv) *Beryllium-Like Ions*

Jordan (1971) and Munro *et al.* (1971) have discussed the density dependent population of the $2s^2\ ^1S$ and $2s2p\ ^3P$ levels of beryllium-like ions, and have computed the statistical equilibrium of these levels as a function of density for C III, N IV, O V, and Si XI. Once the populations of the ground level, and of the three metastable levels have been calculated, the modified coronal excitation equation (Equation (7)) can be used to calculate the populations of the higher excited levels. Gabriel and Jordan (1972) and Sampson and Parks (1974) have discussed the theoretical and experimental studies of collision strengths for the beryllium-like ions. The most extensive theoretical calculations are those of Eissner which are unpublished. However, Gabriel and Jordan quote a number of Eissner's results, and compare then with the experimental observations of excited rates by Tondello and McWhirter (1971) and Johnson and Kunze (1971) for Ne VII and Si XI. The agreement is fair for most transitions, however, there are substantial differences for some transitions. In view of the difficulty of interpreting the experimental observations, which require the determination of the population of the $2s2p\ ^3P$ metastable levels in a transient laboratory plasma, the theoretical cross sections, where available, are likely to be more reliable.

(v) *Neon-Like Ions*

Loulergue and Nussbaumer (1973) and Walker *et al.* (1974c) have recently discussed the coronal spectrum of Fe XVII. Because the $2s^2 2p^5 3s$ levels are populated primarily by cascade, the collisional excitation rates and radiative decay rates for the first 36 excited levels must be known before the relative and absolute intensities of the principal lines of Fe XVII, which lie between ~ 12 and ~ 17 Å, can be calculated. Figure 1 illustrates the complexity of the spectrum of Fe XVII. However, Loulergue (1971, 1973) and Garstang (1969) have calculated all of the required transition rates, and Bely and

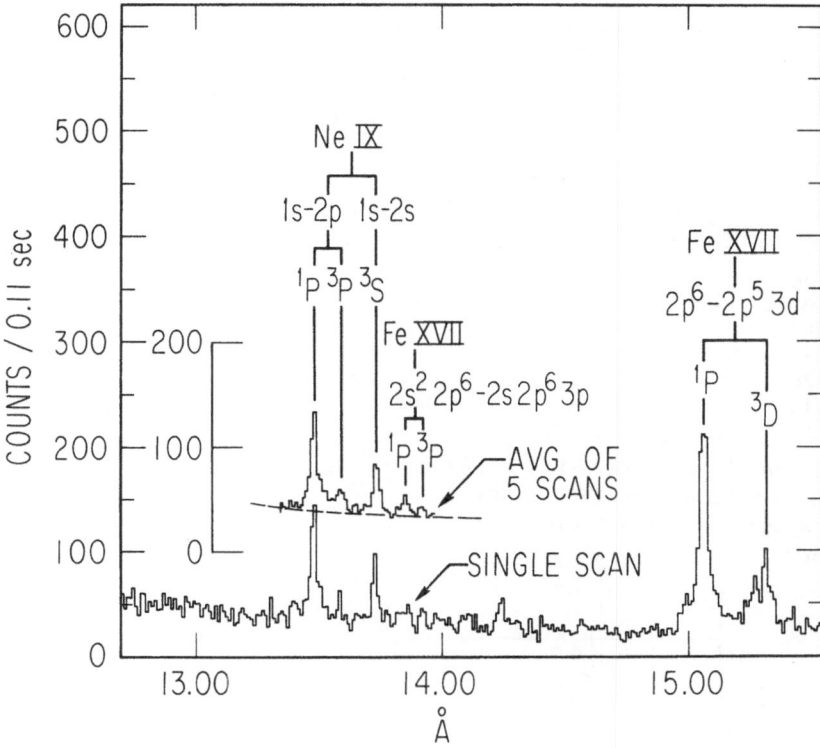

Fig. 2. Portion of the spectrum recorded by the KAP spectrometer on O VI-17 for 1969, March 20, showing the Fe XVII $2s^2 2p^6\ ^1S - 2s2p^6 3p\ ^1P$ and ^3P lines.

Bely (1967) and Flower (1971) have calculated threshold collision strengths for the transitions $2p - 3s$, $3p$, and $3d$. For the $2s - 3s$, p, d singlet excitations, Bely and Bely have estimated collision strengths by using cross sections for lithium-like ions. Using these atomic rate constants, Loulergue and Nussbaumer have calculated the relative intensity of the eight lines observed in the coronal spectrum from $2p^5 3s$, $2s2p^6 3p$, and $2p^5 3d$ configurations of Fe XVII. In table III these results (column 1a) are compared with the line intensities observed by Walker *et al.* (1974c) and Parkinson (1975). The observations of Walker *et al.* are shown in Figures 2 and 3. Loulergue and Nussbaumer have estimated the cross sections for the $2s2p^6 3s$, $3p$ and $3d$ triplet excitations by assuming that the collision strengths are equal to the singlet collision strengths multiplied by the statistical weights of the states. The relative intensities derived with this assumption are given in column 1b of Table III. Walker *et al.* (1974c) have made somewhat different assumptions regarding the $2s - 3s$, $3p$ and $3d$ excitation rates. They made use of the collision strengths for beryllium-like ions calculated by Eissner to derive the $2s - 3s$ and $2s - 3d$ singlet excitation rates, and used the observed intensities of the $2s2p^6 3p$ singlet and triplet lines, and the intensities of the $2s^2 2p^5 3s\ ^1P$ and ^3P lines to derive collision strengths for the $2s2p^6 3s\ ^3S$ and $2s2p^6 3d\ ^3D$ levels.

TABLE III

Comparison of experimental and theoretical line intensities in Fe xvii

Transition		Theory			Experiment		
	(Å)	(1a)	(1b)	(2)	4 Jan. 1967 (2)	20 Mar. 1969 (2)	30 Nov. 1971 (3)
$3s-2p$	17.10	0.23	0.74	0.59	⎰ 1.25	0.75	0.73
$3s-2p$	17.05	0.58	0.97	0.91	⎱	0.80	0.90
$3s-2p$	16.77	0.38	0.54	0.51	0.70	0.72	0.65
$3d-2p$	15.45	0.008	0.08	0.09	0.05	0.09	0.11
$3d-2p$	15.26	0.33	0.37	0.37	0.43	0.56	0.48
$3d-2p$	15.01	1.00	1.00	1.00	1.00	1.00	1.00
$3p-2s$	13.88	0.002	0.10	0.03	–	0.03	0.04
$3p-2s$	13.82	0.15	0.15	0.07	–	0.07	0.07
$4d-2p$	12.26	–	–	–	0.07	–	0.08
$4d-2p$	12.12	–	–	–	0.04	–	0.09

(1a) Loulergue and Nussbaumer ($T=5.5\times10^6$) [$\Omega_{2s^22p^6\ ^1S-2s2p3l\ ^3L}\equiv0$]

(1b) Loulergue and Nussbaumer ($T=5.5\times10^6$)

(2) Walker, Rugge, and Weiss ($T=4.0\times10^6$)

(3) Parkinson

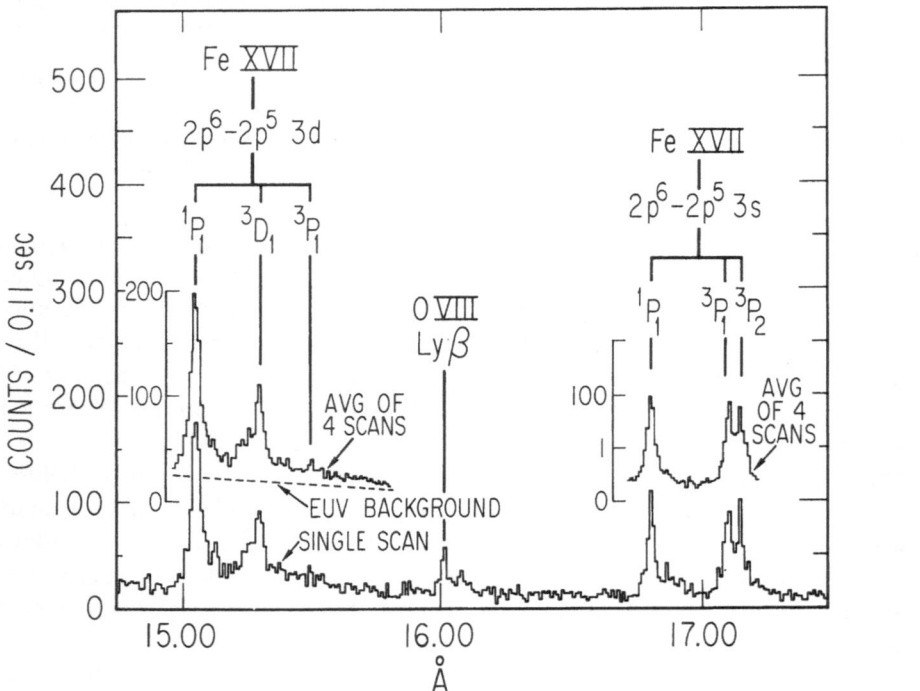

Fig. 3. Portion of the spectrum recorded by the KAP spectrometer on O vi-17 for 1969, March 20, showing the Fe xvii $2s^22p^6\,^1S-2s^22p^53d\,^1P$, ^3D, and ^3P, and $2s^22p^6\,^1S_0-2s^22p^53s\,^1P_1$, $^3P_1$ and $^3P_2$ lines.

TABLE IV

Threshold collision strengths for Fe XVII

Configuration	Level	Collision strength ($\times 10^3$)	
		Bely and Bely	Walker et al.
$2s3p^63d$	$^1D_2$	65.2	73
	$^3D_1$	–	100
$2s2p^63p$	$^1P_1$	22.4	10.1
	$^3P_1$	–	6.5
$2s2p^63s$	$^1S_0$	32.4	30.1
	$^3S_1$	–	45

TABLE V

Parentage of Fe XVII lines observed in the corona

Level	$2s^22p^53s$		$2s^22p\,^53p$								
	$^3P_1$	$^1P_1$	$^3S_1$	$^3D_2$	$^3D_1$	$^1D_2$	$^3D_0$	$^1P_1$	$^3P_1$	$^1P_2$	$^1S_0$
$2p^53s\,^3P_2$	–	–	0.949	0.390	0.048	0.446	–	–	0.068	0.007	–
$2p^53s\,^3P_1$	1.00	–	0.051	0.610	0.952	0.549	0.892	0.451	0.600	0.015	0.500
$2p^53s\,^3P_0$	–	1.00	–	–	–	0.005	0.108	0.549	0.332	0.978	0.500
$2p^53d\,^3P_1$	–	–	–	–	–	–	–	–	–	–	–
$2p^53d\,^3D_1$	–	–	–	–	–	–	–	–	–	–	–
$2p^53d\,^1P_1$	–	–	–	–	–	–	–	–	–	–	–

Level	$2s^22p^53d$							
	$^3P_0$	$^3P_1$	$^1F_3$	$^3P_2$	$^3D_1$	$^3F_2$	$^1D_2$	$^1P_1$
$2p^53s\,^3P_2$	0.947	0.097	0.424	0.621	–	0.007	0.106	–
$2p^53s\,^3P_1$	0.048	0.022	0.576	0.370	0.003	0.423	0.503	–
$2p^53s\,^3P_0$	0.005	–	0.0005	0.009	–	0.570	0.391	–
$2p^53d\,^3P_1$	–	0.881	–	–	–	–	–	–
$2p^53d\,^3D_1$	–	–	–	–	0.997	–	–	–
$2p^53d\,^1P_1$	–	–	–	–	–	–	–	0.999

Level	$2s2p^63s$		$2s2p^63p$		$2s2p^63d$	
	$^3S_1$	$^1S_0$	$^3P_1$	$^1P_1$	$^3D_1$	$^1D_2$
$2p^53s\,^3P_2$	0.594	–	0.186	0.010	0.266	0.204
$2p^53s\,^3P_1$	0.287	0.598	0.139	0.021	0.316	0.369
$2p^53s\,^3P_0$	0.119	0.402	0.054	0.010	0.074	0.149
$2p^53d\,^3P_1$	–	–	–	–	0.197	–
$2p^53d\,^3D_1$	–	–	–	–	0.119	0.073
$2p^53d\,^1P_1$	–	–	–	–	0.013	0.154
$2s2p^63p\,^3P_1$	–	–	0.621	–	0.015	0.005
$2s2p^63p\,^1P_1$	–	–	–	0.959	–	0.046

The relative intensities derived by Walker *et al.* are compared with the observations in Table III. The collision strengths derived by Walker *et al.* are compared with those derived by Bely and Bely in Table IV. The relative contribution of each level to the flux observed in each of the lines of Fe XVII as calculated by Walker *et al.* is shown in Table V. The arrays tabulated in Table V are the products

$$\frac{A_{ij}^r |\Gamma_m^r \delta_{mp} - A_{mp}^r|}{\prod\limits_{k=i}^{N} \Gamma_k^r} \tag{11}$$

required for the evaluation of Equation (6).

(vi) *Calculation of the Coronal Line Spectrum*

Detailed studies of the excitation of ions of the boron-like* through flourine-like sequences have not yet been carried out. However, a number of authors (Tucker and Koren, 1971; Mewe, 1972; and Landini and Fossi, 1970) have calculated the coronal line spectrum below 100 Å. The calculations of Tucker and Koren are the most complete. For ions of the beryllium-like through the fluorine-like isoelectronic sequences, Tucker and Koren have scaled the collision strengths of Bely and Petrini (1970) for the $2p - nl$ excitations in lithium-like ions, using Equation (10). Mewe, on the other hand, has derived effective gaunt factors for a large number of transitions, using the best available theoretical and experimental data for each isoelectronic sequence. Neither author includes the effects of cascades. The general character of the coronal spectrum is shown in Figure 4, which is based on extensive unpublished calculations by the author, in collaboration with Mrs. Kay Weiss at the Aerospace Corporation. Because of the requirement for accurate calculations of the coronal X-ray spectrum in order to interpret the X-ray filtergrams obtained by the S-054 and S-056 experiments on Skylab, extensive calculations of the coronal line spectrum are in progress at American Science and Engineering by Drs Vaiana and Kreiger and their associates, and at the Aerospace Corporation by Drs McKenzie and Walker. The results of these calculations should be available within a year of the date of the present Symposium.

(vii) *Dielectronic Recombination Lines and Satellite Lines*

The major process responsible for the population of doubly excited levels in thermal equilibrium is dielectronic recombination (Gabriel and Paget, 1972) for most isoelectronic sequences. However, for certain isoelectronic sequences, such as neon-like ions (Walker *et al.*, 1974c) inner shell excitation can be an important process. For those cases where dielectronic recombination dominates

$$E_{Zzij}^d = \frac{hc}{\lambda_{ij}} A_{ij}^r n_e^2 a_H A_Z a_z (T_e) \sum_{n'l' > n_i + 1} \frac{\alpha_{gin'l'}^{di} (T_e)}{\Gamma_{in'l'}^r + \Gamma_{in'l'}^a}, \tag{12}$$

---

* Widing (1966) has studied the excitation of some lines of the boron-like ion Si x.

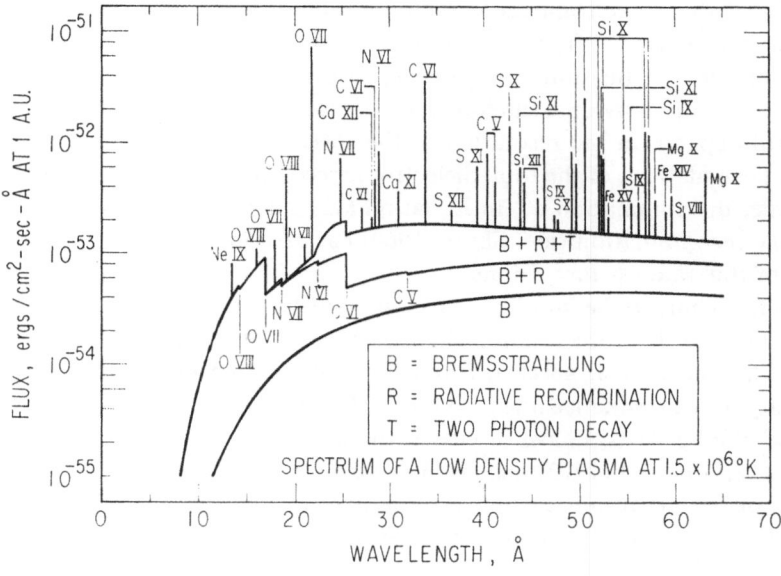

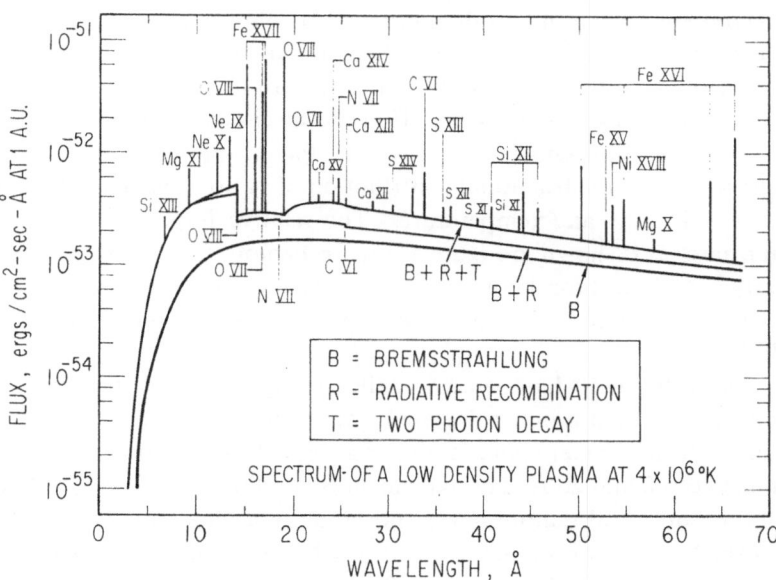

Fig. 4.   Coronal Spectrum for 1.5×10^6 K and 4.0×10^6 K. Only allowed transitions have been shown in order to avoid confusion, so that a number of strong lines (such as the helium-like forbidden lines) are not shown. Note the importance of two photon decay for the continuum at 1.5×10^6 K.

where $\alpha_{\text{g}in'l'}^{\text{di}}$ is the dielectronic recombination rate for the formation of the state $in'l'$ in ionization stage $(z-1)$ from the ground state of ionization stage z, and $\Gamma_{in'l'}^{\text{a}}$ is the autoionizing width of the state. The same equation without the summation is the emission function of an individual satellite line. The lower limit of the summation in equation 12 will depend on the resolution of the spectrometer making the observations. Since most calculations of the total dielectronic recombination rate have assumed that $n'l'$ is large, the summation α^{di} in Equation (12) may be taken equal to the total dielectronic recombination rate. Shore (1969) has carried out the most accurate computation of this rate for low Z ions.

For highly charged helium-like ions the radiative width for states with $n' \sim n_i$ becomes comparable to the autoionizing width, and the satellites with $n' = n_i + 1$, and $n' = n_i$ become prominent spectral features (Walker and Rugge, 1971), approaching the intensity of the resonance line for Fe xxv (Grineva *et al.*, 1973). Tucker and Koren (1971) and Walker (1972) have discussed the evaluation of the emission function in Equation (12). Gabriel (1972) has given a definitive account of the wavelengths and intensities of the important satellites to the lines of the helium-like ions which arise from the $1s2s2p$ and $1s2p^2$ configurations of the lithium-like ion.

(viii) *The X-Ray Continuum*

Three processes make substantial contributions to the coronal X-ray continuum, radiative recombination, bremsstrahlung, and two photon decay of the metastable $2s\,^2S$ and $1s2s\,^1S$ levels of hydrogenic and helium-like ions. Culhane (1969) has given a complete account of the calculation of the spectrum for the first two processes. Walker (1972) and Tucker and Koren (1971) have discussed the calculation of the spectrum for two photon decay of hydrogenic and helium-like ions. Walker *et al.* (1974a) have tabulated the fractional contribution of various elements, and processes for several wavelengths and temperatures. The relative importance of the various continuum processes is shown in Figure 4. Figure 4 emphasizes the importance of two photon decay at temperatures between $\sim 1 \times 10^6$ K and 4×10^6 K, and at wavelengths $\lambda > 20$ Å. The hydrogenic and helium-like ions of oxygen and carbon are chiefly responsible for this flux.

Walker *et al.* (1974a, b) have made detailed comparisons of the observed X-ray continuum, and the continuum calculated for a coronal model derived from line fluxes. Figure 5 (from Walker *et al.*, 1974b) compares the experimental and theoretical continuum between 8 and 25 Å. The agreement is good between 8 and 14 Å, however, above 15 Å, the observed continuum is too large by a factor of ~ 2. Walker *et al.*, suggest that the 'excess' continuum is due to weak unresolved lines from ions such as Fe xviii (~ 16 Å), Cr xv–Cr xi (~ 16–22 Å), Ca xvi–Ca xi (~ 17–25 Å), and Ar xv–Ar xvi (~ 23–25 Å).

(b) DIAGNOSTIC TECHNIQUES FOR THE CORONAL PLASMA

A number of techniques have been developed which, in principle, allow the temperature and density structure of the coronal plasma to be determined without the development of a detailed model. These techniques depend on line intensity ratios which are sensitive to a particular parameter such as temperature or density. However, as we shall see, some care must be exercised in interpreting such intensity ratios in terms of isothermal models.

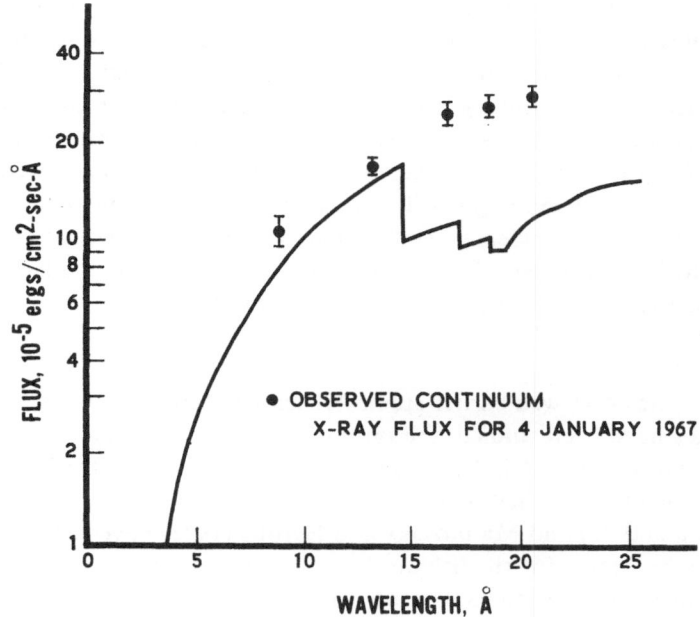

Fig. 5. Continuum fluxes observed on the O VI-10 satellite compared with the fluxes calculated from a coronal model derived from observed line fluxes. The model used is the one shown in Figure 7.

(i) *Density*

The dependence of line intensities on density comes about in two ways:

(1) Metastable excited levels may be competitively de-excited by radiative decay or collisional transfer to an adjacent level which has an allowed decay.

(2) Some ions have compound ground states whose upper level(s) are metastable, or low-lying metastable excited states. These metastable states become populated at sufficiently high densities and the intensities of lines which are preferentially excited from these metastable levels thus become strongly density dependent.

An example of the first situation is the intensity of the $1s^2\,{}^1S - 1s2s\,{}^3S$ transition in helium-like ions. Gabriel and Jordan (1972, 1973) have analyzed the density dependence of this line in detail. A number of authors (Rugge and Walker, 1970, 1971; Acton *et al.*, 1972, Bonnelle *et al.*, 1973) have studied the relative intensity of the forbidden ($1s^2\,{}^1S - 1s2s\,{}^3S$) and intercombination ($1s^2\,{}^1S - 1s2p\,{}^3P$) lines from O VII, Ne IX and Mg XI in the corona. However, only upper limits to the coronal density have been derived from these observations. There are a number of examples of the latter situation. Some ions which have been studied in the case of the solar corona are summarized in Table VI. The line intensity ratios tabulated are sensitive over the range of densities from $10^4-\sim10^{14}$.

The $\Delta n = 1$ transitions in beryllium-like ions will share the same density dependent behavior as the $\Delta n = 0$ transition listed in Table VI. For example, the upper level of the $2p^2\ ^3P - 2p3d\ ^3D$ line will be populated primarily by excitation from the $2s2p\ ^3P$ level. The study of the $\Delta n = 1$ transitions offers certain advantages in comparison with the $\Delta n = 0$ transitions. Line multiplets which have their wavelengths very close together, may be selected thereby reducing the problem of instrument calibration. Furthermore, since the dependence of collisional excitation rates on temperature goes mainly as $\exp(-\Delta E/kT)$, where ΔE is the excitation energy of the upper level, lines which are close together in wavelength will generally have excitation functions with very similar temperature dependence. Unfortunately, relatively few observations of the $\Delta n = 1$ transitions of the beryllium-like ions are available, and detailed studies of the relative intensities of these lines have not been undertaken.

(ii) *Temperature*

Walker (1972) has reviewed earlier studies of temperature sensitive line ratios. Three types of ratios have been studied, the ratio of $2s - 3p$ and $2s - 2p$ lines in lithium-like ions, the ratio of Lα and Lβ in hydrogenic ions, and the ratio of the lithium-like satellite lines and helium-like line intensities.

Heroux *et al.* (1972) and Malinovsky and Heroux (1973) have carried out measurements of the line ratios for the lithium-like ions O VI, Ne VIII, and Mg XI. These authors find that the temperatures derived from the relative intensity of the $2s - 3p$ and $2s - 2p$ lines may be somewhat misleading if the lines are emitted over a broad temperature range.

Malinovsky and Heroux found that the observed Ne VIII and Mg XI line ratios are in good agreement with the theoretical ratios predicted using a coronal thermal model.

Rugge and Walker (1974) have analyzed the relative intensities of the Lα and Lβ lines of O VIII, Mg XII, and Si XIV, using a recent theoretical study of this line ratio carried out by Hutcheon and McWhirter (1973). Using a coronal temperature model Rugge and Walker were able to make a detailed comparison of the observed line ratios with the theory of Hutcheon and McWhirter. The predicted ratio of Lα to Lβ for O VIII derived from the coronal model was 10.8 ± 1.6, while the observed ratio was found to be 10.5 ± 0.5. Good agreement was also found for the observed and predicted Mg XI and Si XIV ratios. However, Rugge and Walker found that the temperature predicted assuming an isothermal plasma differed from the calculated temperature of most efficient emission by $\sim7 \times 10^5$ K for O VIII, by 1.6×10^6 K for Mg XI, and by 3.6×10^6 K for Si XIV. In each case, the 'predicted' temperatures were low.

These results suggest that the prediction of coronal temperatures from 'temperature sensitive' line ratios must be undertaken with considerable caution, unless there is other evidence that the observed lines were emitted from an isothermal structure.

TABLE VI

Compilation of theoretical and experimental studies of density sensitive line ratios

Ion	Ground configuration	Spin	Metastable configuration	Spin	Line ratios studied	References
C III, N IV, O V, Si XI	$2s^2\ ^1S$	0	$2s2p\ ^3P$	0, 1, 2	$[2s2p\ ^3P - 2p^2\ ^3P]/[2s^2\ ^1S - 2s2p\ ^1P]$	Jordan (1970) Munro et al. (1971)
Si X	$2s^22p\ ^2P$	1/2	$2s^22p\ ^2P$	3/2	$[2s^22p\ ^2P_{3/2} - 2s2p^2\ ^2D_{5/2}]/[2s^22p\ ^2P_{1/2} - 2s2p^2\ ^2D_{3/2}]$	Widing (1966)
Fe XIV	$3s^23p\ ^2P$	1/2	$3s^23p\ ^2P$	3/2	$[3s^23p\ ^2P_{3/2} - 3s3p^2\ ^2D_{5/2}]/[3s^23p\ ^2P_{1/2} - 3s3p^2\ ^2D_{3/2}]$	Pottasch (1966)
Fe XIII	$3s^23p^2\ ^3P$	0	$3s^23p^2\ ^3P$	1, 2	$[3s^23p^2\ ^3P_2 - 3s^23p^2\ ^1D_2]/[3s^23p^2\ ^2P_0 - 3s^23p^2\ ^3P_1]$	Flower and Pineau des Forêts (1973)
Fe XV	$3s^2\ ^1S$	0	$3s3p\ ^3P$	0, 1, 2	$[3s3p\ ^3P - 3s3d\ ^3D]/[3s3p\ ^1P - 3s3d\ ^1D]$	Cowan and Widing (1972)

(iii) *Non-Thermal Excitation*

Two spectroscopic techniques have been suggested for the study of non-thermal excitation; the observation of non-autoionizing satellite lines in lithium-like ions, and the observation of K-α lines. Gabriel (1972) has discussed the observations of non-autoionizing satellite lines, and the calculation of the inner shell excitation rates which result in their production.

Phillips and Neupert (1973) have calculated the intensity expected for the K-α lines of sulfur, argon, calcium, and iron for thermal and non-thermal solar X-ray events. They have observed K-α line emission from Fe xvii–xx during a large flare, and find that their theory predicts reasonable values of the emission measure.

(c) CONSTRUCTION OF CORONAL MODELS

Jefferies *et al.* (1972a) have developed a general method for the analysis of line emission from an optically thin gas, and applied this method to coronal forbidden line intensities (Jefferies *et al.*, 1972b, c). Using the formalism developed by Jefferies *et al.*, we may express the flux in a permitted coronal X-ray line as

$$F_{Zzij} = \varepsilon(\lambda_{ij}) \, a_H A_Z \int \int dS N_e \int \int dn_e \, dT_e \mu(T_e, n_e) \, a_z(T_e) \, J(T_e) \, n_e,$$

(13)

We have separated the integrals over the area $S(x_1, x_2)$ which is unresolved by the telescope, and along the line of sight, x_3. Following Jefferies *et al.*, we introduce the distribution function for the coronal plasma, $\mu[T_e(x_v), n_e(x_v)]$ which describes the density and temperature of the plasma, and replace the integral over x_3 by a double integral over this distribution function in a unit column. The quantity $\varepsilon(\lambda_{ig})$ is the efficiency function of the spectroheliograph.

The function μ represents the fraction (dN) of the total number of electrons in a unit column (N_e) which are in neighborhoods where temperature and density are given by T_e and n_e.

$$dN(T_e, n_e, x_1, x_2) = N_e \mu(T_e, n_e, x_1, x_2) \, dn_e \, dT_e.$$

We may also introduce the partial distribution functions $\phi(T_e)$ and $\psi(T_e)$.

$$\phi(T_e) = \int \mu(T_e, n_e) \, dn_e,$$

and

$$\psi(n_e) = \int \mu(T_e, n_e) \, dT_e.$$

The function $J(T_e)$ [or $J(T_e, n_e)$ if the line in question is density dependent] contains the dependence of the excitation function on atomic parameters peculiar to the line being observed, and can be derived from Equations (5) and (12).

Equation (13) is an integral equation, with the kernel $A_z a_z(T_e) J(T_e)$ which we can calculate, and the value F, which we can measure. In order to develop a set of techniques to determine $\mu(T_e, n_e)$, let us assume that there is a functional relationship between temperature and density, i.e.:

$$n_e = n_e(T_e).$$

We may then write Equation (13) as

$$F_{Zzij} = \varepsilon(\lambda_{ij}) \, a_H A_Z \int \int \mathrm{d}S N_e \int \mathrm{d}T_e n_e(T_e) \, \phi(T_e) \, a_z(T_e) \, J(T_e). \qquad (14)$$

The integral equation for F_{Zzij} defined by Equation (14) can, in principle, be solved to determine the temperature distribution of the emission measure; provided that observations including a sufficient number of lines with differing temperature dependence are available. The coronal models I discuss in Section III were derived by solving Equation (14), using various approximation procedures. The spatial integral must be extended over either individual active regions, or over the entire disk for the observations presently available, because of their low spatial resolution.

Pneuman (1972, 1973) and Kopp (1972) have pointed out the critical role that the magnetic field plays in the structure of coronal active regions, because of its predominant influence on the thermal conductivity. If observations of sufficient spatial (as well, of course, as spectral) resolution are available, we can introduce the dependence of ϕ (or of μ and ψ) on the magnetic field B explicitly,

$$F_{Zzij} = \varepsilon(\lambda_{ij}) \, a_H A_Z \int \mathrm{d}u \int \mathrm{d}s \int \mathrm{d}T_e n_e(T_e, B) \, \phi(T_e, B) \, a_z(T_e) \, J(T_e),$$
$$(15)$$

where s is a coordinate along the field line, and u a coordinate transverse to the field lines. We may think of u and s as a set of solar B-L coordinates, much as those used to describe the terrestrial radiation belts. As Pneuman (1972) and Jordan (1975) point out, the coronal plasma will be nearly isothermal along a given field line, allowing simplifying assumptions to be made in the solution of Equation (15).

III. Coronal Models

Coronal models of the structure of active regions developed from X-ray observations have been limited because of the low spatial resolution of the spectroscopically resolved data presently available. Consequently, the models developed have emphasized the temperature structure and composition of the corona. The problem of the coronal energy balance has not been treated. A number of studies using EUV observations have been used to develop models of the energy balance of the transition region (Dupree, 1972; Munro and Withbroe, 1972; Withbroe and Gurman, 1973). Kopp (1972), Withbroe and Gurman (1973), and Gurman and Withbroe (1974) have summarized these results, and discussed the effect of magnetic fields on coronal structure. These authors find an increased 'coronal temperature' overlying active regions, and an increased conductive flux from the corona into the transition region. However, the lack of temperature discrimination for coronal temperatures above $\sim 2.5 \times 10^6$ K with EUV lines prevents these models from distinguishing between an isothermal corona, and a corona with thermal structures extending to higher temperatures beyond the most probable coronal temperature (i.e., the temperature at which the coronal emission measure is at a maximum). The analysis of the coronal X-ray spectra of active regions demonstrates that coronal material is present at considerably higher temperatures than the 'coronal active region temperature of ~ 2.5–3×10^6 K' which is derived from studies of the coronal EUV spectrum.

(a) DEVELOPMENT OF CORONAL THERMAL MODELS

Batstone *et al.* (1970) and Evans and Pounds (1968) developed the first thermal models, using coronal X-ray observations. They assumed that the emission function could be approximated by a step function, and found that coronal material as hot as 6–8×10^6 K must be present in order to explain their observations. Chambe (1971) developed a technique using an analytical expression for the emission measure function which has a maximum at a temperature T_0 (assumed to be near 2×10^6 K) and falls off with temperature as $10^{-T/T_1}$. A study of the heating of the corona by Vil'koveskii (1972) shows that this is a reasonable thermal model based on the assumption that the mechanical heating is proportional to the square of the electron density. Vil'koveskii finds the parameters T_0 and T_1 to be $\sim 1.5 \times 10^6$ and 1×10^6 K for quiet regions, and 2.5×10^6 and 3×10^6 K for active regions. The model found by Chambe from observations requires $T_0 = 1.5 \times 10^6$ K and T_1 between 1.5 and 3.0×10^6 K.

Walker (1972), Acton *et al.* (1972), Bonnelle *et al.* (1973), and Walker *et al.* (1974a, b) have developed models of the coronal emission measure, using the approach of Chambe. Acton *et al.* used the resonance lines of O VII and Ne IX, and were able to derive thermal models for a number of active regions, since their spectrometer was collimated to 1.7' in one direction. They were able to map the emission of individual active regions by scanning the collimator across the solar disk in nearly orthogonal directions. The result of one of these scans is shown in Figure 6. The observations were fitted with an emission measure model $M(T)$:

$$M(T) = C \times 10^{-T/T_1} \text{ cm}^{-3} (10^6 \text{ K})^{-1}.$$

The parameters found for each active region, and for the general corona are presented in Table VII. The total emission measure above 10^6 K, and the 'average' emission temperature of the O VII and Ne IX lines, are also given in Table VII.

Bonnelle *et al.* carried out a similar observation using the resonance lines of Mg XI and Mg XII, and satellite lines of Mg X. These authors developed a thermal model with $C = 5 \times 10^{48}$ cm^{-3} $(10^6$ K$)^{-1}$, and $T_1 = 2.6 \times 10^6$ K.

Walker (1972) and Walker *et al.* (1974a, b) have developed models using the resonance lines of O VII, O VIII, Ne IX, Ne X, Mg XI, Mg XII, Al XII, Al XIII, Si XIII, and Si XIV. They find an excellent fit to the model proposed by Chambe and Vil'koveskii, with $T_0 = 2.5 \times 10^6$ K and $T_1 = 3.5$–4.0×10^6 K. The models developed by Walker

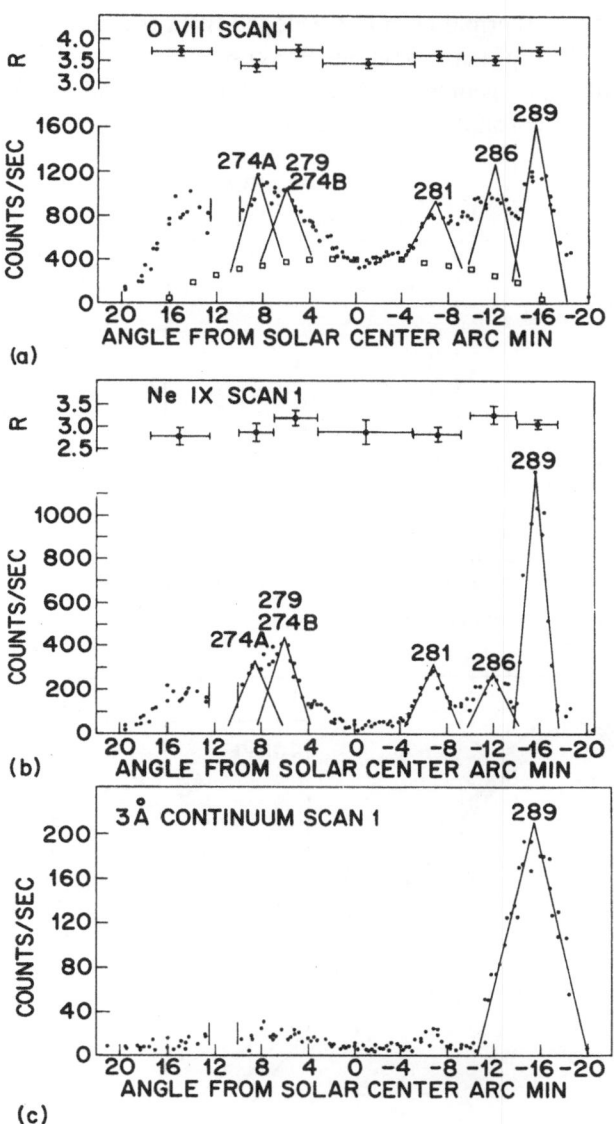

Fig. 6.   Variation of the coronal X-ray intensity across the solar disk observed by Acton *et al.* (1972) with a one dimensional collimator. A second scan approximately orthogonal to the scan shown here allowed a map of the coronal structure to be constructed, and the models given in Table VII.

et al. are shown in Figures 7 and 8. The small values of the parameter χ^2 demonstrate the significance of the fit. From Figure 7, we see that a model with a temperature cutoff will fit the observations since the hottest line included in the model is Mg xi. In Figure 8, we note that the model must include material in excess of 8×10^6 K in order to fit the Si xiv observations.

The model of an isothermal corona developed from the analysis of EUV observations must be modified for active regions. For coronal holes, and for the quiet corona,
where magnetically open geometries introduce the solar wind as an energy loss term
(Pneuman, 1973), the coronal structure can be reasonably fit by an isothermal model,

TABLE VII

Non-isothermal (exponential) model

Plage region	T_1 (10^6 K)	C^a	$EM_{tot}{}^b$ (10^47 cm^-3)	$\langle T_{O\,VII}\rangle$ (10^6 K)	$\langle T_{Ne\,IX}\rangle$ (10^6 K)
289	3.15	3	23	1.95	3.10
279, 274 B	1.80	6	13	1.80	2.70
281	1.50	8	12	1.70	2.55
274 A	1.25	18	16	1.65	2.40
286	1.15	24	16	1.60	2.30
265, 266	0.95	42	15	1.50	2.20
Gen. corona^c	0.60	170	170	1.25	1.70

[a] Units: 10^{48} cm^-3 $(10^6$ K)^-1.
[b] Emission measure above 10^6 K.
[c] All values refer to coronal hemisphere facing the Earth.

Taken from Acton *et al.* (1972).

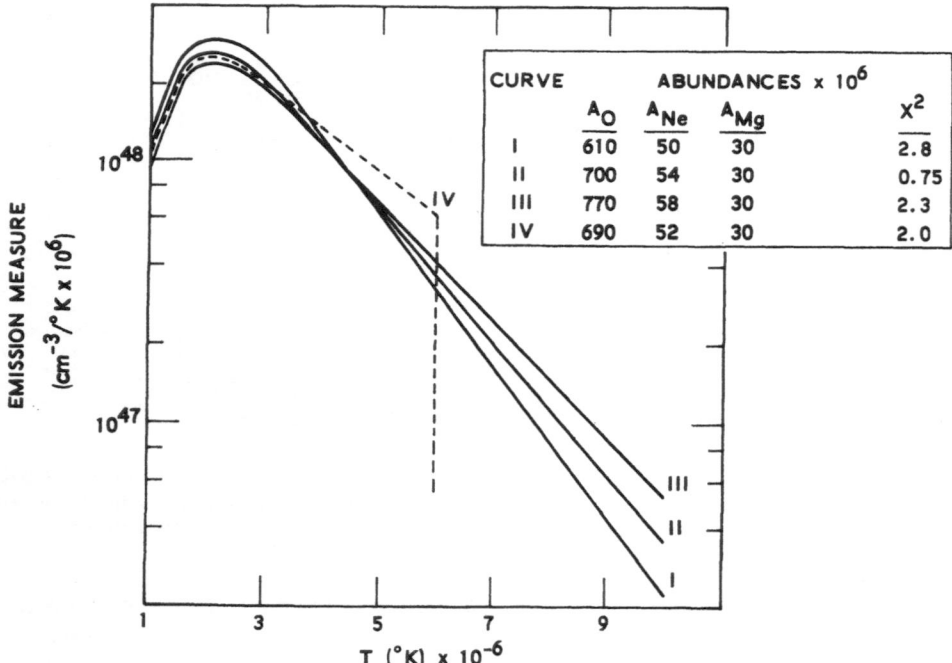

Fig. 7.   Coronal Model constructed from the resonance lines of O VII, O VIII, Ne IX, Ne X, and Mg XI.
Note that a model with no material above 6×10^6 K also fits the observations. A model cutoff above
5.0×10^6 K does not, however, fit the observations.

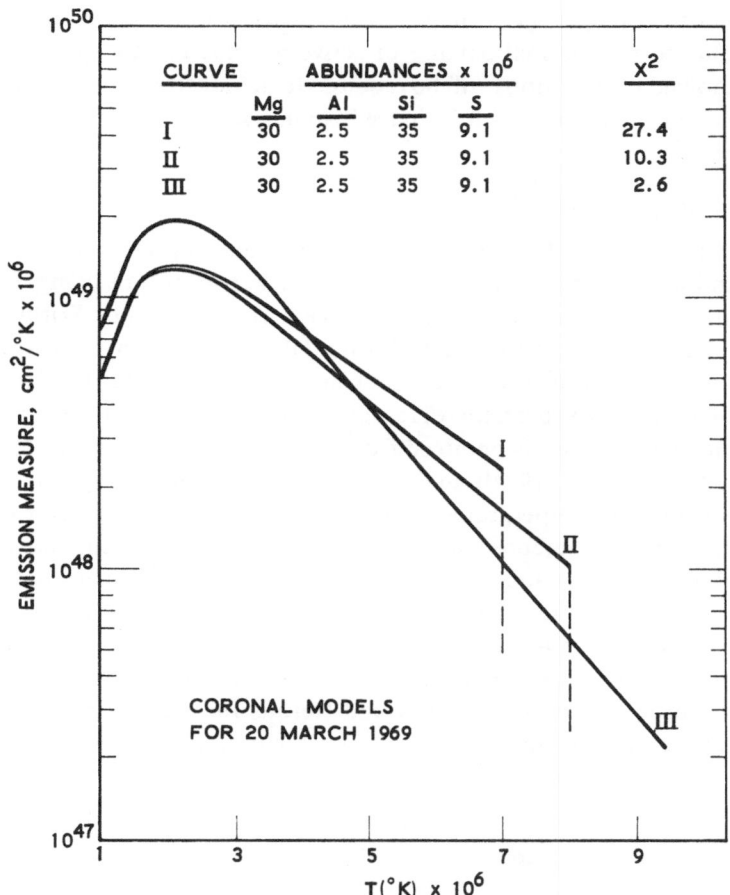

Fig. 8. Coronal Model derived from the resonance lines of Mg IX, Mg X, Al XII, Al XIII, Si XIII, Si XIV, S XV. Note that material at temperatures above 8×10^6 K must be included in the model in order to fit the hot Si XIV and S XV lines.

with $T \sim 1.0 \times 10^6$ K (Munro and Withbroe, 1972) for coronal holes and $\sim 1.5 \times 10^6$ K for the quiet corona (Withbroe and Gurman, 1973). Acton *et al.* find that the general coronal emission measure does indeed fall off very rapidly above the temperature of maximum emission measure ($T_0 \sim 1.5 \times 10^6$ K) with $T_1 \sim 0.3$–1.0×10^6 K. However, for active regions, the emission measure falls off much more slowly, with $T_1 \sim 1.5$–4.0×10^6 K, and models with an isothermal corona, even with temperatures as high as 2.5–3.5×10^6 K (Withbroe and Gurman, 1973; Noyes *et al.*, 1970), cannot account for the X-ray observations.

The high temperatures required to account for the coronal X-ray flux may be an indication of a heating source in addition to the dissipation of mechanical waves, the source generally assumed to be responsible for coronal heating (Kopp and Kupers, 1968). Tucker (1973), Sturrock (1972b), and Syrovatskii and Shmeleva (1972) have suggested that the dissipation of magnetic energy by ohmic heating in current sheets may contribute to the thermal balance of active regions. Highly spatially, as well as spectrally resolved observations will be required to demonstrate that magnetic heating is taking place in the corona, if it is, indeed, an important component in the thermal energy balance.

(b) CORONAL ABUNDANCES

Walker *et al.* (1974a, b, c) have derived coronal abundances from the models given in Figures 7 and 8. The abundances derived by Walker *et al.* are compared with the coronal abundances derived with transition region observations by Withbroe (1971a), Dupree (1972), and Malinovsky and Heroux (1972), and with photospheric abundances compiled by Withbroe (1971a) and Cameron (1974) in Table VIII. In general, the agreement is good; however, there is a significant difference for the neon abundance. Cameron's neon abundance is obtained from cosmic ray observations. Since the neon lines used by Walker *et al.* (1974b) in their analysis were quite strong, and were excited near the mid-range of temperatures in their model I believe their neon results to be quite reliable, and the difference may, therefore, be significant. In the case of the sulfur and argon abundances, the lines analyzed were near the high temperature end of their model, and are thought to be somewhat less reliable than the other abundances. Consequently, we do not regard the difference between the coronal and the photospheric computations to be significant for these elements. The oxygen abundance found by Withbroe and by Malinovsky and Heroux for the transition region is low, however, Dupree's results appear to support a higher oxygen abundance.

TABLE VIII
Comparison of relative abundances
Abundance $\times$ (10⁶)

Element	Active region	Transition region			Photospheric	Solar system
	Walker *et al.*	Withbroe	Dupree	Malinovsky and Heroux	Withbroe	Cameron
C	–	420	350	–	350	370
N	90	89	150	–	115	117
O	700	450	595	350	676	676
Ne	54	28	27	88	–	108
Na	1.7	2.3	1.9	–	1.7	1.9
Mg	30	35	30	50	35	33
Al	2.5	2.3	3.5	–	2.5	2.7
Si	35[a]	35[a]	35[a]	35[a]	35	31.6
S	9	11	20	5	16	16
Ar	6	4.5	–	–	–	3.7
Fe	26	35	20	20	25[b]	26

[a] The coronal abundance values have been normalized relative to the silicon abundance, which was assumed to be 35×10^{-6}.

[b] In a recent analysis, Smith and Whaling (1973) find a value of 25×10^{-6} for the iron abundance, in agreement with the earlier result of Withbroe.

In conclusion, it would appear that the results of Table VIII support a uniform abundance structure for the solar atmosphere.

(c) CORONAL DENSITY STRUCTURE

The X-ray analyses of coronal density structure have, so far, resulted only in upper limits. The analyses of density sensitive EUV lines by Munro *et al.* (1971), Jordan (1971), Widing (1966), and others (see Table VI) have resulted in density models which predict densities of $\sim 5 \times 10^8$ cm^{-3} in the quiet corona, and densities as high as 2.5×10^9 cm^{-3} in active regions. These models are consistent with the X-ray upper limits, and with other observations. Withbroe (1971b) has shown that the intensity of the Mg x resonance line is a good indicator of coronal density, once properly calibrated. Withbroe *et al.* (1971) and Withbroe and Gurman (1973) have developed coronal density maps using this technique, and Gurman *et al.* (1974) have discussed the correlation of coronal density and photospheric magnetic intensity using this technique.

(d) THE STRUCTURE OF CORONAL ACTIVE REGIONS

The poor spatial resolution of the presently available spectroscopically resolved coronal X-ray observations makes it difficult to study the physical structure of active regions. Parkinson (1973a, b) has used two types of observations, high spectral resolu-

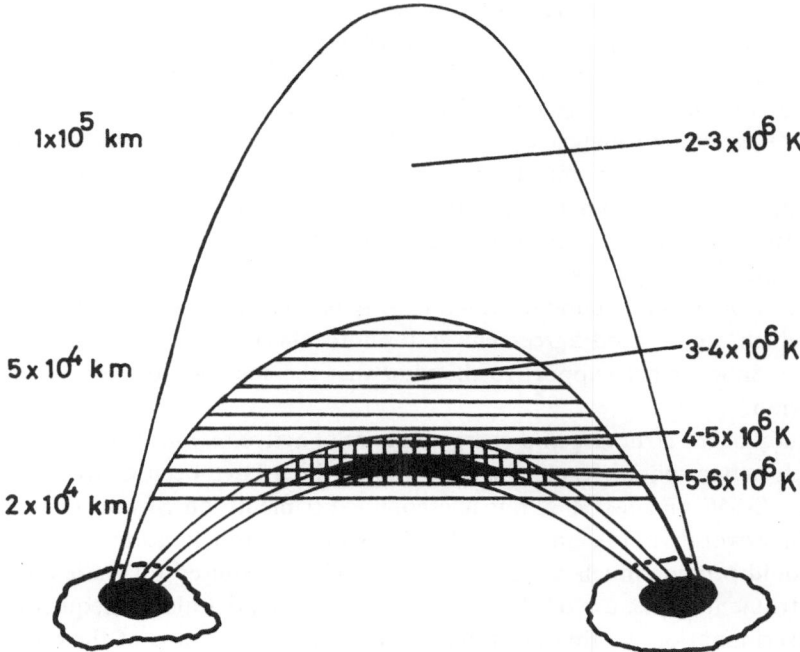

Fig. 9.   Model of a coronal active region derived by Parkinson (1973a, b) from OSO 5 spectrohelio-grams and rocket spectra of active regions.

tion studies with low spatial resolution, and moderate spectral resolution studies of coronal structures from OSO 5 with 1′–2′ resolution (Parkinson and Pounds, 1971) to construct the schematic model shown in Figure 9. This model resembles the X-ray loop structures observed on Skylab (Vaiana, 1974; Underwood, 1974) with high resolution X-ray telescopes.

Pneuman (1972, 1973) has analyzed the structure of coronal active regions in open and closed magnetic structure, and points out that coronal material should be isothermal along magnetic field lines, because of the large thermal conductivity along field lines. This suggests that we may think of an active region as a collection of isothermal flux tubes, with the tubes at lower altitude being higher in temperature. As the flux tube approaches the photosphere, the isothermal approximation breaks down, and the transition region, as described by Kopp (1972) occurs. The temperature structure along each flux tube would indeed have an 'isothermal' corona, coupled to the transition region, however, taken together, the flux tubes result in a thermal structure for the active region which is described by Figures 7 and 8. The study of this structure may require very high (< 10″) spatial resolution. Jordan (1975) has discussed the analytical and observational evidence for the 'multiple flux tube' model of coronal arch (or loop) structures elsewhere in these proceedings.

IV. Conclusions

We have found that the analysis of the coronal X-ray spectrum has resulted in coronal thermal models which show a nearly isothermal corona over quiet regions, in agreement with analyses of the coronal EUV spectrum. However, analyses of active region spectra (or of whole disk spectra which are dominated by active regions) result in models which contain material as hot as 8–10×10^6 K, in contradiction to the 'isothermal' corona derived from the analysis of the EUV spectra of active regions. The X-ray results do predict a maximum in the coronal emission measure near ~ 2–3×10^6 K, for active regions, which corresponds to the temperature of the 'isothermal' corona found by the EUV analyses.

The study of coronal abundances with X-ray observations has found relative abundances which are in good agreement with photospheric and transition region abundances, lending further support to the view that the solar atmosphere is uniform in composition.

The study of the structure and energy balance of active regions will require observations with better spatial resolution than those which have been analyzed up to this time. The S-056 and S-054 X-ray telescope experiments on Skylab (Vaiana et al., 1973; Underwood, 1974), and the OSO-7 X-ray spectroheliograph (Neupert et al., 1974) should provide much of the observational data required for these analyses. All three instruments make use of filters to provide wavelength (and consequently temperature) discrimination, so that accurate calculations of the shape of the coronal spectrum will be required in order to carry out the type of analysis of coronal structure outlined in Section II (c).

References

Acton, L. W., Catura, R. C., Meyerott, A. J., Wolfson, C. J., and Culhane, J. L.: 1972, *Solar Phys.* **26**, 183.
Batstone, R. M., Evans, K., Parkinson, J. H., and Pounds, K. A.: 1970, *Solar Phys.* **13**, 389.
Bely, O.: 1966a, *Ann. Astrophys.* **29**, 683.
Bely, O.: 1966b, *Proc. Phys. Soc.* **88**, 587.
Bely, O. and Bely, F.: 1967, *Solar Phys.* **2**, 285.
Bely, O. and Petrini, D.: 1970, *Astron. Astrophys.* **6**, 318.
Bonnell, C., Senemand, C., Senemand, G., Chambre, G., Guionnet, M., Heroux, J. C., and Michaud, R.: 1973, *Solar Phys.* **29**, 341.
Burgess, A., Hummer, D. G., and Tully, J. A.: 1970, *Phil. Trans. Roy. Soc. London* **A266**, 225.
Burke, P. G., Hubbert, A., and Robb, W. D.: 1972, *J. Phys.* **B5**, 37.
Burke, P. G., Tait, J. H., and Lewis, B. A.: 1966, *Proc. Phys. Soc.* **87**, 209.
Cameron, A. G. W.: 1974, *Space Sci. Rev.* **15**, 121.
Chambe, G.: 1971, *Astron. Astrophys.* **12**, 210.
Cowan, R. D. and Widing, K. G.: 1972, *Astrophys. J.* **180**, 285.
Culhane, J. L.: 1969, *Monthly Notices Roy. Astron. Soc.* **144**, 375.
Culhane, J. L. and Acton, L. W.: 1974, *Ann. Rev. Astron. Astrophys.* **12**, 359, Ann. Rev. Inc. Palo Alto.
Doschek, G. A.: 1972, *Space Sci. Rev.* **13**, 765.
Dupree, A. K.: 1972, *Astrophys. J.* **178**, 527.
Evans, K. and Pounds, K. A.: 1968, *Astrophys. J.* **152**, 313.
Flower, D. R.: 1971, *J. Phys.* **B4**, 697.
Flower, D. R. and Launay, J. M.: 1972, *J. Phys.* **B5**, L207.
Flower, D. R. and Pineau des Forets, G.: 1973, *Astron. Astrophys.* **24**, 181.
Gabriel, A. H.: 1972, *Monthly Notices Roy. Astron. Soc.* **160**, 99.
Gabriel, A. H. and Jordan, Carole: 1972, in E. W. McDaniel and M. R. C. McDowell (eds.), *Case Studies in Atomic Collision Physics*, Vol. II, North-Holland Publ. Co., Amsterdam, p. 211.
Gabriel, A. H. and Jordan, Carole: 1973, *Astrophys. J.* **186**, 327.
Gabriel, A. H. and Paget, T. M.: 1972, *J. Phys.* **B5**, 673.
Garstang, R. H.: 1969, *Publ. Astron. Soc. Pacific* **81**, 488.
Grineva, Y. I., Karev, V. I., Korneev, V. V., Krutov, V. V., Mandelstam, S. L., Varnshtein, L. A., Vaisiljev, B. N., and Zitnik, I. A.: 1973, *Solar Phys.* **29**, 441.
Gurman, J. B., Withbroe, G. L., and Harvey, J. W.: 1974, *Solar Phys.* **34**, 105.
Heroux, L. and Cohen, M.: 1971, *Phil. Trans. Roy. Soc. London* **A270**, 99.
Heroux, L., Cohen, M., and Malinovsky, Monique: 1972, *Solar Phys.* **23**, 369.
Hutcheon, R. J. and McWhirter, R. W. P.: 1973, *J. Phys.* **B6**, 2268.
Jefferies, J. T., Orrall, G. Q., and Zirker, J. B.: 1972a, *Solar Phys.* **22**, 307.
Jefferies, J. T., Orrall, G. Q., and Zirker, J. B.: 1972b, *Solar Phys.* **22**, 317.
Jefferies, J. T., Orrall, G. Q., and Zirker, J. B.: 1972c, *Solar Phys.* **22**, 327.
Johnson, W. D. and Kunze, H. J.: 1971, *Phys. Rev.* **A4**, 962.
Jordan, Carole: 1966, *Monthly Notices Roy. Astron. Soc.* **132**, 463.
Jordan, Carole: 1969, *Monthly Notices Roy. Astron. Soc.* **142**, 499.
Jordan, Carole: 1970, *Monthly Notices Roy. Astron. Soc.* **149**, 1.
Jordan, Carole: 1971, in C. de Jager (ed.), *Highlights in Astronomy*, D. Reidel Publ. Co., Dordrecht, Holland, p. 519
Jordan, Carole: 1975, This volume, p. 109.
Kopp, R. A.: 1972, *Solar Phys.* **27**, 373.
Kopp, R. A. and Kupers, M.: 1968, *Solar Phys.* **4**, 212.
Landini, M. and Monsignori Fossi, B. C.: 1970, *Astron. Astrophys.* **6**, 468.
Landini, M. and Monsignori Fossi, B. C.: 1972, *Astron. Astrophys. Suppl.* **7**, 291.
Loulergue, M.: 1971, *Astron. Astrophys.* **15**, 126.
Loulergue, M.: 1973, Private communication.
Loulergue, M. and Nussbaumer, H.: 1973, *Astron. Astrophys.* **24**, 312.
Malinovsky, Monique and Heroux, L.: 1973, *Astrophys. J.* **181**, 1009.

Mewe, R.: 1972, *Solar Phys.* **22**, 459.
Munro, R. H., Dupree, A. K., and Withbroe, G. L.: 1971, *Sol. Phys.* **19**, 347.
Neupert, W. M., Thomas, R. J., and Chapman, R. D.: 1974, *Solar Phys.* **34**, 349.
Noyes, R.: 1971, *Ann. Rev. Astron. Astrophys.* **2**, 209.
Noyes, R., Withbroe, G. L., and Kirschner, R. P.: 1970, *Solar Phys.* **11**, 388.
Parkinson, J. H.: 1973a, *Solar Phys* **28**, 137.
Parkinson, J. H.: 1973b, *Solar Phys.* **28**, 489.
Parkinson, J. H.: 1975, This volume, p. 45.
Parkinson, J. H. and Pounds, K. A.: 1971, *Solar Phys.* **17**, 146.
Phillips, K. J. H. and Neupert, W. M.: 1973, *Solar Phys.* **32**, 209.
Pneuman, G. W.: 1972, *Astrophys. J.* **177**, 793.
Pneuman, G. W.: 1973, *Solar Phys.* **28**, 247.
Pottasch, S. R.: 1964, *Space Sci. Rev.* **3**, 816.
Pottasch, S. R.: 1966, *Bull. Astron. Inst. Neth.* **18**, 237.
Pottasch, S. R.: 1967, *Bull. Astron. Inst. Neth.* **19**, 113.
Rugge, H. R. and Walker, A. B. C.: 1970, *Solar Phys.* **15**, 372.
Rugge, H. R. and Walker, A. B. C.: 1971, *Solar Phys.* **18**, 244.
Rugge, H. R. and Walker, A. B. C.: 1974, *Astron. Astrophys.* **33**, 367.
Sampson, D. H. and Parks, A. D.: 1974, *Astrophys. J. Suppl. Series* **28**, 323.
Shore, B.: 1969, *Astrophys. J.* **158**, 1205.
Smith, P. L. and Whaling, W.: 1973, *Astrophys. J.* **183**, 313.
Sturrock, P. A.: 1972a, *Solar Phys.* **23**, 438.
Sturrock, P. A.: 1972b, *Progress Astron. Aeronaut.* **30**, 172.
Syrovatskii, S. I. and Shmeleva, O. P.: 1972, *Soviet Astron – A.J.* **16**, 273.
Tondello, G. and McWhirter, R. W. P.: 1971, *J. Phys.* **B4**, 715.
Tully, J. A.: 1974, *J. Phys.* **B7**, 386.
Tucker, W. H.: 1973, *Astrophys. J.* **186**, 285.
Tucker, W. H. and Koren, M.: 1971, *Astrophys. J.* **168**, 283.
Underwood, J. H.: 1974, Paper presented at the Joint *IAU-COSPAR Symp.* **68** on 'Solar Gamma, X- and EUV Radiation', Buenos Aires, Argentina, 11–14 June, 1974 (unpublished).
Vaiana, G.: 1974, Paper presented at the Joint *IAU-COSPAR Symp.* **68** on 'Solar Gamma, X- and EUV Radiation', Buenos Aires, Argentina, 11–14 June, 1974 (unpublished).
Vaiana, G. S., Davis, J. M., Giacconi, R., Kreger, A. S., Silk, J. K., Timothy, A. F., and Zombeck, M.: 1973, *Astrophys. J. Letters* **185**, 247.
Van Regemorter, H.: 1962, *Astrophys. J.* **136**, 906.
Vil'koviskii, E. Ya.: 1972, *Astron Zh.* **49**, 1125 (English translation *Soviet Astron. – A.J.* **16**, 918.)
Walker, A. B. C.: 1972, *Space Sci. Rev.* **13**, 672.
Walker, A. B. C. and Rugge, H. R.: 1971, *Astrophys. J.* **164**, 181.
Walker, A. B. C., Rugge, H. R., and Weiss, K.: 1974a, *Astrophys. J.* **188**, 423.
Walker, A. B. C., Rugge, H. R., and Weiss, K.: 1974b, *Astrophys. J.* **192**, 169.
Walker, A. B. C., Rugge, H. R., and Weiss, K.: 1974c, *Astrophys. J.* **194**, 471.
Walker, D. W.: 1974, *J. Phys.* **B7**, 97.
Widing, K. G.: 1966, *Astrophys. J.* **145**, 380.
Withbroe, G. L.: 1971a, in K. B. Gebbie (ed.), The Menzel Symposium on 'Solar Physics, Atomic Spectra, and Gaseous Nebulae', *NBS Special Publication* **353**, 127.
Withbroe, G. L.: 1971b, *Solar Phys.* **18**, 458.
Withbroe, G. L., Dupree, A. K., Goldberg, L., Huber, C. E., Noyes, R. W., Parkinson, W. H., and Reeves, E. M.: 1971, *Solar Phys.* **21**, 272.
Withbroe, G. L. and Gurman, J. B.: 1973, *Astrophys. J.* **183**, 279.

SOFT X-RADIATION FROM SINGLE ACTIVE REGIONS

D. H. BRABBAN*, E. B. DORLING, W. M. GLENCROSS, and J. R. H. HERRING

*Mullard Space Science Laboratory, Dept. of Physics and Astronomy,
University College London, WC1E 6BT, England*

Summary. The MSSL/Leicester University package on OSO 5 contained proportional counters having fields of view small compared with the area of the solar disk (Herring *et al.*, 1971). Results discussed here were obtained with a detector sensitive in the band 0.3–0.9 nm. This had an entrance window collimated to examine a strip of angular width 2' lying across the Sun.

One mode of operation allowed the field of view to remain stationary across a solar diameter. Observations of single active regions within this field showed that flare-like brightenings could be recognised most of the time, although changes in intensity of radiation at the detector might be as low as 10^{-5} erg cm^{-2} s^{-1} (Glencross *et al.*, 1974). It appears that the bulk of the X-ray emission from any active region might be produced by small flares, even when the area is regarded as 'relatively inactive'. In that case the mechanism responsible for flaring might be the one which is mainly responsible for heating the corona in active regions.

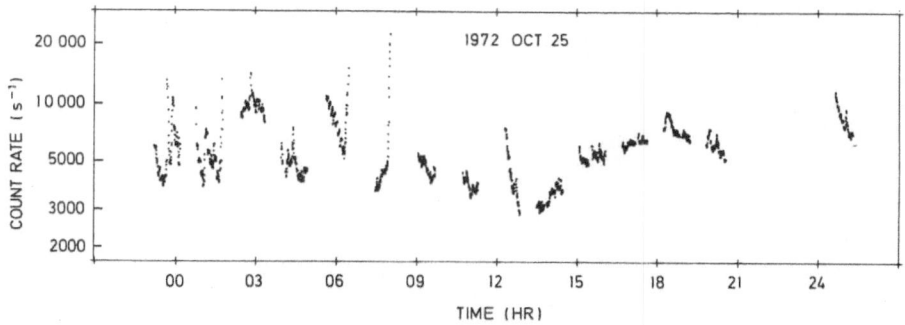

Fig. 1. Soft X-ray emission from McMath region 12094: After a period of considerable flaring, the flux of X-rays in the waveband 0.3–0.9 nm began to increase gradually at 13 00 UT. This suggests there might be a common mechanism for heating coronal plasma during 'quiet' periods as well as during flaring.

Work is now in progress on an examination of long-term changes in flux from single active regions. Figure 1 shows data obtained from McMath region 12094 during one day. Although the region showed frequent flare brightenings until 13 00 UT (optical data and the Solrad satellite show that events peaked at 10 08 UT and 11 36 UT, although OSO 5 was not operating during these periods), this was followed directly

---

* Now at Department of Space Research, University of Birmingham, U.K.

Sharad R. Kane (ed.), Solar Gamma-, X-, and EUV Radiation, 101–102. All Rights Reserved.

by an enhancement of the baseline flux on which no significant short-lived increases developed. It will be argued in a paper being prepared that such changes in the character of the emission suggest once again that there might be a common mechanism for heating coronal plasma during 'quiet' periods as well as during flaring.

References

Glencross, W. M., Dorling, E. B., and Herring, J. R. H.: 1974, *Solar Phys.* **38**, 183.
Herring, J. R. H., Glencross, W. M., Parkinson, J. H., and Pounds, K. A.: 1971, *Proc. Roy. Soc. Lond. A* **321**, 493.

TIME VARIATIONS IN CORONAL ACTIVE REGIONS

A. S. KRIEGER, R. C. CHASE, M. GERASSIMENKO, and S. W. KAHLER

American Science and Engineering, Cambridge, Mass., U.S.A.

A. F. TIMOTHY

NASA Headquarters, Washington, D.C., U.S.A.

and

G. S. VAIANA

*Center for Astrophysics, Harvard College Observatory, Smithsonian
Astrophysical Observatory, Cambridge, Mass., U.S.A.*

Summary. The AS&E X-ray telescope experiment on Skylab has obtained images of the solar X-ray corona with a variety of time resolutions ranging from $2\frac{1}{2}$ s to the regular 12 ± 2 h synoptic observation rate. The form and brightness of coronal active region structures are seen to vary on time scales ranging from seconds, for flare associated changes, to several solar rotations for long term evolution of the regions. The extrapolation of photospheric magnetic fields into the corona, using the potential field approximation, results in a good morphological agreement between the form of the computed coronal field lines and the structure of many of the active regions observed. Thus, in general, the coronal active region structures follow potential field lines and the long term evolutionary changes can be explained on the basis of the spreading of the fields. Short term changes in active region structure frequently take the form of selective brightening or dimming of pre-existing loops due to changes in the pressure of the emitting coronal plasma. In these cases, variations in the non-potential component of the coronal fields supporting and containing the plasma are implied.

The examples shown included the behavior of the X-ray active regions associated with McMath-Hulbert Ca plages 12387 and 12511. McMath 12387 was a magnetically complex, inverted polarity region with complicated X-ray structure. The emergence of new magnetic flux and a class M2 X-ray flare were accompanied by a change in the temperature from less than 2×10^6 to more than 3×10^6 K of some (but not all) of the extensive loop structures of this region without significant change in the form of the structures. McMath 12511 was a two day old simple bipolar region with two compact X-ray emitting loops crossing the neutral line. A subflare in this region was preceded by the gradual brightening of one of these loops. The flare itself consisted of an increase in the surface brightness of this loop of a factor of 20 in less than 2 min and a return to the preflare level in about 3 min. The form of the loop was unchanged through the event.

Sharad R. Kane (ed.), Solar Gamma-, X-, and EUV Radiation, 103. *All Rights Reserved.*
Copyright © 1975 by the IAU.

FLARE-LIKE ULTRAVIOLET SPECTRA OF ACTIVE REGIONS

G. E. BRUECKNER

*E. O. Hulburt Center for Space Research, Naval Research Laboratory,
Washington, D.C. 20375, U.S.A.*

Summary. Some characteristic features of UV spectra which can be seen in flares (Brueckner, this Symposium, p. 135) can also be detected in active regions without the presence of typical other flare phenomena like X-ray enhancement. Another distinct difference between these 'flare-like' spectra and flare spectra is the absence of very high temperature ions like Fe XXI in the 'flare-like' spectra. They occur in small areas and can be detected as 'fluctuating Hα bright points' in broad band Hα. Simultaneously, a strong UV brightening can be seen. As reported by the Skylab crews, these brightenings occur more frequently and quasi-periodic prior to flares. Their spectra show the very broad transition zone lines and often strong line shifts toward the blue or red. Figure 1 shows a selection of typical spectra. One notices a very asymmetric Lα profile in one case. Spectra prior to, during, and after a flare are reproduced in Figure 2. One recognizes that the transition zone instability started intermittently prior to the flare and could be seen during short time intervals long after the flare had ceased.

Sharad R. Kane (ed.), Solar Gamma-, X-, and EUV Radiation. 105–107. *All Rights Reserved.*

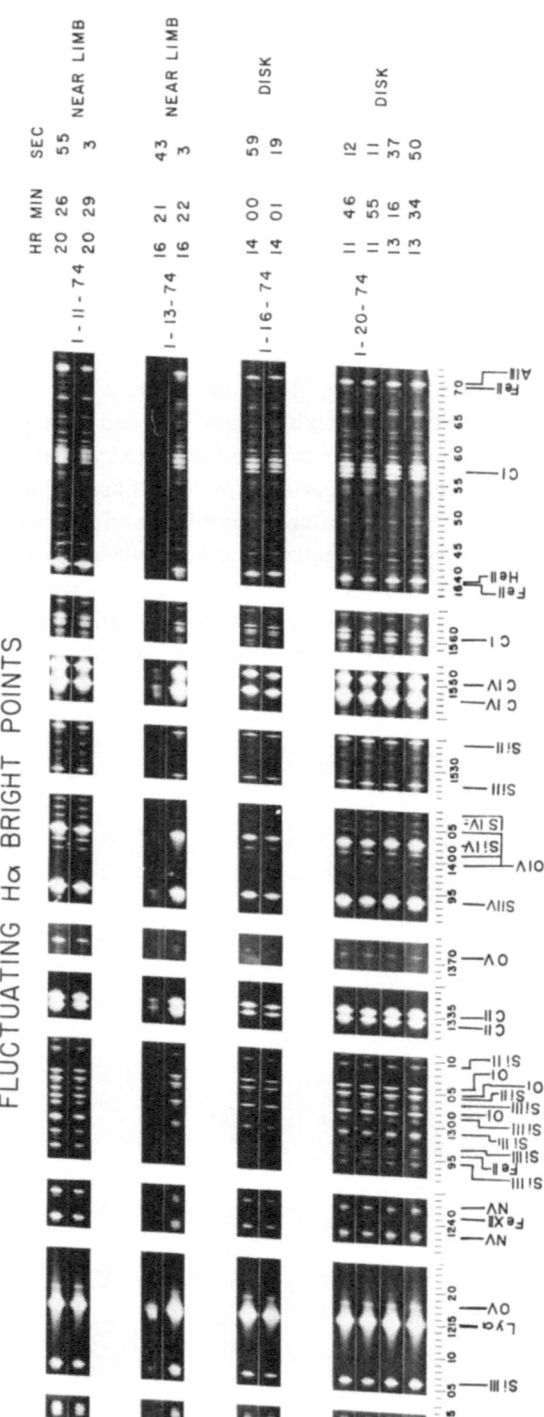

Fig. 1.   Selection of UV spectra of fluctuating Hα bright points.

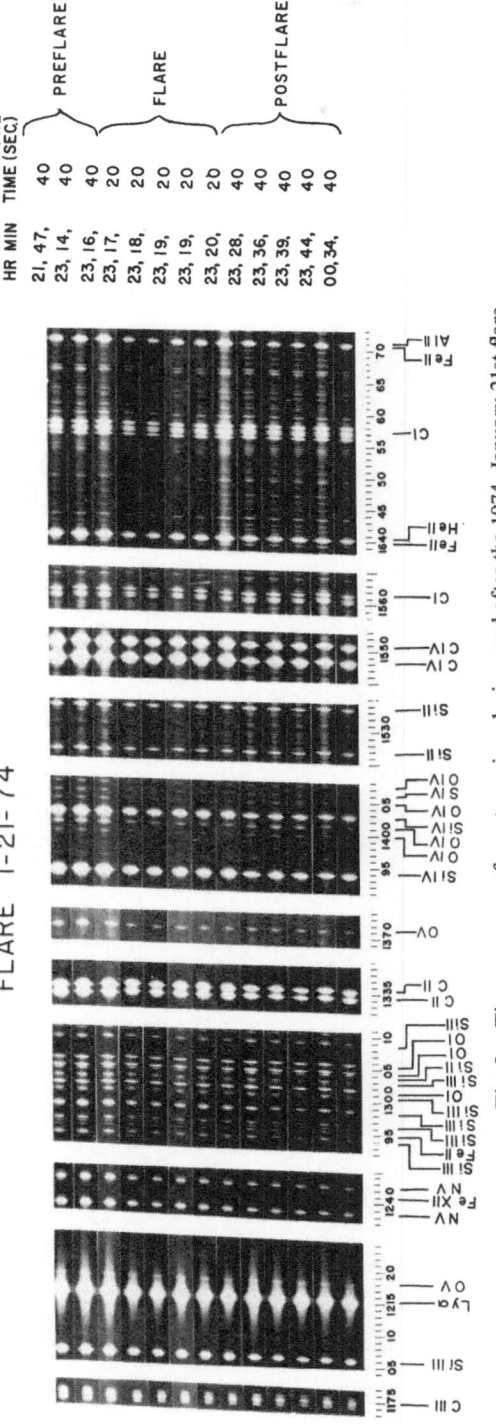

Fig. 2.   Time sequence of spectra prior, during, and after the 1974, January 21st flare.

THE STRUCTURE OF SOLAR ACTIVE REGIONS FROM EUV AND SOFT X-RAY OBSERVATIONS

CAROLE JORDAN*

Center for Astrophysics, Harvard College Observatory and Smithsonian Astrophysical Observatory, Cambridge, Mass. 02138, U.S.A.

Abstract. The structure of solar active regions derived from EUV and soft X-ray observations is reviewed. The methods by which the emission measure as a function of temperature can be interpreted are discussed. The models of density and temperature which can be made from a variety of combinations of the emission measure with information on the spatial distribution of material, are broadly consistent. They show that the plasma at low heights over the central parts of an active region is hotter and denser than that which extends to greater heights. It appears that much of the emitting material exists in the form of loop structures, presumably magnetically controlled flux tubes. Analytical relationships between the physically important parameters describing the properties of the active region at $T_e > 2 \times 10^5$ K are developed and discussed.

I. Introduction

Active regions have been observed for many years at the levels of the photosphere and chromosphere, by making observations in the visible part of the spectrum. The literature on the visible region observations is extensive and good reviews can be found in Zirin (1966), Tandberg-Hanssen (1967), Kiepenheuer (1968) and Howard (1971).

In brief, active regions are characterized at photospheric levels by the presence of sunspots and concentrations of magnetic flux. In the chromosphere, lines such as Hα and Ca H and K are strongly enhanced, and the definition of an active region includes such areas even in the absence of sunspots. Considerable fine structure, mostly of a filamentary nature is observed, for example, in Hα. Because the magnetic field strengths are high these structures are often used to delineate the magnetic field configurations. In addition to the strong enhancement of emission over the areas surrounding the sunspots, features such as Arch Filament Systems (Bruzek, 1967, 1969) are observed extending into the corona when observations are made near the limb.

Before observations from rockets and satellites became possible the structure of the corona above active regions was studied from radio emission and from visible region photographs taken during eclipses. In addition to electron density determinations from the continuum intensities the temperature and density structure could be studied from the forbidden coronal lines formed between $\sim 10^6$ and 3×10^6 K (Aly *et al.*, 1962). Since these observations could be made only at the limb little was known about the three-dimensional structure and usually cylindrical or hemispherical symmetry was assumed, in accordance with the general idea of a 'coronal condensation'. A useful summary of early radio studies of active regions can be found in the book by Kundu (1965).

* On leave of absence from the Appleton Laboratory, Astrophysics Research Division; Culham Laboratory, Abingdon, Berks., England.

Sharad R. Kane (ed.), Solar Gamma-, X-, and EUV Radiation, 109–131. All Rights Reserved.
Copyright © 1975 by the IAU.

The EUV and X-ray observations provide the means of observing the active regions not only on the limb but on the disk, with a spatial resolution of $\sim 5''$. Since the spectrum includes lines which are formed at temperatures between ~ 8000 K and 20×10^6 K (during flares), the whole structure from the chromosphere, through the transition region to the corona can be studied.

The present paper will review the models of the structure of active regions that have developed from the EUV and soft X-ray data.

II. Studies of the Emission Measure

(a) FROM SOFT X-RAY SPECTRA

Some of the early information on the hot components of active regions ($T_e > 2 \times 10^6$ K) came from soft X-ray spectra of the whole sun at times when several active regions were present. Only recently have collimated instruments, which can allow spectra to be taken of individual active regions, been flown (Bonnelle *et al.*, 1973; Brabban and Glencross, 1973). Unless high spatial resolution can be obtained, the interpretation of the emission line intensities is complicated by the presence of the emission from the whole quiet corona ($T_e \sim 1.5 \times 10^6$ K). However, if uncollimated crystal spectrometers are used, the active regions present can appear as components to the observed spectral feature (e.g. See Batstone *et al.*, 1970).

The initial step in analysing soft-X-ray spectra is to find the emission measure $\int_{AV} N_e^2 \, dV$ as a function of temperature. This can be done in the following way.

The emission in a particular line excited only by collisions from the ground level is given by

$$E = 8 \times 10^{-43} \bar{g} f \frac{N(E)}{N(\text{H})} \int_{AV} N_e^2 g(T) \, dV \text{ erg cm}^{-2} \text{ s}^{-1}, \tag{1}$$

where, following Pottasch (1963),

$$g(T) = \frac{N(\text{ion})}{N(E)} T_e^{-1/2} \exp(-W/kT_e) \tag{2}$$

$\bar{g}$ is the average Gaunt factor; f is the oscillator strength; $N(E)/N(\text{H})$ is the abundance of the element relative to hydrogen; $N(\text{ion})/N(E)$ is the relative ion population in equilibrium conditions; W is the excitation potential of the transition; N_e and T_e are the electron density and temperature respectively and AV is the volume of the atmosphere emitting a particular line.

The distribution of $N(\text{ion})/N(E)$ and hence $g(T)$ with temperature can be calculated. A frequent approximation has been to replace $g(T)$ by a constant value over the region where most of the contribution to a given line occurs. Pottasch (1963) replaced $g(T)$ by $0.70 \, g(T_m)$, where $g(T_m)$ is the maximum value of $g(T)$ at T_m. However, the temperature width to which this applies may differ from line to line, and to allow for this Jordan and Wilson (1971) replaced $g(T)$ by $Gg(T_m)$ such that

$$Gg(T_m) = \int_{\log T_1}^{\log T_2} g(T) \, d \log T$$

and the normalization factor G applies to a fixed temperature range $\Delta \log T = \log T_m$ ± 0.15. T_1 and T_2 are chosen well outside the region of the maximum. With either method an average $g(T)$ can be removed from the integral and hence

$$\frac{N(E)}{N(H)} \int_{AV} N_e^2 \, dV \quad \text{can be calculated for each line}.$$

(The abundances may either be derived or assumed). The distribution of the emission measure $\int_{AV} N_e^2 \, dV$ as a function of T_e is thus defined.

However, if the function $g(T)$ for a particular line does not vary steeply with T_e, and the emission measure is varying rapidly with temperature, an iterative procedure must be adopted by recomputing $\int_{AV} N_e^2 \, g(T) \, dV$ until a self-consistent distribution is found. The temperature found for a given line depends strongly on the form of the emission measure, a fact which should be remembered if an empirical distribution is assumed for the emission measure. Also, if whole Sun observations are used it is important to take account of the large value of the emission measure at the temperature of the quiet corona.

Although this review is not primarily concerned with solar abundances it should be emphasized that the form of the emission function between $T_e \sim 1.5 \times 10^6$ K and 6×10^6 K depends heavily on the abundances used and in particular authors differ in the abundance of neon that they use.

Beigman and Vainshtein (1970) analysed the whole Sun intensity data published by Evans and Pounds (1968), Walker and Rugge (1968) and Blake *et al.* (1965), by considering the emission as coming from a quiet coronal component and an active region component. They found that although the data of Walker and Rugge could be fitted by a two component model with little material between T (corona) and T (active), it was not possible for the other data to choose between such a model and one where the emission measure decreased smoothly with increasing T_e.

Batstone *et al.* (1970) have used a least squares fit method to determine the variation of the emission measure with T_e, for three active regions. Their results for one are shown in Figure 1, together with the values that could be obtained using the Pottasch type of analysis, before any iterations. It appears that a least squares method has little advantage over the Pottasch method plus iterations.

Chambe (1971) has also recognized that the $\int_{AV} N_e^2 \, g(T) \, dV$ should be calculated and has obtained a fit to various active region data by using an exponential form such that

$$N_e^2 \frac{dV}{dT} = C \times 10^{-T/T_0},$$

where $T_0 = 1.5 \times 10^6$ K and C is a constant such that $N_e^2 \, dV/dT = 10^{49}$ cm^{-3} 10^6 K^{-1} at 2×10^6 K. His results are shown in Figure 2. The reduction in the temperature at which Ne IX and Ne X are formed is large; T_m for these ions has values of 3.5×10^6 K and 5.0×10^6 K respectively, but the temperatures at which most emission occurs are 2.4×10^6 K and 3.5×10^6 K respectively. Chambe's method as used by him did take account of the large emission measure at $T_e \sim 1.5 \times 10^6$ K, but the results obtained depend strongly on the exponential form assumed. In order to derive N_e and T_e as a function of height Chambe draws on a wide variety of visible region data, and it is

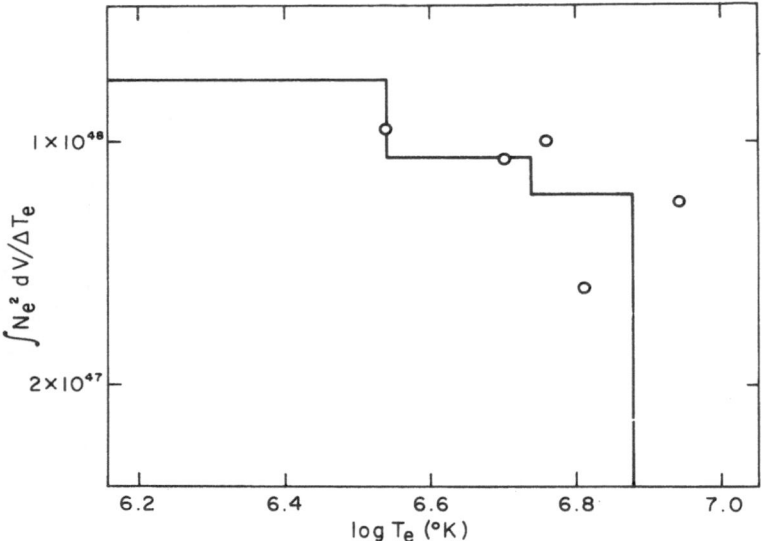

Fig. 1. Differential emission measure for an active region analysed by Batstone *et al.* (1970). The circles show the results of the Pottasch method of analysis.

difficult to realistically compare the significance of his results with others discussed in the present paper. From Figures 1 and 2 it can be seen that Chambe's function falls much more steeply as T_e increases than does the least squares fit.

   Chambe's approach has been used by Acton *et al.* (1972) to analyse emission from O VII and Ne IX. Their spectra were obtained with a collimated instrument and emission from a 1.7′ wide strip was recorded. Although an exponential fit can be made between O VII and Ne IX, extrapolation back to lower temperatures is difficult since the emission measure at lower temperatures is not known for the individual active regions observed. Chambe's fit was to $\int_{AV} N_e^2 \, dV$ for the whole corona plus any active region material present. The function used by Acton *et al.* continues to increase all the way down to $T_e = 10^6$ K, and because of the steep variation of the emission measure with temperature the results will be affected by the assumed variation. This problem illustrates the importance of making observations of lines formed over a wide range of temperature.

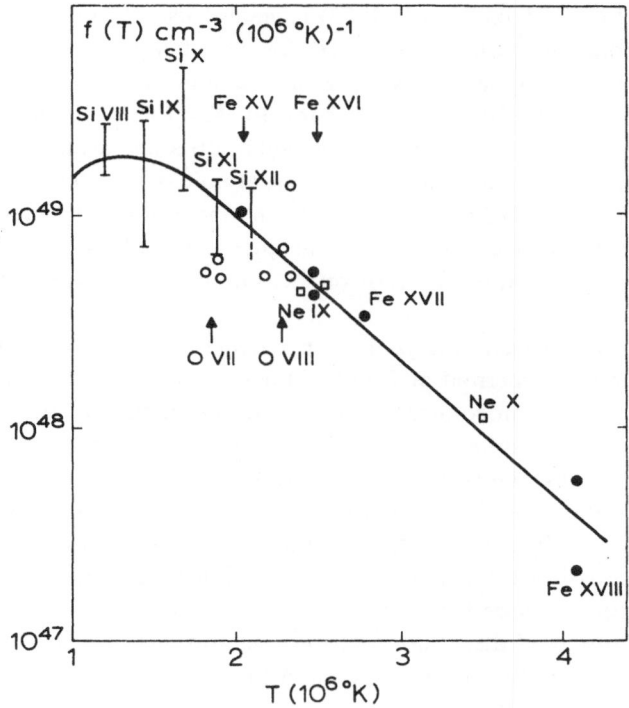

Fig. 2.   Differential emission measure derived by Chambe (1971) using an
exponential form (from Chambe, 1971).

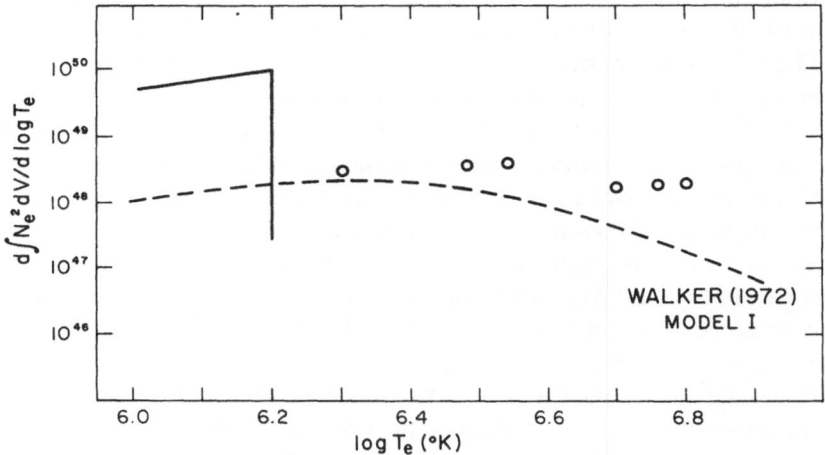

Fig. 3.   Differential emission measure derived by Walker (1972), model I. The circles show the results
of the Pottasch method of analysis. The full curve shows a schematic quiet coronal emission measure.

Walker (1972) and Walker *et al.* (1973) have also used Chambe's approach, combining three exponential forms for $\int N_e^2 \, dV$ (for three temperature ranges) with a least squares fit to the data. These were whole Sun data, and the *form* of the emission function found by Pottasch (1967) was used for $10^6 \, \text{K} < T < 2 \times 10^6 \, \text{K}$. Figure 3 shows Walker's Model I and also the results that would be obtained using $g(T) = 0.70 \, g(T_m)$, with no iteration. (Walker does not tabulate the atomic parameters he used; the values used in Figure 3 are $\bar{g} = 0.30$ and *f*-values from Wiese *et al.* (1966, 1969)). By using only the form of the emission function of the quiet corona between 10^6 and $2 \times 10^6 \, \text{K}$ Walker obtains a variation of the emission measure with temperature that is less steep than that of Chambe.

Walker *et al.* (1973) have analysed further data extending to higher temperature regions using the same method and with similar results. Also Bonnelle *et al.* (1973) have recently analysed their collimated Mg XI and Mg XII data by fitting an exponential fall-off curve and find results consistent with those of other authors.

It is possible to make emission measure studies even from broad band X-ray flux measurements. Landini and Monsignori Fossi (1971) have combined observations made with the Solrad 9 satellite with their theoretical calculations of the temperature dependence of emission integrated over large wavelength bands and have derived a combined emission measure between 10^6 and $3 \times 10^7 \, \text{K}$.

Although the observed distribution of the emission measure can be fitted by functional forms it is not possible to separate $\int_{\Delta V} N_e^2 \, dV$ into density, temperature gradient and volume components without further assumptions or information on the spatial distribution of the emission.

(b) FROM EUV SPECTROHELIOGRAMS AND SPECTRA

The Harvard College Observatory instruments on OSO 4 and OSO 6 have provided observations of the structure of active regions over the temperature range 10^4–$3.5 \times 10^6 \, \text{K}$. Noyes (1971) has reviewed analyses of data from both these experiments. The OSO 4-observations were in the form of whole Sun raster spectroheliograms, made over an interval of 5 min, in selected emission lines. The spatial resolution was $1'$ square. The interpretation of these observations has been discussed by Noyes *et al.* (1970). The main parameter studied was the enhancement in an active region of a given line over its average quiet Sun value. Figure 4 shows the average results from eight active regions. It can be seen that the enhancements vary from about a factor of 3 for transition region ions up to about 40 for Si XII, the highest ion plotted. The enhancement varies from line to line at a given temperature because of differing density dependences, an immediate indication that the density has changed between the quiet and active regions.

The density dependence of the lines of Be I-like ions, in particular of C III, N IV and O V can in principle be used to determine the electron density in the quiet and active regions, (Jordan, 1971a; Munro *et al.*, 1971), but at present the method is limited by uncertainties in the necessary atomic data (Jordan, 1974). However, the data reported by Noyes *et al.* show that the density enhancements for C III and O V are between

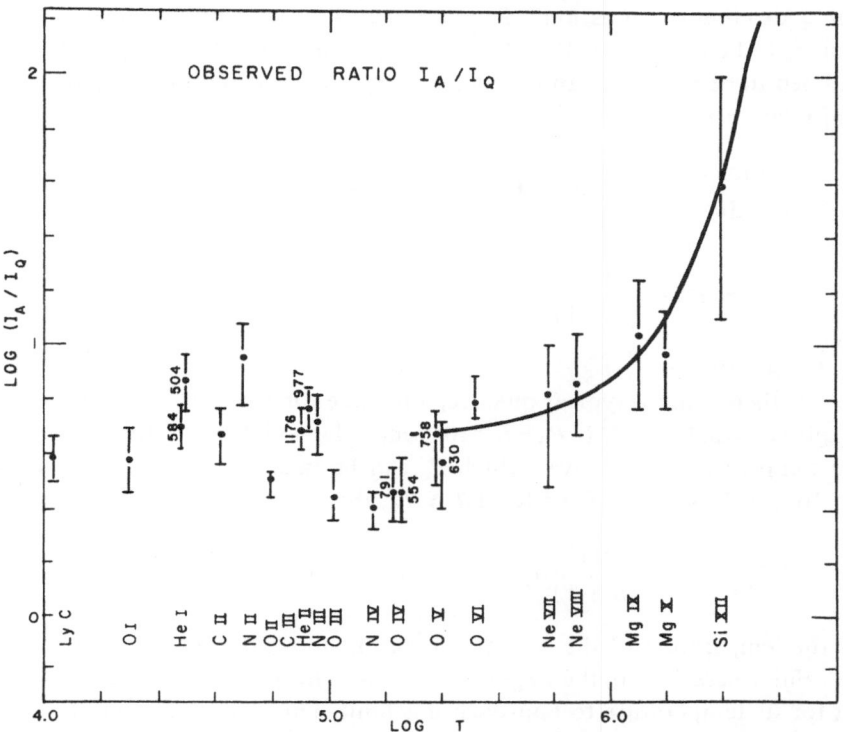

Fig. 4.   Relative intensities of lines between quiet and active regions observed from OSO 4. (From Noyes *et al.*, 1970).

factors of ~2 to 6. Since independent measures of the density allow other parameters such as the conductive flux F_c, and the temperature gradient to be determined, the method is potentially extremely valuable.

The analysis will now be treated in a more general way than used by Withbroe (1970). From spectra of the quiet sun it has been shown that between $T_e \sim 10^5$ to 10^6 K the emission measure has a gradient which is close to that expected if there was constant conductive flux back from the corona (Athay, 1966). This can be seen from the following. The observed emission measure shows that:

$$\int_{\Delta T} N_e^2 \frac{dh}{dT} \, dT = aT^b. \tag{3}$$

The conductive flux is

$$F_c = kT^{5/2} \, dT/dh \text{ erg cm}^{-2} \text{ s}^{-1} \tag{4}$$

Using $P_e = N_e T_e$, Equations (3) and (4) give the relation

$$F_c = \frac{kP_e^2}{ab} T^{-b+3/2} \tag{5}$$

assuming also that the emission is from a spherically symmetric atmosphere. A condition for F_c to be constant is then $b = \frac{3}{2}$ and $P_e = $ const. The regime over which P_e can be assumed to be constant can be examined by using the additional hydrostatic equilibrium relation that:

$$\frac{d \log P_e}{dh} = - 0.86 \times 10^{-4} \frac{1}{T_e} \tag{6}$$

Then

$$P_e^2 = P_0^2 - \frac{Da}{(b+1)} (T_e^{b+1} - T_0^{b+1}), \tag{7}$$

where $D = 4 \times 10^{-4} b$.

If $b = \frac{3}{2}$, then using a typical quiet region value for $\int N_e^2 \, dh$ at 2×10^5 K (Dupree, 1972), gives $a = 3.3 \times 10^{17}$. If $P_0 = 6 \times 10^{14}$ cm^{-3} K, and $T_0 = 2 \times 10^5$ K, the condition on the temperature range over which P_e can be taken as constant (i.e. $P_e^2 = P_0^2$ to within 10%) is $T_0 < T_e < 7 \times 10^5$ K. If $T_e \gg T_0$, then

$$P_e^2 = P_0^2 - \frac{Da}{(b+1)} T_e^{5/2} \tag{8}$$

and if the temperature of the isothermal corona above the transition region is taken as the value where $P_e \to 0$, then $T_c \to 1.8 \times 10^6$ K. The atmosphere must have sufficient height for its temperature to approach this limit. This determination of T_c allows F_c at P_0 to be found directly if T_c is known, since

$$T_c^{5/2} = \left(\frac{P_0^2}{a}\right) \cdot \left(\frac{b+1}{D}\right) \tag{9}$$

and

$$F_c(P_0) = \frac{k}{a} \frac{P_0^2}{b}$$

Thus

$$F_c(P_0) = \frac{k}{b} \cdot \frac{D}{(b+1)} \cdot T_c^{5/2}. \tag{10}$$

Although not expressed in this way the above relations are the physical basis behind Withbroe's (1970) empirical three-parameter description of the structure of the quiet atmosphere. He describes models in terms of F_c, P_0 and T_c. His method has been used by Noyes et al. (1970) to investigate the changes in P_0, F_c and T_c between the quiet and active regions. From the distribution shown in Figure 4 they deduce that to account for the observed variation of both transition region and coronal lines requires an increase in P_0 and F_c of a factor of five, and an increase of T_c to 2.5×10^6 K.

These results are consistent with the analytical method using the increased value of the constant 'a' given by the observed enhancement of a factor of $\simeq 5$ at $T_e = T_0 = = 2 \times 10^5$ K, and a density enhancement of a factor of 5 determined from the O v ratio.

If N_e increases by a factor of 5 at T_0, then (P_0) active $=5(P_0)$ quiet and (F_c) active $=5(F_c)$ quiet. Using Equation (7), the regime where $P_e=$ constant now extends to $T_e=9.6\times10^5$ K, and it is found that $T_c=2.4\times10^6$ K.

Both spectra and spectroheliograms were obtained with the OSO 6 instrument. The spatial resolution for regions where spectra were obtained was 35″. In an analysis of McMath Region 10266 Dupree $et\ al.$ (1973) use Withbroe's arguments and the O v ratio to show that P_0 (active)$/P_0$ (quiet)$=7$, that $F_c=1.5\times10^7$ cm^{-2} s^{-1} (a factor of 12 greater than the quiet value) and $T_c=3.2\times10^6$ K. Proceeding analytically, the density increase of a factor of 7 can be combined with an intensity increase of 5, to say that F_c increases by a factor of 10, and $T_c=3.2\times10^6$ K.

Thus by measuring both N_e and the intensity enhancement at 2×10^5 K, or by determining T_c and N_e the structure of the active region at $T_e>2\times10^5$ K (treated as an average) can be calculated.

(c) STUDIES OF LIMB SPECTRA

Limb observations have the advantage that the variation of the emission measure in lines formed at different temperatures can be studied as a function of height. Prior to the recent Skylab observations active regions observed on spectroheliograms on the disk were usually treated as single features.

The EUV spectroheliograms obtained by the Naval Research Laboratory group can be used to study the relative spatial distribution of material as a function of temperature (See e.g. Tousey, 1967). Boardman and Billings (1969) have made an analysis of an active region seen on the limb during an Aerobee 150 rocket flight by the NRL group on April 28, 1966. They use a Pottasch type of analysis, and from the intensities of a variety of lines they deduce the emission measure as a function of temperature and position. By assuming a cylindrical symmetry about the central point of the region they derive $\langle N_e^2\rangle^{1/2}$ as a function of temperature and height. In view of the loop structures observed from the X-ray photographs (see Section III(a)) the cylindrical approximation may not be appropriate. However, the overall results that they obtain (shown in Figure 5), do not differ substantially from those obtained by Gabriel and Jordan (1974) (see below), if the spatial resolution of the latter results was reduced. Boardman and Billings find that the region has a dense hot 'core' of about 20 000 km in height and 30 000 km radius, which is surrounded by cooler, less dense material. The temperature ranges from 3×10^6 K in the core down to 1.7×10^6 K at 4×10^4 km. The density falls from 2×10^9 cm^{-3} in the core to 1.3×10^8 cm^{-3} at 4×10^4 km.

During the 1970, March 7 total eclipse a rocket was launched from Wallops Island, Virginia into the eclipse path. This experiment and the data obtained have been described by Speer $et\ al.$ (1970) and by Gabriel $et\ al.$ (1971). Some thirty five photographic slitless spectra were obtained over the spectral range 850 Å to 2190 Å during the flight with a grain-limited spatial resolution of about 3″. One of the interesting features observed was a large loop system and intensity enhancement on the north-east limb. From the Fraunhofer Institute maps and the magnetograms of Livingston $et\ al.$ (1970)

this was probably associated with McMath 10623, which was a weak plage region with no sunspots recorded on the maps. The coronal component of the active region is apparent in the forbidden lines of silicon, iron, sulphur and nickel which lie between 1000 Å and 2200 Å, and which were first identified from this data (Gabriel *et al.*, 1971; Jordan, 1971b). The active region can be seen in lines formed at temperatures between

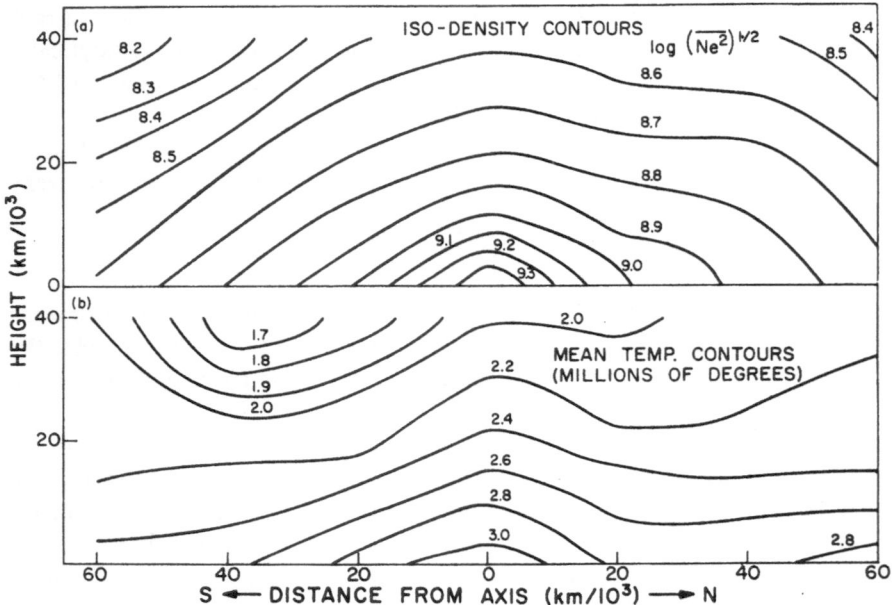

Fig. 5.   The density and temperature as a function of position in a limb active region observed by the Naval Research Laboratory (from Boardman and Billings, 1969).

7×10^5 K and 2.5×10^6 K, and the apparent change in the structure as a function of temperature is shown in Figure 6.

Gabriel and Jordan (1974) have recently made an analysis of the density and temperature structure as a function of height and position in the active region. The spectra were intensity calibrated so that analyses of both absolute and relative intensities have been possible. Frame 30, taken when the Moon's limb was 10950 km above the photosphere, at the latitude of the active region, has been used in the analysis, although a normalization procedure for level populations was made using quiet coronal data over a series of frames.

The method by which the emission measure may be derived has been given in Section II(a). For forbidden lines the approximation that the excited level is populated mainly from the ground is not always valid, so instead of replacing $N_2 A_{21}$ by $C_{12} N_e N_1$, as was done in deriving Equation (1), the upper level population is retained and can be calculated as a function of N_e and T_e. N_1 and N_2 are the populations of the ground and upper levels respectively, A_{21} is the spontaneous transition probability, C_{12} is the

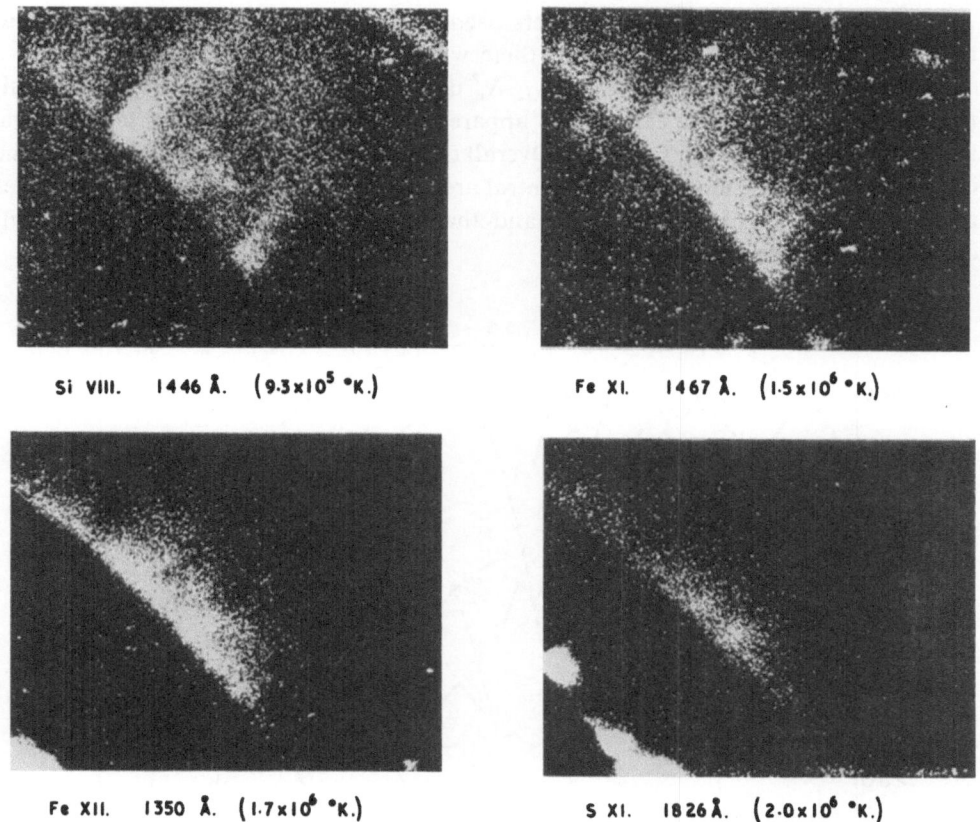

Si VIII.   1446 Å. $(9.3 \times 10^5 \, °K.)$ Fe XI.   1467 Å. $(1.5 \times 10^6 \, °K.)$

Fe XII.   1350 Å. $(1.7 \times 10^6 \, °K.)$ S XI.   1826 Å. $(2.0 \times 10^6 \, °K.)$

Fig. 6.   A limb active region observed in forbidden lines of Si VIII, Fe XI, Fe XII and S XI during the March 1970 total eclipse (from Gabriel *et al.*, 1971).

collisional excitation rate. Hence at a given point in the active region

$$I = \text{const.} \frac{N(E)}{N(\mathrm{H})} \int_{\Delta L} \frac{N_2}{N(\mathrm{ion})} \cdot \frac{N(\mathrm{ion})}{N(E)} \cdot N_e \, \mathrm{d}l \qquad \mathrm{erg \, cm}^{-2} \, \mathrm{s}^{-1} \, \mathrm{sterad}^{-1}$$

where the integration is now over ΔL, the line of sight path length for a particular line.

$N_2/N(\mathrm{ion})$ can be calculated as a function of N_e and T_e, and for the lines of Si IX, Fe XI, Fe XII ($^4S - {}^2P$) and S XI, $N_2/N(\mathrm{ion}) \, N_e \simeq \text{constant}$, and $I \propto N_e^2$, as for many permitted resonance lines. For the lines of Si VIII and Fe XII ($^4S - {}^2D$) a density must be assumed initially and iterations performed as necessary. The values of $N(\mathrm{ion})/N(E)$ have been taken from the calculations by Jordan (1969).
Thus

$$I = \text{const.} \left(\frac{N_2}{N_{\mathrm{ion}}} \frac{1}{N_e} \right) \int_{\Delta L} N_e^2 \, \mathrm{d}l,$$

and the emission measure $\int_{\Delta L} N_e^2 \, \mathrm{d}l$ can be found.

The details of the atomic parameters used are given by Gabriel and Jordan, and only the resulting structure found by them will be discussed at present.

Figure 7 shows the distribution of $\int_{AL} N_e^2 \, dl$ as a function of temperature and position in the active region. (The loops apparent in Si VIII are sketched in to aid the location of the position in Figure 6). Overall the results shown in Figure 7 confirm the impression, from Figure 6, that the central area contains more hot material than does the loop system apparent in Si VIII, and that the fraction of hot material present decreases with increasing height.

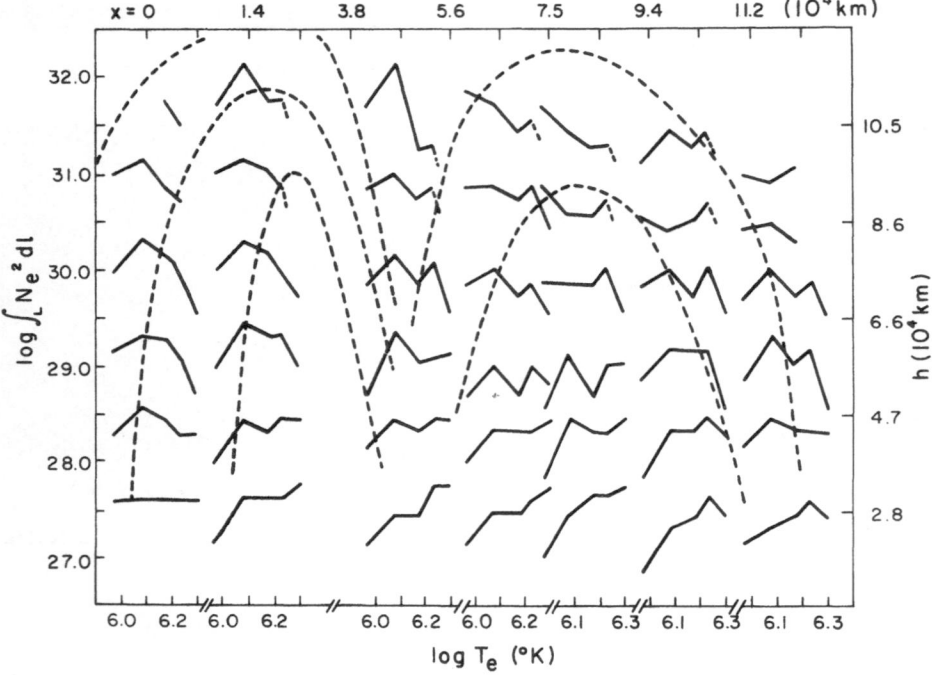

Fig. 7.   The emission measure $\int_{AL} N_e^2 \, dl$ as a function of position in the 1970, March 7 limb active region. x is the distance from the left side of the region, h is the height above the photosphere.

In order to resolve $\int_{AL} N_e^2 \, dl$ into components of N_e and AL for each line, some independent measure of either N_e or AL is necessary. Three methods can be used. Assuming that the loop structures are similar in shape to those seen by Van Speybroeck *et al.* (1970) it can be assumed that $AL = d$, the observed diameter of the loop. This gives an upper limit to AL for the loop may not be filled with the relevant material. However, at 28000 km, (column 1 in Figure 7), $AL = 3.6 \times 10^4$ km leads to $N_e = 1.0 \times 10^9$ cm^{-3} at $T_e = 9.3 \times 10^5$ K $(N_e T_e = 9.3 \times 10^{14}$ cm$^{-3})$. Alternatively, since the Si VIII emission depends on N_e rather than N_e^2 over the density range 10^9–10^{10} cm^{-3}

an average density can be found from the relative intensities of the Si VIII and Si IX lines. This is only a crude method since it assumes that N_e and ΔL are the same for Si VIII and Si IX. However, the method gives the results shown in Table I. The positions correspond to the first, fourth and seventh columns in Figure 7. Densities of $N_e \sim 1$–2×10^9 cm^{-3} are found with ΔL between 3×10^3 and 3×10^4 km. The fact that N_e does not fall by the value expected in hydrostatic equilibrium with $T_e =$ const. – contrary to the results of analyzing the decrease in the absolute intensity with height, presented below – is probably due to the crudeness of the method.

TABLE I

N_e and ΔL from Si VIII/Si IX intensity ratios in 1970, March 7 eclipse active region

Height (km)	Region A		Region B		Region C	
	$\log N_e$ (cm^{-3})	ΔL 10^4 km	$\log N_e$ (cm^{-3})	ΔL 10^4 km	$\log N_e$ (cm^{-3})	ΔL 10^4 km
30 000	9.02	3.5	9.12	1.2	8.98	1.9
49 000	9.17	1.3	9.33	0.45	9.28	0.68
67 000	9.20	0.81	9.25	0.32	9.40	0.26
86 000	9.16	0.74	9.10	0.51	9.04	0.32
105 000	9.05	1.1	9.05	0.57	8.97	0.30

A third method of determining N_e is to use the relative intensities of the Fe XII lines $^4S_{3/2} - {}^2D_{3/2}$ and $^4S_{3/2} - {}^2P_{1/2}$ which are sensitive to N_e over the range $10^{12} > N_e > 10^8$ cm^{-3}. The results are given in Table II. The density, now at $T_e = 1.7 \times 10^6$ K, falls from $N_e = 3 \times 10^9$ cm^{-3} at 30 000 km to $\sim 10^9$ cm^{-3} by 10^5 km. In hydrostatic equilibrium with $T_e = 1.7 \times 10^6$ K, a fall of a factor of 2–3 would be expected. The path length from the Fe XII emission varies between 2×10^3 km to 1.5×10^5 km, but on the whole does not increase substantially with height. However, the path length derived at 1.7×10^6 K is systematically lower than that derived from the Si VIII/Si IX ratio.

TABLE II

Electron densities and ΔL from Fe XII ratios in 1970, March 7 eclipse active regions

Height 10^4 (km)	Region A		Region B		Region C	
	$\log N_e$ (cm^{-3})	ΔL 10^3 km	$\log N_e$ (cm^{-3})	ΔL 10^3 km	$\log N_e$ (cm^{-3})	ΔL 10^3 km
3.0	9.58	2.7	9.56	4.3	9.43	3.9
4.9	9.45	3.5	9.42	4.1	8.99	14.8
6.7	9.44	2.6	9.25	4.5	9.12	5.8
8.6	9.36	2.7	9.27	2.9	9.22	2.6
10.5	9.11	4.3	9.15	5.0		

The results for two points at 30000 km (column 1 and column 4 in Figure 7) are summarized in Table III. The pressure increases from $\sim 10^{15}$ cm^{-3} K at 9×10^5 K to 6×10^{15} cm^{-3} K at 2×10^6 K. (The typical quiet transition region value is $6 \times \times 10^{14}$ cm^{-3} K, and for the quiet corona is $\simeq 4 \times 10^{14}$ cm^{-3} K). Only small magnetic fields would be necessary to balance the gas pressure – between 2 and 5 G. The values of N_e and ΔL have been interpolated between the values for Si VIII and Fe XII.

TABLE III

Models for selected parts of 1970, March 7
eclipse active region

Model for Region A at $h = 30000$ km

$\log T$	Ion	$\log N_e$	ΔL (km)
5.97	Si VIII	9.02	3.5×10^4
6.08	Si IX	9.24	1.3×10^4
6.18	Fe XI	9.44	5.0×10^3
6.23	Fe XII	9.58	2.7×10^3

Model for Region B at $h = 30000$ km

5.97	Si VIII	9.12	1.2×10^4
6.08	Si IX	9.30	6.2×10^3
6.18	Fe XI	9.46	3.1×10^3
6.23	Fe XII	9.56	4.3×10^3

It remains to determine how the material in a given line of sight is distributed in position. Is the hot material within the cool or vice versa? If the active region is considered as composed of adjacent flux tubes, and the hot material is within the cool, the path length for the cool material is sufficiently large for 'gaps' to appear between adjacent regions emitting high temperature lines. This is not observed. At 30000 km the apparent distribution is consistent with a set of loops each with a structure such that the cool material is within the hot. The fraction of the loop material at high temperatures increases towards the central regions. Small-scale helical structures within the loop systems are not ruled out by the observations.

A further study that has been made by Gabriel and Jordan is of the decrease in the observed emission as a function of height. For lines which have intensity $\propto N_e^2$ it would be expected that in the absence of a magnetic field – or along a field line, that the emission per unit volume would eventually decrease according to Equation (6) for hydrostatic equilibrium. The calculated N_e dependence starting from a base density can also be combined with Equation (6) if the line does not have intensity $\propto N_e^2$. If the line of sight distance ΔL increased as a function of height, as might be expected with the expansion of a dipole type field, then this would show as a departure from a hydrostatic fall off with constant T_e. Figure 8 shows the intensity (on a relative scale) observed as a function of height in Si VIII, Si IX and Fe XI, following as well as possible the 'field' direction in the left-hand loop of Figure 7. The hydrostatic fall-off with $T_e = T_m$ for each ion, and the density corrected curves are also given. It can be seen that for heights above 40000 km, the hydrostatic curve is closely followed by the Si IX and Fe XI emission. An increase in ΔL could raise the curve, but a decrease in T_e with height would lower the curve. Unless these two effects cancel out fairly precisely,

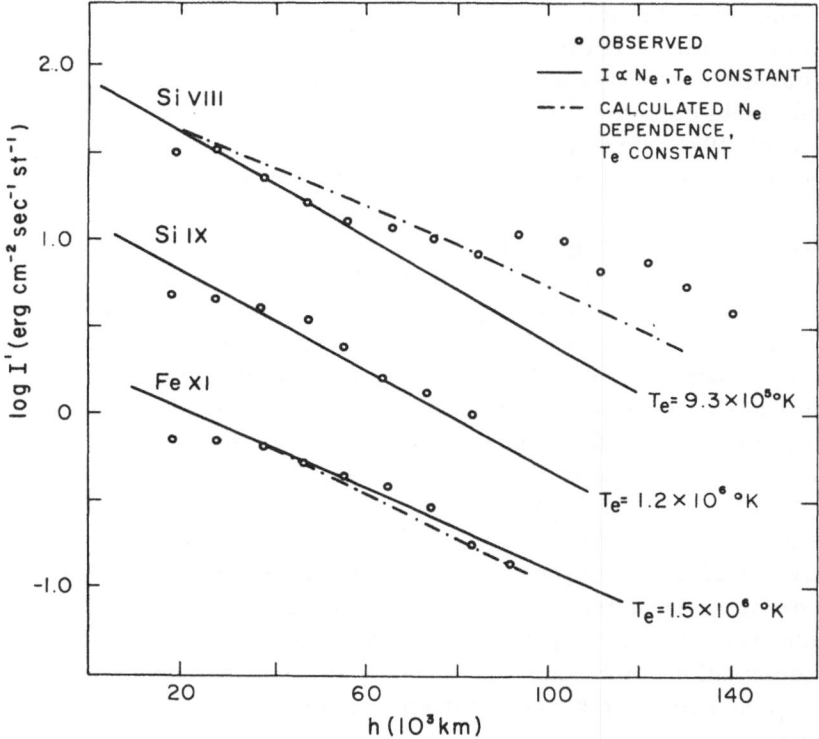

Fig. 8.   The decrease of emission with height in the Si VIII, Si IX and Fe XI, forbidden lines on the left side of the loop structure observed in the 1970, March 7 limb active region. The full lines show the decrease expected in hydrostatic equilibrium at T_m, if the emission is proportional to N_e. The dashed lines take into account the calculated dependence on N_e. The intensities, I', are on relative scales.

Figure 8 shows that ΔL is practically constant with height up to $\sim 9 \times 10^5$ km. The Si VIII emission above 9×10^5 km shows that ΔL eventually increases by a factor of about 2. The values of ΔL can be combined with the emission measures to find N_e as a function of height. There appears to be a systematic departure from the straight line for emission at $h < 4.0 \times 10^4$ km. This could be interpreted as evidence for a non-isothermal medium – i.e., individual loops no longer have a basically isothermal structure. Studies of the transition region material at heights up to 40000 km should allow the type of analysis outlined in the previous section to be made for this active region.

III. The Spatial Distribution of Material in Active Regions

(a) FROM X-RAY IMAGES

The spatial resolution obtainable in images of X-ray emitting regions is now $\sim 2''$. The early photographs obtained at $20''$ resolution (Underwood and Muney, 1967), showed how the material with $T_e \geqslant 2 \times 10^6$ K is concentrated over active regions. By making observations through filters which had different responses as a function of wavelength Underwood and Muney found that the radiation at longer wavelengths (lower T_e) was more diffuse in origin than that at shorter wavelengths. With improved resolution it became possible to observe some of the structure present in the X-ray emitting regions. The photographs obtained by the American Science and Engineering (AS & E) Group (Krieger *et al.*, 1971, and earlier papers) have shown that the material at $T_e > 2 \times 10^6$ K is characteristically in the form of loop structures, connecting regions of opposite polarity observed with ground-based magnetographs. The loops have similarities to those seen earlier in white light photographs but do not coincide with Hα structures. The outline of the X-ray emission follows in general that of the underlying Ca K plage, and is associated with enhanced magnetic fields. The loops have a range of sizes and frequently reach heights of $\sim 1.5 \times 10^5$ km above the photosphere. Figure 9 shows the photograph obtained by Van Speybroeck *et al.* (1970) just after the 1970, March 7 solar eclipse. The form of the loops strongly suggests that the emitting material is confined by the magnetic fields and hence can be used to show the local magnetic field configuration. The results of the AS & E rocket programme over some years has recently been summarized by Vaiana *et al.* (1974). Their analyses have shown that quite frequently loop structures connect adjacent active regions and trans-equatorial arches are also observed. In general they find that the X-ray emission reaches a maximum above the neutral line of longitudinal magnetic field, and that often, where the field gradients are large, a bright hot small core connects the opposite polarities. The AS & E observations on Skylab (Vaiana *et al.*, 1973), made with resolution of up to $2''$, will no doubt considerably extend the present information on the spatial distribution of the X-ray emission.

(b) FROM LIMB-TRANSIT OBSERVATIONS

The spectra and photographs taken of emission on the disk can be supplemented by observations made as an active region rotates beyond the solar limb. Such observations provide useful additional information on the structure as a function of height.

Beigman *et al.* (1969) used observations made from the Cosmos-166 and Cosmos-230 satellites to show that emission in the wavelength bands 2–8 Å and 8–14 Å, and hence at $T_e \gtrsim 4 \times 10^6$ K was formed up to ~ 80000 km above the limb.

Krieger *et al.* (1971) made similar studies of the variation of the emission in the range 2.5–12 Å observed from six active regions with their OSO 4 satellite instrument. They found that the emission ($T_e > 5 \times 10^6$ K) was formed over heights of from 10^4 to 9×10^5 km. The large range of height they obtained probably reflects differences in the

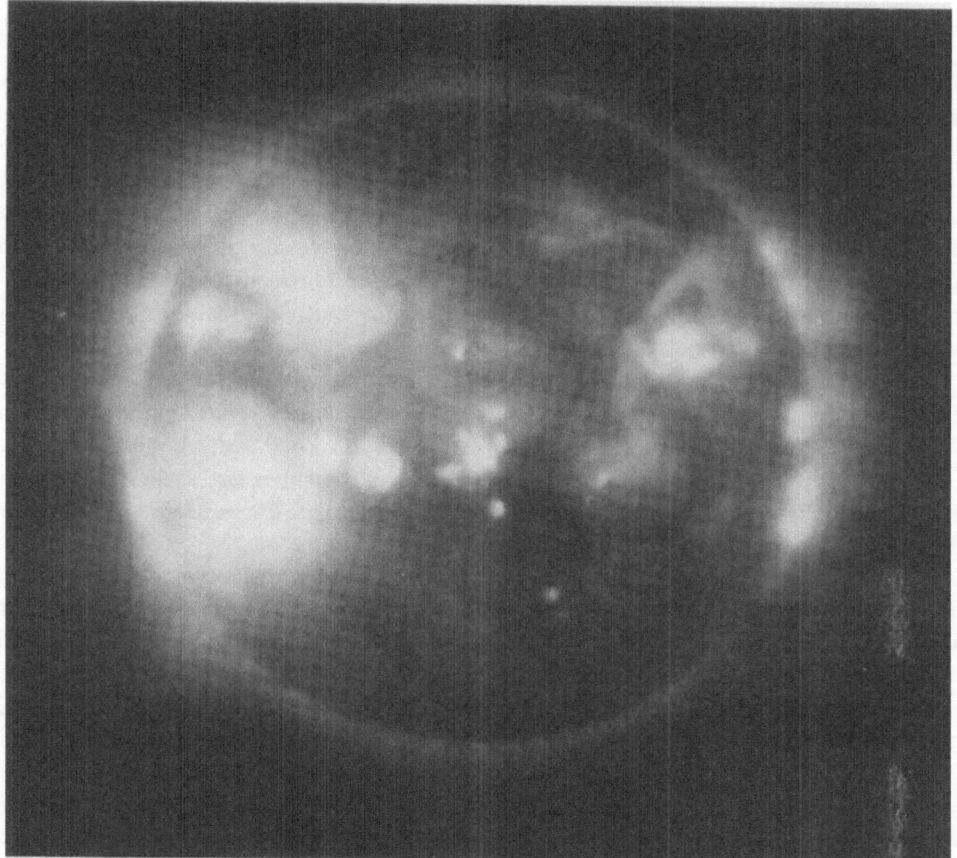

Fig. 9.   The X-ray emission from the corona and active regions observed just after the 1970, March 7 eclipse by the AS & E group (from Van Speybroeck *et al.*, 1970).

temperature structure of the active regions since the height observed by their method will depend on the spectral hardness.

Parkinson (1973) has analyzed flux measurements in the 8.4–9.6 Å wavelength band, (hence $T_e \simeq 6 \times 10^6$ K), obtained from OSO 5, (Herring *et al.*, 1971), and has determined the height of three emitting regions. He finds that the base of the emitting region was $\sim 2 \times 10^4$ km above the photosphere, and that half the emission is formed below 3×10^4 km, with some emission extending to ~ 0.8–1.6×10^5 km.

The limb-transit observations plus the X-ray images show that in general the hottest material lies at the lowest heights and vice versa, with the emission at $T_e \gtrsim 6 \times 10^6$ K lying at $\sim 20\,000$ km above the limb, and that at $T_e \sim 2$–3×10^6 K extending to 10^5 km above the limb. These results are consistent with the spatially resolved EUV data discussed in Section II(c).

IV. Models Using Combined Observations

The analysis by Landini and Monsignori Fossi (1971) was discussed in Section II(a), up to the derivation of the emission measure as a function of temperature. These authors and also Parkinson (1973), and Vaiana *et al.* (1974) have combined X-ray emission measures with models of the geometry in order to find the density and temperature structure of the active regions. Landini and Monsignori Fossi assume equipartition of kinetic and magnetic energy, a spherically symmetric density distribution and hydrostatic equilibrium just at the centre of the 'condensation'. They use as a boundary condition the density distribution given by Saito and Billings (1964), found from visible region observations.

Landini and Monsignori Fossi found that the condensation had a hot core, with $N_e \sim 10^{10}$ cm^{-3}, and $T_e \sim 3$–6×10^6 K, over a radius of some 3×10^4 km. Surrounding the core was a region where the density fell to $\sim 2 \times 10^9$ cm^{-3} and the temperature to 1.5×10^6 K by a height of $\sim 10^5$ km. These electron densities are rather higher than those found by other authors, but the overall model bears similarities to those of Boardman and Billings (1969) and Gabriel and Jordan (1974).

Parkinson (1973) has combined the heights from limb transits discussed in Section III(b) with the emission measures found by Batstone *et al.* (1970). He estimates the emitting volumes at 2–3 $\times 10^6$ K and at 5–6 $\times 10^6$ K as 10^{30} cm^3 and 10^{27} cm^3, respectively and hence deduces $N_e \sim 2 \times 10^9$ cm^{-3} at 2–3 $\times 10^6$ K and $N_e \sim 10^{10}$ cm^3 at $T_e \sim 5$–6×10^6 K. He suggests an empirical model where the hot material lies at the top of a small loop structure ($h \sim 3.0 \times 10^4$ km) below a cooler component extending to $\sim 10^5$ km.

Vaiana *et al.* (1974) have combined their broad-band photographs with theoretical calculations of the dependence of the emission as a function of wavelength and temperature by Tucker and Koren (1971) and Landini (1974, unpublished). They use two models for the geometry, one an isothermal slab model analogous to an isothermal loop and another which has spherical symmetry. The latter gives the now usual result of low lying regions of hot dense material ($T_e \sim 3$–3.6×10^6 K, $N_e \sim 1.6 \times 10^{10}$ cm^{-3}), surrounded by an amorphous less dense, cooler region ($T_e \sim 2.5 \times 10^6$ K, $N_e \sim 6 \times 10^9$ cm^{-3}).

The active region structure found by Aly *et al.* (1962) from visible region eclipse data is broadly consistent with present results for the temperature region 10^6 K–3×10^6 K, although Aly *et al.* used a 'condensation' model. Also, the first active region model derived from radio data (Christiansen *et al.*, 1960) gave results similar to those derived with low spatial resolution EUV data.

The exceptionally high resolution (3″–9″) radio observations recently reported by Kundu *et al.* (1974) are particularly interesting. The active region studied appears to be composed of a succession of bright features along the active region. The brightest regions are the central ones where $T_b \sim 8 \times 10^5$ K, and regions at lower temperatures ($\sim 7 \times 10^4$ K) have a wider separation. It is likely that the radio emission is originating in the dense lower portions of loop structures as seen in lines such as Si VIII during the March 1970 eclipse.

V. The Relation between Observed Structures and Physical Parameters

There has now been accumulated a considerable quantity of data describing the properties of active regions. It is apparent that the different types of observations are more or less consistent and can be explained by a model that has the following properties; enhanced emission over an area comparable to the plage area, plus loop systems extending outwards some 10^5 km above the surface. The X-ray and EUV eclipse data show that larger loops with greater separation of their foot points are cooler than those above the central part of the active region. The density at heights $\sim 10^4$ km increases as T_e increases, but the variation in N_e is not as large at a given height as is the variation in T_e. Above $h \sim 4 \times 10^4$ km the electron density in the loops which extend to $\sim 10^5$ km decreases according to hydrostatic equilibrium in an isothermal region.

The analyses by the Harvard group have shown that the emission in active regions from transition region and coronal ions can be understood in terms of variations in F_c, the conductive flux back from the corona, in T_c, the isothermal coronal temperature, and in P_0 the pressure at the base of the transition region where $T_e \simeq 2 \times 10^5$ K. In Section II(b) of this paper it was shown how these parameters are inter-related. It appears that the observed properties of active regions at $T_e \geqslant 2 \times 10^5$ K can be understood in terms of individual elements each with its own F_c, T_c, P_0 and presumably magnetic field strength. In each flux tube there will be (a) a region where the temperature gradient is given by a constant conductive flux and (b) a region where the temperature is constant at T_c. (An intermediate region lies between.) If T_c can be determined then so can F_c, but an independent measure of N_e in the constant conductive flux part of the region is needed to specify all the structure. The method is strictly valid only if each flux tube has constant diameter.

The behavior of the spatially integrated emission measure at temperatures $T_e > 10^6$ K can also be understood in terms of adjacent magnetically controlled flux tubes each with its own value of F_c and P_0. The observed decrease in the total emission measure as a function of temperature is simply produced by the different weightings of each region.

The summation procedure can be expressed as follows:

Let

$$\int N_e^2 \frac{dh}{dT} \, dT = Y.$$

Below T_q, the minimum value of T_c (which can be *below* the temperature of the quiet corona) the emission function at each temperature T_n is

$$\sum_i Y_n = T_n^b \sum_i a_i,$$

where i refers to an individual flux tube, and T_i is the maximum temperature in that flux tube.

Above T_q, the total emission at each T_n is

$$\sum_i Y_n = T_n^b \sum_{i=n}^m a_i, \qquad (11)$$

where T_m is the maximum temperature present in the whole active region.

At T_q, the lowest maximum temperature in any of the flux tubes,

$$\sum_i Y_q = T_q^{3/2} \sum_{i=q}^m a_i. \qquad (12)$$

Then

$$\sum Y_n / \sum Y_q = (T_n^{3/2}/T_q^{3/2}) \left[1 - \left(\sum_{i=q}^n a_i / \sum_{i=q}^m a_i \right) \right]. \qquad (13)$$

The observed emission function will have a shape determined by $d \sum_i Y_n / d T_n$, which in turn depends on how a_i varies with T_e.

Equation (13) can be compared to the functions that Chambe (1971) and Walker (1972) have used to describe the differential emission measure, i.e. $Y_n = Y_q 10^{-T_n/T_q}$ (Chambe) and

$$Y_n = Y_0 10^{-T_2/T_1} \left(1 + 1 \cdot 1515 \frac{(T_2 - T_0)}{T_1} - 1 \cdot 1515 \frac{(T - T_0)^2}{T_1 (T_2 - T_0)} \right)$$

$$T_0 < T < T_2. \qquad \text{(Walker)}$$

It can be seen from Equation (13) that it is difficult to go back from the observed total emission function to the physical quantities describing the individual parts of the active region.

Some simplification of (13) can be obtained by expressing a_i as a function of T_c only. i.e., let

$$\left(\frac{a_i}{a_a} \right) = \left(\frac{T_{c_i}}{T_a} \right)^x, \text{ then } \sum_i Y_n / \sum_i Y_q = (T_n^{3/2}/T_q^{3/2}) \left[1 - \left(\sum_{i=q}^n T_{c_i} / \sum_{i=q}^m T_{c_i} \right) \right]$$

$$(14)$$

The emission function as a function of T_n can then be calculated for different values of T_q, T_m and χ. Figure 10 shows cases with a range of values of χ and with $T_m = 8 \times 10^6$ K and $T_q = 10^6$ K.

   (i) $\chi = 2.5$
   (ii) $\chi = 1.0$
   (iii) $\chi = 0$
   (iv) $\chi = -1$
   (v) $\chi = -2$

Solutions with lower T_m turn over at lower temperatures, but the overall shapes of the curves are similar. From Figure 10 it can be seen that $1 \leqslant \chi \leqslant -1$ could give a fit to the observations, but it is difficult to determine χ with higher accuracy.

In Section II, it was found that $\frac{5}{2}\Delta \log T_c = \Delta \log (P_0^2/a)$, that $\Delta \log F_c = \Delta \log (P_0^2/a)$ and, more importantly, $\Delta \log F_c (P_0) = \frac{5}{2}\Delta \log T_c$. This latter relationship can be compared with Reimers' (1971) conclusion that $\Delta \log F_c = \frac{7}{2}\Delta \log T_c$, which now appears

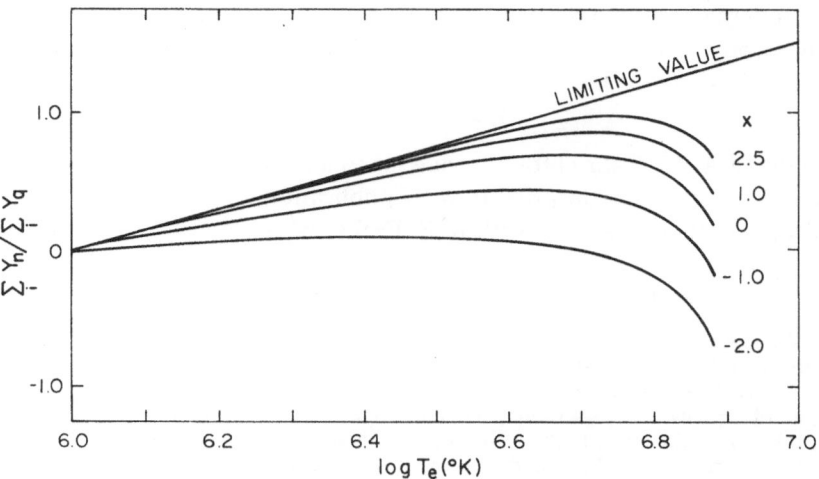

Fig. 10. Theoretical curves for the integrated differential emission measure.

to be incorrect. Reimers derived his relationship between F_c and T as follows; given that,

$$F_c = kT^{5/2} \frac{dT}{dh}, \text{ and } F_c \text{ is constant, then}$$

$$F_c = \frac{2k}{7} \frac{(T_c^{7/2} - T_0^{7/2})}{h_c - h_0} \tag{15}$$

The height difference $h_c - h_0$ was assumed to be constant, by supposing that $h_c - h_0 \propto R$, the damping length for waves heating the corona, and hence $\Delta \log F_c = = \frac{7}{2}\Delta \log T_c$. However, F_c is *not* constant all the way up to T_c, but up to some lower value, say T_*. Thus a relationship $F_c = (2k/7) (T_0^{7/2} - T_*^{7/2})/(h_* - h_0)$ would be valid, and since T_* can be calculated, may provide a useful means of quickly evaluating height changes, as a function of F_c. Reimers also assumed on the basis of earlier work by Unsöld (1960, 1970) that $N(T_0) \propto F_m$, the mechanical energy flux deposited. Hence $\Delta \log N_c = \frac{5}{2}\Delta \log T_c$. But again the basis of this does not seem to be correct, since P_e is not constant from P_0 to P_c.

Provided spatially resolved observations are available, and it is found that Equation (3) is generally valid, the inter-relations found in Section II(b) and above could be used to determine the distribution of F_c and hence F_m over a given active region. When related to the field strength this should provide useful information on the mechanisms heating the active region material. By substituting a range of values of F_c, and P_0 in the equations it can be seen that the observed properties of the loop systems associated with active regions can be reproduced by a model in which loops emerge from below the photosphere, and expand whilst F_m (and hence F_c) is decreasing. With a given F_m the temperature at the top of a loop will first increase, as the loop expands, to a maximum permitted value, T_c, after which an increasing fraction of the loop will become isothermal. These results will be developed further in forthcoming papers.

Acknowledgements

I wish to thank A. H. Gabriel for allowing me to include results of our analyses of the 1970, March 7 active region prior to their publication elsewhere. It is a pleasure to acknowledge useful discussions with R. S. Peckover.

References

Acton, L. W., Catura, R. C., Meyerott, A. J., and Wolfson, C. J.: 1972, *Solar Phys.* **26**, 183.
Aly, M., Evans, J. W., and Orrall, F. Q.: 1962, *Astrophys. J.* **136**, 956.
Athay, R. G.: 1966, *Astrophys. J.* **145**, 784.
Batstone, R. M., Evans, K., Parkinson, J. H., and Pounds, K. A.: 1970, *Solar Phys.* **13**, 389.
Beigman, I. L. and Vainshtein, L. A.: 1970, *Astron. Zh.* **47**, 1030.
Beigman, I. L., Grineva, Yu. I., Mandel'shtam, S. L., Vainshtein, L. A., and Žitnik, I. A.: 1969, *Solar Phys.* **9**, 160.
Blake, R. L., Chubb, T. A., Friedman, H., and Unzicker, E.: 1965, *Astrophys. J.* **142**, 1.
Boardman, W. J. and Billings, D. E.: 1969, *Astrophys. J.* **156**, 731.
Bonnelle, C., Senemaud, C., Senemaud, G., Guionnet, M., Henoux, J. C., and Michard, R.: 1973, *Solar Phys.* **29**, 341.
Brabban, D. H. and Glencross, W. M.: 1973, *Proc. Roy. Soc. London* **A334**, 231.
Bruzek, A.: 1967, *Solar Phys.* **2**, 451.
Bruzek, A.: 1969, *Solar Phys.* **8**, 29.
Chambe, G.: 1971, *Astron. Astrophys.* **12**, 210.
Christiansen, W. N., Mathewson, D. S., Pawsey, J. L., Smerd, S. F., Boischot, A., Denisse, J. F., Simon, P., Kakinuma, T., Dodson-Prince, H., and Firor, J. W.: 1960, *Ann. Astrophys.* **23**, 75.
Dupree, A. K.: 1972, *Astrophys. J.* **178**, 527.
Dupree, A. K., Huber, M. C. E., Noyes, R. W., Parkinson, W. H., Reeves, E. M., and Withbroe, G. L.: 1973, *Astrophys. J.* **182**, 321.
Evans, K. and Pounds, K. A.: 1968, *Astrophys. J.* **152**, 319.
Gabriel, A. H. and Jordan, C.: 1974 (to be published).
Gabriel, A. H., Garton, W. R. S., Goldberg, L., Jones, T. J. L., Jordan, C., Morgan, F. J., Nicholls, R. W., Parkinson, W. H., Paxton, H. J. B., Reeves, E. M., Shenton, D. B., Speer, R. J., and Wilson, R.: 1971, *Astrophys. J.* **169**, 595.
Herring, J. R. H., Glencross, W. M., Parkinson, J. H., and Pounds, K. A.: 1971, *Proc. Roy. Soc. London* **A321**, 493.
Howard, R. (ed.): 1971, 'Solar Magnetic Fields', *IAU Symp.* **43**.
Jordan, C.: 1969, *Monthly Notices Roy. Astron. Soc.* **142**, 501.
Jordan, C.: 1971a, in C. de Jager (ed.), *Highlights of Astronomy*, Vol. 2, D. Reidel Publ. Co., Dordrecht-Holland, p. 519.
Jordan, C.: 1971b, *Solar Phys.* **21**, 381.
Jordan, C.: 1974, *Astron. Astrophys.* **34**, 69.
Jordan, C. and Wilson, R.: 1971, in C. J. Macris (ed.), *Physics of the Solar Corona*, D. Reidel Publ. Co., Dordrecht-Holland, p. 219.
Kiepenheuer, K. O. (ed.): 1968, 'Structure and Development of Solar Active Regions', *IAU Symp.* **35**.
Krieger, A. S., Vaiana, G. S., and Van Speybroeck, L. P.: 1971, in R. Howard (ed.), 'Solar Magnetic Fields', *IAU Symp.* **43**, 397.
Kundu, M. R.: 1965, *Solar Radio Astronomy*, John Wiley-Interscience Publ.
Kundu, M. R., Becker, R. H., and Velusamy, T.: 1974, *Solar Phys.* **34**, 185.
Landini, M.: 1974 (unpublished).
Landini, M. and Monsignori Fossi, B. C.: 1971, *Solar Phys.* **17**, 379.
Livingston, W., Harvey, J., and Slaughter, C.: 1970, *Nature* **226**, 1146.
Munro, R. H., Dupree, A. K., and Withbroe, G. L.: 1971, *Solar Phys.* **19**, 347.

Noyes, R. W.: 1971, in C. J. Macris (ed.), *Physics of the Solar Corona*, D. Reidel Publ. Co., Dordrecht-Holland, p. 192.

Noyes, R. W., Withbroe, G. L., and Kirshner, R. P.: 1970, *Solar Phys.* **11**, 388.

Parkinson, J. H.: 1973, *Solar Phys.* **28**, 487.

Pottasch, S. R.: 1963, *Astrophys. J.* **134**, 347.

Pottasch, S. R.: 1967, *Bull. Astron. Soc. Neth.* **19**, 113.

Reimers, C.: 1971, *Astron. Astrophys.* **10**, 182.

Saito, K. and Billings, D. E.: 1964, *Astrophys. J.* **140**, 760.

Speer, R. J., Garton, W. R. S., Goldberg, L., Parkinson, W. H., Reeves, E. M., Morgan, J. F., Nicholls, R. W., Jones, T. J. L., Paxton, H. J. B., Shenton, D. B., and Wilson, R.: 1970, *Nature* **226**, 249.

Tandberg-Hanssen, E.: 1967, *Solar Activity*, Blaisdell Publ. Co., Waltham, Mass.

Tousey, R.: 1967, *Astrophys. J.* **149**, 239.

Tucker, W. H. and Koren, M.: 1971, *Astrophys. J.* **168**, 283.

Underwood, J. H. and Muney, W. S.: 1967, *Solar Phys.* **1**, 129.

Unsold, A.: 1960, *Z. Astrophys.* **50**, 57.

Unsold, A.: 1970, *Astron. Astrophys.* **4**, 220.

Vaiana, G. S., Davis, J. M., Giacconi, R., Krieger, A. S., Silk, J. K., Timothy, A. F., and Zombeck, M.: 1973, *Astrophys. J.* **185**, L47.

Vaiana, G. S., Krieger, A. S., and Timothy, A. F.: 1974, *Solar Phys.* (in press).

Van Speybroeck, L. P., Krieger, A. S., and Vaiana, G. S.: 1970, *Nature* **227**, 818.

Walker, A. B. C., Jr.: 1972, *Space Sci. Rev.* **13**, 672.

Walker, A. B. C., Jr. and Rugge, H. R.: 1968, *Astron. J.* **73**, S81.

Walker, A. B. C., Jr., Rugge, H. R., and Weiss, K.: 1973, Aerospace Report No. TR-0074 (9260-02)-2.

Wiese, W. L., Smith, M. W., and Glennon, B. M.: 1966, NSROS-NBS4 (U.S. Govt. Printing Office, Washington, D.C.).

Wiese, W. L., Smith, M. W., and Miles, B. M.: 1969, NSROS-NBS22 (U.S. Govt. Printing Office, Washington, D.C.).

Withbroe, G. L.: 1970, *Solar Phys.* **11**, 42.

Zirin, H.: 1966, *The Solar Atmosphere*, Blaisdell Publ. Co., Waltham, Mass.

PART 3

SOLAR FLARES

ULTRAVIOLET EMISSION LINE PROFILES OF FLARES AND ACTIVE REGIONS

G. E. BRUECKNER

E. O. Hulburt Center for Space Research, Naval Research Laboratory,
Washington, D.C. 20375, U.S.A.

Abstract. A preliminary description of ultraviolet spectra of active regions and flares, photographed from Skylab by the Naval Research Laboratory's UV Spectrograph is given. The findings can be summarized as follows: (1) Line profiles of medium ionized lines (transition zone lines) show the most pronounced broadenings and shifts in flares and flare like events. (2) Typical full width at half maximum of these lines correspond to Doppler-velocities of 70 km s^{-1}. (3) Shifts of the same magnitude can be observed. (4) Intersystem lines are not broadened nor shifted. (5) Forbidden coronal lines and intersystem lines become enhanced in the flare spectrum at the moment, when the turbulence seen in the allowed transitions, disappears. (6) A very broad line at 1354.2 Å which appears only in flare spectra, seems to be the forbidden transition $^3P_1 - ^3P_0$ of Fe XXI.

The instrument covered a spectral range from 970 Å to 4000 Å, but in the following we shall concentrate on the wavelength range 1170 Å to 1950 Å, where all the important emission lines are located.

There is no real imaging along the 1′ long slit of the spectrograph, which is only pseudo-stigmatic. A source with even intensity distribution over the whole slit produces a smooth spectrum in the spectrograph's focal plane. But a point source, illuminating the slit only at one point, would produce a different intensity distribution perpendicular to the dispersion in the spectrum. The spectrum would show a distinct peak with long wings measured perpendicular to the dispersion because of some residual imaging capabilities which are inherited in a pseudo-stigmatic imaging system. It seems therefore one can conclude, that spatial structure along the slit exists if one finds an uneven, structured spectrum perpendicular to the dispersion. And vice versa, if the spectrum has a smooth appearance one can conclude that the illuminating source has a fairly even intensity distribution over the 1′ long slit. It is almost impossible to quote numbers. But the experience with the instrument shows, that only very small structures having very high contrast with their surroundings can be distinguished in the focal plane of the pseudo-stigmatic instrument. One has to keep these imaging capabilities of the spectrograph in mind when interpretations of the spectra are made.

A spectral resolution of approximately 0.07 Å could be obtained, including all contributing factors to the instrument profile from the slit width, film and microphotometer effects. This resolution is sufficient, to derive emission line profiles of chromospheric, transition zone and coronal lines.

Figure 1 shows two spectra of the 1973, June 15 flare. The upper spectrum was photographed approximately 1 min after the X-ray peak during the maximum phase of the flare, the lower spectrum for comparison 26 min later when the flare had ceased. Two different exposure times have been selected in order to show spectra of approxi-

Sharad R. Kane (ed.), Solar Gamma-, X-, and EUV Radiation, 135–151. All Rights Reserved.
Copyright © 1975 by the IAU.

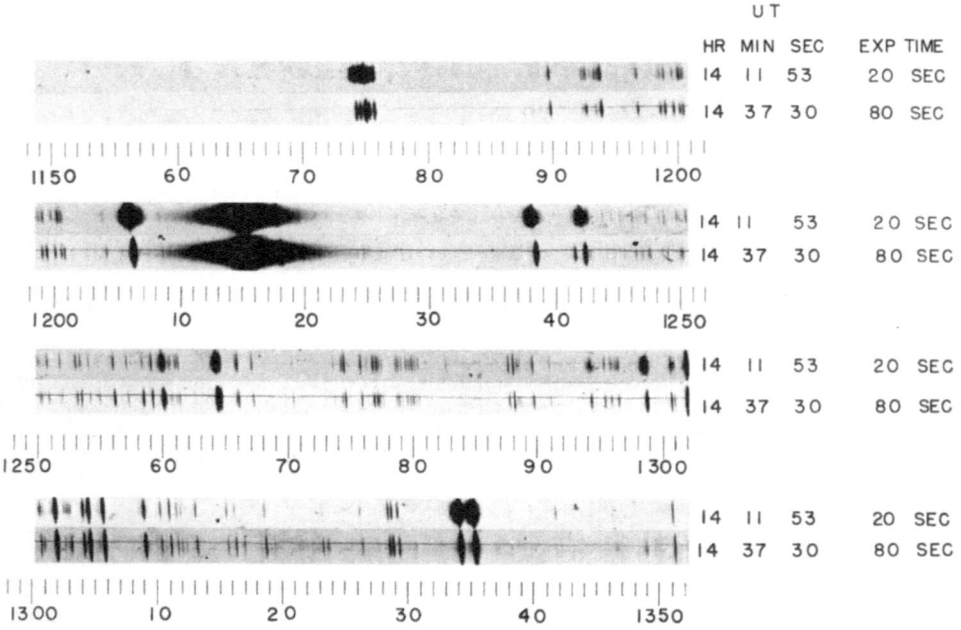

Fig. 1.   Spectra of the 1973, June 15th flare, 1150 Å to 1350 Å. Upper spectrum: Flare during maximum phase. Lower spectrum: Same pointing, after flare has ceased.

mately the same photographic density. (The spectra as photographed from Skylab
consist always of a series of four different exposures to bridge a wide dynamic range
from the faintest to the strongest lines.)

At 1175 Å we see the C III multiplet resolved. It is strongly enhanced in the flare and
all lines are broadened.

At 1204 Å we observe the weak intersystem line from S v. This line should be density
sensitive if the local density exceeds a certain value. It should be noted that this line,
like all other intersystem lines does not show any broadening in the flare spectrum.
We may therefore conclude that it does not originate in the flare plasma itself. The
intersystem lines seem to come from the surrounding area, where they are enhanced
like all other lines.

Lα is overexposed, but it is enhanced approximately $3 \times$ in the flare.

The O v $^1S - ^3P$ intersystem line at 1218 Å behaves similar like the S v line. No
broadening of the line profile nor any enhancement during the flare phase can be
observed.

At 1240 Å follows the N v resonance doublet, which is enhanced and broadened in
the flare spectrum. Both components are blue shifted contrary to the Si III resonance
line at 1206 Å which is reshifted during this phase of the flare. The Fe XII forbidden
line at 1241.7 Å is not enhanced in the flare spectrum during the early phase.

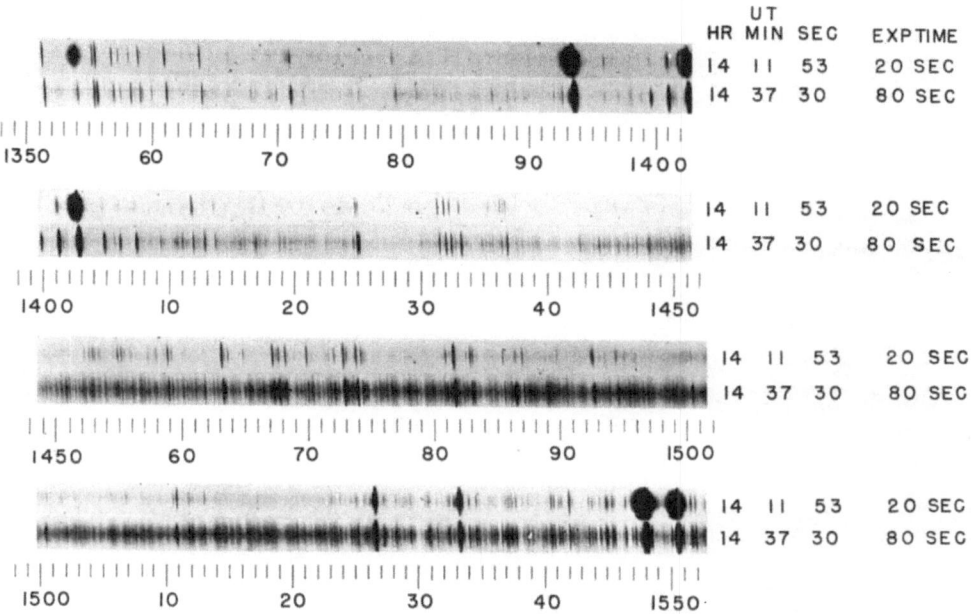

Fig. 2.   Spectra of the 1973, June 15th flare, 1350 Å to 1550 Å. Upper spectrum: Flare during maximum phase. Lower spectrum: Same pointing, after flare has ceased.

The Si II doublet at 1260 Å and 1265 Å shows less broadening in the flare spectrum than the Si III line at 1206 Å.

O I at 1300 Å is enhanced, but its profile is only slightly changed. Weak lines near the oxygen triplet which show flare broadening are from Si II and Si III.

C II at 1334 and 1335 Å shows a blue shifted flare profile. At 1349 we find the forbidden Fe XII line, which is not enhanced in the flare spectrum during the early phase. An unidentified line of similar behavior can be seen at 1323.8 Å. Another unidentified sharp line at 1342 Å shows a very strong enhancement.

The most peculiar line in flare spectra can be found at 1354 Å (Figure 2). This line becomes very bright in flares, it is very broad and has no counter part in active regions. Feldman *et al.* (1974) have calculated the wavelength of the forbidden transition $^3P_1 - {}^3P_0$ of Fe XXI in the carbon sequence to be $1355\,\text{Å} \pm 3$ Å. An ion temperature of approximately 20×10^6 K can be derived from the halfwidth of this line. The intensity change of this line with time is distinctly different from the intensity change of the transition zone lines. We also do not see any structure of the line profile along the slit. All these observations indicate that the 1354 line indeed comes from a region of the flare, where different physical conditions prevail than those in the area, where the transition zone lines are formed. It is therefore very likely that the identification will hold.

At 1355.6 and 1358.5 the O I intersystem doublet $^3P - {}^5S$ does show only a very slight enhancement in the flare spectrum. Because all other neutral lines from allowed

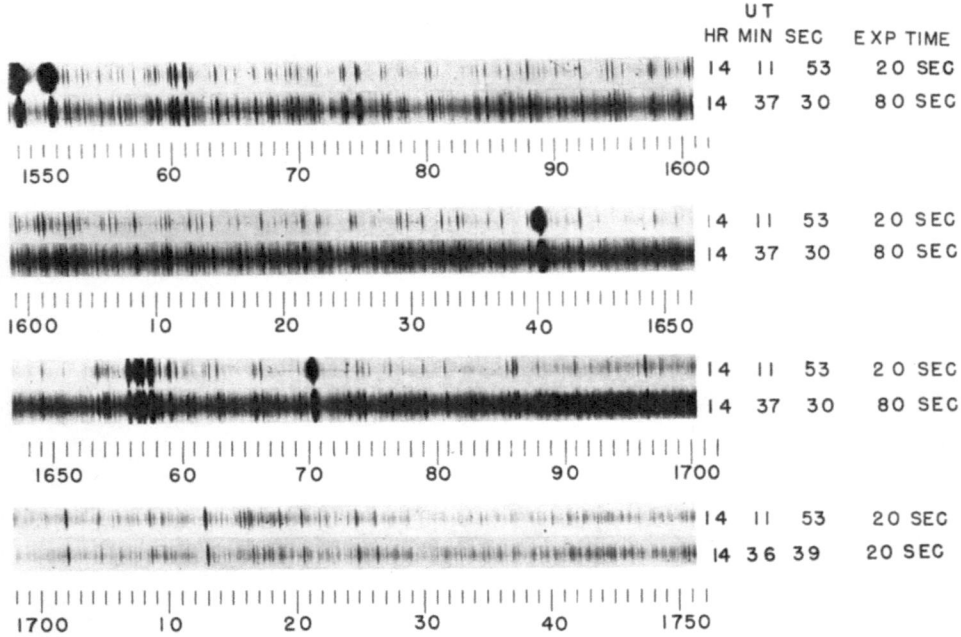

Fig. 3.   Spectra of the 1973, June 15th flare, 1550 Å to 1750 Å. Upper spectrum: Flare during maximum phase. Lower spectrum: Same pointing, after flare has ceased.

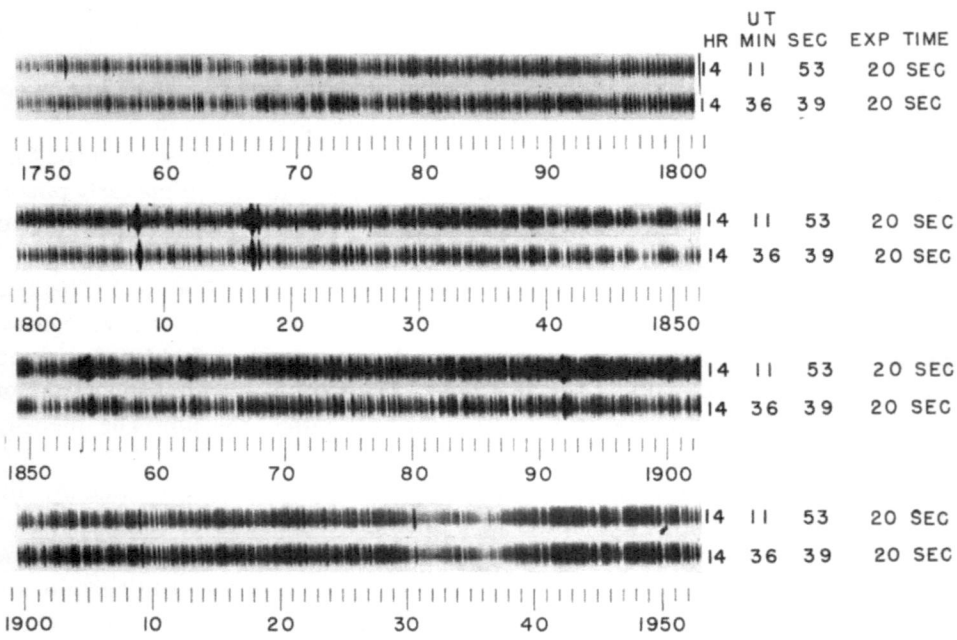

Fig. 4.   Spectra of the 1973, June 15th flare, 1750 Å to 1950 Å. Upper spectrum: Flare during maximum phase. Lower spectrum: Same pointing, after flare has ceased.

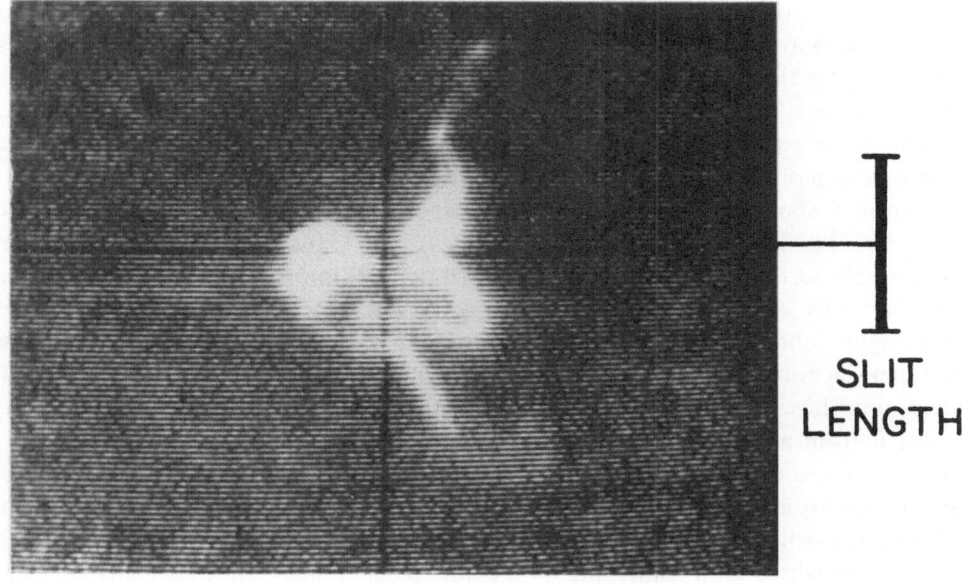

FLARE, JUNE 15th 1973

Fig. 5.   Hα image of the June 15th flare, transmitted from the display on board Skylab. The spectrograph slit was positioned along the vertical crosshair.

transitions in this region of the spectrum show a stronger enhancement, we can conclude that in the chromosphere where O I is formed, the collision rate during a flare is enhanced.

In the flare spectrum the most broadened lines are the Si IV lines at 1400 Å. Also around 1400 Å we find the $^2P - {}^4P$ intersystem lines of O IV, which are not enhanced and not broadened in the flare spectrum. The same applies to the S IV $^2P - {}^4P$ intersystem lines at 1406 and 1416.9 Å.

The Si II doublet at 1526 and 1533 Å does not show significant broadening in the flare, contrary to the Si II doublet at 1260 Å.

The C IV resonance doublet at 1540 Å shows a very broad line profile and strong enhancement in the flare spectrum.

Hα of He II (Figure 3) at 1640 Å seemed to be broadened, but the line is blended with an Fe II line, therefore it is difficult to determine the contribution of He and Fe to the composite line profile.

The same applies to Al II at 1670 Å. Fe II does not show signs of significant broadening, which can be seen from the appearance of the many isolated Fe II lines in this region of the spectrum.

The N III intersystem line at 1754 Å is slightly weakened in the flare spectrum (Figure 4). This is not the case for the Si III intersystem line at 1892 Å which does not show any intensity change or broadening.

Finally we are pointing out that the continuum below 1520 Å, (Si triplet continuum) does show only a very slight enhancement in the flare spectrum, while we cannot see any change in the continuum above 1520 Å. Also, the Al autoionization lines at 1934 Å do not change. This is not the case in all spectra of flares. There are spectra where both continua are enhanced, particularly the Si triplet continuum. We have found at least one case where the Al-autoionization lines seem to disappear.

Figure 5 shows the slit position of the spectrograph as it was pointed onto the June 15th Hα flare along the vertical cross hair line. It is obvious that the slit covered only small portions of the very bright Hα ribbons. Widing (1975) has found, that the very hot flare emission, as seen in the Fe XXIV spectroheliograms, is located between the bright ribbons. Therefore the slit covered the hottest area of the flare. This makes the identification of the 1354 Å line with Fe XXI even more likely. The line at 1354 Å does not show structure perpendicular to the dispersion, which means, that it is emitted from an area comparable with the slit area, contrary to the appearance of the broadened transition zone lines. Widing measured the size of the Fe XXIV emission in the spectroheliograms to be approx. 30″. It is therefore safe to assume, that the Fe XXI line originates from the same hot coronal plasma above the explosive region of the flare where the Fe XXIV emission comes from. While a time sequence of spectra was taken the pointing had not been changed.

We shall now discuss the time dependence of the spectra. Figure 6 shows a series of spectra covering a time interval of 25 min, taken all with 20 s exposure time. The turbulent phase of the flare, which is distinguished by the very broad line profiles of the medium ionized lines like Si II at 1260 Å, Si III at 1206 Å and N V at 1240 Å lasts approximately 7 min. During this phase, the broadening gradually decreases in all the lines mentioned. But we also see changing shifts from predominantly blue shifts at the beginning to red shifts during the later turbulent phase. We also note, that the Si III line at 1206 exhibits a red shift at the same time when Si II at 1260 is blue shifted. (Upper two spectra in Figure 6.) After the turbulent phase has ceased at 14 h, 22 min, 26 s abruptly the forbidden Fe XII lines at 1240 Å appears very strong and then decreases gradually toward the end of the sequence. While intensity and line width of the C II lines at 1335 Å (Figure 7) are decreasing with time, beginning with the first spectrum, the Fe XXI 1354 Å line increases in intensity during the first 3 min, to disappear gradually toward the end of the sequence. C II behaves very similar like all other transition zone lines. 1354 Å should therefore not originate in the transition zone. It also cannot be attributed to the chromosphere, because of its disappearance after the flare has ceased and its very broad line profile.

The same behavior of the medium ionized lines can be seen in Figure 8, which shows the broadening of Si IV. At 14 hr, 22 min, 26 s when the Si IV line broadening stops abruptly, the intersystem lines of O IV and S IV become very intense.

Figure 9 demonstrates another example of the selective appearances of the line shifts in different ions. While C IV at 1550 Å shows a significant red shift, at 14 hr, 18 min, 23 sec, the Si II doublet remains unshifted. The slight enhancement of the Si triplet continuum below 1520 Å lasts very long through 14 h, 28 min. Another notice-

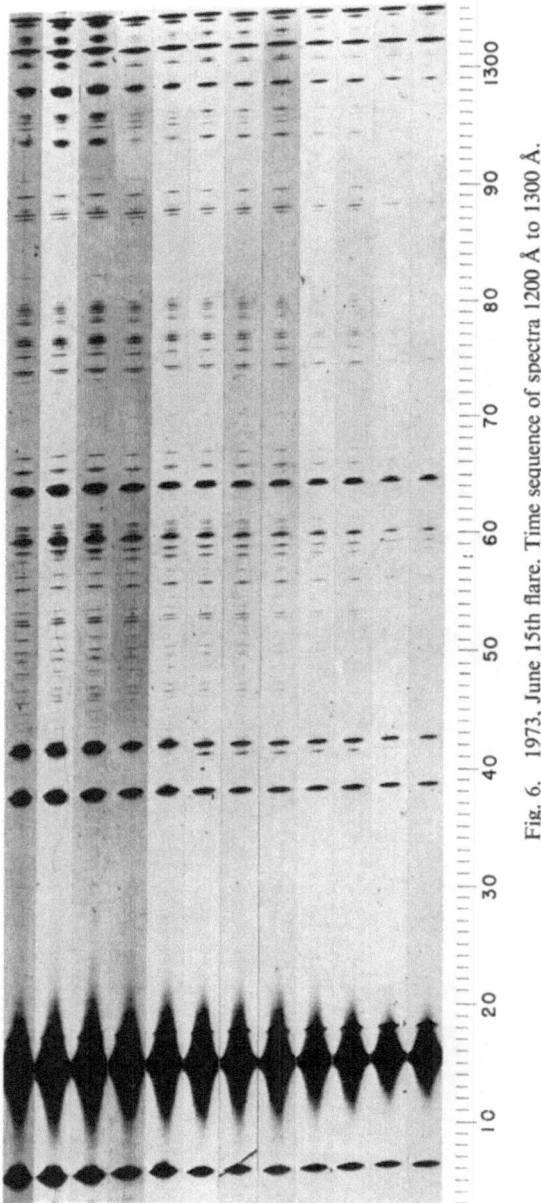

Fig. 6.   1973, June 15th flare. Time sequence of spectra 1200 Å to 1300 Å.

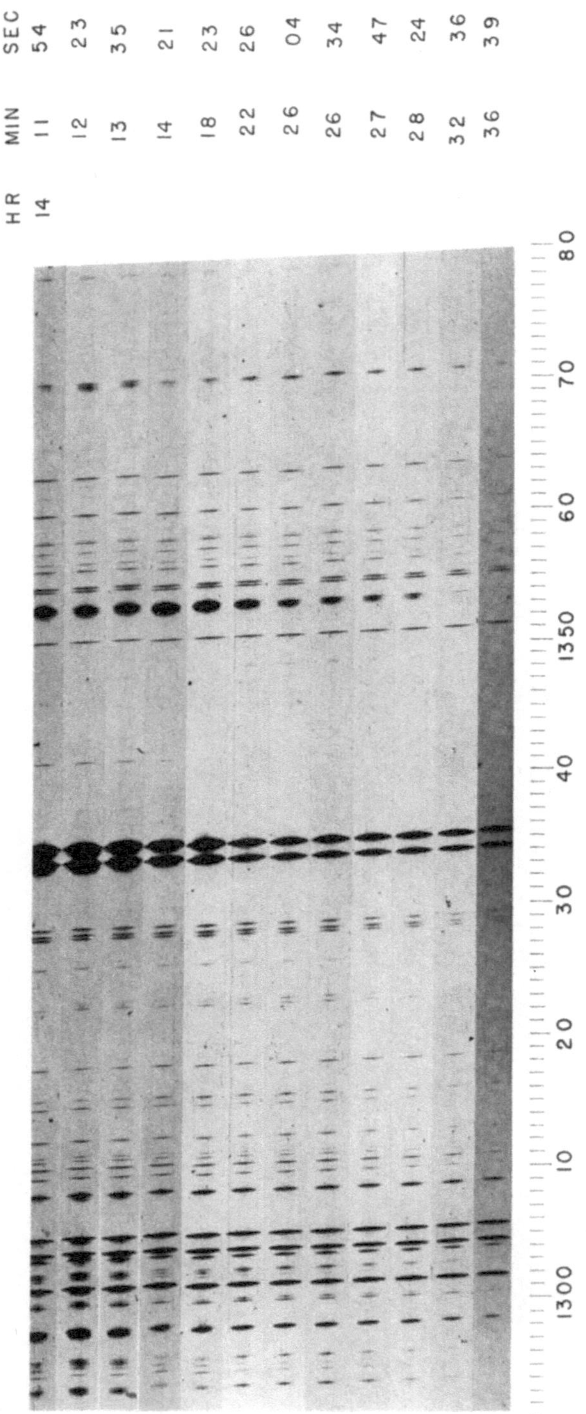

G. E. BRUECKNER

Fig. 7. 1973, June 15th flare. Time sequence of spectra 1290 Å to 1380 Å.

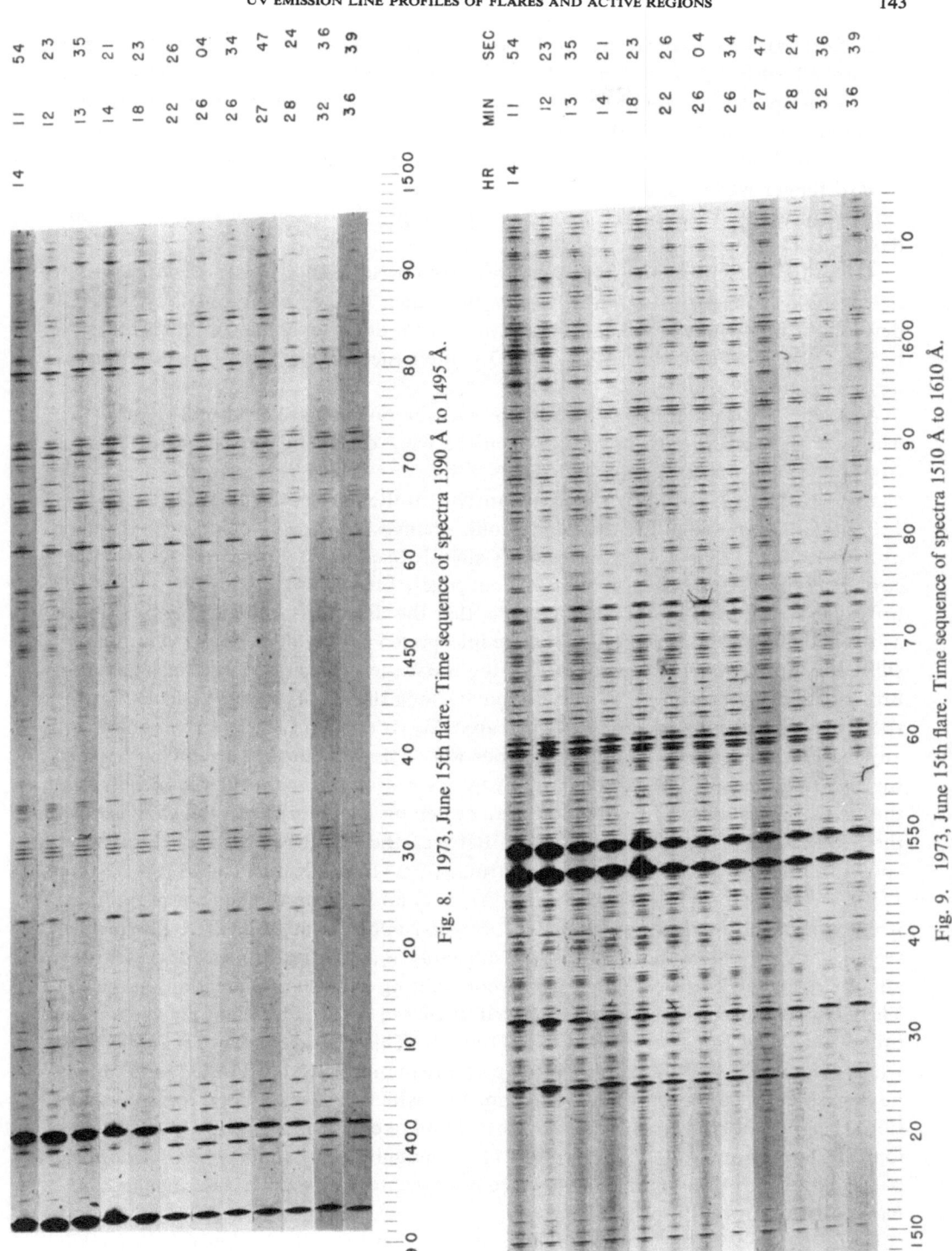

Fig. 8. 1973, June 15th flare. Time sequence of spectra 1390 Å to 1495 Å.

Fig. 9. 1973, June 15th flare. Time sequence of spectra 1510 Å to 1610 Å.

able effect can be demonstrated in this portion of the spectrum. At the beginning of the flare all neutral and 1 × ionized lines are strongly enhanced in the short wavelength part of the spectrum around Lα. This is not the case for the lines of the same origin at longer wavelength. It should be noted that there is no continuum at shorter wavelength, while above 1520 Å the Si singlet and triplet continuum becomes stronger toward longer wavelength.

In Figures 10 through 15 we show some profile tracings of a small limb flare.

At the limb, considerable forground absorption causes a self reversal of the Si III line at 1206 Å (Figure 10), which cannot be treated here as being optically thin. Lα develops strong wings in the flare spectrum, while it does not change its core intensity, because of foreground absorption. The O v intersystem line at 1218 Å does not show any change of its profile.

At 1240 Å at the limb one still sees the Lα wing in the flare spectrum (Figure 11). The two absorption lines near 1240 Å in the wing of Lα are Mg II resonance lines from the transition $3s\ ^2S - 4p\ ^2P^0$. The N v lines are strongly broadened, their profile changes from a Gaussian to a Lorentzian type in the flare spectrum.

Figure 12 shows the intensity and profile changes of the O I, Si II and Si III lines around 1300 Å. O I is not enhanced, only slightly broadened. Si II at 1309 Å becomes strongly selfreversed. Si III exhibits the strongest broadening. Figure 13 shows the broadened Si IV lines at 1400 Å. We note, that the flare line profile is not as smooth as the active region profile. This could be interpreted as a profile composited of many shifted components. The appearance of the broadened transition zone lines (see Figures 1–4 and 6–9) seems to be very irregular along the slit. Despite the fact, that the spectrograph has no stigmatic imaging capability, it does not smooth out completely any features, which cover only a small portion of the slit. But the irregular shape of the Si IV lines may also be caused partially by absorption of enhanced unknown lines in the wing of these lines. The latter point can be demonstrated in Figure 14, which shows the profiles of the C IV lines at 1550 Å. Between the C IV lines we see several weak emission lines in the active region spectrum. (C IV 1548.20 Å, Si I 1548.694 Å, Fe II 1550.27 Å, C IV 1550.77 Å, Si I 1551.199 Å.) Si I and Fe II appear in absorption in the wings of C IV of the flare spectrum. The irregular appearance of the flare line profiles seems to be caused by both effects, the appearance of enhanced absorption features and the irregular Doppler shifts of different emitting flare plasma areas along the slit. The absorption effect can be demonstrated further in Figure 15, where the wing of the broadened He II line is absorbed by CO bands. This would indicate, that the He II wings originate very deep below the temperature minimum. On the other hand, it may be that spicules contain a very cool component which causes the foreground absorption. It should be noted here, that any absorption of cooler ions in the wings of higher ionized flare lines seems to occur only close enough to the limb, where the optical pathlength is large enough. It is therefore not easy, to locate exactly the height of the flare emission in the transition zone lines.

In most cases, the unshifted but enhanced component of the background is equally

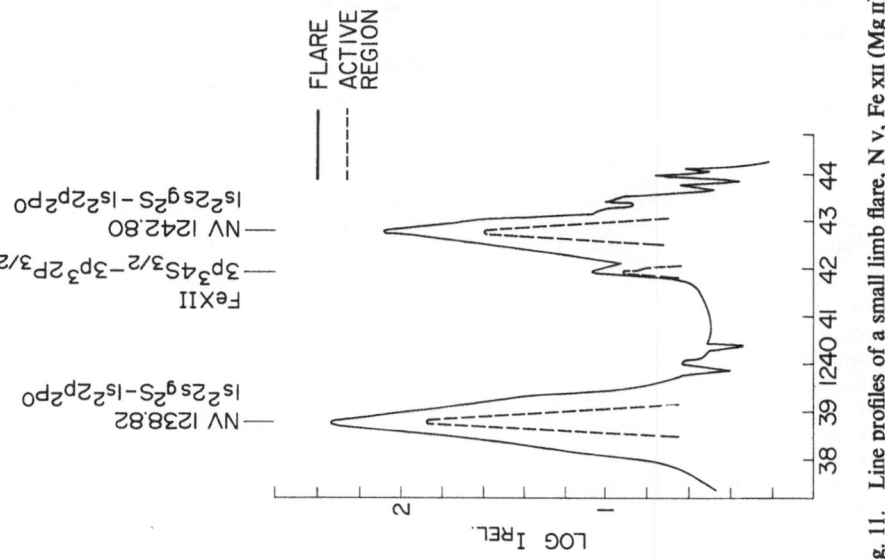

Fig. 11. Line profiles of a small limb flare, N v, Fe xii (Mg ii).

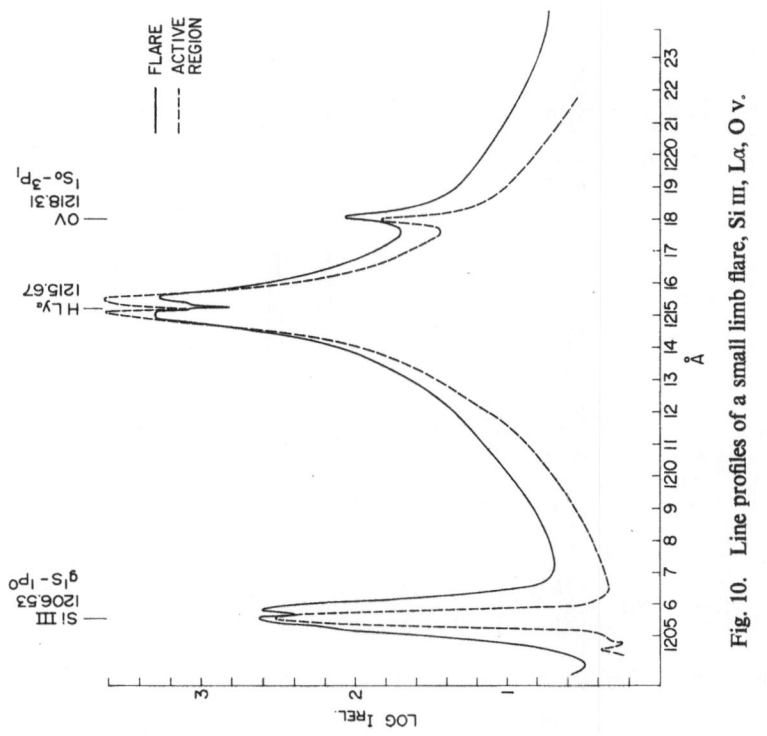

Fig. 10. Line profiles of a small limb flare, Si iii, Lα, O v.

strong or stronger than the pure flare profile. An isolation of the flare profile can be done, if the latter shows a large shift.

In Figure 16 we demonstrate an example of strongly shifted N v lines. Assuming, that the blue wing of the composite profile is representative of the background profile,

TABLE I

Typical line-widths of optically thin lines in active regions and flares

λ	Ion	T_e	$T_K \times 10^6$ K		ξ_0 km s^{-1}	
			active region	small flare	active region	small flare
1296	Si III	5×10^4	1.0	3.1	24	43
1393	Si IV	7×10^4	1.1	8.4	26	70
1402	Si IV	7×10^4	1.1	8.3	26	70
1548	C IV	1×10^5	1.4	3.0	43	65
1550	C IV	1×10^5	1.4	2.8	43	65
1399	O IV	1.25×10^5	1.5	1.6	39	41
1238	N v	1.6×10^5	1.4	3.2	41	62
1242	N v	1.6×10^5	1.5	2.6	43	56
1371	O v	2.2×10^5	1.5	2.4	35	50

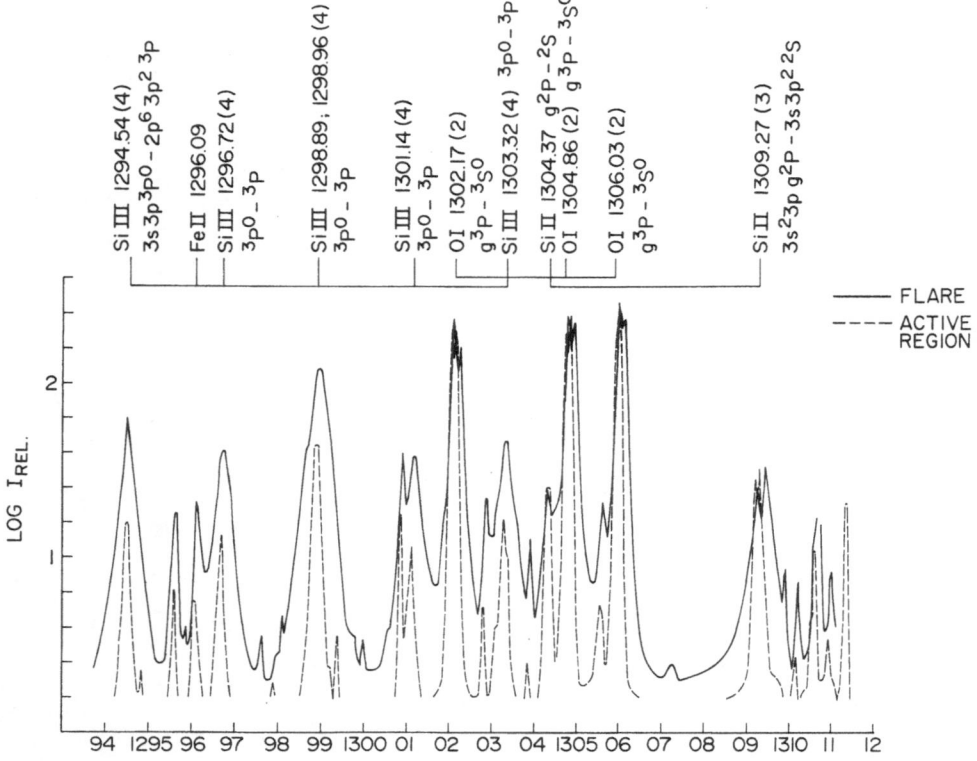

Fig. 12.   Line profiles of a small limb flare, O I, Si II, Fe II, Si III.

we determine the shift of the flare profile to be 68 km s^{-1} to the red and obtain for its halfwidth a Doppler velocity of 68 km s^{-1}.

In Table I typical line width of optically thin lines in a flare spectrum compared with an active region are given. These lines are listed in order of increasing excitation temperature, covering a range from 50 000 to 220 000 K. We notice the strong discrepancy between excitation and ion temperature, which exist already in the active region spectrum. (The same applies to the quiet Sun transition layer.) This discrepancy is greatly

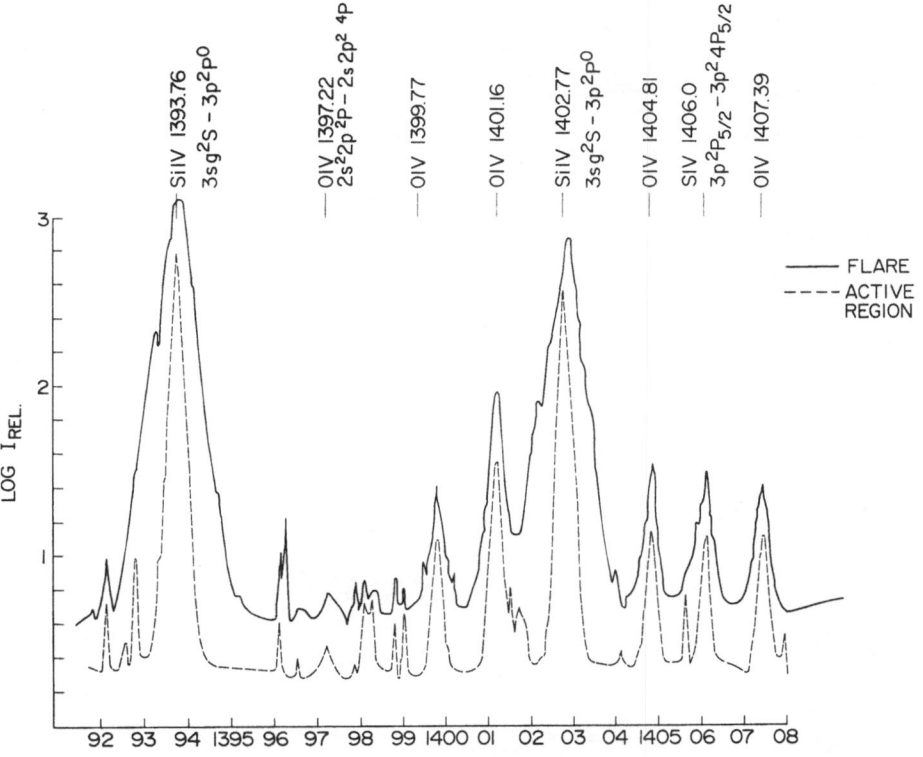

Fig. 13. Line profiles of a small limb flare, Si IV, O IV, S IV.

enhanced in the flare spectrum, but we also note that the maximum ion temperature can be found at an excitation temperature of 70 000 K. At higher excitation temperatures, the ion temperatures decrease. The list shows that there is no change in the halfwidth of the O IV intersystem line indicating, that this line does not originate in the turbulent flare plasma. If we express the halfwidth of the lines in terms of Doppler velocities, we find 70 km s^{-1} as a maximum value for Si IV in this particular case. The term 'ion temperature' is used here in a very broad sense only as an expression for the measured halfwidth of the lines. It is perhaps not related to the temperature of the emitting gas, as will be pointed out below.

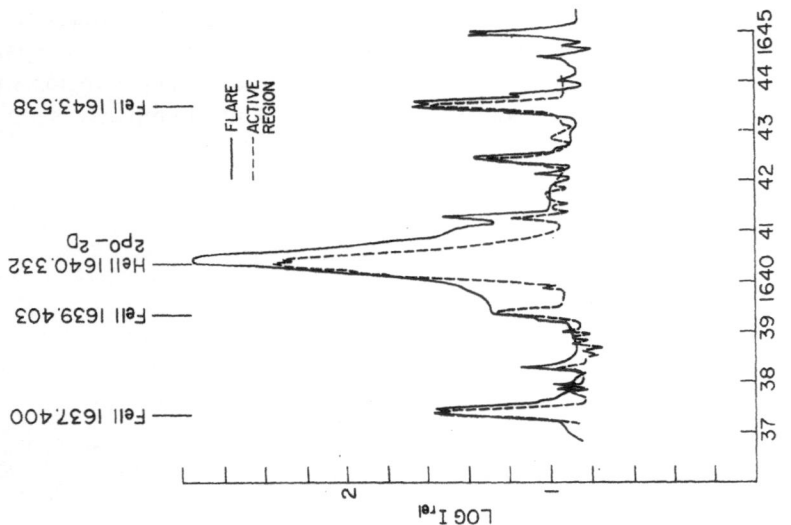

Fig. 15.   Line profiles of a small limb flare, Fe II, He II.

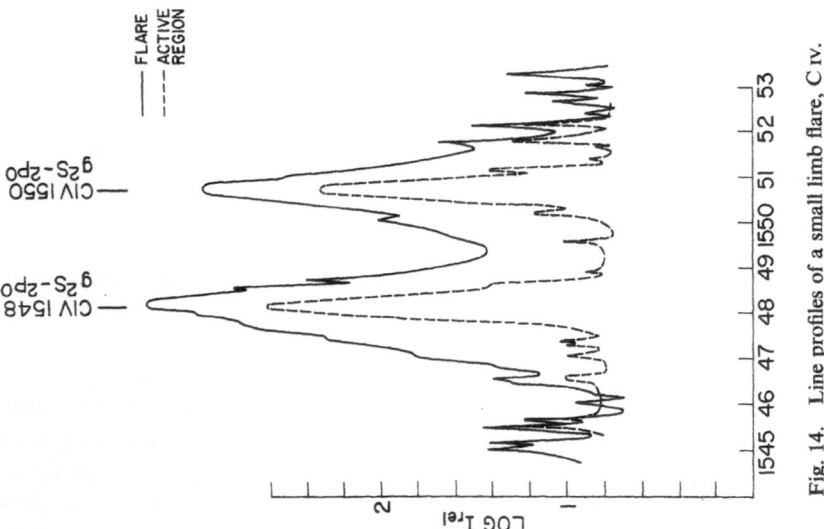

Fig. 14.   Line profiles of a small limb flare, C IV.

Starck broadening, microturbulence and macroturbulence are possible mechanisms which could be used to explain the very broad profiles of the transition zone lines in flares. In addition we must consider the possibility, that local strong electrical fields cause the high ion velocities. We believe, that the latter explanation is very unlikely, because if such fields would exist, the electrons should be accelerated to comparable energies like the ions and all atoms should be completely ionized. On the other hand, this possible mechanism should not be discounted completely until exact calculations disprove it. If such a mechanism would be working we would expect that the line width are proportional to the z/m ratio of the specific ion. Such a dependence cannot be found. The z/m ratio of C IV is twice as large as that of Si IV and yet the full width at half maximum of both lines is almost identical.

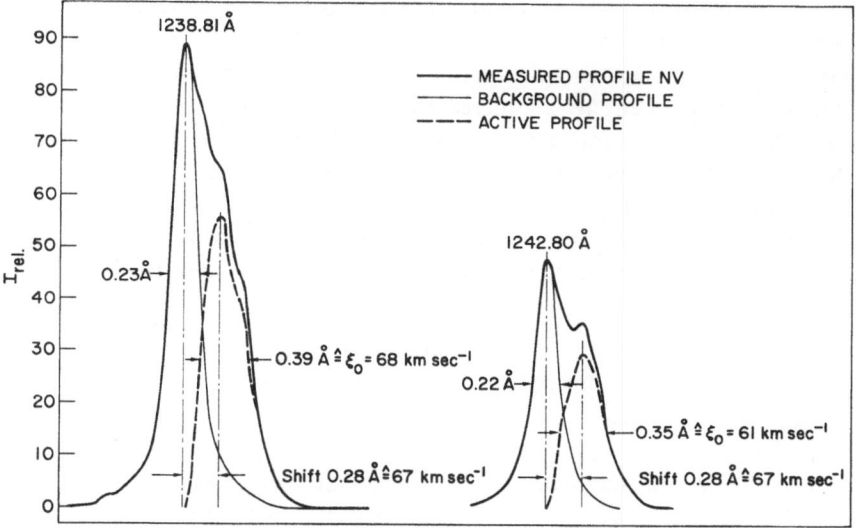

Fig. 16.   Separation of N v flare line profiles from background line profiles.

Starck broadening by random electrons would yield to extreme high densities, larger than 10^{18} cm^{-3}. Such densities can only be found in deep photospheric layers. A mechanism would have to be invented to ionize photospheric material locally to such high ion temperatures without the accompanying adiabatic expansion. If highly energetic electron beams can penetrate to such depth, than local ionization could take place. But calculations have shown, that even 100 KeV electrons will only penetrate to layers, where the local density is approximately 10^{12} cm^{-3}. These estimates have been done under the assumption, that the present models of the upper photosphere and chromosphere are valid, which is not necessarily the case.

Our observations do not show any correlation between line shifts and line broadening. Also, different ions can show different line shifts at the same time. Furthermore, the irregular structure of the line profiles along the slit (our spatial resolution is very marginal) indicate, that different areas of the flaring regions exhibit line shifts in different directions and of different magnitude. These fact seem to indicate, that local Doppler motions are causing the line broadening and the observed profiles are averaged over areas where these motions are not homogeneous. The size of these elements remains unknown. A mechanism still must be found, which is able to accelerate material to such high velocities without heating it up to very high temperatures. Sprays and surges are common features connected with solar flares. But a simple identification of our line broadening mechanism with sprays and surges, as they are observed in Hα, cannot be made, because we observe blue shifts, red shifts and in many cases the lines are only broadened and not shifted.

However, if we explain the broadening of the transition zone lines by macroscopic Doppler effect, we must find an explanation for the fact, that the intersystem lines appear very strong at the very moment, when the turbulence, observed in the allowed transitions ceases. If the flare plasma as a whole exhibits an instability during the turbulent phase, an increased collision rate could de-excite the intersystem transitions. (This would be equivalent to an increased density during the turbulent phase.) If however, the line broadening mechanism is macroturbulence, then it is not easy to understand, why the sudden stop of the large scale movements should increase the intensity of the intersystem lines. It is therefore likely, that the broad lines of the allowed transitions reflect both, a strong turbulent motion within the flare plasma and local anisotropic acceleration of material. At the moment, when the Doppler broad-

TABLE II

Intersystem lines in flare spectrum

λ	Ion	Transition	$A\,(\mathrm{s}^{-1})$	Remarks
1487	N IV	$2s^2\,{}^1S_0 - 2s2p\,{}^3P_1$	4.71 (2)	enhanced after 'turbulent' phase
1218	O V		1.88 (3)	no change during flare
1754.1	N III	$2s^2p\,{}^2P_{3/2} - 2s2p^2\,{}^4P_{1/2}$	?	very slightly weaker during turbulent phase
1399.8	O IV	$2s^2p\,{}^2P_{1/2} - 2sp^2\,{}^4P_{1/2}$	4.8 (3)	enhanced after 'turbulent' phase
1401.1	O IV	${}^2P_{3/2} - {}^4P_{5/2}$	7.5 (3)	enhanced after 'turbulent' phase
1404.8	O IV	${}^2P_{3/2} - {}^4P_{3/2}$	2.6 (3)	enhanced after 'turbulent' phase
1407.4	O IV	${}^2P_{3/2} - {}^4P_{1/2}$	2.9 (3)	enhanced after 'turbulent' phase
1660.8	O III	$2s^22p^2\,{}^3P_1 - 2s2p^3\,{}^5S_2$	?	slightly enhanced after 'turbulent' phase
1666.1	O III	${}^3P_2 - {}^5S_2$	?	slightly enhanced after 'turbulent' phase
1355.6	O I	$2s^22p^4\,{}^3P_2 - 2s2p^5\,{}^5S_2$	1.3 (3)	enhanced during flare
1358.5	O I	${}^3P_1 - {}^5S_2$	3.8 (2)	enhanced during flare
1892.1	Si III	$3s^2\,{}^1S_0 - 3s3p\,{}^3P_1$	8.3 (4)	no change during flare
1204.3	S V		?	enhanced during early flare phase
1406.0	S IV	$3s^23p\,{}^2P_{3/2} - 3s3p^2\,{}^4P_{1/2}$	?	slightly enhanced after 'turbulent' phase
1416.9	S IV	${}^2P_{3/2}\,{}^4P_{3/2}$	?	slightly enhanced after 'turbulent' phase
1485.6	S I	$3p^4\,{}^3P_0 - 3p^33p^1\,{}^5D_1$	2.3 (6)	not identified

ening of the allowed transitions ceases not only the acceleration of material but also turbulent motion within the remaining flare plasma comes to a halt. It could also be, that the intersystem lines are emitted only as a recombination spectrum of the cooling corona above the explosive plage region. After the turbulence in the plage spectrum has ceased, no additional heating of the corona overlaying the plage takes place and cooling starts, which is reflected in the appearance of lines of cooler ions. The intersystem lines are all originating from relative 'cool' ions compared with the very hot ions (Fe XXI, Fe XXIV) emitting during the explosive phase of the flare.

Table II summarizes the observations of the intersystem lines in the 1973, June 15 flare. The list of lines has been compiled from Jordan (1973). Jordan's identifications and A-values have been listed together with remarks about the behavior of these lines in the flare spectrum.

In principle, one should be able, to derive lower limits for densities from the strength of these lines. However, the uncertainties are too large at the moment to give meaning-ful values. There is a wide margin of error in the atomic parameters. But worse, the collision rates in a highly turbulent plasma are a completely unknown quantity. For these reasons we rather refrain at the time being from trying to quote quantitative density values.

Acknowledgements

The spectra used for this analysis were taken by the Skylab Astronauts P. Conrad, Dr J. Kerwin, P. Weitz (Skylab I) and A. Bean, Dr O. Garriott, J. Lousma (Skylab II). I am indebted to K. Nicolas, who did the microphotometry of the limb flare spectrum and to Dr G. Doschek, O. K. Moe, K. Nicolas and Dr C. Moore Sitterly for helpful discussions. Dr R. Tousey is the Principal Investigator of the NRL Skylab Project. This work was supported by NASA DPR S-60404G.

References

Feldman, U., Doschek, G. A., Cowan, R. D., and Cohen, L.: 1975, *Astrophys. J.*, in press.
Jordan, C.: 1973, *Nuclear Instr. Methods* **110**, 373.
Widing, K. G., 1975. This volume, p. 153.

Fe xxiv EMISSION IN SOLAR FLARES OBSERVED WITH THE NRL/ATM XUV SLITLESS SPECTROGRAPH

KENNETH G. WIDING

*E. O. Hulburt Center for Space Research, Naval Research Laboratory,
Washington, D.C. U.S.A.*

Abstract. During the Skylab Mission, the NRL slitless spectrograph photographed a number of flares in the 170–600 Å region with a spatial resolution approaching 2″. At flare maximum the $2s\,^2S_{1/2}-$ $-2p\,^2P_{1/2,\,3/2}$ transitions of Fe xxiv are present, and show the location of the (approx.) 20×10^6 deg plasma with respect to the surface magnetic field and chromospheric (He ii) emissions. Three examples are discussed (two only briefly).

In the small, intense disk flare of 1973, August 9 the high temperature region appears at the foot of a low altitude arch. The estimated electron density is 5×10^{11} cm^{-3}.

In the limb flare of 1974, January 15 the hot X-ray emitting component is at a very low altitude compared to the flare loops.

In the impulsive double ribbon flare of 1973, June 15 the Fe xxiv emission is centered over the neutral line, forming a bridge-like structure between magnetic regions of opposite polarity. The estimated electron density is 5×10^{10} cm^{-3}.

The Fe xxiv emission was visible 8 to 10 min as compared with a calculated cooling time by conduction of only 5 min. The lengthened life of the emission may be associated with the observed 'turbulence', which inhibits the heat conduction, or alternatively, with a slower energy release prolonged beyond the end of the burst phase.

I. Introduction

The NRL slitless objective grating spectrograph (SO82A) was part of the Apollo Telescope Mount (ATM) carried aboard the Skylab Earth Orbiting Laboratory, which was launched by the National Aeronautics and Space Administration on 14 May 1973.

The optical layout of the instrument is shown in Figure 1. Sunlight entering the instrument at normal incidence is both dispersed and focused by a single concave grating to produce a series of monochromatic images of the Sun approximately 18.5 mm in diameter. The only other optical element is a thin aluminum filter in front of the film which excludes long wavelength stray light. For more detail on the instrument together with examples of spectroheliograms photographed with it during the Skylab mission see Tousey *et al.* (1973).

Photographing the wavelength range 170–630 Å at normal incidence, this instrument demonstrated unique capabilities for the observation of solar flares. The wide field of view enabled it to photograph a solar flare anywhere on the disk or limb without the time lost in repointing an instrument with a narrower field of view. Stigmatic images of the flare are formed in each emission line with a spatial resolution approaching 2″. This means that the spatial distribution of the various flare emissions is visible almost at a glance. No significant overlap of images occurred even for the largest flare (40″ × 40″) observed during the mission.

The spectral region 170–630 Å contains flare radiations in a range from chromo-

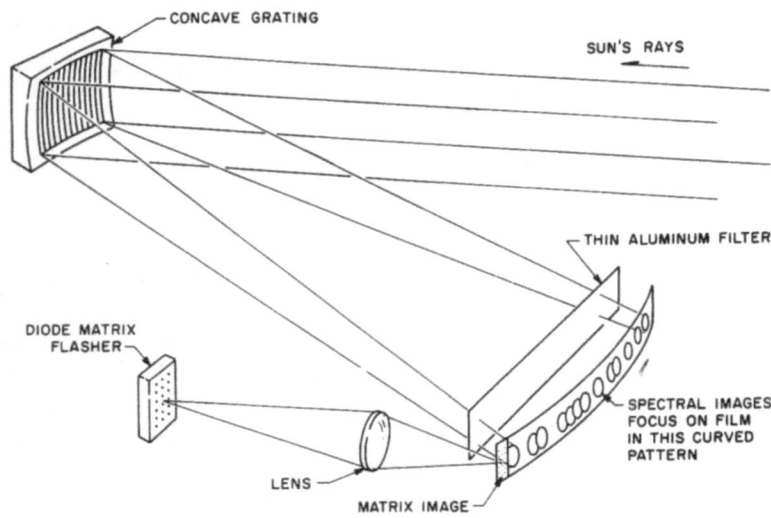

Fig. 1.   Optical layout of S082A, the NRL slitless spectrograph. The single concave grating is rotated
between 2 positions to select either the short wavelength range 170–335 Å, or
the longer range 320–630 Å.

spheric and transition zone ions to the highest stages of iron. Of particular interest is
the comparison of the Fe XXIV image at 255.2 Å with the He II 256 Å image; this shows
at a glance the spatial location of the high temperature flare relative to the chromospheric ribbons (Figure 10).

The $2s\,^2S_{1/2} - 2p\,^2P_{1/2,\,3/2}$ transitions of Fe XXIV at 192.1 and 255.2 Å were previously observed and identified by Neupert (1971) and Purcell and Widing (1972). In
the Skylab observations the Fe XXIV images are nearly always present at flare maximum and during the early post-maximum phase. The emissivity of Fe XXIV maximizes
at a temperature of 17×10^6 deg (Jordan, 1970), which is in the range of typical peak
temperatures observed during the Skylab mission (Dere and Kreplin, 1974). The
emissivity on the low temperature side drops off sharply and essentially vanishes below
10×10^6 deg. This agrees with the flare behavior of the Fe XXIV images, which disappeared quickly once the cooling trend was well underway.

In the following sections we illustrate these remarks with a sample of the flare
observations obtained. We choose 3 contrasting examples: a small but intense disk
flare (1973, August 9), a limb flare with flare loops (1974, January 15), and the impulsive double ribbon flare of 1973, June 15. The last flare showed the location of the
high temperature flare relative to the chromospheric ribbons better than any other
example.

II. The Disk Flare of 1973, August 9

Figures 2 and 3 illustrate a series of events (possibly interconnected) taking place
between 1500 and 1600 UT on August 9. The two bright regions near the limb which

CORONAL TRANSIENT IN He II

AUG 9, 1973 $15^h 21^m$ U.T.

Fig. 2.   Pre-flare active region on disk. The arrow points to the bright foot-point region in the image of Fe xv 284 Å, which subsequently flared (see Figure 3).

flared in succession 30 min apart appear to be associated with a complex coronal event. Sub-flare activity at the limb prior to 1521 UT presumably was responsible for the material ejected into the corona visible in Figure 2. We leave to a later study the problem of relating the transient to events at the surface, and here focus on the structure of the disk region which flared at 1554 UT (Figure 3).

In the pre-flare photograph of Figure 2 this region (McMath plage 12474) is already bright in the Fe xv 284 Å image at the place pointed to by the arrow. In particular, the arrow points to a small bright Knot which appears to be situated at the foot-point of a low-altitude arch or loop. This same small feature subsequently flared 30 min later, as may be seen in Figure 3. Note also that the field lines are closed over the whole low-altitude emission structure, i.e. no evidence of an over-lying current sheet associated with this flare (see also Cheng and Spicer, 1975).

The disk flare that followed at 1554 UT was one of the more energetic events observed by the NRL instruments. The flare rise in X-rays coincided with a strong

DISK FLARE IN PROGRESS

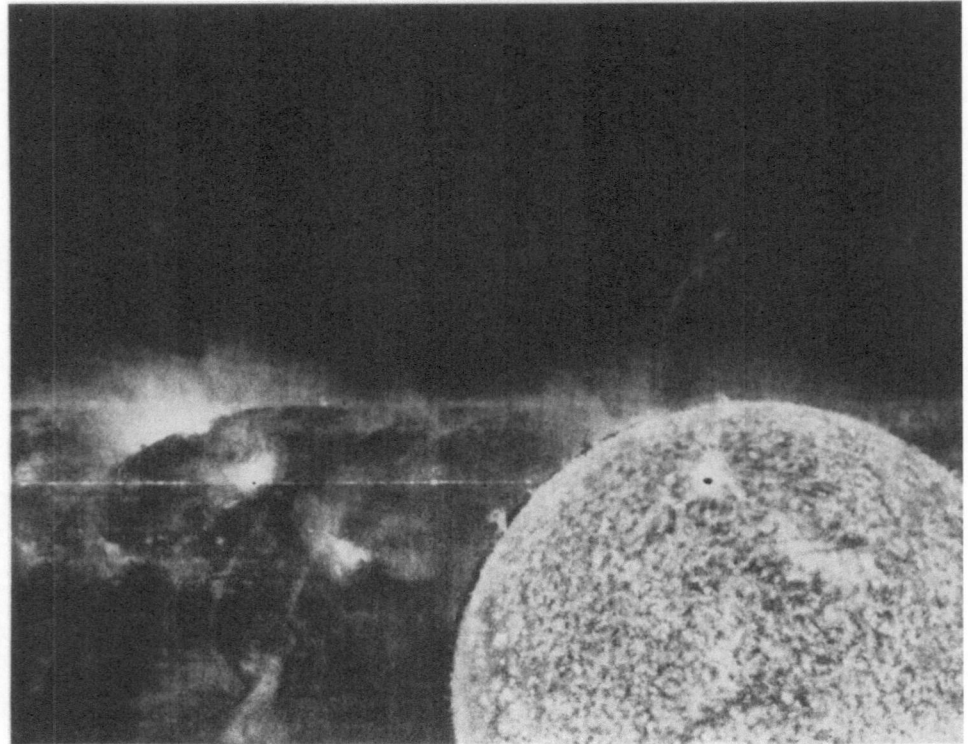

AUG 9, 1973   15^h 54^m U.T.

Fig. 3.   The disk flare two minutes after microwave burst maximum. The Fe xv and He ɪɪ images are
solarized (dark spots) at the place of greatest intensity.

microwave burst maximum at 1552 UT and a type III event at about the same time.
The exposure at 1554 UT shows the flare already cooling down and saturating the
film image in Fe xv (to produce the film reversal). It will be noted that the intense core
is the same foot-point region already bright 30 min earlier.

To derive an estimate of electron density from the flare observations we used the
Solrad-9 fluxes (Dere and Kreplin, 1974) to obtain the emission measure at flare
maximum and measured the volume on the spectroheliograms. The basic assumption
is that the 0–3 Å and 1–8 Å fluxes are emitted by the same volume defined by the
Fe xxɪv image. The intense core of the August 9 flare has about the same size in all
the flare images, and it is fairly small. The maximum dimension of the Fe xxɪv image
does not exceed 5000 km.

The observed X-ray intensity therefore requires a high density of the order of
5×10^{11} cm^{-3}. Values of this order may turn out to be typical when the source region
is at a low altitude in the limb of a flux loop.

III. The Limb Flare of 1974, January 15

Figure 4 shows the coronal loops associated with a flaring region at the limb at 1425 UT near the time of X-ray maximum. The loops were visible eight hours later, although changed in aspect.

The prominent features are the high altitude loops together with the foreground ribbons in which they are rooted. As for the high temperature flare itself, it is relatively inconspicuous at a much lower altitude. It is visible in the 2 or 3 small, bright, crescent-shaped images (of very high stage ions) aligned at the level indicated by the arrows in Figure 4.

FLARE LOOPS

Fe XIV 274 Å   Fe XIV   Fe XVI 263 Å

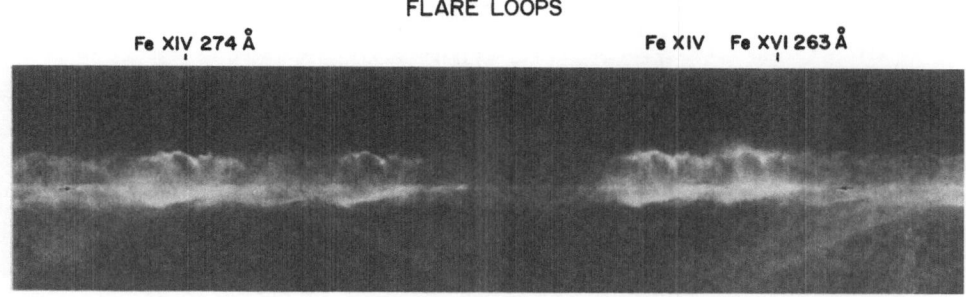

Fig. 4.   Limb flare of 1425 UT, 1974, January 15. The X-ray emitting component is visible in 2 or 3 bright images at the level of the arrows.

The shape of this feature suggests that it is basically the top of a low arch, but whether it is rooted in the foreground, or beyond the limb, is not clear. Alternatively, the high temperature feature could be a foot-point region, similar to the foreground ribbons. In any case, the electron density of the emission region appears to be high.

IV. The Double Ribbon Flare of 1973, June 15

This was probably the largest flare observed with good coverage by the ATM/NRL spectrographs during the three Skylab missions. Except for the impulsive phase, the observational coverage is quite complete, and the various phases in the spectral evolution are fairly easy to identify. *It also showed the location of the high temperature flare relative to the chromospheric ribbons better than any other example.*

The ATM/S-082A observations essentially cover the time from the end of the burst phase at 1411 over flare maximum (X-ray) and through the subsequent cooling phase till 1438. The significant changes in the spectral development and in the spatial and intensity changes of individual images may be followed through the approximate 30 min of observation with the aid of Figures 6 and 7. These figures compare selected spectral ranges at four different times in the flare development at sufficient magnification to illustrate the major changes. A sequence showing the changes in the Fe XXIV

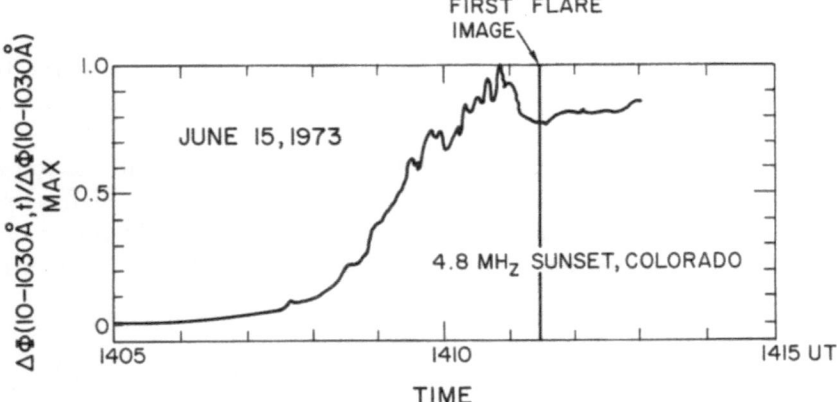

Fig. 5.   The first flare spectroheliogram in relation to the SFD observed for this event (communicated by R. F. Donnelly).

image at still greater magnification is given in Figure 8, where it is compared with the Fe xv image.

(a) POST-IMPULSIVE PHASE (1411–1415)

Figure 5 illustrates the Sudden Frequency Deviation (SFD) associated with the impulsive phase of the June 15 Event (Donnelly, 1974). Microwave bursts at 2.7 and 5.0 GHz were also recorded between 1409 and 1411, as well as a Dekameter type III burst at 1410.

The first flare observation at 1411.7 coincided with the end of the burst phase and the rise to peak flux in the 0–3 Å band, as recorded from Solrad 9 and Solrad 10 (Dere and Kreplin, 1974). On the first exposure the Fe xxiv images are present at their maximum extent and approaching peak brightness (Figures 7 and 8). Evidently, the

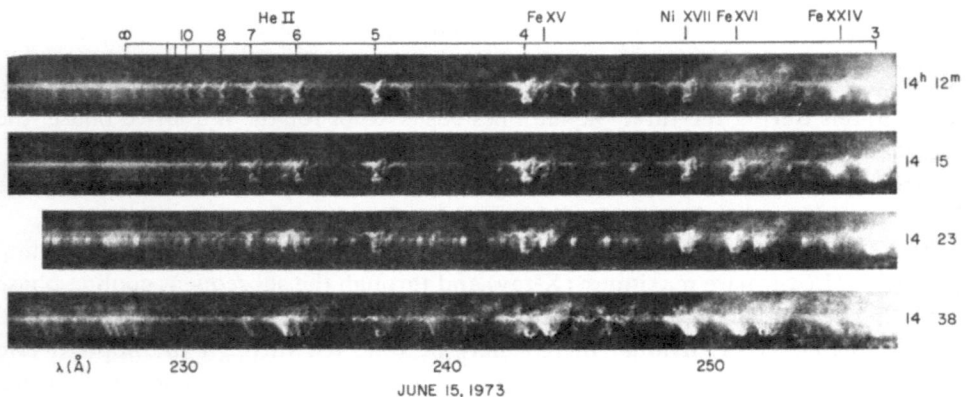

Fig. 6.   The enhanced brightness of He ii relative to coronal lines in the early phase may be traced by comparing the adjacent images of He ii ($n=4$) with Fe xv.

flare temperature peaked at approximately 1412, a conclusion confirmed by the analysis of the Solrad 10 observations (Dere, 1974).

Although the burst phase ended at 1411, significant impulsive effects remain in the images. Alternatively, these effects may indicate a continuation of the energy release, but on a slower scale, or represent a longer-lived 'turbulence' associated with the flare instability.

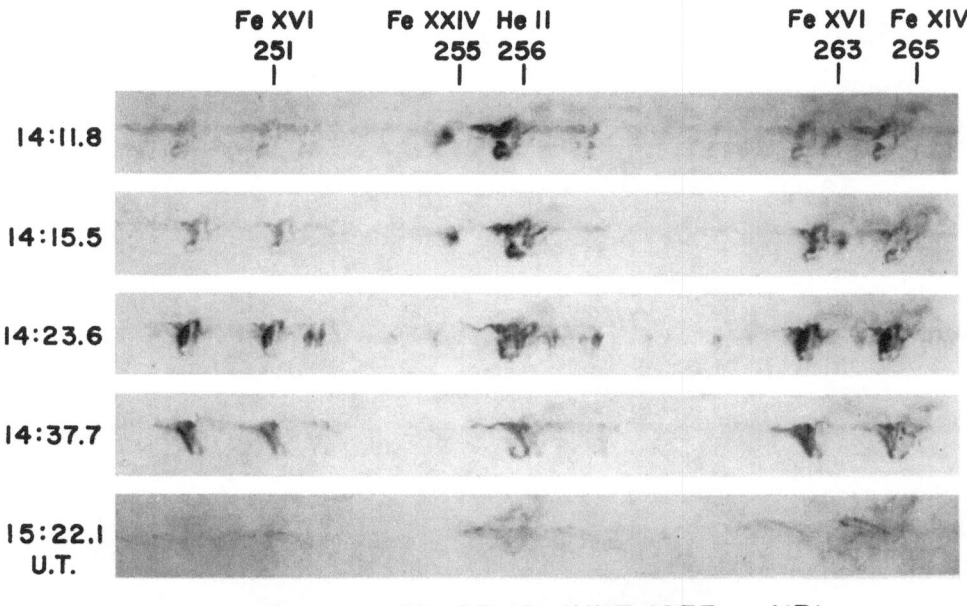

Fig. 7. The cooling of the flare is shown by the disappearance of the Fe XXIV image after 1415 followed by the brightening of lower stage iron ions at 1423.

One sign of post-burst heating and dynamical activity is shown by the enhanced brightness and distortion of the He II images visible in the first two frames of Figure 6. The enhanced emission of He II occurs primarily in the chromospheric ribbons overlying areas of strong vertical field. For the orientation of the He II emission ribbons with respect to the neutral line see Figure 10; compare also the Hα image and magnetogram in Figure 9.

The He II images are distorted by Doppler shifting and mass motion which continued till 1415, or later.* The most dramatic effect, however, is shown by the blueshifted spike of the erupting filament at 141140 (Figure 8). This absorption filament occupies the channel between the emission ribbons, and is best seen in the He II 256 Å and Fe XV 284 Å images. In Figure 7 the filament as a whole is greatly broadened in the Fe XIV 265 Å image. The Doppler spike appears to erupt more or less at the place

* Dr Brueckner (this symposium, p. 135) has noted a similar phenomenon in simultaneous slit spectra of this flare.

where the filament intersects the neutral line. The length of the spike corresponds to a projected velocity of approximately 460 km s^{-1}.

The Fe XXIV image at 141140 in Figure 8 also shows a broadened emission knot with extensive wings near the center of the image. Whether this excess broadening is thermal or non-thermal, it reflects the instability over the neutral line.

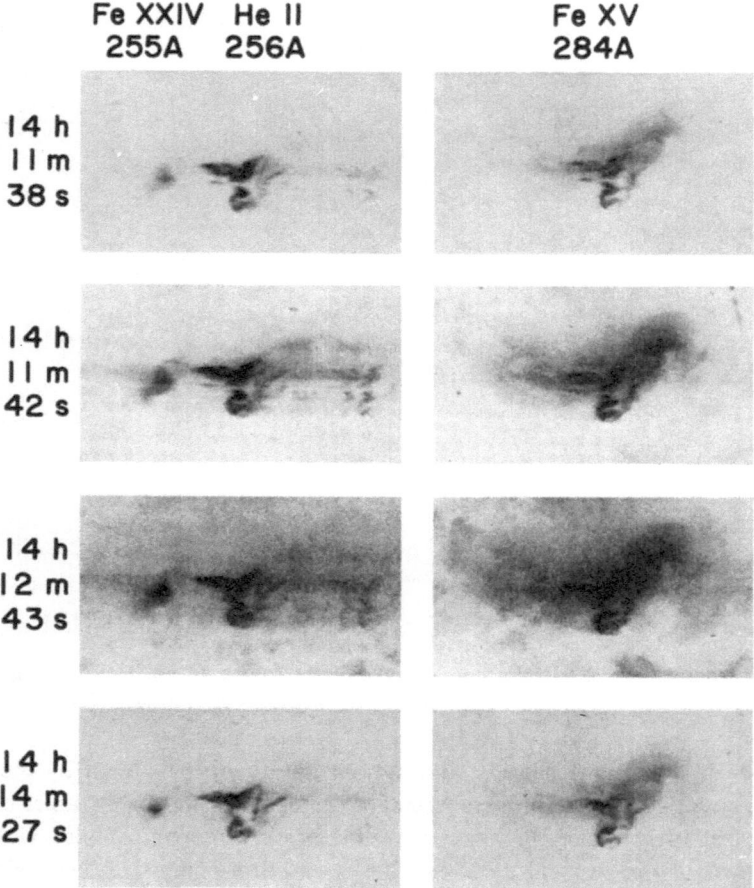

Fig. 8.   A sequence showing the changes in the intensity and structure of the Fe XXIV image.

The previous observations suggest an explanation in terms of a primary energy release over the neutral line.

(1) As one effect, it produces the broadening and eruption of the filament.

(2) At a different altitude (perhaps lower) the energy release produces heating as its primary effect in a flux loop joining opposite polarities across the neutral line. The emitting region – as seen in the Fe XXIV image – appears stabilized and contained.

Flare June 15, 1973

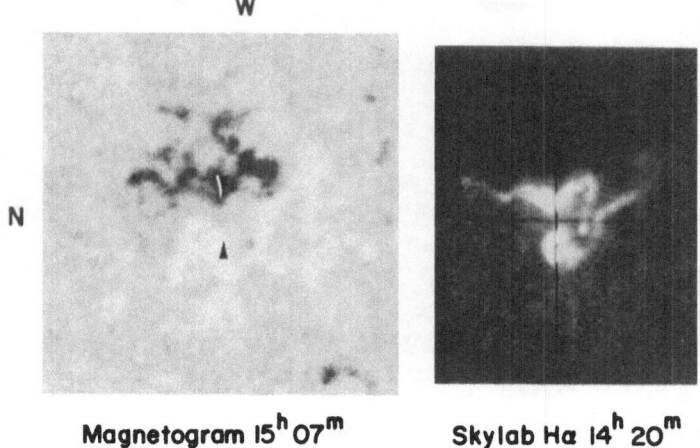

Fig. 9.   The Fe XXIV image crosses the gap between the Hα ribbons at the vertical cross wire. In the magnetogram the Fe XXIV image occupies the gap between the arrow tips.

(3) Simultaneously, the energy flow downward into the denser foot-point regions produces the enhanced brightness and mass-motion visible in the He II ribbons.

(b) COOLING PHASE (After 1415)

The cooling of the flare is particularly evident in the disappearance of the Fe XXIV images after 1420; this happens when the temperature in the main body of the flare has dropped below 10×10^6 deg. This is followed by the rapid brightening of typical coronal lines such as Fe XV, XVI, and Ni XVII to produce the rich spectrum observed at 1423 in Figure 6. These lines appear at optimum brightness when the temperature has dropped into the range of 5×10^6 deg.

(c) SPATIAL LOCATION OF THE Fe XXIV EMISSION

In Figure 9 the Hα image has been enlarged to approximately the same scale as the Kitt Peak Magnetogram. The double ribbon nature of the Hα flare is clear. The magnetogram shows a relatively simple bipolar structure with the neutral line passing through the region in a N-S direction, turning slightly to the SW in the filament channel.

In Figure 10 the region of the Fe XXIV 255 Å line has been enlarged from the exposure at 1415.5. The approximate position of the neutral line has been transferred to the He II 256 Å image with the aid of an overlay of the magnetogram enlarged to the same scale.

The blurred appearance of the He II image in Figure 10 is caused by the Doppler shifting and mass-motion still going on some 4 min after the burst phase; but the general 2 ribbon structure is still clear (compare with the more stable Hα image in Figure 9).

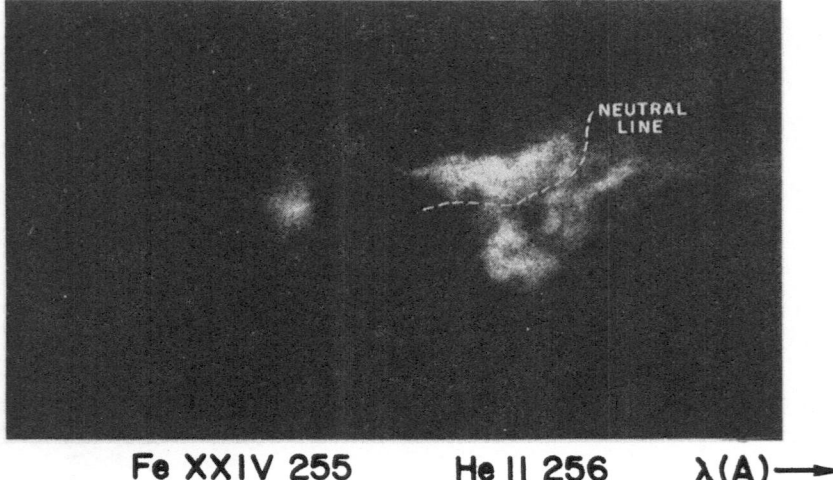

Fe XXIV 255   He II 256   λ(A)⟶

Fig. 10.   He II ribbons and neutral line. When placed in register with the He II image, the Fe XXIV emission lies over the neutral line in the narrow waist of the gap between the ribbons.

The general location of the Fe XXIV emission with respect to both the neutral line and the chromospheric ribbons is evident by inspection of Figure 10. A more refined position of the Fe XXIV image may be determined by translating it along the direction of dispersion by the known difference in wavelength (Purcell and Widing, 1972). This puts the Fe XXIV image approximately over the bend in the neutral line in the narrow waist of the gap between the ribbons. A similar comparison of the Fe XXIV image at 1411.8 with the magnetogram shows that the main body of the Fe XXIV image spanned the gap in the magnetogram between the points shown in Figure 9. This suggests that the basic structure of the Fe XXIV image is a magnetic loop crossing the neutral line at right angles and with its foot-points rooted in fields of opposite polarity.

TABLE I

Properties of the Fe XXIV emission region
1973, June 15

Volume	10^{27} cm³
$N_e^2 V$	3×10^{48} (Dere and Kreplin, 1974)
N_e	5×10^{10} cm⁻³
Altitude (estimated)	13 000 km

(d) PROPERTIES OF THE Fe XXIV EMISSION REGION

Table I summarizes the properties of the Fe XXIV emission region observed on 1973, June 15. As the absolute calibration of the Fe XXIV image is not yet available, we determined the electron density by using the Solrad 10 fluxes (Dere and Kreplin, 1974) to obtain the emission measure at flare maximum, and measuring the volume on the spectroheliogram.

The volume of the Fe XXIV emitting region was derived under the plausible assumption that the image basically represents a cylindrical arch with a projected length equal to 26 000 km. The altitude was roughly estimated by assuming the arch is circular in shape.

Finally, we discuss the cooling of the flare, and in particular, the observed duration of the Fe XXIV emission, which lasted about 8 or 9 min after the first observation at 1411.8 UT. Fe XXIV emits only in the temperature range above 10×10^6 deg where thermal conduction is the most efficient cooling process. Following a method of Culhane *et al.* (1970) we consider the conductive cooling of a cylindrical loop initially heated to some peak temperature. If we put in the dimensions and electron density derived from the Fe XXIV image, we find that a 20×10^6 deg plasma will cool below 10×10^6 deg in 5 min.

This is somewhat too short compared to the Fe XXIV observed lifetime of 8 or 9 min. But as noted previously, the activity observed in the He II images suggests that appreciable energy release may have occurred as late as 1415. In support of this, the Fe XXIV images in Figure 8 show considerable changes in structure and intensity till 1415 but fade rapidly thereafter. We could therefore regard this as the time when the final cooling phase begins.

Alternatively, we may take the observed mass-motions at face-value, and assume that the observed 'turbulence' effectively inhibits the heat conduction as late as 1415. This effectively lengthens the calculated cooling time of 5 min by another 3 min, which is in agreement with observation.

Acknowledgments

The author is grateful to the members of the NRL Skylab Mission Team, headed by Dr R. Tousey, for the opportunity to discuss these observations. He is also grateful to K. P. Dere and R. W. Kreplin for providing the Solrad-9 and Solrad-10 observations and preliminary analysis in advance of publication. The author thanks Dr R. F. Donnelly for communicating the SFD observations of the June 15 flare.

References

Cheng, C. C. and Spicer, D. S.: 1975, This volume, p. 423.
Culhane, J. L., Vesecky, J. F., and Phillips, K. J. H.: 1970, *Solar Phys.* **15**, 394.
Dere, K. P.: 1974, private communication.
Dere, K. P. and Kreplin, R. W.: 1974, private communication.
Donnelly, R. F.: 1974, private communication.
Jordan, C.: 1970, *Monthly Notices Roy. Astron. Soc.* **148**, 17.
Neupert, W. M.: 1971, *Phil. Trans. R. Soc. London A* **270**, 143.
Purcell, J. D. and Widing, K. G.: 1972, *Astrophys. J.* **176**, 239.
Tousey, R., Bartoe, J.-D. F., Bohlin, J. D., Brueckner, G. E., Purcell, J. D., Scherrer, V. E., Sheeley Jr., N. R., Schumacher, R. J., and VanHoosier, M. E.: 1973, *Solar Phys.* **33**, 265.

X-RAY AND EUV SPECTRA OF SOLAR FLARES
AND LABORATORY PLASMAS

G. A. DOSCHEK

*E. O. Hulburt Center for Space Research, Naval Research Laboratory,
Washington, D.C. 20375, U.S.A.*

Abstract. Recent laboratory work relevant to solar flares on the spectroscopy of highly ionized atoms is reviewed. Much of this work has concerned the X-ray and EUV spectrum of iron ions, Fe xviii–Fe xxiv, which produce prominent emission lines in the spectra of solar flares. Also discussed are recently obtained laboratory X-ray spectra of emission lines of hydrogen-like and helium-like ions, and associated satellite lines due to transitions of the type, $1s2l-2p2l$, $1s^2 2l-1s2p2l$, and $1s^2 2l--1s2l3p$. Satellite lines have also been identified in spectra of solar flares, and can be used to determine the electron temperature of the plasma. The laboratory work is important in the planning of future experiments in solar flare X-ray and EUV spectroscopy.

I. Introduction

During the last decade, a considerable amount of phenomenological information has been obtained on the X-ray and EUV radiations emitted by solar flare plasmas. This extension of our knowledge of flares to the very short wavelengths is a product of unmanned and manned space vehicles; the most recent highlight being the successful Skylab Mission. X-ray crystal spectrometers and broadband X-ray detectors aboard unmanned satellites such as the Orbiting Solar Observatories (OSO) and the Solrad satellites, have revealed the existence of high electron temperature ($T_e \approx 20 \times 10^6$ K) plasmas associated with solar flares, that emit intense line and continuum radiation in the X-ray region below 100 Å. These plasmas appear to be largely thermal in nature; in contrast to the hard X-ray emission at energies greater than about 7 keV, which appears to have a non-thermal origin (Kane, 1974). The Soft X-ray plasmas are temporally associated with the Hα event, and recent NRL and AS & E Skylab observations indicate that these plasmas are confined to loop-like structures (Widing, 1975; Krieger *et al.*, 1975).

This review will consider some of the details of the soft X-ray emission line spectrum, and in particular will compare the solar flare spectra with recently obtained laboratory spectra of similar high temperature plasmas. The results of the laboratory work make possible future solar flare X-ray and EUV observations that offer considerable promise for determining the physical conditions in high temperature flare plasmas. This information is important for testing various models of solar flares.

II. Flare Spectra

At electron temperatures of 20×10^6 K, most of the abundant solar elements are ionized to the hydrogen-like and helium-like ionization stages. The emission line spectra of the elements carbon through calcium consist primarily of the resonance lines and principal series lines of the hydrogen-like and helium-like ions. These lines fall between ~ 2.7 Å and ~ 41 Å, and are produced primarily by electron impact excitation of ions in the ground state.

Iron is the only appreciably abundant solar element that is significantly heavier than the other elements. The relatively larger ionization potential of iron ions insures that iron ions other than hydrogen-like and helium-like ions will also be present in high temperature plasmas near 10×10^6 K. This is indeed observed in flare spectra, and lines from iron ionization stages from Fe XVII through Fe XXV are prominent. In contrast to flares, only the Fe XVII lines are strong in the X-ray spectra of active regions.

The intense flare iron lines emit in three relatively narrow spectral ranges. The resonance transitions $2s^2 2p^k - 2s 2p^{k+1}$ fall in the EUV region from ~ 80 Å to ~ 260 Å. Actually, only the two Fe XXIV lines ($^2S - {}^2P$) fall near the long wavelength end of this region (192 Å and 256 Å). Most of the other iron lines from Fe XVIII through Fe XXIII are confined between ~ 80 Å and ~ 150 Å.

Iron lines from Fe XVII through Fe XXIV involving $\Delta n \geqslant 1$ type transitions fall between ~ 7.0 Å and 17 Å. These lines are due to transitions of the type, $2s^2 2p^k - 2s^2 2p^{k-1} \, nl$, and were first observed in flare spectra by Neupert *et al.* (1967). The nd transitions are predicted to be considerably stronger than the ns transitions (Fawcett *et al.*, 1974).

Finally, a prominent spectral feature has been observed between 1.85 Å and 1.93 Å in solar flare spectra (Neupert *et al.*, 1967; Meekins *et al.*, 1968, 1970; Doschek *et al.*, 1971; Grineva *et al.*, 1973). This feature is due to emission from helium-like Fe XXV, and innershell emission from lower iron ionization stages, i.e., $1s^2 2s^2 2p^k - 1s 2s^2 2p^{k+1}$. Most of the innershell emission is due to lines of lithium-like Fe XXIV. All three of these iron-line groupings are important for investigating high temperature plasmas.

The innershell transitions, or satellite lines, have also been observed for elements lighter than iron. Most of these lines have been identified as transitions in helium-like and lithium-like ions (Gabriel and Jordan, 1969a; Walker and Rugge, 1971; Neupert, 1971; Doschek, 1972; Feldman *et al.*, 1974a). The wavelengths of most of the satellite lines fall close to the wavelengths of the resonance lines of the hydrogen-like and helium-like ions, and are stronger relative to the resonance lines in spectra of the heavier elements. Satellite lines due to transitions in beryllium-like ions and lower ionization stages have not been conclusively identified for any element other than iron.

Most of the available X-ray flare spectra have been obtained from groups at the Naval Research Laboratory (OSO 4, 6), Goddard Space Flight Center (OSO 3, 5) and at the Lebedev Institute in Moscow (Mandelstam and his colleagues). Neupert and his colleagues at Goddard are the only group to have obtained high resolution EUV iron-line spectra of flares near 100 Å.

Figures 1–3 show X-ray and EUV spectra from several large flares. These spectra illustrate the major spectral features discussed above. In addition, Figures 2 and 3 show active region and quiet Sun spectra for comparison with the flare spectrum. Emission of iron ions from Fe XVIII through Fe XXIV is either weak or absent in the active region and quiet Sun spectra, but strong during flares.

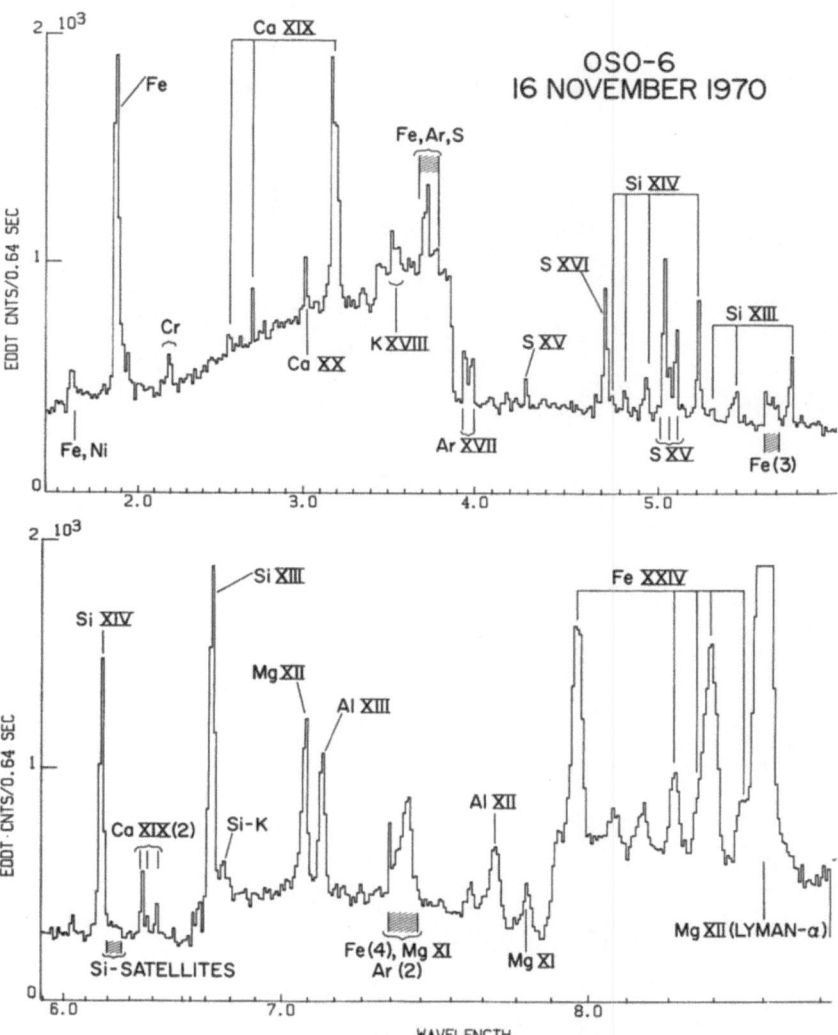

Fig. 1. A solar flare spectrum obtained from an NRL spectrometer on OSO 6. The spectrum is dominated by emission from the resonance lines of hydrogen-like and helium-like ions. Lines of Fe XXIV (2l–4l') are also prominent. The edges in the continuum are instrumental.

In Figures 1 and 2 the emission from the strongest lithium-like satellite lines occurs close to, and on the long wavelength side of, the helium-like resonance lines, and is blended with the intercombination $(1s^2 \, ^1S_0 - 1s2p \, ^3P_{1,2})$ and forbidden lines $(1s^2 \, ^1S_0 - 1s2s \, ^3S_1)$ of the helium-like ions. A detailed list of flare emission lines and identifications for the wavelength region from ~ 1 Å to 8.5 Å that summarizes the observations up to 1971 has been published by Doschek (1972). Since then Doschek *et al.* (1973) and Neupert *et al.* (1973) have also published wavelengths and identifications for flare spectra in the region from 8.5 Å to 21 Å. Wavelengths and identifica-

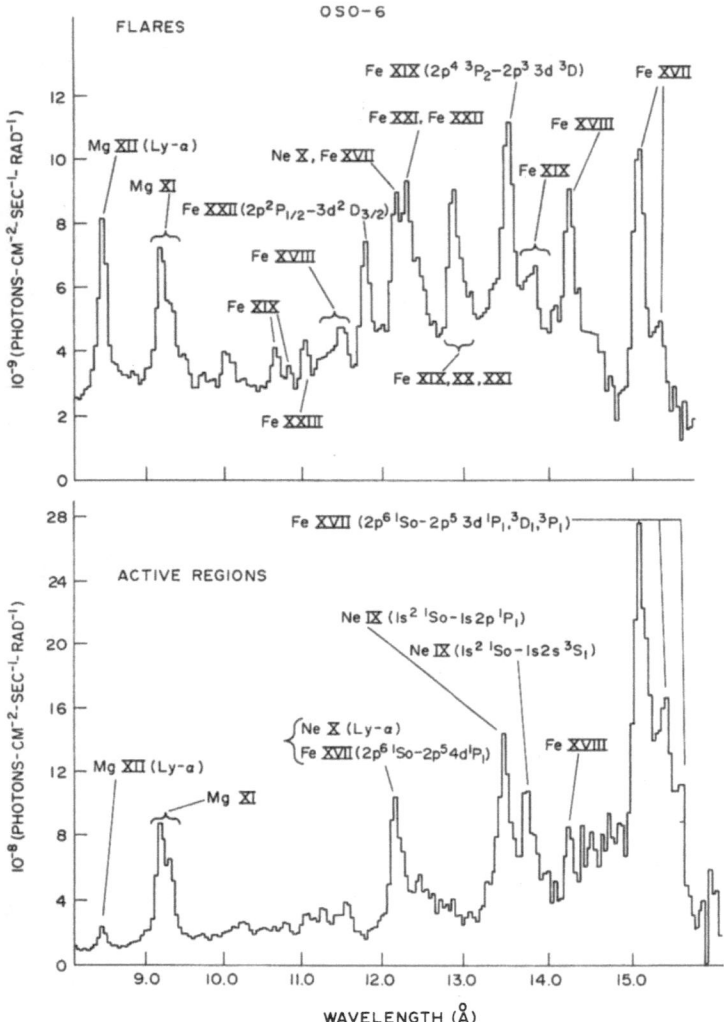

Fig. 2.   Solar flare and active region spectra obtained from an NRL spectrometer on OSO 6. Emission from Fe XVIII–Fe XXIV is strong in the flare spectrum but weak or absent in the active region spectrum.

tions for the flare lines near 100 Å (Figure 3) have been given recently by Kastner *et al.* (1974). The 100 Å lines have also been recently observed in laboratory spectra of considerably higher resolution by Feldman *et al.* (1973a, 1974b) and Doschek *et al.* (1974a, b), and detailed classifications have been published. These results will be discussed further below.

Almost all of the existing flare spectra have low resolution compared with the resolution that is needed to use the full diagnostic capabilities of the spectrum. This will

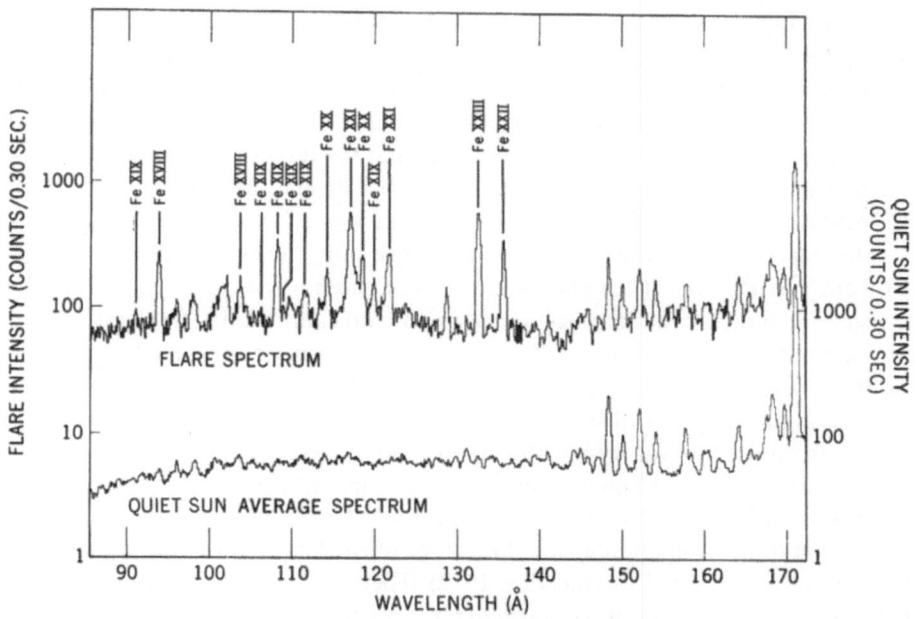

Fig. 3.   Solar flare and quiet Sun spectra obtained from a GSFC spectrometer on OSO 5. Emission from Fe XVIII–Fe XXIII is strong in the flare spectrum. Some of the identifications are incorrect. (Spectrum courtesy of S. O. Kastner and W. M. Neupert.)

become clear upon inspecting typical laboratory spectra to be shown below. The majority of the existing solar spectra were obtained by experiments that sacrificed spectral resolution for a large total wavelength coverage with reasonable temporal resolution. The subsequent analysis of the solar data and the analysis of the laboratory data have shown that the full diagnostic possibilities of the spectrum may be exploited by employing much higher wavelength resolution and much higher temporal resolution, over narrow selected wavelength bands.

III. Recent Results from Laboratory Plasmas

(a) LABORATORY AND SOLAR PLASMAS

Generating high temperature laboratory plasmas is motivated by the thermonuclear fusion program and the desire to support astrophysical programs. Laboratory sources that are capable of producing highly ionized atoms include the low inductance vacuum spark, thetapinch sources, Tokamak-type magnetic traps, plasma focus machines, and the focussed pulse of a high power laser. These devices are described in detail in the literature, e.g., Elton and Lee (1972), Burgess (1972), and Fawcett (1973). The low inductance spark devices described by Cohen *et al.* (1968) at Goddard, Lee and Elton (1971) at NRL, and Schwob and Fraenkel (1972) at the Hebrew University are capable of producing and exciting Fe XXV and Fe XXVI.

Recently, high power lasers have been used to generate high temperature plasmas. A high power laser pulse is focused onto a solid target of the material under investigation. Typically, the NRL glass laser can deliver about 60 J onto a target in 0.9 ns. This energy is converted into thermal energy of the target material and a high temperature plasma is formed that expands outward from the target. The dense regions of the NRL plasmas are about 100 μm in size, and exist on the order of nanoseconds.

Although there are fundamental differences between laboratory and flare plasmas, e.g., the electron densities of vacuum spark and laser-produced plasmas are $\approx 10^{19}$ cm^{-3}–10^{21} cm^{-3} (Lee and Elton, 1971; Feldman *et al.*, 1974a), the similarities between flare and laboratory plasmas are useful and significant. The electron temperatures, or excitation conditions if thermal equilibrium is not assumed, are nearly the same in laboratory and flare plasmas. Therefore many of the lines that are observed in solar plasmas are also observed in the laboratory. The additional permitted lines that are strong in laboratory spectra due to the statistical populating of levels near the ground state are an advantage in making line identifications.

Another significant manner in which laboratory plasmas resemble solar flare plasmas is in the overall phenomenological appearance of the EUV and X-ray radiations from both types of plasmas (Elton and Lee, 1972). There is a remarkable similarity regarding the sequence of emission of hard X-rays, microwaves, and soft X-rays. There is also a close similarity in the time-flux profiles of the various types of radiations. These points have been emphasized most recently by Lee (1974), and Figure 4 shows a typical hard X-ray spectrum of a low inductance vacuum spark plasma. As with solar flare plasmas, a combination thermal spectrum and nonthermal power law spectrum can be fitted to the data.

(b) IRON LINES IN LABORATORY AND FLARE SPECTRA

The spectra shown in Figures 1–3 indicate the importance of iron line emission in solar flare spectra. Line emission from a large number of iron ionization stages offers the possibility of treating the iron line fluxes in terms of a Pottasch-type analysis, i.e., each ion is assumed to emit at its temperature of maximum emitting efficiency. Simultaneous observations of iron lines from different ionization stages with collimated instruments will enable the emission measure-temperature distribution to be determined as a function of position and time over the flare area. A similar analysis using the resonance lines of hydrogen-like, helium-like, and lithium-like ions alone is more difficult because of the high temperature tails of their excitation functions (Jordan, 1969, 1970).

Iron-line emission near 1.85 Å will be discussed in the next section. In this section iron-line emission from ~ 7 Å to 17 Å and line emission near 100 Å are discussed.

The solar flare spectra obtained to date do not have sufficient resolution to resolve the blends of iron lines due to the transitions, $2s^2 2p^k - 2s^2 2p^{k-1} nl$, $n \geqslant 1$, in the ~ 7 Å to 17 Å region. Laboratory spectra of iron in this wavelength range are quite complex (Feldman and Cohen, 1968; Cohen and Feldman, 1970; Connerade *et al.*, 1970). There is considerable overlapping of lines of different ionization stages. The complexity of

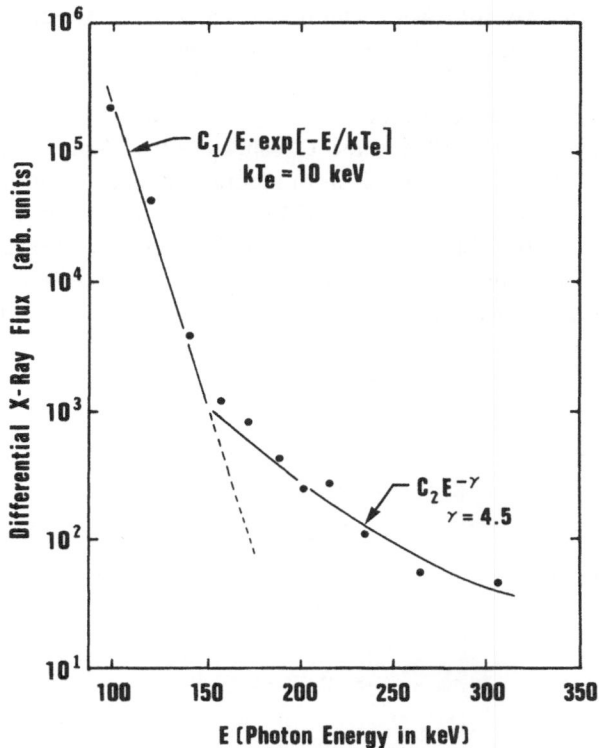

Fig. 4. Hard X-ray spectrum of a vacuum spark plasma. Both a thermal and non-thermal spectrum are combined to fit the data. (Courtesy of T. N. Lee.)

the laboratory spectra is also due in part to statistical populating of the low lying levels of the $2s^2 2p^k$ and $2s 2p^{k+1}$ configurations at the high electron densities in laboratory plasmas, which leads to a large number of lines in the laboratory spectrum.

Considerable work on the laboratory spectra is still necessary. However, some recent progress has been reported. Feldman et al. (1973b) have identified or reclassified 27 lines of Fe XVIII between 13.95 Å and 16.27 Å, and Fawcett et al. (1974a) have recently discussed the expected solar flare spectrum of iron due to Fe XIX through Fe XXIV. Fawcett et al. point out that the solar flare spectrum should be considerably simpler than the laboratory spectrum, because most of the iron ions are in the ground state. Doschek et al. (1972, 1973) have identified lines of Fe XXIII and Fe XXIV near 8 Å in flare spectra, and have shown that the Fe XXIV lines near 10.6 Å ($2s - 3p$) are blended (at the low resolution of the available flare spectra) with lines that are probably due to Fe XIX ($2p - 4d$).

Most of the recent line identifications of highly ionized iron have relied in part on ab initio theoretical wavelength calculations, as well as on the extrapolation of identified lines of lighter elements along the iso-electronic sequences. Many of the recent theoretical calculations have been done by R. D. Cowan at the Los Alamos Scientific

Laboratory. Cowan's ab initio wavelengths are accurate to about ± 0.03 Å at X-ray wavelengths, and to about ± 1–5 Å for iron lines due to $2s^2 2p^k - 2s2p^{k+1}$ transitions near 100 Å.

Recently, there has been considerable progress in identifying the iron lines near 100 Å (Feldman et al., 1973a, 1974b; and Doschek et al., 1974a, Fawcett et al., 1974b). Most of this work is based on the analysis of laboratory spectra obtained at NRL from laser-produced plasmas. Kastner et al. (1974) have also recently identified some of these lines in flare spectra obtained by Goddard instruments on OSO 5. Their identifications are based on some of the laboratory work, and on iso-electronic extrapolations. The low resolution of the flare spectra make the line identifications difficult.

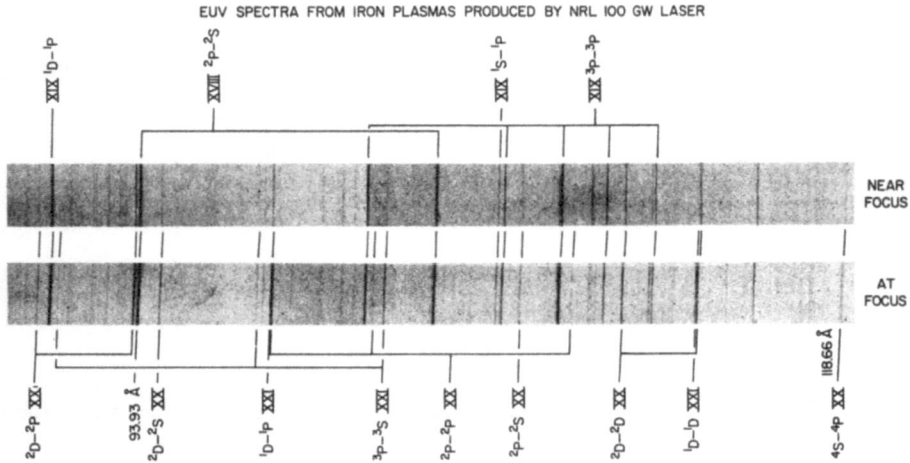

Fig. 5.   Iron-line spectra of laser-produced plasmas. Lines of Fe xx and Fe xxi are substantially stronger in the 'at focus' spectrum than in the 'near focus' spectrum.

Figure 5 shows iron line spectra obtained from NRL 100 GW laser-produced plasmas. By changing the position of the focusing lens (see p. 170), the excitation conditions of the plasma could be varied. The higher ionization stage lines are stronger in the spectrum obtained with the lens well-focused. In this manner, lines of Fe xix, Fe xx, and Fe xxi can be distinguished from one another by visual inspection.

The detailed identifications of the lines in Figure 5 relied on the earlier work on these iso-electronic sequences given in Moore's Atomic Energy Levels, and on the extension of these data to elements up through $\sim$ vanadium by Fawcett and colleagues at Culham. A summary of the data before the recent work at NRL is given by Fawcett (1971), and Figure 6 shows three spectra of titanium, scandium, and vanadium obtained by Fawcett. The work at NRL confirms most of the identifications of the $2s^2 2p^k - 2s2p^{k+1}$ lines of the fluorine, oxygen, nitrogen, and carbon iso-electronic sequences in these lighter elements, and in some cases extends the identifications up through nickel.

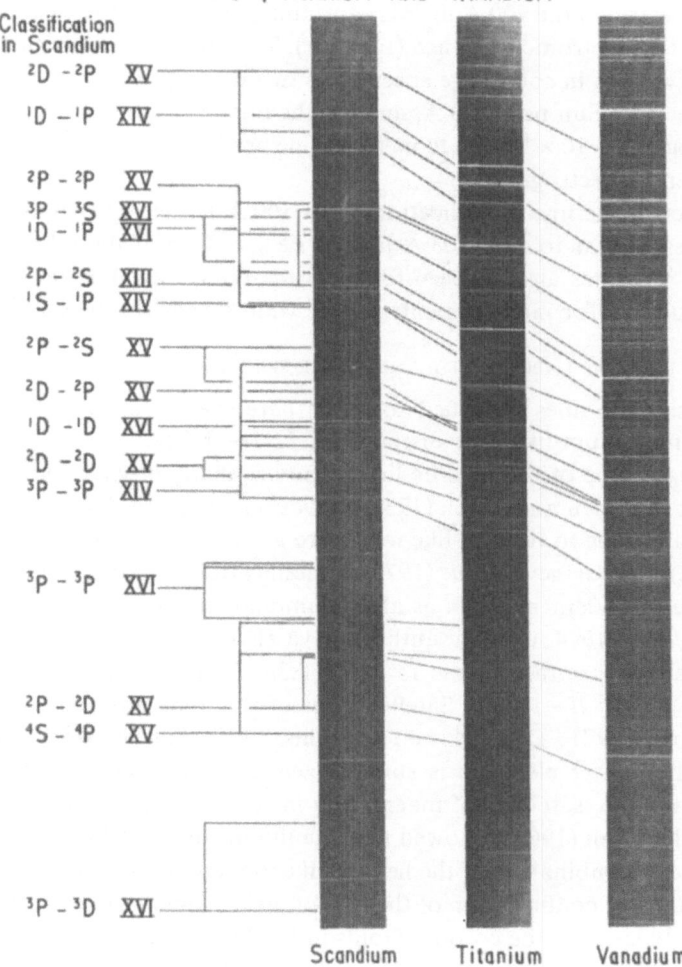

Fig. 6. Laboratory spectra showing lines due to transitions of the type $2s^2 2p^k - 2s 2p^{k+1}$, in ions of the elements, titanium, scandium, and vanadium. Identifications of lines in these elements is important for identifying the same transitions in iron and nickel. (Spectra courtesy of B. C. Fawcett.)

It is interesting to compare the laboratory iron spectrum of Figure 5 with the solar flare spectrum shown in Figure 3. Because electron densities of flare plasmas $(\approx 5 \times 10^{10} \text{ cm}^{-3}$, Cowan and Widing, 1973) are orders of magnitude less than the densities of laboratory plasmas, some of the lines that are strong in laboratory spectra are weak in the flare spectrum. An example is Fe XIX $(2s^2 2p^4 \, {}^1D_2 - 2s2p^5 \, {}^1P_1)$. This line is weak in the flare spectrum because the ${}^1D_2$ level is not appreciably populated at flare electron densities.

Laboratory work on the 100 Å lines is continuing, and needs to be extended through the beryllium iso-electronic sequence (Fe XXIII). The 100 Å lines are particularly suitable for observations in solar flare spectra for two reasons: (a) there are few spectral lines from the quiet Sun near 100 Å, and (b) the lines are sufficiently wide at coronal flare temperatures ($\sim 10 \times 10^6$ K) to measure line profiles with existing high resolution grazing incidence spectrographs.

A byproduct of the line identifications near 100 Å is the wavelength prediction of forbidden lines similar to the 5303 Å line of Fe XIV. One of these lines (1355 ± 3 Å, Fe XXI $({}^3P_0 - {}^3P_1)$) has already been found in the NRL Skylab data. These lines are particularly suitable for measurements of line widths and wavelength shifts.

(c) SATELLITE LINES IN LABORATORY AND FLARE SPECTRA

Most of the satellite lines identified so far in flare spectra are due to transitions in lithium-like and helium-like ions of the type, $1s^2 2l - 1s2p2l$, and $1s2l - 2p2l$, respectively. Although some of the satellite lines were originally observed long ago in laboratory spectra by Edlén and Tyrén (1939), it was only recently that detailed classifications for the lines due to lithium-like ions were given by Gabriel and Jordan (1969a), Gabriel (1972) and Grineva et al. (1973). Classifications for the helium-like satellite lines of the heavier elements such as aluminum and silicon have recently been given by Feldman et al. (1974c). These authors have also observed and classified satellite lines in laboratory spectra near the $1s^2 \, {}^1S_0 - 1s3p \, {}^1P_1$ helium-like lines due to transitions of the type, $1s^2 2l - 1s2l3p$. Satellite lines of this type are also observed in flare spectra (Neupert, 1971; Doschek and Meekins, 1973). Earlier work on helium-like satellites of the lighter elements is summarized in Walker and Rugge (1971), who identified helium-like satellites of magnesium in solar active region spectra.

Gabriel and Jordan (1969a) showed that the lithium-like satellite lines were formed by dielectronic recombination of the helium-like ion, and therefore the observation of these lines is a direct confirmation of the role of dielectronic recombination in shaping the ionization balance in the corona. Goldsmith (1969) had reached a similar conclusion concerning the formation of the helium-like satellites. Gabriel (1972) and Grineva et al. (1973) subsequently showed that the satellite lines could be used to determine the electron temperature of a plasma, and the departure of the plasma from ionization equilibrium. The impetus for solar and laboratory observations of the satellite lines arose in part from the identification by Gabriel and Jordan (1969a) of the forbidden line ($1s^2 \, {}^1S_0 - 1s2s \, {}^3S_1$) of the helium-like ions in solar spectra of active regions and flares. The forbidden line may have diagnostic value for flares because the intensity ratio of the forbidden to intercombination line has been shown by Gabriel and Jordan (1969b) to be density sensitive.

Recent laboratory observations of satellite lines have been made using vacuum sparks (Schwob and Fraenkel, 1972; Elton and Lee, 1972; Lee, 1974; Margalit et al.,

1974; Holz *et al.*, 1974), the plasma focus (Peacock *et al.*, 1971), and laser-produced plasmas (Peacock *et al.*, 1973; Feldman *et al.*, 1974a, c). Satellite lines of iron have so far only been observed in vacuum spark spectra. Only the satellite lines of lithium-like iron are strong in these spectra. Emission from Fe XXVI and associated satellite lines is quite weak.

Figure 7 shows a vacuum spark spectrum of iron obtained by Schwob and Fraenkel (1972), and Figure 8 shows a recently obtained solar flare iron spectrum reported by Grineva *et al.* (1973). (Recent results concerning high resolution solar satellite line spectra of magnesium and lighter elements are summarized in this Symposium by Parkinson (1975).) The spark spectra still do not have the resolution necessary to resolve the closely spaced lines near 1.86 Å. Furthermore, the solar and spark spectrum differ in that the solar spectrum is indicative of a plasma that is close to ionization equilibrium, while the spark spectrum indicates extreme transient ionization of the laboratory plasma (Gabriel, 1972). The presence of substantial innershell emission from the lower ionization stages such as Fe XVIII in the spark spectrum is indicative of extreme transient ionization. The forbidden line observed in the solar spectrum

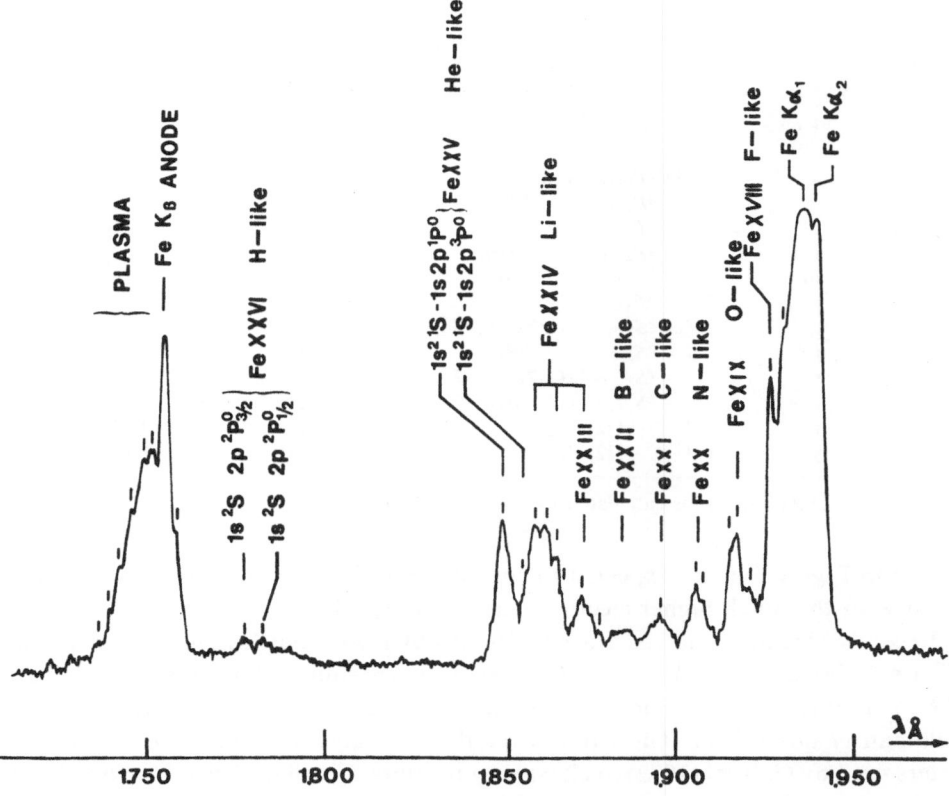

Fig. 7.   A vacuum spark spectrum of iron. Many of the blended lines near 1.85 Å are also prominent in solar flare spectra. (Spectrum courtesy of B. S. Fraenkel.)

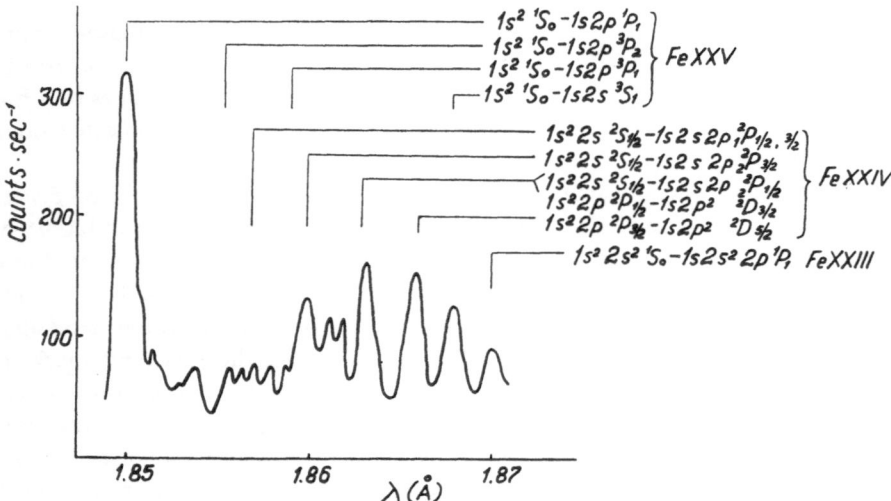

Fig. 8.   Solar flare spectrum of iron obtained by Grineva *et al.* (1973). Most of the lines of Fe xxɪv are produced by dielectronic recombination of Fe xxv. Compare with Figure 7.

TABLE I

Magnesium and calcium satellite lines[a]

Key	Transition	λ(Å) Mg x	λ(Å) Ca xvɪɪɪ
a	$1s^2 2p\ {}^2P_{3/2} - 1s2p^2\ {}^2P_{3/2}$	9.296	3.203
d	${}^2P_{1/2} - \qquad {}^2P_{1/2}$	9.295	3.203
j	${}^2P_{3/2} - \qquad {}^2D_{5/2}$	9.321	3.210
k	${}^2P_{1/2} - \qquad {}^2D_{3/2}$	9.318	3.207
m	${}^2P_{3/2} - \qquad {}^2S_{1/2}$	9.221	3.189
n	${}^2P_{1/2} - \qquad {}^2S_{1/2}$	9.218	3.185
q	$1s^2 2s\ {}^2S_{1/2} - 1s(2s2p\ {}^3P)\ {}^2P_{3/2}$	9.283	3.200
r	${}^2S_{1/2} - \qquad {}^2P_{1/2}$	9.286	3.202
s	${}^2S_{1/2} - 1s(2s2p\ {}^1P)\ {}^2P_{3/2}$	9.235	3.191
t	${}^2S_{1/2} - \qquad {}^2P_{1/2}$	9.236	3.192
w	$1s^2\ {}^1S_0 - 1s2p\ {}^1P_1$	9.168	3.176
y	$1s^2\ {}^1S_0 - 1s2p\ {}^3P_1$	9.231	3.192

[a] Wavelengths calculated by Gabriel (1972).

shown in Figure 8 is not expected to contribute to the blend of emission in Figure 7, because of the much higher electron densities of spark plasmas.

Recently, satellite line spectra have been obtained from laser-produced plasmas. Figure 9 shows lithium-like satellite lines of magnesium obtained by Peacock *et al.* (1973). Figure 10 shows similar satellite spectra for heavier elements obtained by Feldman *et al.* (1974c). The letters over the lines and multiplets are the same key letters used by Gabriel (1972) in classifying the lines. Finally, Figure 11 shows helium-like satellites of aluminum obtained by Feldman *et al.* (1974c). Also shown are satellite lines near the $1s^2\ {}^1S_0 - 1s3p\ {}^1P_1$ aluminum line.

Table I gives Gabriel's (1972) theoretical wavelengths for lithium-like satellite lines of magnesium and calcium. The satellite line laser spectra are interesting because the relative intensities of the three pairs of lines, (a, d), (q, r), and (j, k) in Figures 9 and 10 are not in accord with theoretical expectations (Gabriel, 1972). However, similar anomalous intensity behavior is not observed in spectra obtained with vacuum sparks (Elton, 1974) or the plasma focus (Peacock *et al.*, 1971), and therefore some aspect of the laser-produced plasmas may be responsible for the anomalous relative intensities. *(See Note added in proof.)*

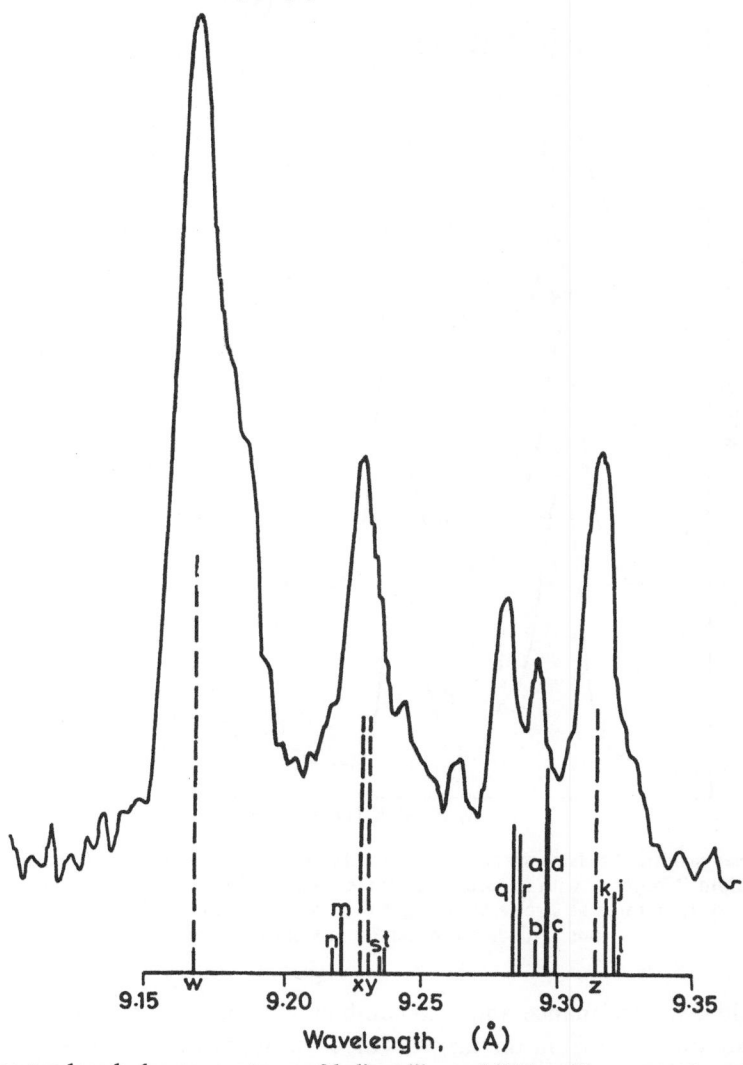

Fig. 9.   Laser-produced plasma spectrum of helium-like and lithium-like magnesium. The line z is the forbidden line ($1s^2\,{}^1S_0 - 1s2s\,{}^3S_1$) of Mg XI and may not be present in the laboratory spectrum because of the high electron density. The same remark may also apply to line x, a magnetic quadrupole transition ($1s^2\,{}^1S_0 - 1s2p\,{}^3P_2$). (Spectrum courtesy of N. J. Peacock.)

Care must be exercised in interpreting laboratory spectra obtained photographically, because different types of lines can be formed at different times during the lifetime of the plasma and within different regions of the plasma. Peacock *et al.* (1973) report that the lithium-like satellite lines are formed close to the target surface in laser-produced

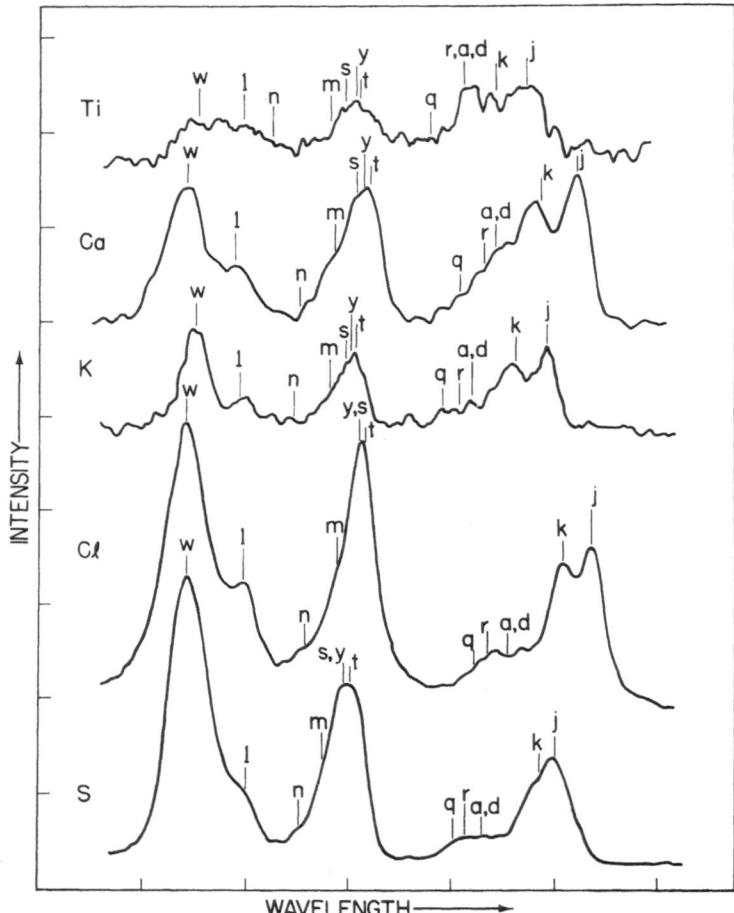

Fig. 10.   Laser-produced plasma spectra of helium-like and lithium-like ions of heavier elements up through titanium (Compare with Figure 9.) In these laser-produced plasmas, satellite line emission for calcium and titanium is as strong or stronger than resonance line emission from the helium-like ions. This is not observed in solar flare spectra.

plasmas, while the resonance and intercombination lines are emitted in the plasma plume expanding away from the target. This is dramatically illustrated in a pinhole-type spectrum of magnesium recently obtained by Burkhalter (1974), and shown in Figure 12. Emission from the resonance and intercombination lines can be seen as far as seven millimeters from the target surface.

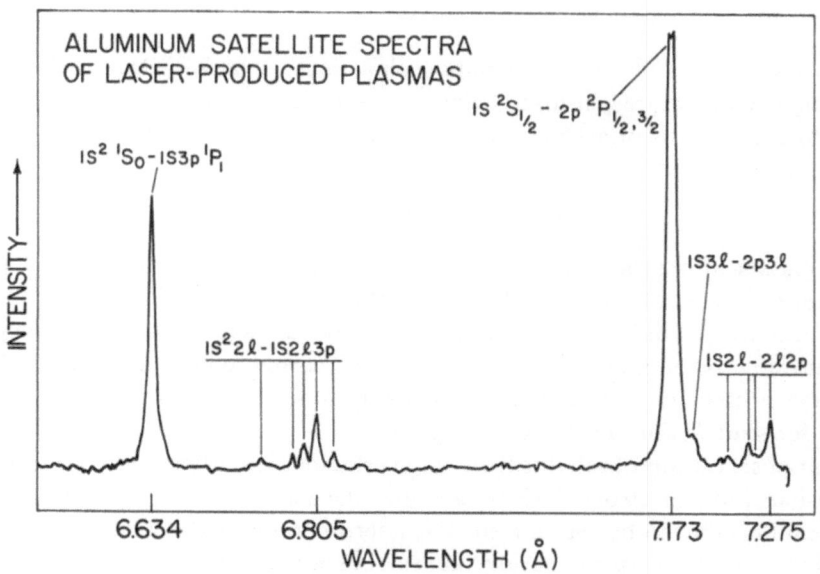

Fig. 11. Laser-produced plasma spectrum of satellite lines due to transitions of the type $1s^2 2l-$ $-1s2l3p$, and $1s2l-2p2l$. These types of satellite lines are also found in solar flare spectra.

Mg TARGET

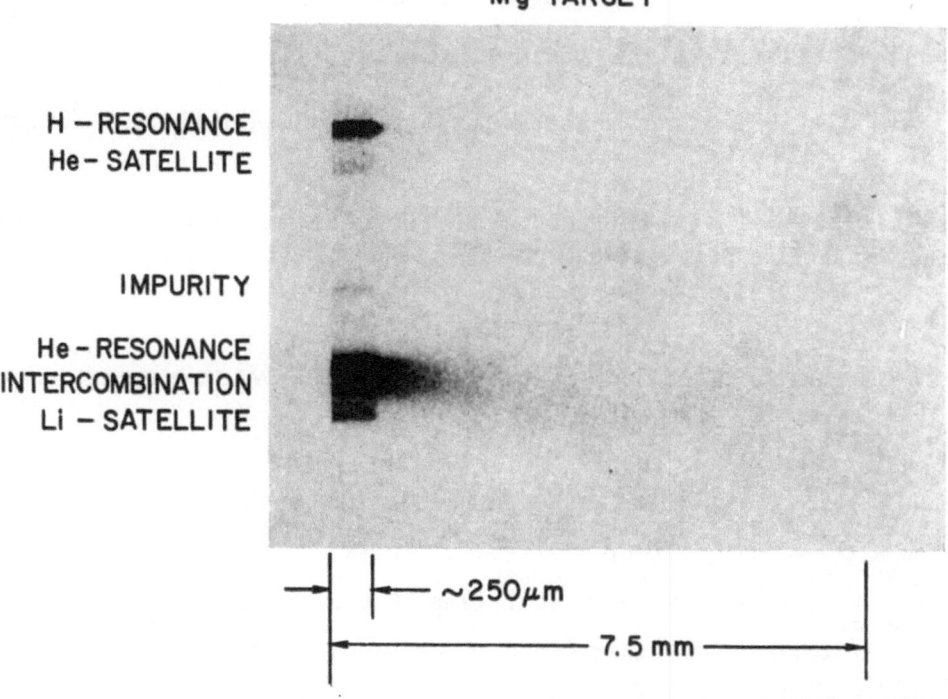

Fig. 12. Laser-produced plasma pinhole-type spectrum of hydrogen-like, helium-like, and lithium-like magnesium obtained from a laser-produced plasma. Emission from satellite lines is confined near the target surface. Emission from the other lines is observed as much as 7 mm from the target surface. Compare with Figure 9. (Spectrum courtesy of P. G. Burkhalter.)

It is from the analysis of laboratory data such as in Figures 7–12 that the atomic physics in high temperature plasmas is clarified; and can therefore be applied with confidence to the analysis of astrophysical plasmas.

IV. Conclusions

The recent classifications of iron lines in X-ray and EUV laboratory spectra, and recent ab initio wavelength and oscillator strength predictions for iron lines in the X-ray region, make possible new experiments in solar flare spectroscopy that enable the temperature-emission measure distribution of the plasma to be determined in flare plasmas. Line profiles can be measured for highly ionized iron lines in the 100 Å region, and for similar lines in the X-ray region.

From solar observations of satellite lines of the principal series lines of hydrogen-like and helium-like ions, electron temperatures and departures of flare plasmas from ionization equilibrium can be determined. The interpretation of the solar satellite line observations are made possible in part by analysis of spectra obtained from laboratory plasmas. In overall phenomenology, laboratory plasmas show a striking resemblance to solar flare plasmas.

Acknowledgements

I would like to thank P. G. Burkhalter, B. C. Fawcett, B. S. Fraenkel, S. O. Kastner, T. N. Lee, and N. J. Peacock for kindly sending me examples of their data.

I would also like to acknowledge helpful discussions with Uri Feldman, J. Meekins, and R. W. Kreplin.

Note added in proof: The anomalous satellite line intensities have been explained by a revision of the theoretical line intensities by Bhalla and Gabriel (1974).

References

Bhalla, C. P. and Gabriel, A. H.: 1974, *Proceedings of IAU Colloquium No. 27*, held at Harvard University, 9–11 September 1974.
Burgess, D. D.: 1972, *Space Sci. Rev.* **13**, 493.
Burkhalter, P. G.: 1974, private communication.
Cohen, L. and Feldman, U.: 1970, *Astrophys. J. Letters* **160**, L105.
Cohen, L., Feldman, U., Swartz, M., and Underwood, J. H.: 1968, *J. Opt. Soc. Am.* **58**, 843.
Connerade, J. P., Peacock, N. J., and Speer, R. J.: 1970, *Solar Phys.* **14**, 159.
Cowan, R. D. and Widing, K. G.: 1973, *Astrophys. J.* **180**, 285.
Doschek, G. A.: 1972, *Space Sci. Rev.* **13**, 765.
Doschek, G. A., Meekins, J. F., Kreplin, R. W., Chubb, T. A., and Friedman, H.: 1971, *Astrophys. J.* **170**, 573.
Doschek, G. A., Meekins, J. F., and Cowan, R. D.: 1972, *Astrophys. J.* **177**, 261.
Doschek, G. A., Meekins, J. F., and Cowan, R. D.: 1973, *Solar Phys.* **29**, 125.
Doschek, G. A. and Meekins, J. F.: 1973, in R. Ramaty and R. G. Stone (eds.), *High Energy Phenomena on the Sun*, NASA SP-342, p. 262.
Doschek, G. A., Feldman, U., Cowan, R. D., and Cohen, Leonard: 1974a, *Astrophys. J.* **188**, 417.
Doschek, G. A., Feldman, U., and Cohen, L.: 1974b, *J. Opt. Soc. Am.*, in press.

Edlén, B. and Tyrén, F.: 1939, *Nature* **143**, 940.

Elton, R. C.: 1974, private communication.

Elton, R. C. and Lee, T. N.: 1972, *Space Sci. Rev.* **13**, 747.

Fawcett, B. C.: 1971, special publication, S.R.C. Astrophysics Research Unit, Culham Laboratory, England.

Fawcett, B. C. 1973, preprint: Review paper on The Identification of Emission Lines of Highly Ionized Atoms in the Solar Spectrum, to be published in *Advances in Atomic and Molec. Physics*.

Fawcett, B. C., Cowan, R. D., and Hayes, R. W.: 1974a, *Astrophys. J.* **187**, 377.

Fawcett, B. C., Galanti, M., and Peacock, N. J.: 1974b, *J. Phys. B* **7**, 1149.

Feldman, U. and Cohen, L.: 1968, *Astrophys. J. Letters* **151**, L55.

Feldman, U., Doschek, G. A., Nagel, D. J., Behring, W. E., and Cohen, Leonard: 1973a, *Astrophys. J. Letters* **183**, L43.

Feldman, U., Doschek, G. A., Cowan, R. D., and Cohen, Leonard: 1973b, *J. Opt. Soc. Am.* **63**, 1445.

Feldman, U., Doschek, G. A., Nagel, D. J., Behring, W. E., and Cowan, R. D.: 1974a, *Astrophys. J.* **187**, 417.

Feldman, U., Doschek, G. A., Cowan, R. D., and Cohen, Leonard: 1974b, *Astrophys. J.*, in press.

Feldman, U., Doschek, G. A., Nagel, D. J., Cowan, R. D., and Whitlock, R. R.: 1974c, *Astrophys. J.* **192**, 213.

Gabriel, A. H.: 1972, *Monthly Notices Roy. Astron. Soc.* **160**, 99.

Gabriel, A. H. and Jordan, C.: 1969a, *Nature* **221**, 947.

Gabriel, A. H. and Jordan, C.: 1969b, *Monthly Notices Roy. Astron. Soc.* **145**, 241.

Goldsmith, S.: 1969, *J. Phys. B* **2**, 1075.

Grineva, Yu. I., Karev, V. I., Korneev, V. V., Krutov, V. V., Mandelstam, S. L., Vainstein, L. A., Vasilyev, D. N., and Zhitnik, I. A.: 1973, *Solar Phys.* **29**, 441.

Holz, E. Ya., Kononov, E. Ya., Mandelstam, S. L., Sidel'nikov, Yu. V., Zitnik, I. A.: 1974, paper presented at the XVII Plenary Meeting of COSPAR, June-July 1974, Sao Paulo, Brazil.

Jordan, C.: 1969, *Monthly Notices Roy. Astron. Soc.* **142**, 501.

Jordan, C.: 1970, *Monthly Notices Roy. Astron. Soc.* **148**, 17.

Kane, S. R.: 1974, in Gordon Newkirk, Jr. (ed.), 'Coronal Disturbances', *IAU Symp.* **57**, 105.

Kastner, S. O., Neupert, W. M., and Swartz, M.: 1974, *Astrophys. J.* **191**, 261.

Krieger, A. S., Chase, R. C., Gerassimenko, M., Kahler, S. W., Timothy, A. F., and Vaiana, G. S.: 1975, This volume, p. 103.

Lee, T. N.: 1974, *Astrophys. J.* **190**, 467.

Lee, T. N. and Elton, R. C.: 1971, *Phys. Rev. A* **3**, 865.

Margalit, G., Goldsmith, S., and Feldman, U.: 1974, preprint.

Meekins, J. F., Kreplin, R. W., Chubb, T. A., and Friedman, H.: 1968, *Science* **162**, 891.

Meekins, J. F., Doschek, G. A., Friedman, H., Chubb, T. A., and Kreplin, R. W.: 1970, *Solar Phys.* **13**, 198.

Neupert, W. M.: 1971, *Solar Phys.* **18**, 474.

Neupert, W. M., Gates, W. J., Swartz, M., and Young, R.: 1967, *Astrophys. J. Letters* **149**, L79.

Neupert, W. M., Swartz, M., and Kastner, S. O.: 1973, *Solar Phys.* **31**, 171.

Parkinson, J. H.: 1975, This volume, p. 45.

Peacock, N. J., Hobby, M. G., and Morgan, P. D.: 1971, *Plasma Physics and Controlled Fusion*, Vol. 1, IAEA, Vienna, p. 537.

Peacock, N. J., Hobby, M. G., and Galanti, M.: 1973, *J. Phys. B* **6**, L298.

Schwob, J. L. and Fraenkel, B. S.: 1972, *Space Sci. Rev.* **13**, 589.

Walker, A. B. C., Jr. and Rugge, H. R.: 1971, *Astrophys. J.* **164**, 181.

Widing, K. G.: 1975, This volume, p. 153.

ASSOCIATION OF X-RAY FLARES WITH SOLAR CORONAL ACTIVE REGIONS

P. R. SENGUPTA

Institute of Applied Manpower Research New Delhi-1, India

Summary. It is known that the flare component of solar X-ray emission is sensitive to the level of solar activity. The location of the X-ray flaring region is not directly known but can be inferred from the location of the associated Hα flare. A detailed study of more than 4000 solar X-ray flares recorded by UI and the NRL detectors during past eight years has shown that 85% of these flares definitely occurred in the active regions. For the rest 15% no definite conclusion was possible because either Hα flare data was not available or no Hα flare was reported within ± 15 min of these X-ray flares.

An analysis of soft X-ray flares associated with three active regions, viz. McMath Regions 10607, 10618 and 11128, which occurred in February–March, 1970 and January, 1971, has led to the following conclusions:

(i) X-ray flares generally occur in the active regions.

(ii) The frequency of occurrence, intensity and the spectral hardness of the soft X-ray flare emission are related to solar activity level and are consistent with the index of activity derived by Sengupta (1971, 1974).

(iii) Flare X-ray emission increases with the activity level and is maximum during the final phase of the growth. Flare emission declines markedly during the decay phase of the activity.

(iv) Eruptive flares and hard X-ray bursts generally occur during the final phase of the growth.

References

Sengupta, P. R.: 1971, 'A Study of Solar X-ray Emission and its Effect on the Earth's Ionosphere', D.Sc. Thesis, Department of Electronics, I.I.T., Kharagpur, India.
Sengupta, P. R.: 1974, *Space Res.* **14**, 461.

STUDIES OF THE DYNAMIC STRUCTURE AND SPECTRA OF SOLAR X-RAY FLARES

S. W. KAHLER, A. S. KRIEGER, J. K. SILK, and R. W. SIMON,

American Science and Engineering, Cambridge, Mass., U.S.A.

A. F. TIMOTHY

NASA Headquarters

and

G. VAIANA

Center for Astrophysics, Harvard College Observatory – Smithsonian Astrophysical Observatory, Cambridge, Mass., U.S.A.

Summary. Data from the AS & E X-ray spectrographic telescope on Skylab were used for a preliminary study of two solar flares on 1973, August 9 and September 5. Photographic images taken through broad band filters during the early flare onsets and subsequent evolution of the flares were analyzed by microdensitometry and subsequent conversion to energy maps of the flare areas.

Three exposures of 4, 16, and 64 s taken of McMath plage region 12474 were used to study the preflare configuration of the August 9 flare. Although all three images were taken before the onset of the flare as determined by Solrad, microwave, and Hα reports, a significant X-ray flux enhancement is apparent in a bright coronal X-ray spot. The center of the flux increase coincides with the center of the X-ray spot and has a linear dimension first of 6″–8″ as determined from the 4 and 16 s exposures and later of 10″–12″ as determined from the 16 and 64 s exposures.

The flare of September 5 occurred in one of three bright X-ray spots surrounding a large negative sunspot. Two frames taken after the onset of type III bursts but before the X-ray onset as seen in the Solrad data indicated a flux increase in a region with dimensions of about 8″ × 20″. During the rising phase of the flare event the shape of the emitting region was changing with X-ray fluxes increasing fastest in the northwest part of the flare where a small appendage to the main flare region appeared. A comparison of preflare and postflare images indicated that the X-ray spot in which the flare occurred nearly disappeared after the flare, but a second spot which did not flare showed enhanced intensity.

Future work will relate the positions of the flares and flare onsets to the pre-flare X-ray structures, the magnetic fields, and the Hα features of the active regions. The photographic images will also be used to derive flare volumes and rates of change of the X-ray flux in various regions of the flares.

Sharad R. Kane (ed.), Solar Gamma-, X-, and EUV Radiation, 185. All Rights Reserved.
Copyright © 1975 by the IAU.

ON THE THERMAL STRUCTURE OF THE
FLARE-PRODUCED PLASMA

IAN CRAIG

Department of Physics and Astronomy, University College London, England

Summary. Recent flare studies have shown that soft X-ray data are not compatible with simple isothermal models of the source (Herring and Craig, 1973; Craig, 1973; Neupert *et al.*, 1973). With this in mind, the emitting flare plasma has been represented by the temperature-emission measure distribution function,

$$\zeta(T)\,dT = \left(\sum_{i=1}^{n} A_i T^{-\alpha_i}\right) dT \quad \text{for} \quad T \geqslant T_0, \tag{1}$$

where $\zeta(T)$ is the differential emission measure (cm^{-3} per 10^6 K), T is the electron temperature in units of 10^6 K, T_0 is a low temperature cut-off for the distribution, α_i are real positive numbers, and A_i are positive coefficients determined from data (for appropriate values of T_0 and α_i) by a least squares fitting procedure. Such a distribution is suggested by results obtained by the present author using simple delta-function representations for $\zeta(T)$ (with $n \leqslant 4$); these discreet multi-temperature models usually indicate that the emission measure decreases with increasing temperature. Also, as discussed by Brown (1974), a power law distribution for $\zeta(T)$ is consistent with the observed bremsstrahlung emission in the hard X-ray (> 10 keV) domain. In attempting to find a suitable form for the differential emission measure, a simple empirical function of the type assumed by Chambe (1971) for active regions was also tried, but the fit, as evidenced by the χ^2 test was unsatisfactory.

The present analysis indicates that the flare regions above about 3×10^6 K may be described adequately (as defined by the χ^2 test), though not uniquely, by a single power law for $\zeta(T)$; viz. $\zeta(T) = AT^{-\alpha}$ for $T > 3 \times 10^6$ K. Other terms in the expansion for $\zeta(T)$ may be neglected above this temperature. Figure 1 shows the differential emission measure for three points in a flare previously discussed by Craig (1973). Note the tendency for α to increase as the spectrum softens during the decay phase of the flare.

In considering the decay of the high temperature flare regions, it has been assumed that the dominant energy loss mechanisms are radiation and conduction. It is further assumed that the hot region has an approximately cylindrical structure, and that, at any instant, constant pressure prevails throughout the source. Thus $N_e(z)\,T(z) = P$ is constant, where $N_e(z)$ (cm^{-3}) is the electron density at some position z (cm) in the source, and $T(z)$ (10^6 K) is the associated local temperature (cf. Brown, 1974). Under the above conditions, the ratio of the conductive energy flux to the radiative energy loss, for unit volume of plasma, at each point in the tube is given by

$$R(T) = \left|\frac{\nabla \cdot \mathbf{F}_{\text{cond}}}{\nabla \cdot \mathbf{F}_{\text{rad}}}\right| = 9.2 \times 10^{14} \frac{S^2 P^2}{A^2 f(T)} \left(\alpha + \tfrac{1}{2}\right) T^{2\alpha - 1/2}, \tag{2}$$

Sharad R. Kane (ed.), Solar Gamma-, X-, and EUV Radiation, 187–189. All Rights Reserved.

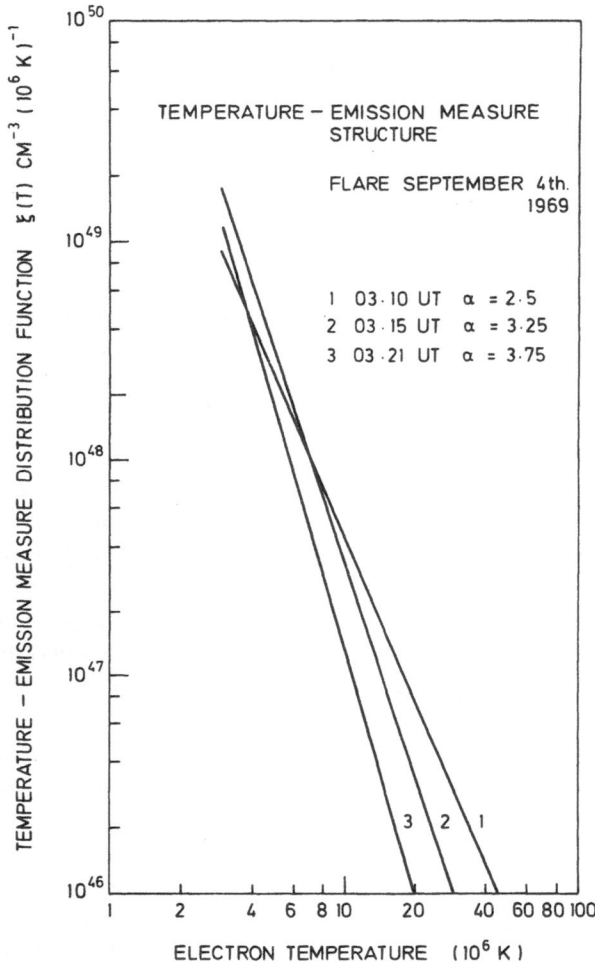

Fig. 1.   The temperature structure at three instants during the decay phase of a flare
on 1969, September 4.

where **F** represents the appropriate energy flux vector, $f(T)$ is the radiative energy loss function calculated by Tucker and Koren (1971), and $S(\text{cm}^2)$ is the cross-sectional area of the filament.

Inserting typical values for the parameters in the above equation ($\alpha = 3$, $A = 10^{51}$, $f(T) = 10^{-23}$, $S = 10^{17}$ and $P = 10^{12}$) yields $R(T = 10) \simeq 1$, which implies that conductive effects will dominate strongly for those regions of the plasma above 10^7 K. This preliminary analysis suggests that both radiative and conductive mechanisms are required to explain the decay of the multithermal plasma. However, bearing in mind that conduction is an energy redistribution process, radiation provides the dominant energy loss mechanism over the temperature structure ($> 3 \times 10^6$ K) as a whole.

References

Brown, J. C.: 1974, in Gordon Newkirk, Jr. (ed.), 'Coronal Disturbances', *IAU Symp.* **57**, 395.
Chambe, G.: 1971, *Astron. Astrophys.* **12**, 210.
Craig, I. J. D.: 1973, *Solar Phys.* **31**, 197.
Herring, J. R. H. and Craig, I. J. D.: 1973, *Solar Phys.* **28**, 169.
Neupert, W. M., Swartz, M., and Kestner, S. O.: 1973, *Solar Phys.* **31**, 171.
Tucker, W. H. and Koren, M.: 1971, *Astrophys. J.* **168**, 283.

THE RELATIONSHIP BETWEEN HARD AND SOFT X-RAY
BURSTS OBSERVED BY OSO 7

DAYTON W. DATLOWE

University of California, San Diego, La Jolla, Calif., U.S.A.

Abstract. Solar X-rays in the energy range 1–100 keV originate in hot plasmas and streams of energetic electrons in solar flares, and since these phenomena may represent a significant fraction of the energy in a flare, an understanding of them is important for any flare theory. This paper presents the results of the University of California, San Diego, solar X-ray instrument on the OSO-7 satellite. Study of the time evolution of the emission measure in a typical burst indicates that the growth of soft X-ray emission is due to the addition of new hot material to the flare plasma, and the study of the time evolution of the temperature of the plasma indicates that conduction is the dominant cooling mechanism. Comparison of the hard (10–100 keV) and soft (5–10 keV) data indicates that the main heat input to the flare plasma is not collisions by the electrons which make the hard X-rays. The fraction of soft X-ray bursts observed by the instrument which also have a detectable hard X-ray component is $\frac{2}{3}$; this result is the same for bursts which occured near the center of the disk ($\theta < 60°$) and for those bursts believed to have been partly occulted by the limb, indicating that hard X-ray emission comes at least part from high in the corona. For a sample of 62 hard X-ray bursts which occurred near or beyond the limb, the spectral index of the hard X-ray power law was significantly larger, as compared with the spectra of a comparable number which occurred at solar longitudes less than 60°.

1. Introduction

This paper is a review of the observations made by the University of California, San Diego, solar X-ray instrument on the OSO-7 satellite. The instrument is the first to observe simultaneously both the hard and soft components of an X-ray burst in separate detector systems; thus it is uniquely suited to examine the relationship between the two spectral components. Solar X-rays give information about hot plasmas and streams of energetic electrons in solar flares, and since the amount of energy in these phenomena are comparable to the total energy of the flare, understanding their origin is of considerable importance for any flare theory. This paper will explore the relationship between the hard and soft X-ray components and what can be learned about the physics of solar flares from a detailed comparison of them.

II. Soft X-Rays

Soft X-rays in the 4–10 keV range observed by the instrument come from hot (10^7 K) plasmas associated with solar flares. Previous measurements of the soft X-ray continuum have been described in White (1964), Neupert (1967), Hudson *et al.* (1969), Culhane and Phillips (1970), Teske (1971), Kahler *et al.* (1970) and Valnicek *et al.* (1973). The only previous systematic study of a number of bursts was by Horan (1971), uing 17 large bursts observed by a broad band counter on the OSO-4 satellite. Little observational effort has been directed toward the statistical study of the temperatures and emission measures of bursts, nor has much effort been directed toward direct correlative observations with hard X-ray bursts. The OSO-7 observations have

Sharad R. Kane (ed.), Solar Gamma-, X-, and EUV Radiation, 191–208. All Rights Reserved.
Copyright © 1975 by the IAU.

made possible the first systematic study of a large number of bursts; the set contains 197 events which occurred between October, 1971, and June, 1972, over which time about 40% data coverage was obtained.

The OSO-7 soft X-ray detector (Datlowe *et al.*, 1974a) consists of a 2.54 cm diameter proportional counter with a $Xe(CO_2)$ fill. Pulses from the detector are analyzed into channels approximately 1.4 keV wide. A thermal spectrum is fitted to the channels from 5.1–10.6 keV to determine a temperature and emission measure, such that when the spectrum is folded through the detector response the best fit to the pulse height distribution is obtained. The thermal spectrum is approximated by the formula of Culhane and Acton (1970), which takes into account both free-free and free-bound emission; line emission, particularly from iron at 6.65 keV, is neglected. The estimated total systematic error in the temperature is 2×10^6 K, and in the emission measure a factor of 2.

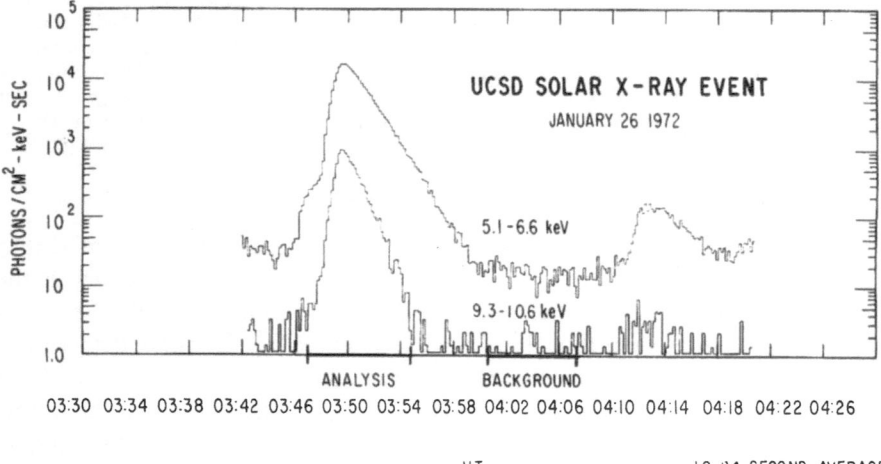

Fig. 1.   Soft X-ray flux from a $-$N flare at 0346 UT on 1972, January 26. The two energy channels shown are 5.1–6.6 and 9.3–10.6 keV. The time over which the analysis was performed and the time used to estimate the background are indicated by heavy lines at the bottom of the figure. Each point represents 10.24 s of data.

A typical burst observed by the OSO-7 solar X-ray instrument is shown in Figure 1. The event occurred in 1972, January 26 at 0348 UT and was associated with a $-$N subflare. The figure shows the 5.1–6.6 keV and 9.3 to 10.6 keV channels. The flux shows a rapid rise, lasting for about 2 min, and a slow decay lasting for about 10 min. The time intervals used for the spectral analysis and for the background estimate are indicated by heavy lines at the bottom of the figure.

Although solar X-ray bursts are sometimes characterized by a single temperature and emission measure pair, these quantities evolve throughout the course of an event. Figure 2 shows this evolution for the sample burst, where each point represents a

30.72 s average. The maximum temperature, 15×10^6 K, was achieved early in the burst and thereafter the temperature declined slowly. The emission measure grew exponentially for about 2 min, reaching a maximum value of 10^{49} cm^{-3}. For an assumed density of 5×10^{10} cm^{-3} (Hudson and Ohki, 1972), the thermal energy in the hot flare plasma is 1×10^{30} erg.

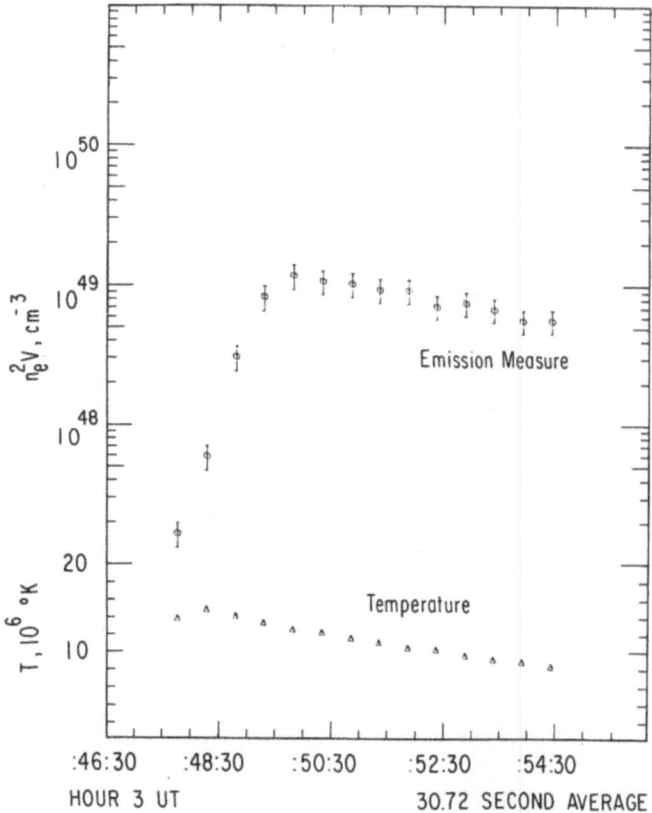

Fig. 2.   The temperature and emission measure for the burst shown in Figure 1. The temperature (triangles) is plotted on a linear scale and the emission measure (circles) is plotted on a logarithmic scale. Error bars shown for the emission measure represent an assumed 20% systematic error, as well as the statistical error.

This kind of analysis was carried out systematically for all bursts during the observing period, for which the flux in the 5.1–6.6 keV channel exceeded 10^3 photons (cm^2 s keV)$^{-1}$ and for which there was sufficient data coverage. In all, 197 bursts were analyzed, 17 associated with class 1 flares, 135 associated with subflares, and the remainder were unreported in *Solar Geophysical Data*. The evolution of the emission measure was largely the same in each case. There is a rising phase which is quasi-exponential and which lasts for a few minutes. The distribution of risetimes is shown in Figure 3c, showing that half of the risetimes are between 30 and 100 s. The figure

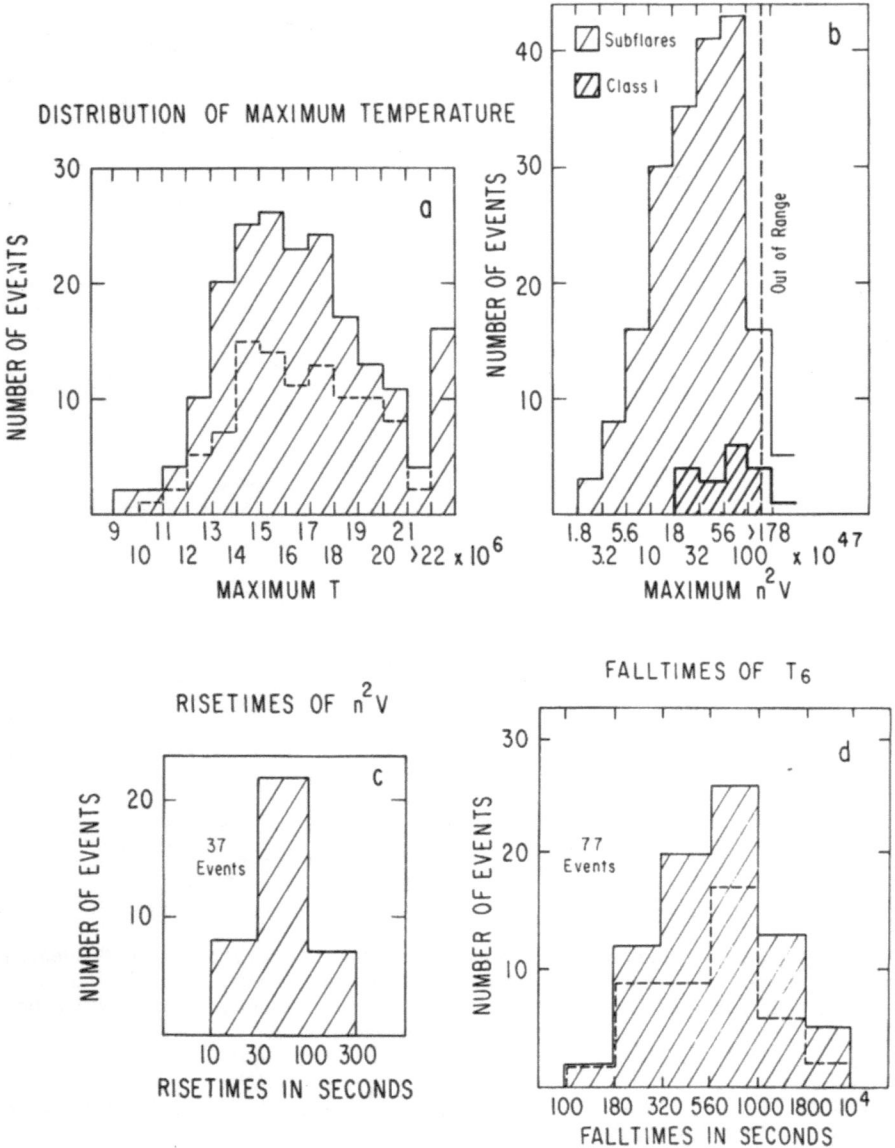

Fig. 3.   Distributions of parameters characterizing soft X-ray events from 197 examples. (a) Distribution of maximum temperature for an individual burst. (b) Distribution of maximum emission
measure. Importance I flares are shown separately, and instrument saturation causes the cutoff for
large bursts. (c) Distribution of risetimes of emission measure for the 37 cases in which an exponential growth rate gave a good fit to the data. (d) Distribution of falltimes of temperature for 77 subflares in which an exponential decay gave a good fit to the data. Events below the dotted lines in (a)
and (d) are associated with analyzed hard X-ray bursts.

contains only 37 events because a good fit to an exponential was required in preparing the distribution; including events with deviations from an exponential would produce a similar distribution. The median value of the peak emission measure is 30×10^{47} cm^{-3}, and for the bursts in this sample the maximum emission measure usually fell between 10 and 100×10^{47} cm^{-3}. The median temperature at the time of peak emission measure was 11×10^6 K.

The temperature evolution is somewhat different. The maximum temperature is reached early in the event; frequently the observations are consistent with $dT/dt < 0$ throughout the event. The median decay time for the temperature is 600 s, and the distribution of decay times is shown in Figure 3d. The distribution of maximum temperatures is shown in Figure 3a. This latter distribution is relatively flat for temperatures of $13 \leqslant T_{max} \leqslant 19 \times 10^6$ K, and the median value is 17×10^6 K. Since the time of maximum temperature is early in the burst, the emission measure at that time is much smaller than the peak value, and in most of these events it lies between 10^{47} cm^{-3} and 5×10^{47} cm^{-3}.

Inference about the cooling mechanism of the hot flare plasma can be made from a study of the cooling time. The two principal cooling mechanisms (Culhane *et al.*, 1970; Roy and Datlowe, 1974) which must be considered are radiation and conduction. These possibilities can be examined on the assumption that the heat flow is determined by changes in temperature only; this is probably valid during the decay phase but, as well be shown later, cannot be valid during the rising phase of the burst. For free-free radiative cooling, the decay time is related to the temperature by $\tau \propto T^{1/2}$. Free-bound and line emission can be taken into account (Culhane *et al.*, 1970) but this does not make a significant change in the numerical value of the exponent in the present context. Conductive cooling on the other hand predicts a relationship between the cooling time and the temperature of the form $\tau \propto T^{-2.5}$. It can be seen that there should be a correlation between the temperature and the cooling time, and that a positive correlation would favor radiative cooling while an anticorrelation would favor conductive cooling. This correlation has been tested using both T_{max} and the average temperature over the time interval used to determine the cooling time. The result is that the sign of the correlation is negative although it is not a strong correlation. This favors conduction as the dominant cooling mechanism in these subflares.

Many of the soft X-ray bursts were accompanied by hard X-ray bursts simultaneously observed by the instrument. To see if there is a correlation between the soft X-ray parameters and the presence or absence of an accompanying hard X-ray burst, Figure 3a shows the maximum temperatures for bursts with a hard X-ray component *below* the dotted line. The distribution is substantially the same for bursts above and below the line. For emission measures, similar results are obtained except that a greater fraction of bursts with large emission measures have hard components, possibly a threshold effect. In general, when the distribution of soft X-ray parameters for those bursts with a hard X-ray component is compared to the same distribution for those in which no hard component was detected, there is no important difference.

These observations place a number of constraints on the physics of the hot flare plasma and its heating mechanism. The amount of energy stored in the hot plasma may be comparable to the total energy of the flare and thus its origin is very important for flare theory. The X-ray data make it possible to estimate the energy content of the plasma and to observe its time variations.

The key experimental facts are that the temperature and emission measure appear to evolve independently, inasmuch as the maximum of one is not reflected in the maximum of the other, and that the emission measure typically grows by a factor of 100 throughout a burst while the temperature steadily declines. The following are possible models for the evolution of the plasma:

(1) A fixed amount of plasma is directly heated by nonthermal electrons. The emission measure stays constant and the temperature increases in proportion to the energy input.

(2) A constant amount of plasma is compressed by a factor of 100. In this case the number of particles nV stays the same but the emission measure n^2V increases.

(3) New plasma is injected into a relatively constant volume, such as a pre-existing arch system. The amount of matter in the volume nV, and the plasma thermal energy $3\,nVkT$ each increase by a factor of 10, but the emission measure n^2V increases by a factor of 100.

(4) The volume of the emission region and the amount of matter and thermal energy in it grow by a factor of 100, while the density stays relatively constant. This might occur if successive flux tubes are heated to temperatures of 10^7 K but no mass motion occurs.

Case number 1 can be eliminated immediately, since it predicts precisely the converse of the observed evolutions. The independence of the temperature and the emission measure evolution rules out case 2; during the growth phase the heat of compression must be removed from the plasma, since the temperature is slowly falling. Once the compression is complete the same cooling mechanism would cause a rapid temperature drop in the plasma, which is not observed. In fact, the transition from the growth phase of the emission measure to the decay phase is not detectable in the temperature evolution.

Cases 3 and 4 are both consistent with the OSO-7 observations. In both cases new plasma is added to the soft X-ray emitting region. While the two possibilities cannot be distinguished on the basis of the present data alone, it should be noted that case 3 (constant volume) involves the addition of significantly less energy to the plasma than case 4 (constant density). Spatially resolved or imaging soft X-ray experiments can distinguish between the two models.

A second consequence of the independence of the temperature and emission measure evolution is that heat flows cannot be inferred from temperature changes alone. The new matter which is added to the flare plasma brings in heat, while from the declining temperature one might infer that heat is flowing out of the region. Thus the question of conductive cooling versus radiative cooling must be replaced by a realistic model of the total thermal evolution of the plasma.

III. Hard X-Rays

Hard solar X-rays come from bremsstrahlung due to collisions of 10–100 keV electrons with the solar atmosphere. The spectrum of this component is characteristically a power law, and its duration is typically 1 minute for bursts associated with subflares, although variability in the flux on time scales of the order of 1 second has been observed (Frost, 1969). Several satellite experiments have observed hard solar X-radiation since its discovery by Peterson and Winckler (1959); a good historical summary will be found in Kane (1974). The OSO-7 observations (Datlowe *et al.*, 1974b) have provided the first systematic study of hard solar X-ray burst morphology.

The hard X-ray detector on the OSO-7 solar X-ray instrument consists of a 9.57 cm^2 Na I(T1) scintillation counter which is actively shielded by a Cs I (Na) collimator. The entrance window consists of 0.041 g cm^{-2} of aluminum. Pulses from the detector are analyzed into 9 logarithmically spaced channels from 10.6–323 keV. In the analysis a power law X-ray spectrum is fitted to the hard X-ray pulse height distribution in the same way that a thermal spectrum is fitted to the soft X-ray data. Since the photons which the instrument observes are typically at energies of 20 keV, spectra are normally referred to that energy, given in the form of $A(E/20)^{-\gamma}$ photons (cm^2 s keV)$^{-1}$; normalization at 1 keV is sensitive to small errors in the slope and gives less reliable results.

The behavior of a typical event is shown in Figure 4, which gives the soft

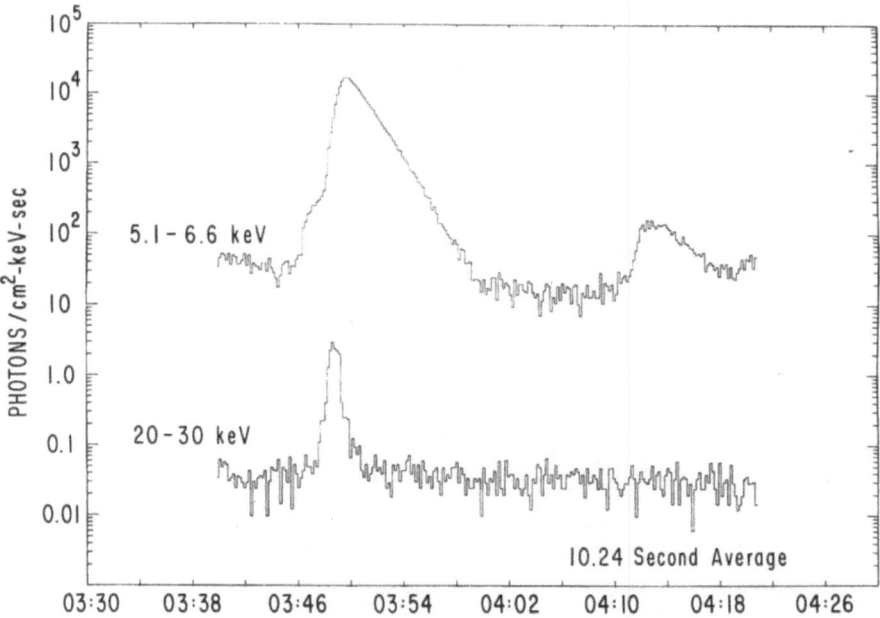

Fig. 4.   Hard and soft X-ray fluxes from the sample event, Figure 1. The lower trace gives the 20–30 keV channel of the Na I (T1) detector, representative of the hard X-ray flux. For comparison, the 5.1–6.6 keV channel of the soft X-ray detector is shown again in the upper trace. Hard X-ray analysis was carried out from 034731 to 034924 UT.

DAYTON W. DATLOWE

(5.1–6.6 keV) and hard (20–30 keV) fluxes for the 1972, January 26 sample event. The soft and hard X-ray components have comparable time histories during the rising phase, with the greatest flux increase in both channels occurring in the same data sample. The hard X-ray flux peaks earlier and dies away rapidly compared to the soft X-ray flux. This figure shows clearly the ability of the OSO-7 solar X-ray instrument to separate the hard and soft components of an X-ray burst. The evolutions of the spectral parameters A and γ are shown in Figure 5. Each data point represents a 10.24 s average, the maximum time resolution of the instrument. The maximum value of the 20 keV flux A was ≈ 4 photons $(cm^2 \, s \, keV)^{-1}$ and the flux remained at this value for five data samples, 51 s. The hardest individual spectrum, $\gamma = 2.6$, occurred at 20 keV flux maximum, and the average spectral index was $\bar{\gamma} = 3.1$.

This kind of analysis has been carried out for 123 bursts. Two-thirds of the bursts observed by the proportional counter with peak fluxes greater than 10^3 photons

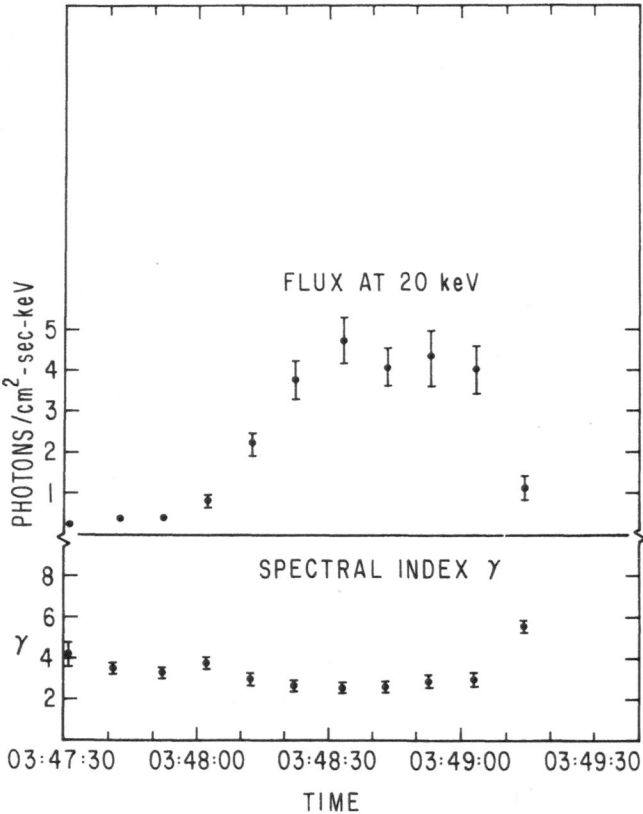

Fig. 5.   Hard X-ray emission parameters for the sample event in 1972, January 26. The spectral index γ and the flux at 20 keV are plotted on a linear scale. Statistical errors in the flux are shown, while for the spectral index the error shown is the combination of the statistical error and an assumed systematic error of $\Delta\gamma = 0.3$.

$(cm^2 \, s \, keV)^{-1}$ at 5 keV also had a detectable hard X-ray component at 20 keV. Eighty percent of these events reached a maximum flux $A_m > 0.2$ photons $(cm^2 \, s \, keV)^{-1}$ and could be analyzed, while for the remainder the flux was too small to fit a spectrum. The analysis yielded a measurement of the peak 20 keV flux, the spectral index at maximum, the average spectral index, and the duration of the burst.

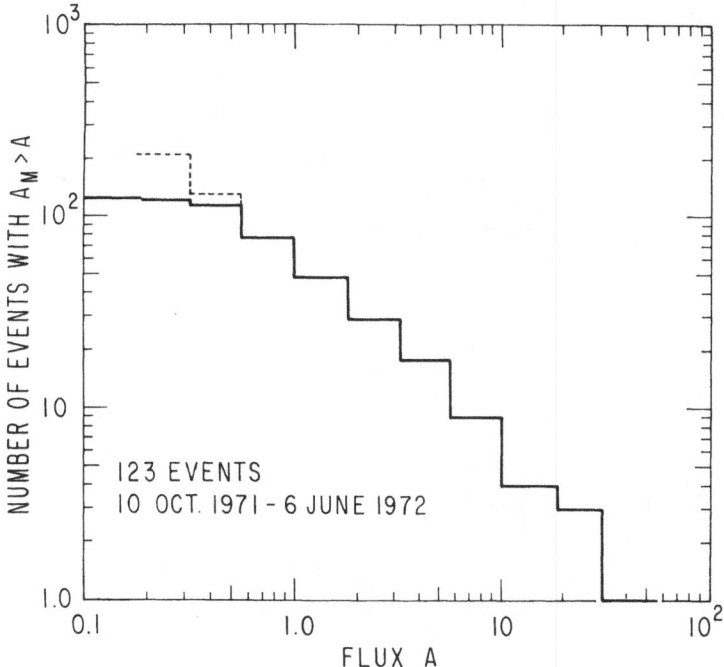

Fig. 6. Integral distribution of frequency of events vs peak hard X-ray flux. The dotted line represents an extrapolation of this distribution until N equals the number of soft X-ray bursts observed. If the distribution continued smoothly down to $A = 0.2$ photons $(cm^2 \, s \, keV)^{-1}$ then the number of hard and soft X-ray bursts in the sample would be the same.

The distribution of peak hard X-ray flux at 20 keV, A_m, is shown in Figure 6. The distribution is relatively smooth between $1 < A_m < 10$ photons $(cm^2 \, s \, keV)^{-1}$ and flattens below a flux of 0.5. This flattening appears to be a threshold effect; if this is the case, then the fraction of soft X-ray bursts with a hard X-ray component may be greater than $\frac{2}{3}$. Nonetheless, it does appear that there is a distinct class of soft X-ray bursts with large fluxes which do not exhibit any hard X-ray emission.

The quantity γ_m denotes the spectral index in the 10.24-s data sample which had the largest 20 keV flux, A_m, in a particular burst. Figure 7a shows the distribution of the frequency of occurrence of γ_m. The median value is $\gamma_m = 4.0$ and values outside the range 2.5–6.0 rarely occur. The distribution of the average spectral index $\bar{\gamma}$ computed as described in Datlowe *et al.* (1974b), is shown in Figure 7b. The median is $\gamma = 4.6$ and

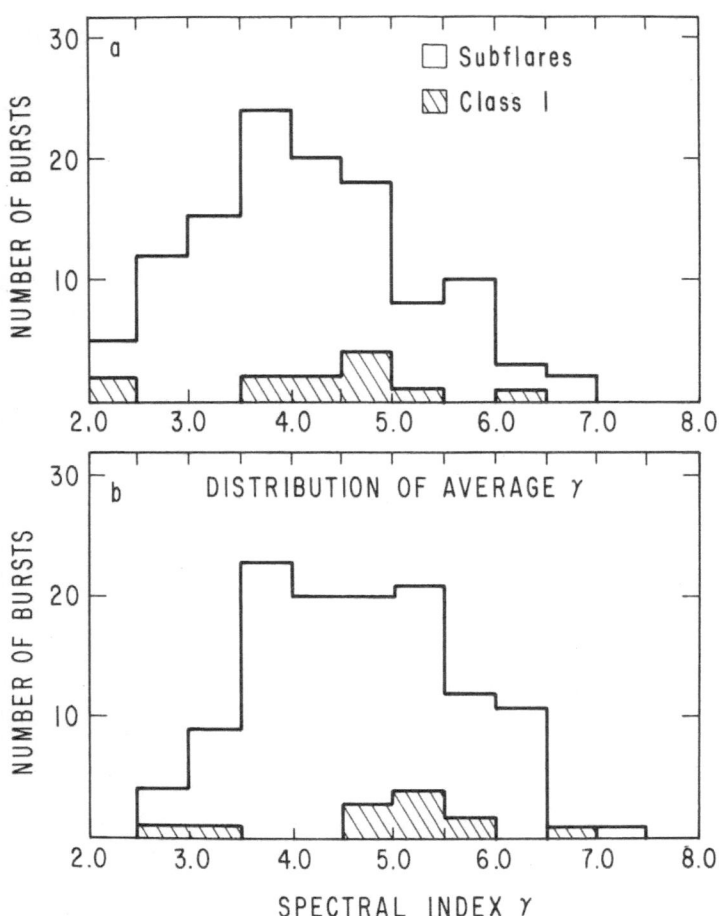

Fig. 7. Distributions of the frequency of occurrence of the spectral indices for 123 bursts. (a) The spectral index at the time of maximum 20 keV flux, γ_m. (b) Average spectral index, $\bar{\gamma}$, computed as described in Datlowe *et al.* (1974). Events associated with importance 1 flares are crosshatched.

the width of the distribution is $\sigma_\gamma = 1.0$. The two spectral indices are highly correlated and related by $\bar{\gamma} \approx \gamma_m + 0.5$.

Within a given burst, individual samples of the spectral index γ are highly correlated. Kane and Anderson (1970) found in their OGO-5 data that the spectral index tended to vary from soft to hard to soft as the burst evolved. The sample burst in Figure 5 also shows this kind of evolution. However, in the OSO-7 data a much more common case is an evolution in which the spectral index becomes progressively larger with each data sample; thus the spectrum softens or stays nearly constant throughout the course of the burst. There is no correlation between the spectral index and the flux in an individual sample using 1111 individual spectral in the OSO-7 data, the coefficient of correlation between A and γ is 0.04.

Finally, the experiment yields statistics on the durations of hard X-ray bursts, the length of time that a given burst remains above the flux threshold of 0.1 photons $(\text{cm}^2\,\text{s\,keV})^{-1}$ at 20 keV. Durations observed by the instrument range from the shortest time resolution of the instrument, 10.24 s, to 800 s. For the subflares the median duration is 92 s, while for bursts associated with class 1 flares the median is 50% longer. There is a weak correlation between burst size and duration in the sense that events with a larger peak flux A_m tend to last longer. These durations are notably longer than those of Kane and Anderson (1970) but are in agreement with the observations of Frost and Dennis (1971).

IV. Burst Energetics

From hard X-ray spectra it is possible to estimate the energy of the nonthermal electrons in the X-ray emitting region. The X-ray spectrum depends only on the instantaneous electron spectrum, $N(E)$, but the total collision losses depend on the spectrum of the accelerated electrons. In the thin target case electrons predominantly escape, so that the accelerated and instantaneous spectra are essentially the same; in the thick target case electrons lose all of their energy to collisions, and the injected spectrum $F(E)$ differs from $N(E)$. In the present analysis the formulae of Brown (1972) are used as the model for X-ray emission. Slightly different results would be obtained under different assumptions about the chemical composition of the X-ray region, about inhomogeneities in the X-ray region, and about the directivity of the radiation due to electron streaming. In all of the OSO-7 solar X-ray calculations, it is assumed that the electron spectrum is a power law above 20 keV, corresponding to the electron energies to which the experiment is sensitive; however, Kahler (1974) and Peterson et al. (1973) have given evidence that the solar electron spectrum may at times extend down to 5 keV or below.

Historically the first question in energetics has been whether or not the nonthermal electrons have enough energy to provide the heating of the thermal flare plasma. This computation depends on assumptions about the ambient density of the hot flare plasma and about the low energy cutoff of the electron spectrum. Kahler and Kreplin (1971) and McKenzie et al. (1973) compared the two energetic quantities and found that nonthermal electrons do have sufficient energy to heat the plasma. For the OSO-7 hard X-ray events, the median thermal energy is $U = 3 \times 10^{29}$ erg if a density of 5×10^{10} cm^{-3} is assumed. For a 20 keV electron spectrum cutoff, the median thin target and thick target collision losses were respectively 5×10^{27} erg and 2×10^{28} erg, which is insufficient; however, if a cutoff of 10 keV is assumed, the collision losses are increased by an order of magnitude and they become comparable to the plasma thermal energy. If, following the above mentioned authors, a 10 keV cutoff is adopted, then the nonthermal collision loss energy and the plasma thermal energy are comparable in magnitude.

Since the OSO-7 instrument simultaneously observes hard and soft X-ray spectra, it is possible to compare not only the total amounts of thermal and nonthermal energy, but also make a detailed time comparison of these forms of energy. Figure 8 gives such a comparison. The top trace in the figure gives the product of the thermal energy of the flare plasma and the ambient density, obtained from the product of the observed temperature and emission measure, $Q = 3 n_e n_i V k T = (3 NkT) n_i$. Q is taken to be representative of the thermal energy input, which as stated earlier must be due to the addition of new hot material to the soft X-ray emitting region. If this region maintains a constant volume throughout the evolution (case 3), then Q will be proportional to

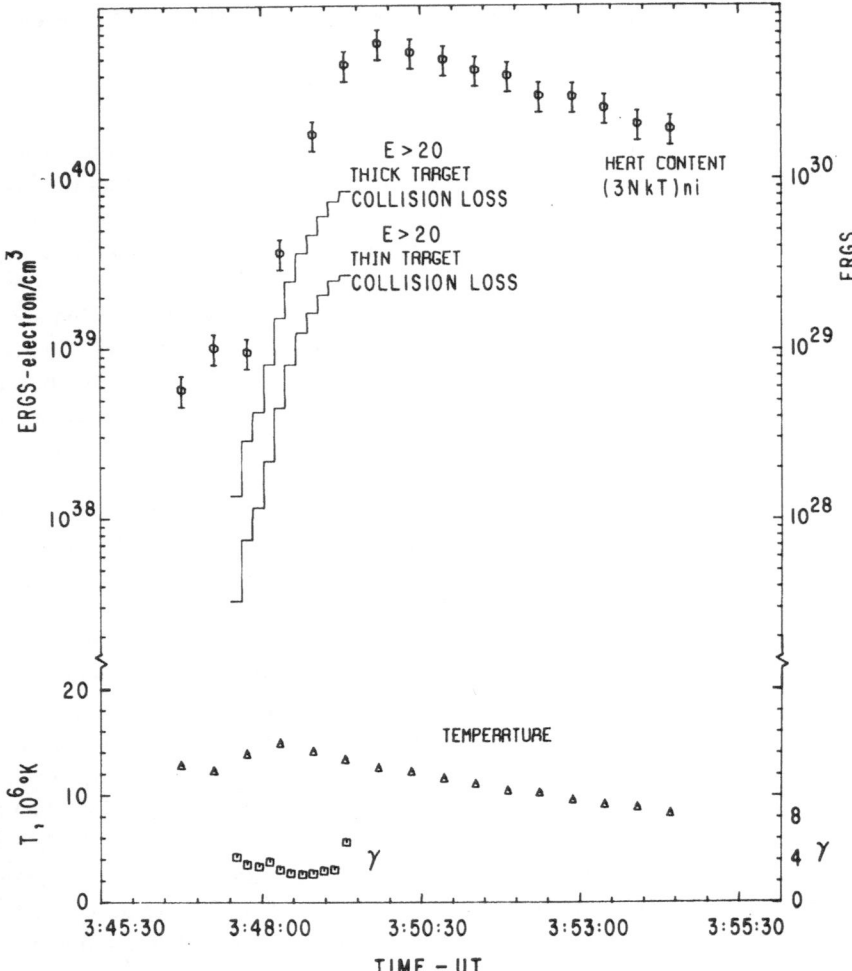

Fig. 8.   Energetics of the sample X-ray burst in 1972, January 26. The top trace gives the heat content of the thermal plasma, the product of the total thermal energy and the ambient density. The peak value of this quantity is 6×10^{40} erg electron cm^{-3}, so that if the ambient density is assumed to be 5×10^{10} cm^{-3}, then the thermal energy is 1.2×10^{29} erg. The next two traces give the cumulative thick and thin target collision losses in ergs, which are the time integrals of the thick and thin target power. The triangles give the successive measurements of the temperature of the flare plasma, and the boxes give the values of the hard X-ray spectral exponent γ. Note that the input of nonthermal energy drops below threshold 30 s before the peak in Q.

the square of the energy input, whereas for constant density (case 4) Q will be directly proportional to the thermal energy growth. The solid line traces the cumulative collision loss energy in both the thick and thin target approximations, $W = \Sigma P \Delta t$. The lowest two traces give the plasma temperature as derived from the soft X-rays and the spectral index of the hard X-rays, γ.

Figure 8 presents the comparison for the sample event of 1972, January 25. The growth of the quantity Q continues for 30 s after the end of the collisional energy input; this behavior is typical of the 123 events analyzed. This behavior is more strongly pronounced in the event of 1972, May 14 at 1504 UT, shown in Figure 9; in

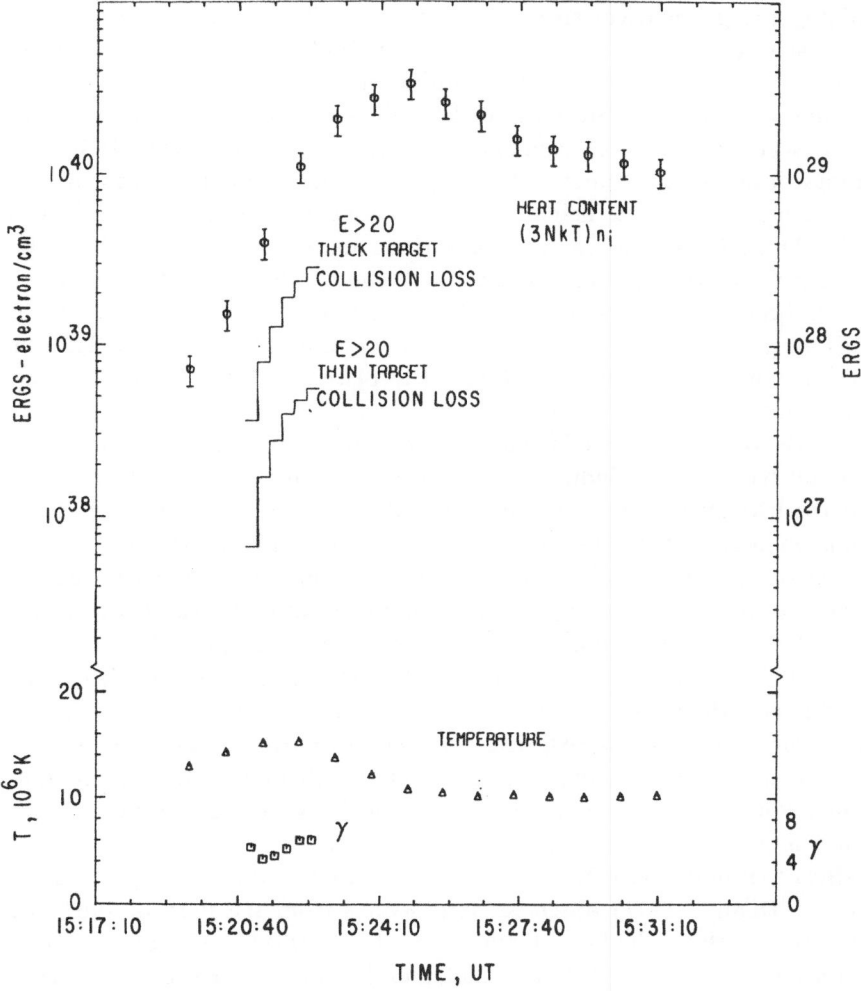

Fig. 9.   The same as Figure 8, but for the event of 1972, May 14 at 1504 UT. When the nonthermal energy input had ceased, the heat content has reached only 25% of its largest value, which occurred 60 s later.

this case the heat content Q attained only 25% of its final value when the collision losses were no longer detectable. This detailed comparison of the timing of the thermal and nonthermal energies in solar X-ray bursts indicates that the collision losses do not come at the correct time to directly energize the hot flare plasma. Thus an alternative plasma heating source is required. The existence of another energy source is also consistent with the fact that in one third of the soft X-ray bursts no hard X-rays are detected.

V. Center-to-Limb Effects

The study of longitude variations of solar X-ray emission is a powerful technique for examining the geometrical structure of X-ray emitting regions. For events on the visible disk of the sun, variations from center to limb will reveal any anisotropy which may be present in the emission. Occultation of the lower portions of a burst by the limb will give additional information about the spatial extent of the emitting regions.

In the soft X-ray region, Catalano and Van Allen (1973) successfully applied the technique of observing a burst with two spacecraft at different heliocentric longitudes: they determined the scale height for soft X-ray emission in the 2–12 Å band to be 1.1×10^4 km. Kreplin and Taylor (1971) used lunar occultation of an X-ray burst and measured its horizontal extent in the 1–8 Å band, 140000 km. Imaging experiments have also been flown (Vaiana *et al.*, 1973) to measure the sizes of soft X-ray emitting regions. In the hard X-ray region the only direct observation is that of Takakura *et al.* (1971); they used a modulation collimator to place a 1' upper limit on the size of the X-ray source.

Evidence for the height of hard X-ray emission comes from the observations of Wood and Noyes (1972) indicating the EUV radiation is well correlated in time with nonthermal X-ray emission; they identified the origin of EUV emission to be heating by collision losses of the X-ray emitting electrons. Since the spectral lines observed were characteristic of ionic species found in the transition region, they concluded that the energetic electron collision losses were taking place at that level. Because of the high density, the thick target approximation is expected to be applicable. Vorpahl (1973), Hudson (1973), and Kane and Donnelley (1971) have all given evidence to support chromospheric thick target models. In these cases we would expect that since the height of emission is below 5000 km, X-ray emission from flares more than 7° beyond the limb could reach us only by scattering off of higher layers in the solar atmosphere, and thus would be strongly suppressed. Accordingly, no burst which is well behind the limb should give detectable hard X-ray fluxes.

On the other hand, hard X-ray emission from high in the corona has been reported. Frost and Dennis (1971) observed hard X-rays from the flare of 1969, March 30 which was far behind the limb. Datlowe and Lin (1973) have reported correlated observations of X-rays, type III bursts, and interplanetary electrons, which they interpret as good evidence for escaping electrons and X-ray emission high in the corona.

Thirty-seven OSO-7 soft X-ray bursts have been identified as having come from behind the limb of the Sun. The method of identification, described in detail in Roy and Datlowe (1974), uses full-disk Hα pictures to select bursts with no accompanying optical counterpart and to track prolific bursts producing regions over the limb. Locations of over-the-limb bursts are estimated using the time of limb passage and the solar rotation rate. Of the 37 bursts, 24 (two-thirds) had a detectable hard X-ray component. This is the same fraction as exhibited by events near the center of the disk. Moreover, of the eight events with expected longitudes $\theta > 100°$, corresponding to minimum visible heights from 10^4–10^5 km, five had a detectable nonthermal component. This is entirely inconsistent with the chromospheric emission model or with any model of localized X-ray emission. At least some of the hard X-ray emission must therefore come from considerable heights in the corona, and the evidence supports the contention that the vertical distribution of this emission is similar to that of the soft X-ray emission.

For events on the solar disk, there is no center-to-limb variation in the relative frequency of either soft or hard X-ray bursts. Since the optical counterparts are used to identify the location of the X-ray bursts, the Hα longitude variation must be taken into account. To remove this effect the longitude distribution of the X-ray bursts was normalized by the longitude distribution of 2011 confirmed subflares reported during the observing period. The data were collected into three groups 0–29°, 30–59° and 60–87° longitude, with east and west summed together. No statistically significant longitude variation was found for either the hard or soft X-ray bursts. For the soft X-rays one standard deviation in each of the bins was respectively 13%, 13%, and 19%; the corresponding numbers for the hard X-rays were 19%, 19%, and 28%. Since the instrument operates at a fixed 20 keV threshold, these observations mean that there is no longitude variation in the intrinsic brightness of hard X-ray bursts, no limb brightening or darkening at 20 keV comparable to the effect seen in Hα.

In the hard X-ray spectra there is however a significant center-to-limb effect. Figure 10 shows the distribution of X-ray spectra at the time of maximum flux, γ_m, in a manner similar to Figure 7. In this case however the data have been divided into two groups, bursts occurring between $0 \leqslant \theta \leqslant 59°$ (upper part) and $\theta \geqslant 60°$ (lower part). The limb events are subdivided into the ones which occurred on the disk (shaded) and those which were beyond the limb (black). These two distributions differ principally in the region of small spectral index, $\gamma_m \leqslant 3.5$. The difference in the medians of the two distributions is $\Delta\gamma = 0.7$. Using the $\theta < 60°$ events as the reference distribution, the probability that the over-the-limb event distribution would occur by randomly choosing events is 6×10^{-3}; the probability that the limb event distribution would result from randomly choosing events from the reference distribution is 5×10^{-3}, and the equivalent number for the 62 events with $\theta \geqslant 60°$ is 8×10^{-4}. The difference in the two hard X-ray spectral distributions has a high degree of statistical significance.

One immediate consequence of these observations is that thermal interpretations of the hard X-ray emission from the Sun can be ruled out. Anisotropic emission from a thermal plasma can occur if the pitch angle distribution is highly anisotropic, but on the Sun the relaxation times should be quite short. Therefore no thermal plasma would be expected to yield the observed longitude effect.

There is at present no satisfactory quantitative model for the origin of the center-to-

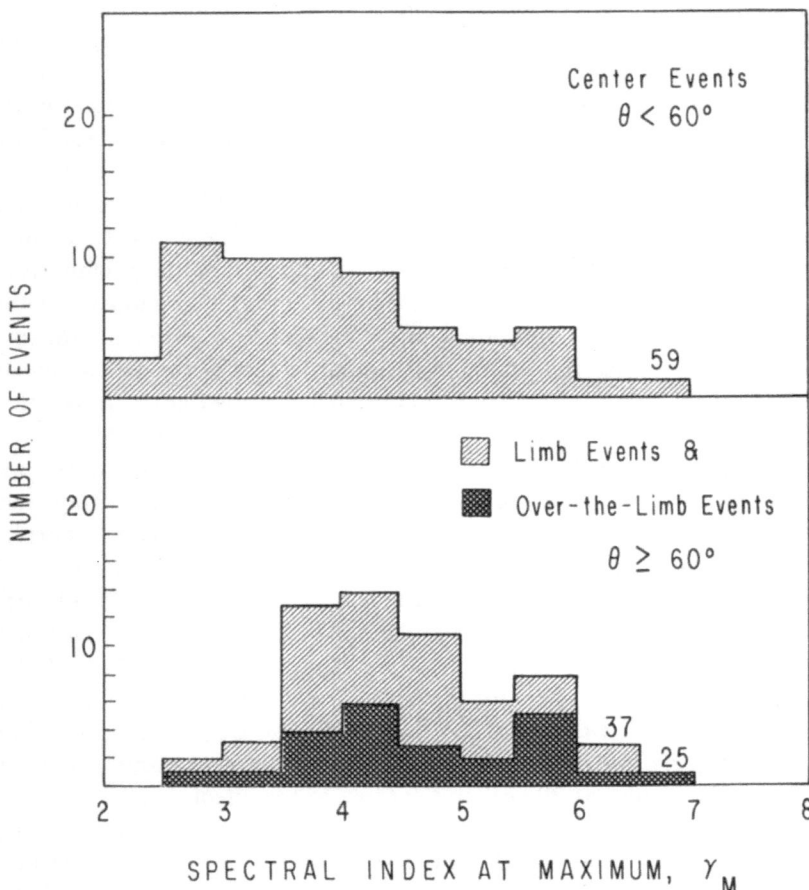

Fig. 10. Variation of spectral index at minimum with solar longitude. The upper panel gives the distribution of frequency of occurrence of γ_m for events associated with flares occurring at longitudes of 0–59°. The lower panel gives the same distribution for limb (60–87°) and over-the-limb events. The two distributions differ primarily in the absence of events in the range $2.5 \leqslant \gamma_m \leqslant 3.5$ in the $\theta \geqslant 60°$ group.

limb spectral variation. Occultation of an inhomogeneous source with spectra that soften with altitude cannot explain the effect, because it occurs on the disk as well as behind the limb. Compton backscattering of X-rays from the solar atmosphere (Tomblin, 1972) will produce a hardening of spectra from bursts near disk center at some energies, but according to the calculations of Santangelo *et al.* (1973), in the 20–50 keV energy range on which the OSO-7 spectra principally depend, the alteration of the spectral slope will be small. Thick target streaming models (Brown, 1972; Petrosian, 1973) predict a small variation of the spectrum with longitude but they also predict a large limb brightening which is not observed. None of these models gives a quantitative explanation of the origin of the effect.

VI. Summary

The OSO-7 solar X-ray experiment has provided a wealth of new information about bursts. The principal results of the data analysis so far are the following:

(1) The growth of the soft X-ray emission in a burst must be due to the addition of new hot material to the flare plasma.

(2) The anticorrelation between the cooling time and the maximum plasma temperature gives new evidence to support conduction as the principal cooling mechanism of the hot flare plasma.

(3) Although nonthermal electrons may have sufficient energy to heat the flare plasma, the relative timing between hard and soft X-ray observations indicates that some other source of energy must also heat the plasma.

(4) The observations of hard X-ray emission from bursts which occurred behind the limb indicate that hard X-ray emission comes at least in part from considerable heights in the corona.

(5) The spectra of hard X-ray bursts which occur near or beyond the limb are significantly softer than those occurring near the center of the disk.

Acknowledgments

Professor Laurence Peterson of the University of California, San Diego, is the principal investigator for the OSO-7 Solar X-ray Experiment. D. L. McKenzie of the Aerospace Corp. played an important role in the design and construction of the instrument and in the early analysis of the data. This presentation would not have been possible without numerous valuable discussions with Hugh S. Hudson and Michael J. Elcan of UCSD. This work was supported under National Aeronautics and Space Administration Contract NAS-5-11081 and U.S. Air Force Contract F19628-74-C-0046.

References

Brown, J. C.: 1972, *Solar Phys.* **26**, 459
Catalano, C. P. and Van Allen, J. A.: 1973, *Astrophys. J.* **185**, 335.
Culhane, J. L..: 1969, *Monthly Notices Roy. Astron. Soc.* **144**, 375.
Culhane, J. L. and Acton, L. W.: 1970, *Monthly Notices Roy. Astron. Soc.* **151**, 141.
Culhane, J. L. and Phillips, K. J. H.: 1970, *Solar Phys.* **11**, 117.
Culhane, J. L., Vesecky, J. F., and Phillips, K. J. H.: 1970, *Solar Phys.* **15**, 394.
Datlowe, D. W. and Lin, R. P.: 1973, *Solar Phys.* **32**, 459.
Datlowe, D. W., Hudson, H. S., and Peterson, L. E.: 1974a, *Solar Phys.* **35**, 193.
Datlowe, D. W., Elcan, M. E., and Hudson, H. S.: 1974b, *Solar Phys.* **39**, 155.
Frost, K. J.: 1969, *Astrophys. J. Letters* **158**, L159.
Frost, K. J. and Dennis, B. R.: 1971, *Astrophys. J.* **165**, 655.
Horan, D. M.: 1971, *Solar Phys.* **21**, 188.
Hudson, H. S.: 1973, in R. Ramaty and R. G. Stone (eds.), *High Energy Phenomena on the Sun*, NASA SP-342, p. 207.
Hudson, H. S. and Ohki, K.: 1972, *Solar Phys.* **23**, 155.
Hudson, H. S., Peterson, L. E., and Schwartz, D. A.: 1969, *Astrophys. J.* **157**, 389
Kahler, S. W.: 1973, in R. Ramaty and R. G. Stone (eds.), *High Energy Phenomena on the Sun*, NASA SP-342, p. 124.

Kahler, S. W. and Kreplin, R. W.: 1971, *Astrophys. J.* **168**, 531.

Kahler, S. W., Meekins, J. F., Kreplin, R. W., and Bowyer, C. S.: 1970, *Astrophys J.* **162**, 293.

Kane, S. R.: 1974, in G. Newkirk, Jr. (ed.), 'Coronal Disturbances', *IAU Symp* **57**, 105.

Kane, S. R. and Anderson. K. A.: 1970, *Astrophys. J.* **162**, 1003.

Kane, S. R. and Donnelley, R. F.: 1971, *Astrophys. J.* **164**, 151.

Kreplin, R. W. and Taylor, R. G.: 1971, *Solar Phys.* **21**, 452.

McKenzie, D. L., Datlowe, D. W., and Peterson, L. E.: 1973, *Solar Phys.* **28**, 175.

Neupert, W. M.: 1967, *Solar Phys.* **2**, 294.

Peterson, L. E. and Winckler, J. R.: 1959, *J. Geophys. Res.* **64**, 1969.

Peterson, L. E., Datlowe, D. W., and McKenzie, D. L.: 1973, in R. Ramaty and R. G. Stone (eds.), *High Energy Phenomena on the Sun*, NASA-SP-342, p. 132.

Petrosian, V.: 1973, *Astrophys. J.* **186**, 291.

Roy, J. R. and Datlowe, D. W.: 1974, *Solar Phys.,* to be published.

Solar Geophysical Data, CRPL-FB 332-340, 1972.

Santangelo, N., Horstman, H., and Horstman-Moretti, E.: 1973, *Solar Phys.* **29**, 143.

Takakura, T., Ohki, K., Shibuya, N., Fujii, M., Matsuoka, M., Miyamoto, S., Nishimura, J., Oda, M., Ogawara, Y., and Ota, S.: 1971, *Solar Phys.* **16**, 454.

Teske, R. G.: 1971, *Solar Phys.* **19**, 356.

Tomblin, F. F.: 1972, *Astrophys. J.* **171**, 377.

Vaiana, G. S., Davis, J. M., Giacconi, R., Krieger, A. S., Silk, J. K., Timothy, A. F., and Zombeck, M.: 1973, *Astrophys. J. Letters* **185**, L47.

Valcinek, B., Farnik, F., Horn, J., Letfus, V., Sudova, J., Komarek, B., Engelthaler, P., Ulrych, J., Moucka, L., Fronka, O., Vasek, T., Beranek, I., Plch, J., and Zderadicka, J.: 1973, *Bull. Astron. Inst. Czech.* **24**, 362.

Vorpahl, J. A.: 1973, in R. Ramaty and R. G. Stone (eds.), *High Energy Phenomena on the Sun*, NASA SP-342, p. 221.

White, W. A.: 1964, in W. N. Hess (ed.), AAS-NASA Symposium on the *Physics of Solar Flares*, NASA SP-50, p. 131.

Wood, A. T. and Noyes, R. W.: 1972, *Solar Phys.* **24**, 180.

RELATIONSHIP BETWEEN HARD AND SOFT SOLAR X-RAY SOURCES OBSERVED BY OSO-7

D. W. DATLOWE and H. S. HUDSON

University of California, San Diego, La Jolla, Calif., U.S.A.

Summary. The UCSD experiment on the OSO-7 satellite has provided hard and soft X-ray observations of a large number of solar flares. Of these a sample of 123 had sufficiently large fluxes to permit analysis of their spectra (Datlowe *et al.*, 1974). The locations of these flares upon the solar disk have been obtained by comparison with Hα flare listings. We find that the soft (5.1–6.6 keV) X-ray bursts, above a threshold of 1000 photons $(cm^2 \ s \ keV)^{-1}$, have a relatively flat distribution from center to limb. The frequency of occurrence of hard (20–30 keV) X-ray bursts, as normalized to the longitude distribution of the soft X-ray bursts, shows a statistically insignificant excess of $19 \pm 34\%$ in the longitude range 80–90°. Furthermore, the limb flares exhibit a small but statistically significant spectral softening.

Brown (1972) and Petrosian (1973) have shown that a thick-target source model for the hard X-ray emission, under the assumption of a uniformly vertical magnetic field, should produce a strong limb-brightening pattern of hard X-ray burst occurrence. Such a model would explain the polarization observations of Tindo *et al.* (1972). The lack of limb brightening in the OSO-7 hard X-ray data suggests the incorrectness of this simple thick-target model with vertical streaming.

Further evidence against the simple thick-target model comes from the OSO-7 observations of hard X-ray bursts originating in solar flares which occurred beyond the limb, as identified by Roy and Datlowe (1974). The characteristics of these bursts were similar to those seen near the limb on the visible disk. For the 8 soft X-ray bursts which occurred at minimum visible heights greater than 10^4 km, 5 had detectable nonthermal components. Since this fraction is the same as that for bursts near the center of the disk, we conclude that significant hard X-ray fluxes originate high in the corona and that the soft and hard X-ray sources may have similar geometrical distributions.

References

Brown, J. C.: 1972, *Solar Phys.* **26**, 459.
Datlowe, D. W., Elcan, M., Hudson, H. S.: 1974, *Solar Phys.* **35**, 193.
Petrosian, V.: 1973, *Astrophys. J.* **186**, 291.
Roy, J. R. and Datlowe, D. W.: 1974, *Solar Phys.* **40**, 165.
Tindo, I. P., Ivanov, V. D., Mandelstam, S. L., and Shurygin, A. I.: 1972, *Solar Phys.* **24**, 429.

THERMAL AND NONTHERMAL INTERPRETATIONS
OF FLARE X-RAY BURSTS

S. KAHLER

American Science and Engineering, Cambridge, Mass., U.S.A.

Abstract. Various authors have presented arguments for either the thermal or the nonthermal interpretations of impulsive $E > 20$ KeV X-ray bursts and slowly varying $E < 10$ keV X-ray bursts. In this
review the arguments for and against the prevailing opinion that the impulsive bursts are nonthermal
and the slowly varying bursts are thermal are presented.

For the impulsive bursts we discuss the spectra, electron mean free paths, center-to-limb distributions of both the numbers of events and spectra of events, and polarization data as relevant criteria.
For the slowly varying events we discuss electron self collision times, distribution of X-ray temporal
parameters, associated gradual rise and fall radio bursts, spectral and time profiles of special events
and center-to-limb distributions of numbers of events as the relevant criteria.

I. Introduction

The prime goal of solar flare observations is to deduce the physical mechanisms which
give rise to the release of a large quantity of energy in a short time. It is generally
assumed that the magnetic field of the active region in which the flare occurs is the
source of that energy and that the energy is released by some plasma process or processes (Sturrock, 1973). As the flare evolves, we can expect that the released energy
will eventually manifest itself largely as heat. Observations late in the course of the
flare should then be predominantly thermal in nature, but earlier in the event one may
hope to observe phenomena of a nonthermal nature which will reflect the processes
by which the energy conversion is carried out. In particular, the observations of certain kinds of X-ray events are thought by some to give good insights into the non-
thermal phases of flares. Others have disagreed, arguing that these events are usually
or always thermal in nature. In this paper we will discuss the different kinds of observed flare X-ray events and their properties and review the various interpretations
given them in the literature.

In the interpretation of the X-ray spectra one must decide whether the electrons
producing the emission are thermally relaxed or not. By thermally relaxed we understand that the electron distribution at any point in the X-ray source region has a Maxwellian velocity distribution and a well defined temperature. The temperature may
vary from point to point and the ions may not have the same temperature due to the
longer relaxation times (Spitzer, 1962). From the point of view of studying the non-
thermal flare process the thermal interpretation is a pessimistic one since it says that
from observations we can only hope to calculate the amount of energy converted to
heat and the rate of that conversion. In the nonthermal point of view we are seeing
X-ray bursts from energetic electrons which are a direct product of the nonthermal
process. Because the differences between the two interpretations are so critical to our

Sharad R. Kane (ed.), Solar Gamma-, X-, and EUV Radiation, 211–231. All Rights Reserved.

understanding of flare phenomena it is important that we understand the bases for these two interpretations and their consequences.

II. X-Ray Event Morphology

The two kinds of X-ray events that we shall discuss have been described by Kane and Anderson (1970). An example is shown in Figure 1 (Kane, 1975). At high energies ($E > 20$ keV) the X-ray event is impulsive, with a total time duration of about 1 min, and has a relatively hard spectrum. The other component is seen best at low energies ($E < 10$ keV) and is characterized by a much longer time duration and a relatively soft spectrum. It is also closely associated with the main phase of the optical flare and is slowly varying in time. When it occurs, the impulsive event almost always takes place during the rising phase of the slow event. These two kinds of events have been widely

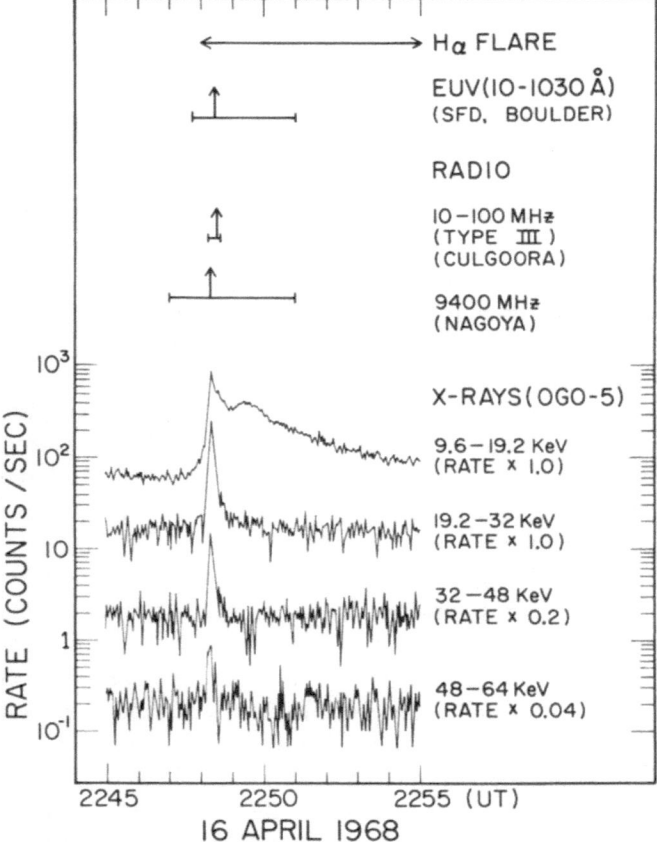

Fig. 1.   The impulsive and low energy X-ray bursts observed with the OGO-5 scintillator (Kane, 1975). The impulsive burst is observed best at energies above 20 keV and is well correlated with the impulsive microwave bursts, while the low energy component peaks later and has longer time constants.

discussed in the literature, and various authors have argued thermal and nonthermal origins for each. In addition to these two kinds of events, a third class is known which is characterized by high energy events with relatively flat power law spectra out to 250 keV or more and decay times on the order of minutes or tens of minutes. They are usually associated with type II or type IV radio bursts (Peterson *et al.*, 1973; Frost and Dennis, 1971). Since everyone seems to agree that the latter events are nonthermal in nature, we shall consider thermal and nonthermal arguments only for the slowly-varying and the impulsive events.

III. High Energy ($E > 10$ keV) Events

(a) NONTHERMAL ARGUMENTS

In their study of the properties of thirteen impulsive X-ray bursts Kane and Anderson (1970) advanced several reasons for choosing a nonthermal interpretation for the bursts. These were:

(1) The short rise time. For ~ 40 keV X-rays the *e*-folding rise times were 2–5 s. The decay times were slightly longer at 3–10 s.

(2) The power law energy spectrum. They showed that a power law of the form:

$$dJ/dE = K E^{-\gamma} \text{ photons cm}^{-2} \text{ s}^{-1} \text{ keV}^{-1}, \tag{1}$$

where E is the photon energy and K and γ constants is a good fit to the observed X-ray spectrum at the peak of the event. They further showed that a bremsstrahlung interpretation of the X-ray spectrum yielded an instantaneous electron spectrum of the form:

$$dJ_e/dE_e = A \cdot E_e^{-\delta} \text{ electrons cm}^{-2} \text{ s}^{-1} \text{ keV}^{-1}, \tag{2}$$

where E_e is the electron energy and A and δ are constants. They concluded that the observations were not consistent with a thermal spectrum which can be represented by a *single* temperature. This is shown in Figure 2.

(3) The occurrence of the burst before the Hα maximum of the flare. The time of the X-ray maximum preceded the Hα maximum by 0.5 to 3 min.

These three arguments, especially the second one, have led most workers in the field to accept the nonthermal interpretation.

More recent work by Kane (1972) has tended to confirm this interpretation. He found a very close association between some impulsive X-ray bursts and type III radio bursts. Although only about one third of all impulsive X-ray bursts he observed are associated with reported type III bursts, it was found that when the X-ray and type III bursts are associated, the time correlation between the two emissions is very good. Figure 3 from Kane (1973b) shows an example of this close association. In general, not only the times of maxima of the bursts but also the total durations of the bursts were in close agreement. Since the type III radio bursts are produced by streams of non-thermal electrons (Lin *et al.*, 1973) this close association again argues for the nonthermal interpretation of the impulsive X-ray bursts. A close temporal association also

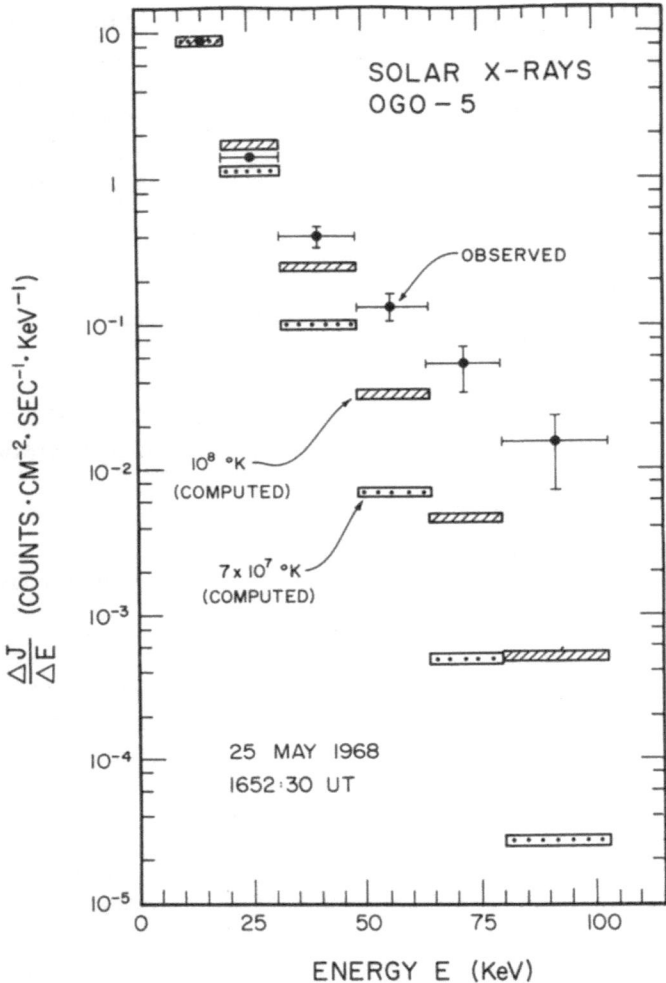

Fig. 2.   The comparison of an impulsive burst spectrum compared with the calculated values of the spectra for plasmas with $T = 7 \times 10^7$ K and $T = 10^8$ K. This shows the problem of trying to fit power law spectra with a single temperature thermal model (Kane and Anderson, 1970).

exists between the impulsive X-ray bursts and impulsive microwave bursts (Kane, 1972) which are believed to be due to gyrosynchrotron emission from nonthermal electrons (Ramaty, 1973). Kane (1973a) further found an approximate proportionality between the peak microwave flux and the peak X-ray flux in the correlated cases.

Another basis for the nonthermal interpretation is the observation of multiple or periodic impulsive X-ray bursts. One such burst was observed by Frost (1969) in 1969, March 1 and another burst with a 16 s periodicity was reported by Parks and Winckler (1969). In each case each individual burst was found to be correlated with impulsive microwave emission.

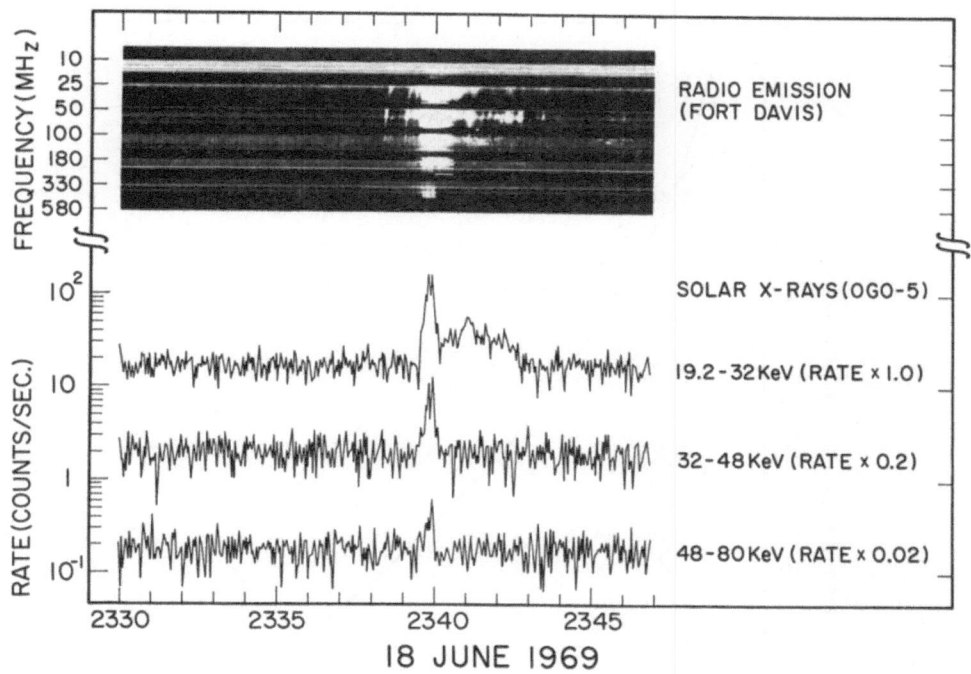

Fig. 3.   An example of time correlated type III and impulsive X-ray bursts (Kane, 1973b).

(b) THERMAL ARGUMENTS

The general preference for the nonthermal interpretation is based largely on choosing the simplest interpretation of the observed spectrum. Since most impulsive bursts can be fitted by a simple power law energy spectrum but not a single temperature thermal spectrum, the nonthermal power law distribution is preferred. Chubb (1970), however, has discussed impulsive bursts in terms of a multithermal interpretation. He points out that X-ray flare measurements with crystal spectrometers have indicated the multithermal character of cooler flare plasmas and that it is only reasonable to expect this interpretation to be required at higher temperatures, i.e., shorter wavelengths. Chubb also points out that the impulsive bursts cannot be described by a simple power law over the entire observable energy range but instead they show a cutoff or steepening in the power law spectrum at ~ 100 keV which is characteristic of thermal emission. The thermal interpretation of impulsive bursts typically requires temperatures of 10^8 K or more and emission measures of some 10^3–10^4 times smaller than those characterizing cooler ($\sim 10^7$ K) plasmas observed with crystal spectrometers. Figure 4 shows his thermal interpretation of the 1969, March 1 impulsive event discussed by Frost (1969). The temporal behavior of the typical impulsive spectrum also is consistent with the thermal interpretation since it either first grows harder during the rise phase and then softens during the decay phase or it continuously

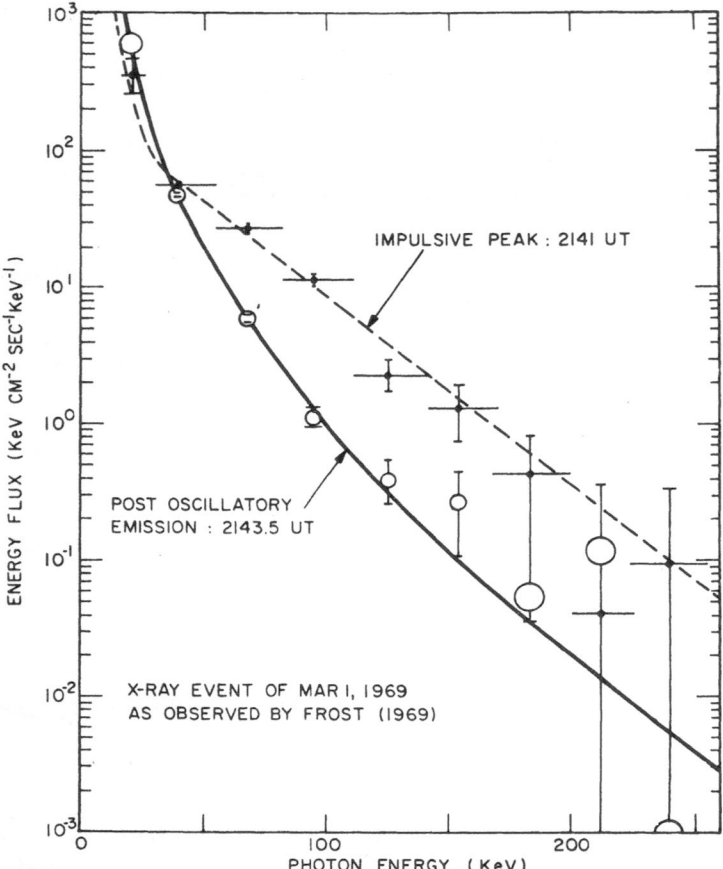

Fig. 4.   The thermal interpretation by Chubb (1970) of an impulsive burst in 1969, March observed
by Frost (1969). Chubb interprets the impulsive data at 2141 UT as free-free emission
from a 3.7×10^8 K plasma.

softens throughout the burst (Datlowe *et al.*, 1974) corresponding to a plasma which is
heated and subsequently cools. Milkey (1971) similarly has pointed out that the impulsive X-ray bursts can be interpreted in terms of emission from a volume with a
strong temperature gradient. In some cases the temperature distribution of the emitting region may result in a spectrum which mimics the power law non-thermal
spectrum.

Brown (1974) has taken the next logical step in the multithermal analysis by deriving
an analytic expression for the thermal source of any observed hard X-ray spectrum.
For this he uses an emission measure $\mu(T)$ per unit temperature such that:

$$\mu(T)\,\mathrm{d}T = n_\mathrm{e}^2\,\mathrm{d}V. \tag{3}$$

His result then yields $\mu(T)$ for any given observed X-ray spectrum.

Kahler (1971a, b) used a simplified model of a flare region with a peak temperature of 10^8 K to argue against Chubb's (1970) multithermal interpretation. Assuming electron streaming losses with no heat sources or sinks along the path and taking a range of emission measures given by Chubb, he calculated that the total energy needed to sustain the plasma against heat losses is prohibitively large for $n_e \gtrsim 10^9$ cm^{-3}. Lower electron densities require volumes larger than observational upper limits (Takakura et al., 1971). In addition, the mean free path of the electrons averaged over the $T \gtrsim 10^8$ K Maxwellian distribution is greater than the longest linear dimension of the flare volume, so that electrons are more likely to escape or be reflected by mirroring in a flux tube than to undergo binary collisions which will result in a Maxwellian velocity distribution.

Brown (1974) has countered Kahler's arguments by pointing out that his simple model of a 10^8 K region is not valid. One must first unfold the actual distribution of the emission measure $\mu(T)$ per unit temperature and then, assuming constant pressure, calculate the thermal gradient as a function of temperature in order to calculate the conductive cooling at any point. Brown has used these calculations to obtain τ_c, the characteristic time for the conductive temperature redistribution within the hot plasma. If this time is long, say 50 s, then the entire decay of an impulsive burst can be explained by the conductive relaxation and the required energy input is small. If the time is short, say 2 s, a much larger amount of energy is needed to sustain the burst. τ_c is inversely proportional to $n_{10}^3 \Sigma^2$ where n_{10} is the density of the surrounding plasma in which $T = 10^7$ K and Σ is the cross section of the flux tube. A small change in either n_{10} or Σ results in a large change in τ_c, and since these quantities are not well known, the values of τ_c are therefore very tentative. Brown has not answered one of the basic points raised by Kahler, that the mean free path of the electrons is excessively long. Brown calculates that for one large observed burst the scale distance for change in T is comparable to the mean free path. This calculation was done only to decide whether the dominant cooling mechanism is conduction or electron streaming, but one must still consider whether the plasma can be thermally relaxed in the first place. Using Equation (8) of Takakura and Kai (1966) for the electron deflection time:

$$t_D = \frac{2.4 \times 10^{12} \, \varepsilon^{3/2}}{n_0} \text{ s}, \tag{4}$$

where ε is the ratio of the electron kinetic energy to its rest mass and n_0 the ambient density, we find a mean free path of

$$1 = v t_D \approx 5 \times 10^{10} \text{ cm} \tag{5}$$

for a 50 keV electron in a density of $n_0 = 10^{10}$ cm^{-3}. This is on the order of one solar radius and more than an order of magnitude larger than the upper limit for an impulsive burst deduced by Takakura et al. (1971). Therefore, it still appears unlikely that the electrons will be thermally relaxed at densities of 10^{10} cm^{-3} or less. Avoiding this problem by assuming high densities then leads to the new problem of excessively high energy requirements in the treatments of both Kahler and Brown. Brown has pointed out that all the preceding arguments assume that the classical formulas for conductive cooling or electron streaming apply. If we have plasma turbulence then wave-particle interactions will constrict the heat loss and the thermal interpretation may then be valid.

(c) ANISOTROPY AND POLARIZATION RESULTS

If the impulsive bursts are produced by bremsstrahlung from nonthermal electrons, it is reasonable to expect that the bursts will be anisotropic in their distribution of emission and polarized due to the anisotropic nature of the electron velocity distribution. Pinter (1969) and Ohki (1969) each used the forty-six hard X-ray events listed by Arnoldy *et al.* (1968) to study the distribution of the events as a function of solar longitude. In each case the listed X-ray event was associated with an Hα flare which gave the position of the X-ray source on the disk. Ohki found a sharp limb darkening which could not be explained by the center-to-limb variation of Hα flares while Pinter found a peak in the longitude distribution at 40–50°. Pinter interpreted his results in terms of the model of Takakura and Kai (1966) in which electrons are trapped in and travel along a magnetic flux tube which is essentially parallel to the solar surface. The angular distribution function for bremsstrahlung (Sommerfeld, 1951) is

$$J(\theta) = \sin^2 \theta (1 - \beta \cos \theta)^{-4}, \tag{6}$$

where θ is the angle between the direction of motion of the electron and the direction of radiation and βc is the electron velocity. Since higher energy bremsstrahlung peaks at smaller angles, the Takakura-Kai model and an east-west orientation for the magnetic field predicts that low energy bremsstrahlung will peak near central meridian and higher energy bremsstrahlung will peak closer to the limb, in accord with Pinter's findings. Pinter's calculations were improved by Elwert and Haug (1971) who used a power law distribution for electron energies and considered different pitch angles for the electrons. Their results were in qualitative agreement with Pinter's. Subsequently, Brown (1972) considered a model (de Jager and Kundu, 1963) in which electrons with a power-law distribution are continuously injected into a *vertical* magnetic field where they travel down into the chromosphere and rapidly decay in a thick target situation. Limb brightening is predicted, but the electron scattering in the thick target diminishes the center-to-limb effect, particularly for low energies. Phillips (1973) reanalyzed Ohki's data in terms of the de Jager-Kundu vertical field model by considering the distribution with respect to the Sun's apparent center and found the results to be consistent with a probability of occurrence which is independent of the distance from the Sun's center.

Kane (1973b) has analyzed the distribution of over 300 impulsive X-ray events observed on OGO 5 and finds that there is no significant center-to-limb variation in the frequency of occurrence of X-ray bursts associated with small solar flares. His data are shown in Figure 5. Datlowe *et al.* (1974) found a similar result using a smaller sample of events from their scintillation detector on OSO 7. These results probably present no difficulty to nonthermal electron models since non-vertical magnetic fields and various pitch angle distributions at the injection point of the de Jager-Kundu model (Brown, 1972) or departures from the east-west direction for the magnetic field of the Takakura-Kai model (Shaw, 1972) may substantially reduce any expected nonuniform distribution of events. For our purposes, however, it is important to note that Kane's result may not be consistent with a thermal origin of the impulsive bursts.

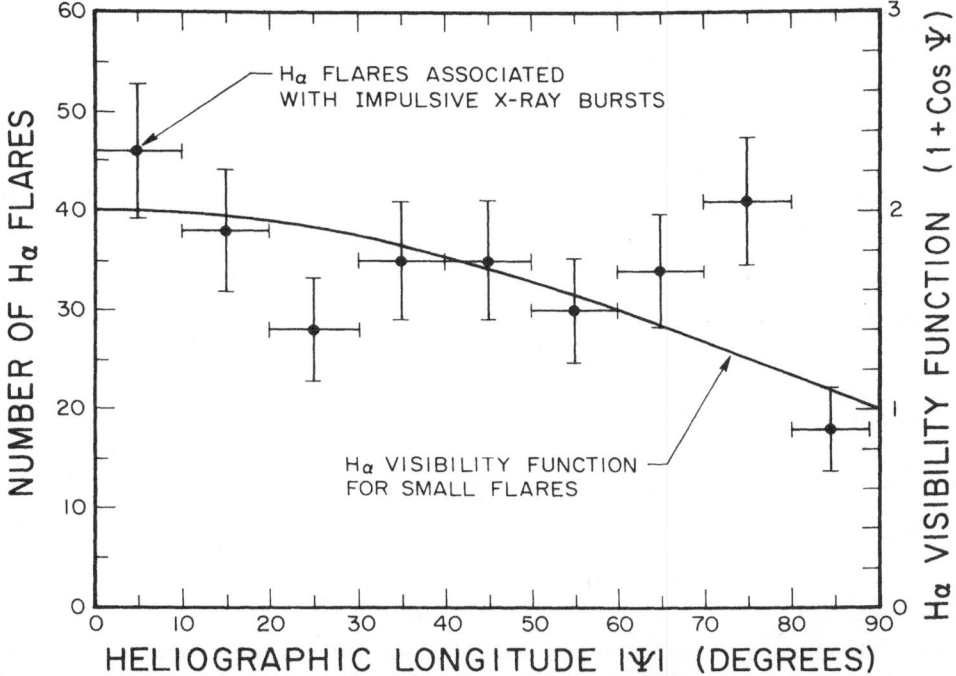

Fig. 5. The center-to-limb distribution of ∼300 impulsive X-ray bursts measured by Kane (1973b). With the Hα visibility function taken into account, the data are consistent with an isotropic source for the X-rays.

With an isotropic electron velocity distribution, no anisotropy in the X-ray emission is expected, but Santangelo *et al.* (1973) have shown the Compton backscattering from the solar surface can modify the intensity and spectral shape of the burst as observed at the Earth. Qualitatively, one expects some limb darkening, but the quantitative value depends on the energy spectrum and the effective passband of the X-ray detector.

In a recent result Datlowe *et al.* (1974) found an asymmetric distribution of X-ray burst spectral indices with solar longitude. Figure 6 shows how the average spectral index of a burst, $\bar{\gamma}$, becomes larger for limb and over-the-limb events. According to the authors, this result is statistically significant at the 99% confidence level and is too large to be explained by the center-to-limb variation of spectral hardening by Comp-

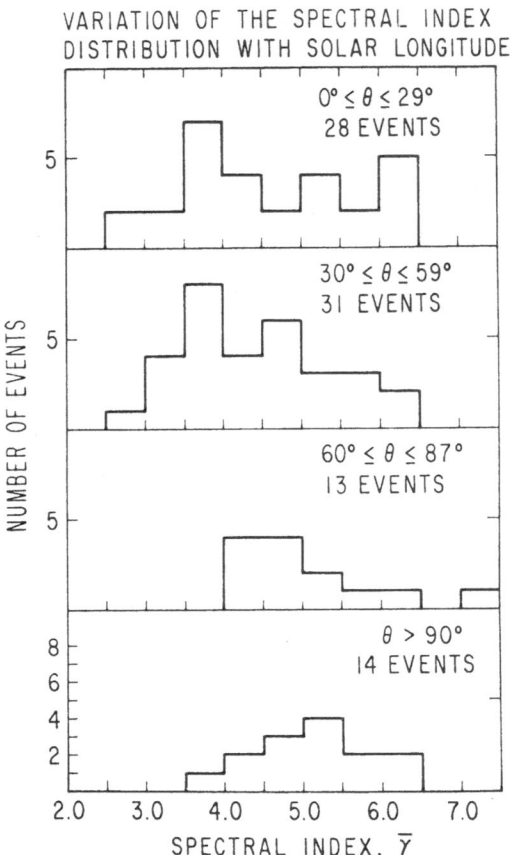

Fig. 6.   The variation with solar longitude of the distribution of the average spectral indices for impulsive events measured by Datlowe *et al.* (1974). Events in the top three panels are identified with Hα flares or subflares while events in the bottom panel are from unreported events believed to be from over the limb.

ton back-scattering as discussed by Santangelo *et al.* (1973). They attribute the variation to directivity of the electron pitch angle distribution in the solar magnetic fields. Their result certainly appears to conflict with the thermal interpretation.

The polarization observations of flares reported by Tindo and his colleagues (Tindo *et al.*, 1972a, b; 1973) have provided one of the strongest arguments for the nonthermal interpretation of hard X-ray bursts. Polarimeters consisting of beryllium scattering blocks and three pairs of counters were flown on the Intercosmos 1, 4 and 7 satellites to measure the polarization and polarization angles of flare X-rays. The Intercosmos 4 instrument was used to determine these parameters for two flares in 1970, October 24 and November 5 (Tindo *et al.*, 1972b). The maximum polarization of the two events was 0.16 and 0.21 respectively and in each case the plane of polarization is found to be near the plane of the projected radius vector of the flare on the solar disk. Tindo *et al.* (1972b) compared the flux profile of their broad band counters with a scintillation counter on board and found the closest correlation with the 15 keV energy channel. The fact that a significant polarization has been measured for a number of hard X-ray bursts is not a conclusive argument against the thermal interpretation since some polarization will be produced by the albedo X-rays from flares (Santangelo *et al.*, 1973). The reflected radiation will be linearly polarized along a line perpendicular to the projected radius vector of the flare on the disk rather than parallel to it as found by Tindo *et al.* Brown *et al.* (1974) have criticized the method of calibrating the polarization data by assuming a zero polarization late in the event on the grounds that some residual polarization will always be present in the albedo flux. Further polarization measurements of flares, especially those near sun center where the reflected radiation should be unpolarized and those near the limb where the plane of the reflected radiation should be parallel to the limb, could determine the validity of the thermal interpretation.

An additional problem from the point of view of Chubb's (1970) interpretation is whether the large X-ray flares observed by Tindo *et al.* correspond to a larger scale of

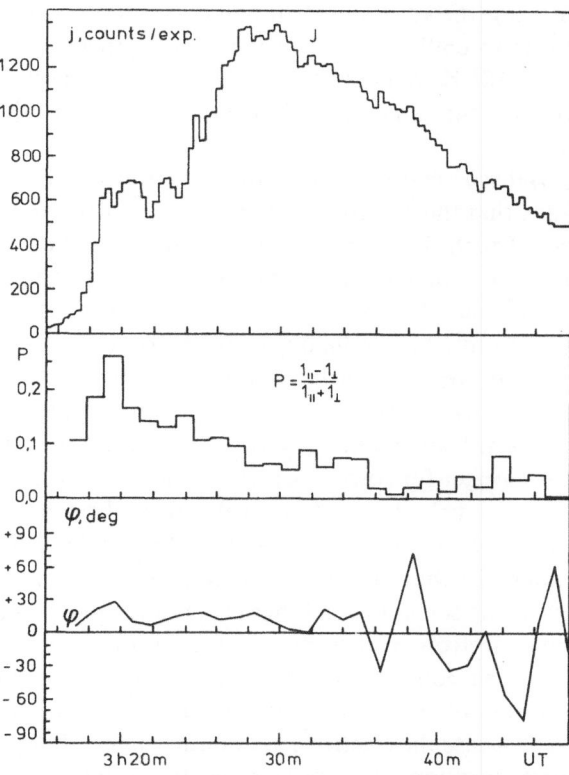

Fig. 7.   The measurement of linear polarization by Tindo *et al.* (1972a) of a flare in 1970, November 5. (a) is the flux of the $E \sim 15$ keV X-rays; (b) is the degree of polarization; and (c) is the angle of polarization.

the impulsive bursts of Kane and Anderson (1970) and Frost (1969) with a characteristic spectral cutoff at ~ 100 keV or whether they are more like the second phase of the
two phase events reported by Frost and Dennis (1971). The event of 1970, November
5, is shown in Figure 7. The facts that the flares observed on Intercosmos 4 were large
Hα events, with polarization persisting for 5–10 min, and that the existence of a
spectral cutoff around 100 keV is in question, suggest the possibility that these may
not be the impulsive events which Chubb interpreted as thermal.

IV. Low Energy Events

(a) THERMAL ARGUMENTS

In contrast to the impulsive X-ray bursts, the low energy ($E < 10$ KeV) X-ray events
are almost always treated as thermal in origin. This is due in large part to the analysis
of data from Bragg crystal spectrometers in which flare spectra can be observed with
high wavelength resolution from ~ 0.6 to ~ 20 Å (Doschek *et al.*, 1972). Temperatures
may be estimated by (1) comparing the ratios of resonance lines of hydrogenic ions
to helium-like ions of the same element; (2) measurement of the slope of the continuum; and (3) measuring the ratio of satellite lines due to dielectronic recombination
to resonance lines due to collisional excitation. The temperatures obtained by these
methods range from $\lesssim 10^7$ K to 3.4×10^7 K. The various methods used generally give
varying temperatures for the same event, indicating that a multithermal origin exists
for the flare plasma.

Kahler *et al.* (1970) discussed two X-ray flares observed with a proportional counter
on OGO 5 and argued that the $E < 10$ keV radiation was thermal in origin because (1)
the flux-time profiles of their detector were slowly varying and not impulsive as would
be expected from a comparison with the high energy impulsive events; and (2) for a flare
density of $n_e \sim 10^9 - 10^{10}$ cm^{-3} the self collision time, i.e., the time for binary collisions
to relax any deviation from a Maxwellian velocity distribution, is on the order of 10^{-1}
to 10^0 s. These arguments are not conclusive for the following reasons. First, the time
scales of low energy nonthermal bursts, if they exist, do not have to be similar to those
of the high energy bursts. Kane and Anderson (1970) found that the rise times of the
impulsive events were longer for lower energies. In addition, there are numerous low
energy bursts that have much shorter time scales than the two events presented by
Kahler *et al.* (1970). For example, Culhane and Phillips (1970) discussed an event observed in 1968, January 23 that lasted only 4 min. The problem with Kahler *et al.*'s
second argument involving self collision times is that it is model dependent; it assumes
a volume of energetic electrons which interact among themselves and have lifetimes
much greater than the self collision time for relaxation. One could postulate a thick
target model of continuous injection similar to that suggested for impulsive bursts
(Brown, 1972). In such a case the electrons are nonthermal with lifetimes much shorter
than the X-ray burst duration and comparable to or less than the self collision
time.

Drake (1971) has statistically analyzed over 4000 solar X-ray bursts using data from the 2 to 12 Å Geiger counter experiments on Explorers 33 and 35. He plotted the differential distributions of the X-ray bursts with respect to rise time, decay time, total duration, ratio of rise time to total duration, and ratio of rise time to decay time. Figure 8 from his paper shows the distribution of the ratios of the rise time to the decay time. In this distribution, as in the others, there is no evidence for the existence of more than one class of X-ray burst in the 2–12 Å (1–6 keV) range. More classes may exist but cannot occur for a substantial number of bursts. For the great majority

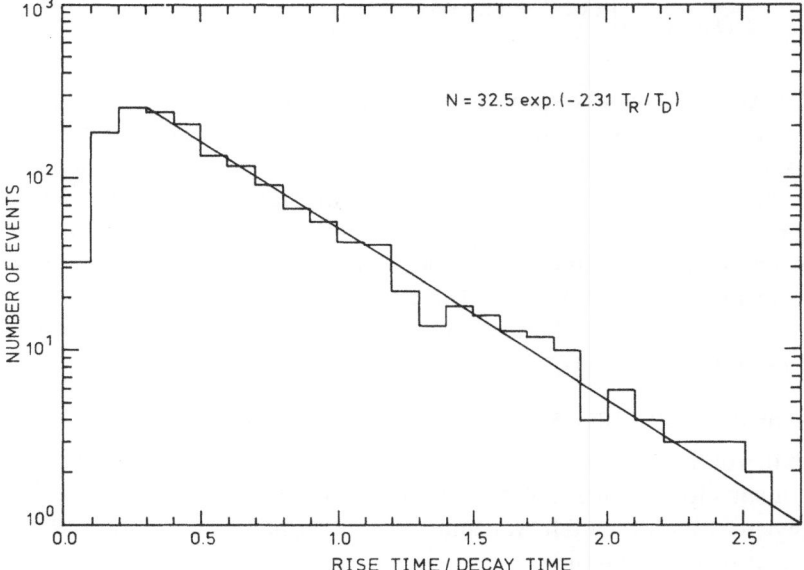

Fig. 8. The differential distribution of 2–12 Å X-ray bursts with respect to the ratio of the rise time to the decay time (Drake, 1971). The straight line is a least-squares fit.

of soft X-ray bursts then we can conclude that (1) they are all thermal or ((2) they are all nonthermal or (3) some are thermal and some are nonthermal but the time constants characteristic of each are similar.

Hudson and Ohki (1972) used the temperatures and emission measures derived from sixteen X-ray flare events observed with broad band detectors on the Solrad and Vela satellites to calculate the intensity of the associated 16 or 17 GHz microwave event assuming an isothermal plasma. The measured intensity of the 'gradual rise and fall' or 'post burst increase' radio event near the time of the maximum X-ray emission measure was compared to the calculated value and good agreement was found. Hudson and Ohki concluded that the correlation confirms the thermal model of these phenomena.

We can show, however, that their data are also consistent with a nonthermal model. If we observe free-free bremsstrahlung from an isothermal plasma with a temperature

T then the ratio of the X-ray flux F_X to the radio flux F_R measured in the same units is:

$$\frac{F_X}{F_R} = \exp\left\{\frac{E_R - E_X}{kT}\right\} \approx \exp -\left(\frac{E_X}{kT}\right), \tag{7}$$

where E_X and E_R are the X-ray and radio energies and $E_R \ll E_X$. For $E_X \approx 2$ keV and $kT \approx 2$ keV this yields a value of $F_X/F_R \approx 0.4$. Suppose on the other hand one were observing thin target bremsstrahlung given by the Bethe-Heitler formula (Jackson, 1962)

$$F(\hbar\omega) \propto \ln\left[\frac{(E^{1/2} + (E - \hbar\omega)^{1/2})^2}{\hbar\omega}\right], \tag{8}$$

where E is taken to be 4 keV and ω is the observed frequency. Then F_X/F_R is ~ 0.1, within a factor of 4 of the value given above for the thermal case. We have assumed the radio region is optically thin and have neglected free-bound emission. Thin target bremsstrahlung from a monoenergetic electron distribution would not give a thermal X-ray spectrum but one could choose an incident electron spectrum which would result in an X-ray spectrum that would be consistent with a thermal spectrum as observed in broad band detectors. The point is that while the data of Hudson and Ohki are consistent with a thermal source for flares, they cannot be used to exclude possible nonthermal models.

(b) NONTHERMAL ARGUMENTS

Since the thermal interpretation for low energy X-ray events is so widely accepted, the only arguments for nonthermal emission have come from authors who have presented specific events which they felt warranted a nonthermal interpretation. We shall now discuss several such examples.

Blake and House (1971) analyzed iron line emission from a rocket crystal spectrometer flown in 1966, October 4. Their analysis of the intensity of the iron line complex at 1.9 Å due to Kα line emission and the intensity of the optical $(2p - 31)$ transitions of iron from 10 to 17 Å yielded values of the ratio F_0/F_K of less than 50. They computed the expected ratio as a function of the electron temperature and found that an electron temperature of 200×10^6 K or a monoenergetic electron distribution peaked at $E \gtrsim 15$ keV was required to account for the observation. The emission measure of these electrons was calculated to be $\approx 3 \times 10^{47}$ cm^{-3}. If we take a density of $n_H = 10^9$ cm^{-3}, then the total energy of the quasithermal or nonthermal electrons is $\sim 7 \times 10^{30}$ erg, substantially larger than the energy of a small flare, estimated at 10^{29} erg (Lin, 1971). In view of the fact that there were no Hα flares, radio bursts or ionospheric disturbances reported at the time of the observations, their deductions would appear questionable.

Another event was analyzed by Landini et al. (1972). Their work with a small event in 1969, January 7 observed with the 0.5 to 3 Å and 1–8 Å detectors on Solrad indicated that if the event were thermal the temperature first decreased early in the event and then increased late in the event, opposite to the results obtained by Horan and others.

The authors interpreted this behavior as indicative of nonthermal emission. As in the event analyzed by Blake and House, there was no accompanying sign of nonthermal activity; however, because it was a small event none might be expected. The January 7 event occurred during the decay phase of an earlier, larger X-ray burst which may have presented problems for the data analysis. If such events do take place on the sun, one would like to see several more examples in order to define them as a class. Their existence constitutes proof of low energy nonthermal events only as long as one assumes that the cooling time is substantially longer than the heating time. If this is not the case, then the temperature and emission measure profiles may simply reflect the time variation of the heating mechanism.

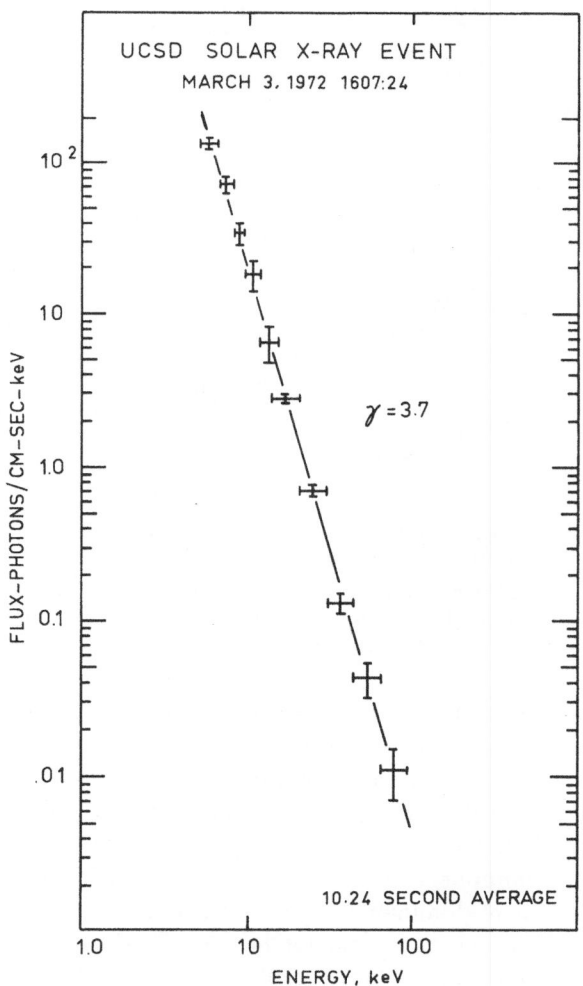

Fig. 9.   The power-law spectrum at the peak of a hard X-ray burst observed on OSO 7 (Peterson *et al.*, 1973). The power law fit extends from 5 to 100 keV.

Peterson *et al.* (1973) have classified over 200 events observed on OSO-7 into four general categories. One of their categories, which consisted of only one event, has a single power law fit over both the hard and soft X-ray spectrum early in the event, followed by a steeper low energy spectrum. This event observed in 1972, March 3 could be fitted by a single power law from 5 keV to $\sim$ 100 keV with an exponent of $\gamma = 3.7$, as shown in Figure 9. If we accept the nonthermal interpretation of the hard X-ray burst, then this event may indeed constitute evidence for the existence of low energy nonthermal bursts. The nature of the bremsstrahlung spectrum is such that it will

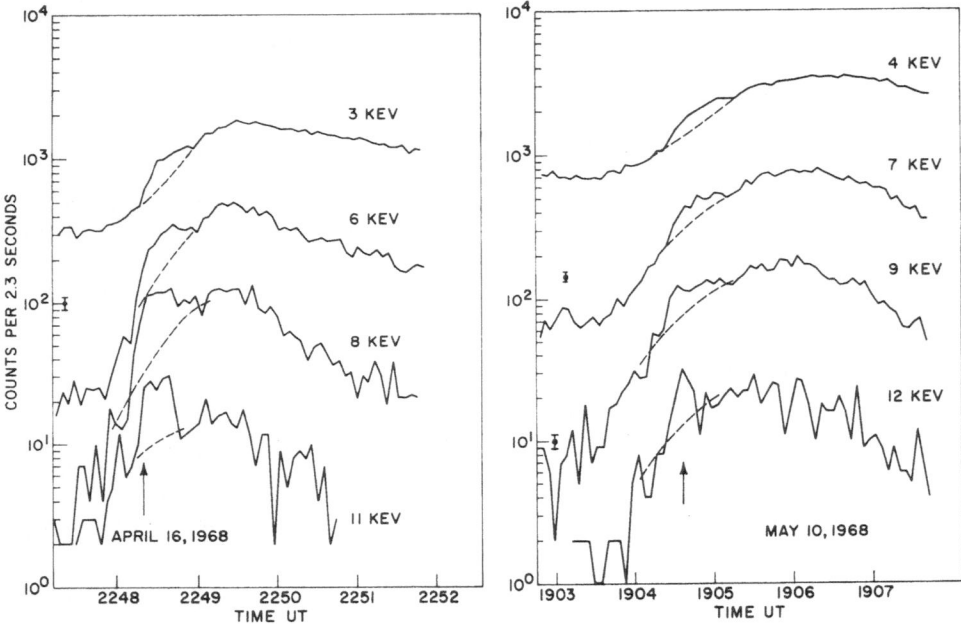

Fig. 10. Counting rate profiles of two impulsive events from the NRL detector on OGO 5 (Kahler and Kreplin, 1971). The dashed lines represent the estimated profiles of the thermal components during the impulsive bursts.

extend down to the lowest measurable energy for any electron spectrum. The rarity of the March 3 event may simply be due to the requirement that the development of the low energy thermal event be sufficiently retarded in time relative to the impulsive burst for the low energy end of the impulsive burst to dominate the spectrum. Kahler and Kreplin (1971) found that on only two out of twelve impulsive hard X-ray events could the impulsive component be traced down to 3 keV. In the other ten cases the slowly varying component dominated the low energy spectrum. In their analysis they assumed that they could accurately estimate the intensity-time profile of the $E < 10$ keV slowly varying fluxes and subtract those profiles from the profiles of the total emission to yield the profiles of the impulsive component alone. Figure 10 shows the flux profiles of these two events. The fact that Peterson *et al.* and Kahler and Kreplin both

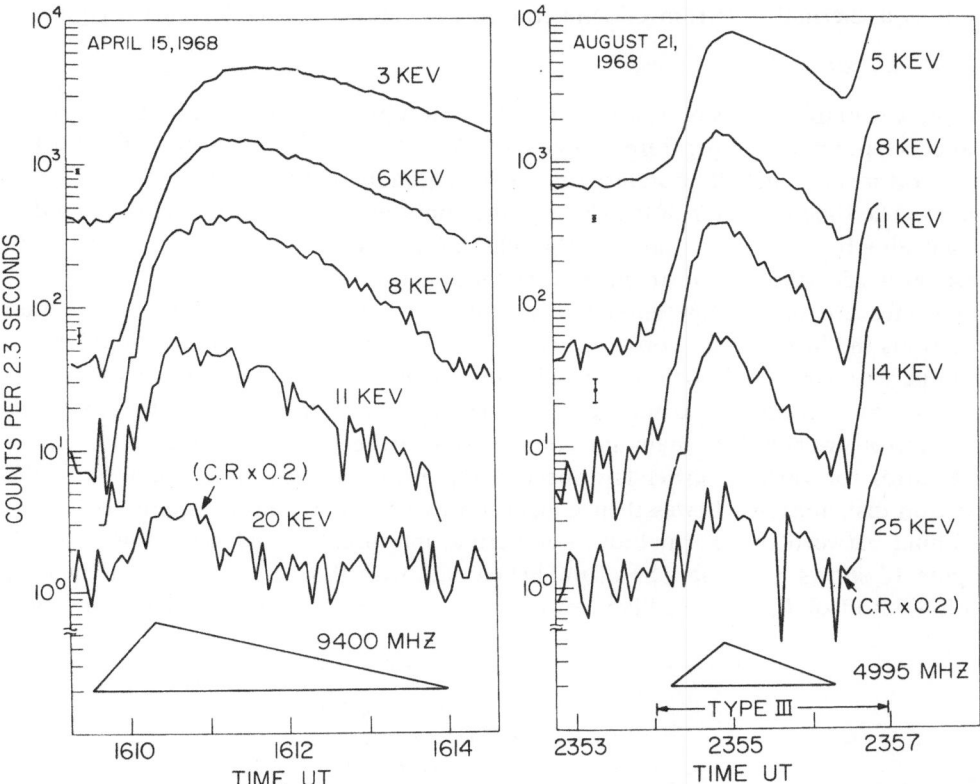

Fig. 11. Two low energy impulsive events which were accompanied by impulsive events in the
$E > 20$ keV range and were closely correlated with microwave bursts (Kahler, 1973).

were able to trace the low energy end of the impulsive bursts down to 5 keV using
different analytical methods seems strong evidence that in at least some cases the low
energy emission is dominated by a nonthermal component. This statement, of course,
assumes a nonthermal interpretation for the impulsive burst.

Kahler (1973) has used the method of tracing the intensity-time profiles down to
low energies to argue that three events he observed on OGO 5 in 1968 might be purely
nonthermal in nature down to 3 keV. In each case an impulsive hard X-ray burst with
an accompanying microwave burst could be traced down to 3 keV where only a single
simple rise and fall profile was seen, as shown in Figure 11. He pointed out, however,
that the low energy profiles were quite consistent with the behavior usually seen in
slowly varying bursts which are almost always treated as thermal.

(c) SPECTRAL INTERPRETATIONS

Kahler and Kreplin (1971) have discussed the ambiguity involved in the interpretation of low energy continuum spectra. One can measure the slope of the continuum at some energy E and interpret it either as a power law spectrum of exponent

γ or as a thermal spectrum of temperature T. The two are related by the equation

$$kT = E/(\gamma - 1).\tag{9}$$

Using reasonable values of γ such as those obtained by Kane (1971) and $2 \lesssim E \lesssim 8\,\mathrm{keV}$, the corresponding temperatures range from 7 to 55×10^6 K, in agreement with the temperatures one usually obtains from various broadband and spectrometer measurements. The time behavior of the slope of the continuum also provides no help in distinguishing between power law and thermal interpretations since the spectra typically first get harder and then near the peak begin to get softer again.

One feature of the low energy spectrum which might be used for nonthermal criteria is the line emission observed with crystal spectrometers. Landini et al. (1973) have performed calculations of the ratios of H-like to He-like ions for a plasma with a power law electron spectrum. They assumed a quasi-steady state situation in which the electrons were able to maintain a power law distribution in time and calculated the line ratios for various power law spectral exponents and low energy cutoffs in the electron distributions. It was their conclusion that there was no possibility of distinguishing between the thermal and nonthermal interpretation by using this method. Figure 12 shows a comparison of published data with their calculations for the case with a cutoff of $E_1 = 1$ keV. Since lower values of the cutoff imply lower values of γ

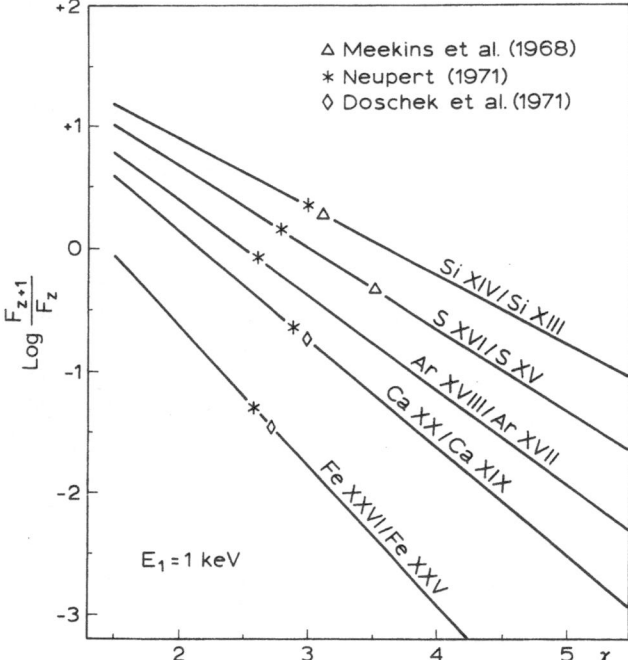

Fig. 12. Computed flux ratios between the resonance lines of H-like and He-like ions as a function of the electron power law spectral index γ (Landini et al., 1973). A low energy cutoff of 1 keV was used. Some experimental values are shown for comparison.

than those derived by Kane (1971) for higher energies, Landini *et al.* (1973) suggest that a higher energy cutoff with associated larger values of γ represent a more reasonable interpretation of the data.

A more obvious method is to look for enhanced Kα line emission either during the impulsive burst or during the slowly-varying component. Phillips and Neupert (1973) have calculated the Kα line emission expected from S, Ar, Ca, and Fe for both the thermal and nonthermal cases. The electron spectra deduced by Kane and Anderson (1970) for impulsive bursts were used for the nonthermal case, but it was found that in most cases the expected line emission was relatively weak. Phillips and Neupert did their nonthermal calculations for an assumed ion temperature of 2×10^6 K. They point out that the proportions of S, Ar, and Ca atoms ionized to the B-like stage or lower, which are necessary for Kα line emission in the solar atmosphere, are very small at temperatures above 10×10^6 K, hence no significant Kα line emission would be expected. It would appear, then, that the requirement of low ion temperatures and energetic electrons for Kα line emission would imply a possible nonthermal event whenever the electron temperatures deduced from the continuum measurements were high but yet Kα line emission was simultaneously detected. This procedure would be more useful in working with the hard impulsive bursts than with low energy events.

(d) ANISOTROPY AND POLARIZATION RESULTS

As discussed in Section II, one expects to see a heliographic longitude dependence of the number of observed soft X-ray bursts for certain nonthermal models. The 2–12 Å X-ray bursts observed with the University of Iowa Geiger counter experiments on Explorers 33 and 35 were used by Pinter (1969) and Ohki (1969) to obtain the longitude dependence of the bursts. Ohki found no longitude dependence of a set of 232 such bursts, but Pinter found a peak at 30–40° longitude using a larger sample of 490 bursts. Neither author took into account the Hα flare longitude distribution.

Drake (1971) plotted the longitude distribution of a total of 2698 2–12 Å bursts and took into account the Hα flare distribution for the period he studied. His result is shown in Figure 13. No statistically significant directivity is indicated with the possible exception of the 70° to 90° region. The results in this region are due to the fact that the Hα distribution during the time of X-ray coverage had a sharp drop in the 70°–80° longitude region followed by a peak in the 80° to 90° region. The 2–12 Å data should not be subject to substantial modification through the Compton backscattering process (Tomblin, 1972) because of the dominance of photoelectric absorption over Compton scattering. The lack of directivity can be taken as evidence for a thermal interpretation of the bulk of soft X-ray events.

Unfortunately, there do not appear to be any polarization measurements made on the low energy X-ray events probably due partly to the assumed thermal nature of these events. Brown (1974) has pointed out that Compton scattering will introduce a small amount of polarization in the detected signal even if the source is thermal. Wolff (1973) has calculated the minimum polarization measurable over a 100 s period with a lithium scattering block experiment as a function of energy for a flare spectrum.

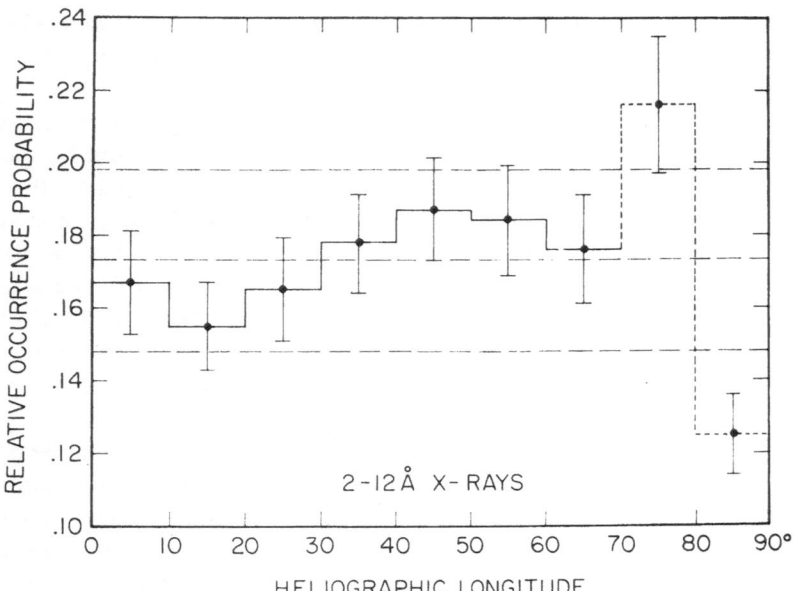

Fig. 13. The relative probability for the occurrence of 2–12 Å solar X-ray bursts as a function of solar longitude (Drake, 1971). Data are from Explorers 33 and 35 from July 1966 to September 1968.

The detectable polarization is very low ($<1\%$) in the lowest energy range (5–9 keV) and increases with energy due to the poorer statistics at high energy. Polarization measurements in the low energy X-ray range are facilitated by higher fluxes and longer characteristic time scales than occur with the high energy impulsive bursts. Two polarization experiments are planned for the Solrad 11 satellite to be launched in 1975, a Bragg crystal mounted to observe a region of the continuum at 2.8 Å and a Compton scattering experiment with two energy channels at about 8 to 15 and 15 to 50 keV. The experiments also have the advantage that they are mounted around the spin axis of the spacecraft which points at the Sun. This experiment should provide measurements very useful in establishing the thermal or nonthermal nature of low energy X-ray events.

References

Arnoldy, R. L., Kane, S. R., and Winckler, J. R.: 1968, *Astrophys. J.* **151**, 711.

Blake, R. L. and House, L. L.: 1971, *Astrophys. J.* **166**, 423.

Brown, J. C.: 1972, *Solar Phys.* **26**, 441.

Brown, J. C.: 1974, in G. Newkirk (ed.), 'Coronal Disturbances', *IAU Symp.* **57**, D. Reidel Publ. Co., Dordrecht, Holland, p. 395.

Brown, J. C., McClymont, A. N., and McLean, I. S.: 1974, *Nature* **247**, 448.

Chubb, T. A.: 1970, in E. R. Dyer (General Ed.), *Solar Terrestrial Physics*, Vol. 1, D. Reidel Publ. Co., Dordrecht, Holland, p. 99.

Culhane, J. L. and Phillips, K. J. H.: 1970, *Solar Phys.* **11**, 117.

Datlowe, D. W., Elcan, M. J., and Hudson, H. S.: 1974, *Solar Phys.* **39**, 155

de Jager, C. and Kundu, M. R.: 1963, *Space Res.* **3**, 836.

Doschek, G. A., Meekins, J. F., Kreplin, R. W., Chubb, T. A., and Friedman, H.: 1972, in K. Schindler (ed.), *Cosmic Plasma Physics*, Plenum, p. 165.

Drake, J. F.: 1971, *Solar Phys.* **16**, 152.

Elwert, G. and Haug, E.: 1971, *Solar Phys.* **20**, 413.

Frost, K. J.: 1969, *Astrophys. J.* **158**, L159.

Frost, K. J. and Dennis, B. R.: 1971, *Astrophys. J.* **165**, 655.

Hudson, H. S. and Ohki, K.: 1972, *Solar Phys.* **23**, 155.

Jackson, J. D.: 1962, *Classical Electrodynamics*, Wiley, New York.

Kahler, S.: 1971a, *Astrophys. J.* **164**, 365.

Kahler, S.: 1971b, *Astrophys. J.* **168**, 319.

Kahler, S. W.: 1973, in R. Ramaty and R. G. Stone (eds.), *High Energy Phenomena on the Sun*, NASA SP-342, p. 124.

Kahler, S. W. and Kreplin, R. W.: 1971, *Astrophys. J.* **168**, 531.

Kahler, S. W., Meekins, J. F., Kreplin, R. W., and Bowyer, C. S.: 1970, *Astrophys. J.* **162**, 293.

Kane, S. R.: 1971, *Astrophys. J.* **170**, 587.

Kane, S. R.: 1972, *Solar Phys.* **27**, 174.

Kane, S. R.: 1973a, in R. Ramaty and R. G. Stone (eds.), *High Energy Phenomena on the Sun*, NASA SP-342, p. 55.

Kane, S. R.: 1973b, in G. Newkirk (ed.), 'Coronal Disturbances', *IAU Symp.* **57**, D. Reidel Publ. Co., Dordrecht, Holland, p. 105.

Kane, S. R.: 1975 (to be published).

Kane, S. R. and Anderson, K. A.: 1970, *Astrophys. J.* **162**, 1003.

Landini, M., Monsignori Fossi, B. C., and Pallavicini, R.: 1972, *Solar Phys.* **27**, 164.

Landini, M., Monsignori Fossi, B. C., and Pallavicini, R.: 1973, *Solar Phys.* **29**, 93.

Lin, R. P.: 1971, *Acceleration of 10–100 keV Electrons in Solar Flares*, Seminar on the Acceleration of Particles in Near-Earth and Interplanetary Space, Galaxy, and Metagalaxy, Leningrad.

Lin, R. P., Evans, L. G., and Fainberg, J.: 1973, *Astrophys. Letters* **14**, 191.

Milkey, R. W.: 1971, *Solar Phys.* **16**, 465.

Ohki, K.: 1969, *Solar Phys.* **7**, 260.

Parks, G. K. and Winckler, J. R.: 1969, *Astrophys. J.* **155**, L117.

Peterson, L. E., Datlowe, D. W., and McKenzie, D. L.: 1973, in R. Ramaty and R. G. Stone (eds.), *High Energy Phenomena on the Sun*, NASA SP-342, p. 132.

Phillips, K. J. H.: 1973, *Observatory* **93**, 17.

Phillips, K. J. H. and Neupert, W. M.: 1973, *Solar Phys.* **32**, 209.

Pinter, S.: 1969, *Solar Phys.* **8**, 142.

Ramaty, R.: 1973, in R. Ramaty and R. G. Stone (eds.), *High Energy Phenomena on the Sun*, NASA SP-342, p. 188.

Santangelo, N., Horstman, H., and Horstman-Moretti, E.: 1973, *Solar Phys.* **29**, 143.

Shaw, M. L.: 1972, *Solar Phys.* **27**, 436.

Sommerfeld, A. J. F.: 1951, *Atombau und Spectrallinien*, Ungar, New York.

Spitzer, L.: 1962, *Physics of Fully Ionized Gases*, Interscience, New York.

Sturrock, P. A.: 1973, in R. Ramaty and R. G. Stone (eds.), *High Energy Phenomena on the Sun*, NASA SP-342, p. 3.

Takakura, T. and Kai, K.: 1966, *Publ. Astron. Soc. Japan* **18**, 57.

Takakura, T., Ohki, K., Shibuya, N., Fujii, M., Matsuoka, M., Miyamoto, S., Nishamura, J., Oda, M., Ogawara, Y., and Ota, S.: 1971, *Solar Phys.* **16**, 454.

Tindo, I. P., Ivanov, V. D., Mandel'stam, S. L., and Shuryghin, A. I.: 1972a, *Solar Phys.* **24**, 429.

Tindo, I. P., Ivanov, V. D., Valniček, B., and Livshits, M. A.: 1972b, *Solar Phys.* **27**, 426.

Tindo, I. P., Mandel'stam, S. L., and Shuryghin, A. I.: 1973, *Solar Phys.* **32**, 469.

Tomblin, F. F.: 1972, *Astrophys. J.* **171**, 377.

Wolff, R. S.: 1973, in R. Ramaty and R. G. Stone (eds.), *High Energy Phenomena on the Sun*, NASA SP-342, p. 162.

HIGH TIME RESOLUTION ANALYSIS OF SOLAR FLARES OBSERVED ON THE ESRO TD-1A SATELLITE

PETER HOYNG, JOHN C. BROWN*, GERARD STEVENS and H. F. VAN BEEK

Space Research Laboratory, University of Utrecht, The Netherlands

Summary

(a) INSTRUMENT AND DATA REDUCTION

The Utrecht Hard Solar X-Ray Spectrometer on board the ESRO TD-1A satellite (launch March 1972) is permanently Sun-pointed and measures the solar radiation between 30 keV and 1000 keV, in 12 logarithmically spaced energy channels, with a continuous fine time resolution, viz. 1.2 s for the four lowest energy channels and 4.8 s for the rest. The detector has a 5 cm² Cs I (Na) crystal; counts due to particles are rejected and even during the largest solar flares saturation effects (e.g. pulse pile-up) are absent. For further details see Van Beek (1973), and Van Beek and De Feiter (1973).

The instrument has successfully operated during the two years' lifetime of the satellite and has observed a number of solar flares.

A method was developed that actually reconstructs the photon spectrum from the measured pulse height distribution and yet is much faster than the usual two parameter χ^2-fit (Hoyng and Stevens, 1974). This permitted conversion of large amounts of pulse height distributions into photon spectra and thus full utilization of the 1.2 s time resolution. In addition, a single power law fit to the photon spectrum is routinely computed.

(b) OBSERVATIONAL MATERIAL; THICK TARGET ANALYSIS

A discussion is presented of our analysis of two smaller events, 1972, May 18, UT 1406, class 1B/M4 and 1972, August 7, UT 0252, class SB/M2, and the large event of 1972, August 4, UT 0620, class 2B/X5 (see Hoyng and Stevens, 1973, and Van Beek *et al.*, 1973).

The May 18 event shows a regular single spike structure lasting $\sim 10^2$ s, not unlike the spikes seen by Kane and Anderson (1970); the August 7 event was similar, except for its more detailed time structure. Above 100 keV no flux was detected. The August 4 event is among the largest events ever recorded. The flare still emitted hard X-rays amply when data coverage ended after 21 min from the onset. In addition to rapid time variations the count rates appear to fluctuate periodically, the period increasing from some 20 s through 60 s to 120 s. The photon spectra of the two small events

* On leave from Department of Astronomy, University of Glasgow, Scotland.

Sharad R. Kane (ed.), Solar Gamma-, X-, and EUV Radiation, 233–235. All Rights Reserved.

are simple power laws with γ increasing linearly (May 18) or constant (August 7) through the entire event, as also found by McKenzie *et al.* (1973) but contradicting Kane and Anderson's (1970) results. All authors however seem to find that towards the end of *small* events the spectrum softens. This feature, together with the detailed (spiky) structure of the time profile itself, are the main arguments for the existence of continuous electron acceleration.

The spectra of the August 4 event, detectable up to ~ 400 keV, are not simple power laws. They have a break, not necessarily sharp, around ~ 60 keV after which the spectrum steepens by $\Delta\gamma \sim 1$. (In the single power law fits, $\gamma(t)$ *decreases* towards the end of the event). This break (cf. also Frost, 1969) is probably too large to be interpreted as an anisotropy effect (cf. Petrosian, 1973) and so should be present in the spectrum of the X-ray emitting electrons. However, as long as no good theoretical explanation for pure power laws exists, observation of a power law with a break must not be considered as more peculiar than a simple power law itself (cf. the cosmic ray spectrum). Using the single power law fits, we determined for all three events with full time resolution the thick target parameters F_{25} and P_{25}, being the total required number flux and energy flux of fast electrons $\geqslant 25$ keV into a target region (Brown, 1971). Table I summarizes typical values.

TABLE I

Total required number and energy flux of electrons $\geqslant 25$ keV
in a thick target X-ray source

	Smaller event (lasting 10^2 s)	Large event (lasting 10^3 s)
Mean F_{25} (s^{-1})	0.5–1.0×10^{36}	4×10^{36}
$\int F_{25}dt$	4×10^{37}	5×10^{39}
$\int P_{25}dt$ (erg)	2×10^{30}	2×10^{32}
Mean $n_0 N_{25}$ (cm^{-3})	3–8×10^{45}	3.5×10^{46}

The interpretation of these numbers poses great problems: the energy put into fast electrons about equals the total flare energy spent and the total number of electrons ever accelerated is about equal to the number of electrons in the whole flare, taking some reasonable density (10^{10-11} cm^{-3}). Moreover these numbers are lower limits in the sense that one could well extrapolate below 25 keV and/or the target might be thin!

By using reasonable inflow velocities into some flare region we find that in particular F_{25} is too high. We argue that the only practical way to bring F_{25} down is by accelerating an electron not once but many times. In this case F_{25} has no meaning and from the data one derives the nonthermal emission measure $n_0 N_{25} =$ ambient density × total number of fast electrons $\geqslant 25$ keV at the given time. One could term this a thick target with containment. There is however no way around the large values of $\int P_{25}\, dt$ and we have to face the situation that nearly all flare energy is channeled through fast electrons into other forms. An interesting new feature of the data is the marked correlation of γ and F_{25} in the August 4 event. During the last 7 min (decay phase) the correlation is particularly good and γ and F_{25} decrease monotonically with time. This could indicate the slow collisional decay of a large, low density cloud. From elementary considerations we find a density $n_0 \sim 10^8$ cm^{-3} and a volume $V \sim 10^{33}$ containing $\sim 10^{38}$ fast electrons. These electrons could be due to escape of a small fraction of the total number available, viz. 5×10^{39} (Table I).

(c) FOURIER ANALYSIS

Power spectra have been computed of the raw count rates of channel 2, F_{25} and γ for the three events (the assistance of Drs R. Rutten and G. Geytenbeek is gratefully acknowledged). The results can be summarized as follows:

(i) The power spectrum of the raw count rates of channel 2 of the August 4 event shows *significant* periodicities at 120, 60 and 33 s. The same periodicities are present in the power spectra of F_{25} and, to a lesser degree, of γ. No periodicities were found in the two other events.

(ii) In all three events, the power spectrum of the raw count rates of channel 2 shows an average high frequency power level *significantly* above the level expected from Poisson noise alone. We drew the conclusion that the solar flare must have intrinsic white noise in its emission at all periods from 15 s down to the Nyquist period, 2.4 s.

The periodicities in the August 4 event must correspond to *coherent* changes of the whole flare region. Directly from the time profile we infer physical timescales $\tau = |(1/c)\,(dc/dt)|^{-1}$, $c =$ count rate, for these changes and find $\tau \gtrsim 10$ s for the large event and $\tau \gtrsim 2$ s (non periodic) for the small events – so that neither case shows changes in the source *as a whole* in times down to 1.2 s.

References

Brown, J. C.: 1971, *Solar Phys.* **18**, 489.

Frost, K.: 1969, *Astrophys. J.* **158**, L159.

Hoyng, P. and Stevens, G. A.: 1973, Proceedings First European Astronomical Meeting, Athens 1972, vol. 1, 97.

Hoyng, P. and Stevens, G. A.: 1974, *Astrophys. Space Sci.* **27**, 307.

Kane, S. R. and Anderson, K. A.: 1970, *Astrophys. J.* **162**, 1003.

McKenzie, D. L., Datlowe, D. W., and Peterson, L. E.: 1973, *Solar Phys.* **28**, 175.

Petrosian, V.: 1973, *Astrophys. J.* **186**, 291.

Van Beek, H. F.: 1973, Ph.D. Thesis, Utrecht.

Van Beek, H. F. and de Feiter, L. D.: 1973, Proceedings First European Astronomical Meeting, Athens 1972, vol. 1, 103.

Van Beek, H. F., Hoyng, P., and Stevens, G. A.: 1973, in H. E. Coffey (ed.), *Collected Data Reports on the August 1972 Solar-Terrestrial Events*, Report UAG-28, part II, 319.

RISE TIME OF HARD X-RAY BURSTS

JOAN VORPAHL

University of California, San Diego and Sacramento City College, California, U.S.A.

and

TATSUO TAKAKURA

Dept. of Astronomy, University of Tokyo, Tokyo, Japan

Summary. A study was made of the hard X-ray component in the impulsive phase of solar flares. In 36 randomly chosen events the value for the slope in the differential electron power spectrum $E^{-\delta}$ electrons cm^{-2} s^{-1} keV^{-1}, was related to the 20–32 keV spike rise time (*e*-folding) as $t_{\text{rise}} = 0.56 \exp(0.88\delta)$ in the thin target model and $t_{\text{rise}} = 0.10 \exp(0.88\delta)$ in the thick target picture. In the thin target model, the above empirical relation would imply that the acceleration of electrons can be longer when the acceleration rate is smaller. An alternative interpretation would be that an impulsive hard X-ray burst is a superposition of two components emitted from thin and

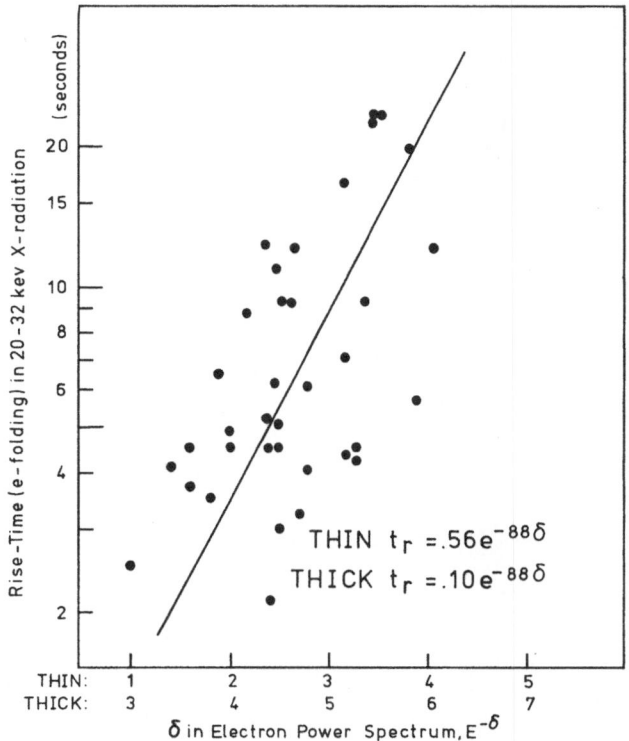

Fig. 1.   Rise-time in 20–32 keV X-radiation versus electron hardness.

Sharad R. Kane (ed.), Solar Gamma-, X-, and EUV Radiation, 237–238. All Rights Reserved.
Copyright © 1975 by the IAU.

thick targets; when the former predominates, the duration is longer and the photon spectral index is larger, while when the latter predominates, the duration is shorter and the photon spectral index is smaller; $3 \lesssim \delta \lesssim 4$ is required (Figure 1). The uncertainty in δ is 0.5 while that in the rise time is 1 s.

DETERMINATION OF THE HEIGHT OF HARD X-RAY SOURCES IN THE SOLAR ATMOSPHERE BY MEASUREMENT OF PHOTOSPHERIC ALBEDO PHOTONS

JOHN C. BROWN* and H. F. VAN BEEK

Space Research Laboratory of the Astronomical Institute, Utrecht, The Netherlands

Summary **. The importance and difficulties of determining the height of hard X-ray sources in the solar atmosphere, in order to distinguish source models, have been discussed by Brown and McClymont (1974) and also in this Symposium (Brown, 1975; Datlowe, 1975). Theoretical predictions of this height, h, range between $\lesssim 10^3$ and 10^5 km above the photosphere for different models (Brown and McClymont, 1974; McClymont and Brown, 1974). Equally diverse values have been inferred from observations of synchronous chromospheric EUV bursts (Kane and Donnelly, 1971) on the one hand and from apparently behind-the-limb events (e.g. Datlowe, 1975) on the other.

Direct resolution of the height of a source at the limb for sources at the low end of this height range (corresponding to angles $\zeta \lesssim 1\overset{''}{.}4$ at the Sun) is certainly impossible for any planned hard X-ray heliograph while, even for sources in the 10^4–2×10^4 km range ($\zeta \simeq 14\overset{''}{.}-28''$), this method does not distinguish low sources exactly at the limb from high sources near the limb. Thus it is of interest to consider whether heights might be inferred for sources *on the disk* and particularly for those at low altitudes. This is made potentially possible only by virtue of the fact that the dense photosphere effectively provides a 'mirror' behind the source due to Compton backscattering of X-rays (Tomblin, 1972; Santangelo *et al.*, 1973). Though several means may be considered for utilising this (e.g. Brown *et al.*, 1974) the most promising one in terms of planned instrumentation is, in our view, spatial resolution of the patch of albedo X-rays behind the primary source. For simplicity here we consider only the case of a small primary source (diameter $\lesssim 10''$) at the disc center, for which the albedo patch consists of circular isobrightness contours out to the source's solar horizon at a distance $r \simeq \sqrt{2Rh}$ (R = solar radius) corresponding roughly to angular radii of 50", 160" and 350" for $h = 10^3$, 10^4 and 5×10^4 km respectively (i.e. $\zeta \sim 1\overset{''}{.}3$, 13" and 65"). Clearly, therefore, observation of this patch requires much less resolution than needed for the source height itself – the essential advantage of the method. Furthermore, the evidence is that the primary source does indeed have a small horizontal extent (Takakura *et al.*, 1971), so the method is unlikely to be vitiated by obscuration of the albedo distribution by the primary emission.

Actual detection of the albedo patch depends on the capabilities of the hard X-ray heliograph employed. In particular, the brightness of the patch is easily shown to drop

---

* On leave from: Dept. of Astronomy, University of Glasgow, Scotland.
** Paper to appear in *Astronomy and Astrophysics*.

Sharad R. Kane (ed.), Solar Gamma-, X-, and EUV Radiation, 239–241. All Rights Reserved.

off from the subsource point roughly like

$$dI \text{ (cts per cm}^2 \text{ per square arc sec)} = \frac{f I_0 \zeta}{2\pi (\varrho^2 + \zeta^2)^{3/2}}, \tag{1}$$

where ϱ is the angular distance from the subsource point, I_0 is the primary source photon flux (here assumed isotropic) and f is the *total* backscattered fraction. Distribution (I) has to be compared with the instrumental response profile to a point source to determine the actual angular radius ϱ_0 over which the albedo is detectable. The Hard X-ray collimating heliograph under development in Utrecht forms real-time pictures in the form of a 4.3×4.3 array of square elements of $8'' \times 8''$ each in 5 energy bands from 3.5–20 keV. Each element has a sensitive area of 4 mm^2. The highest (15–20 keV) channel of this instrument is about ideal for observing the albedo, being above the range where f is reduced by photospheric absorption (Santangelo *et al.*, 1973) but low enough to maintain reasonable count rates. Figure 1 shows the angular

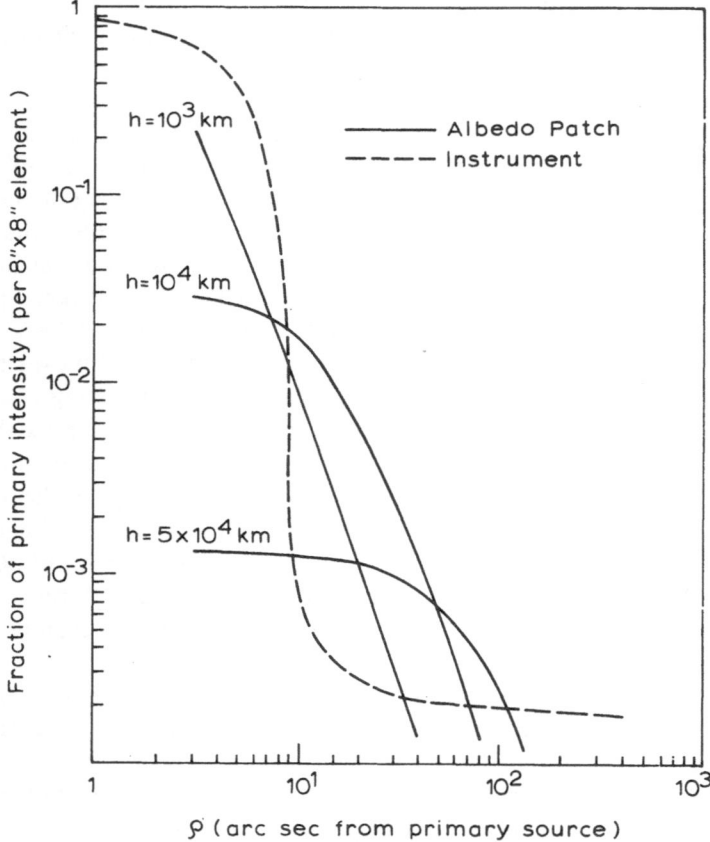

Fig. 1.   Albedo brightness distribution as a function of the angular radius ϱ'' from a primary source at the disc center, for various source heights h, compared to the instrumental profile for a point source.

response profile of this instrument to a point source, mainly due to collimator characteristics, compared to the albedo profile for $h = 10^3$, 10^4, and 5×10^4 km. It is evident that the albedo patch can indeed be measured with this instrument for such source heights (see also Table I).

Finally it is necessary to consider the integration time necessary to determine the albedo distribution above the statistical noise which depends of course, on the actual primary source intensity. However, even for a fairly modest event, with $I_0 = 10^3$ photons cm^2 s^{-1} in the 15–20 keV range, it is found that a 30 sec integration time will give ample precision in the albedo measurement for $h \lesssim 5 \times 10^4$ km (Table I).

TABLE I

Visibility of X-ray albedo above the instrument profile

Source Height		Albedo visible above instrument profile over		Estimated variance in albedo counts after 30 s integration for a flare producing 10^3 photons cm^{-2} s^{-1} between 15 and 20 keV
h (km)	ξ	Radius ϱ_0	No. of image elements	
10^3	1″37	34″	53	16%
10^4	13″7	71″	244	6%
5×10^4	68″7	110″	590	10%

References

Brown, J. C.: 1975, This volume, p. 245.
Brown, J. C., McClymont, A. N., and McLean, I. S.: 1974, *Nature* **247**, 448.
Brown, J. C. and McClymont, A. N.: 1974, *Solar Physics*, in press.
Datlowe, D. W.: 1975, This volume, p. 191.
Kane, S. R. and Donnelly, R. F.: 1971, *Astrophys. J.* **164**, 151.
McClymont, A. N. and Brown, J. C.: 1974, in preparation.
Santangelo, N., Horstman, H., and Horstman-Moretti, E.: 1973, *Solar Phys.* **29**, 143.
Tomblin, F. F.: 1972, *Astrophys. J.* **121**, 377.
Takakura, T., Ohki, K., Shibuya, N., Fujii, M., Matsuoka, M., Miyamoto, S., Nishimura, J., Oda, M., Ogawara, Y., and Ota, S.: 1971, *Solar Phys.* **16**, 454.

INFERENCE OF THE HARD X-RAY SOURCE DIMENSIONS IN THE 1972, AUGUST 7 WHITE LIGHT FLARE

DAVID M. RUST

Sacramento Peak Observatory, Air Force Cambridge Research Laboratories, Sunspot, N.M., U.S.A.

Summary*. Broadband photographs and spectra of the white light flare of 1972, August 7 have been compared with hard X-ray spectra from the same event. There is a very close temporal correspondence between the hard X-ray and white light emission curves, and these emissions come from layers that are separated by a height of less than 2000 km. The flare shows at least two distinct particle acceleration phases: the first, occurring at a stationary source, gave very bluish continuum emission from 4 bright stationary knots while the X-ray ($E > 60$ keV) spectrum hardened and reached peak intensity. This phase occurred between 1520 and 1523 UT. In the second phase (1524–1537 UT) the bright knots dissolved and a faint wave moved out from the flare center at 40 km s^{-1}. The spectrum of the wave was nearly flat in the range 4950–5900 Å and analysis of the spectrum indicates that the emission was probably due to heating and ionization by 20–100 keV electrons. The X-ray spectrum, as derived from Interkosmos 7 and ESRO TD-1A satellite data, becomes softer during the wave phase. The close correspondence between the X-ray and continuum emission events shows that, in effect, the hard X-ray source has been resolved. It consists of several changing patches approximately $3'' \times 5''$ in area, consistent with the upper limit of 1' from balloon observations (Takakura *et al.*, 1971).

References

Takakura, T., Ohki, K., Shibuya, N., Fuji, M., Matsuoka, M., Miyamoto, S., Nishimura, J., Oda, M., Ogawara, Y., and Ota, S.: 1971, *Solar Phys.* **16**, 454.

---

* For the full text of the paper see *Solar Phys.* **40** (1975), 141.

THE INTERPRETATION OF SPECTRA, POLARIZATION, AND DIRECTIVITY OF SOLAR HARD X-RAYS

JOHN C. BROWN

Space Research Laboratory of the Astronomical Institute, Utrecht, The Netherlands *

Abstract. The importance of interpretation of hard X-ray burst spectra, polarisation and directivity to the flare process as a whole is emphasised. After critically reviewing observations of these and related burst characteristics, the problems of analytic and numerical inversion of the X-ray spectrum to give the flare electron spectrum are discussed and it is concluded that electron spectra cannot be accurately and unambiguously inferred from their bremsstrahlung emission. Consideration of directional, albedo, and model-dependent effects, on the other hand, shows that none of the X-ray data are at present inconsistent with a power-law electron acceleration spectrum.

Characteristics of thick-target, thin-target and electron-trap models of hard X-ray sources are discussed quantitatively and their ability to fit the observations is examined. Selection of a satisfactory model is precluded by lack of both sufficient observations and of adequate theoretical description of models. Nevertheless, it is suggested that redistribution of the flaring atmosphere and the effects of collective energy losses may reconcile even behind-the-limb burst observations and interplanetary electron spectra with a thick-target description (which fits other data well). This is attractive since a thick-target X-ray source makes the minimal demand on flare energy. Even a thick-target, however, requires an embarrassingly large number and energy of fast electrons. Therefore the review is completed by discussing how these requirements might be reduced if thermal emission extended to hard X-ray energies or if multiple reacceleration of electrons occurred.

I. Introduction

Interesting as some problems of the hard X-ray flare may be *per se*, their basic importance lies in their relationship to the fascinating and long-standing conundrum of the solar flare as a whole, reviews of which include those by Sturrock and Coppi (1966) and by Sweet (1969, 1971). Of the several basic requirements, listed in these reviews, for any flare model, two of the most demanding are the sufficiently rapid conversion of energy from preflare (magnetic) storage into both energetic particle streams and heating of the thermal flare plasma to produce thermal radiation and mass motions. The importance of hard X-ray bursts in this connection is, firstly, to give the most direct view of accelerated flare particles *in situ*, radio burst and interplanetary particle observations being relatively difficult to interpret due to propagation effects. Secondly the electrons responsible for the bursts lie at the low energy end of the steep spectrum of accelerated particles and so comprise by far the bulk of the total number and energy of these particles (Neupert, 1968; Brown, 1971, and many subsequent authors). Furthermore, there is considerable evidence that this total energy may in some flares equal or exceed the total requirements for the thermal flare phenomena (e.g. Neupert, 1968; Brown, 1971, 1972a; Kahler and Kreplin, 1971; Syrovatskii and Shmeleva, 1972; McKenzie *et al.*, 1973).

Consequently there has been wide interest lately in the detailed development of flare

---

* On leave from: Dept. of Astronomy, University of Glasgow, Scotland.

Sharad R. Kane (ed.), Solar Gamma-, X-, and EUV Radiation, 245–282. All Rights Reserved.
Copyright © 1975 by the IAU.

models in which the thermal flare plasma is heated by the energetic electrons responsible for the hard X-ray burst (e.g. Strauss and Papaggianis, 1971; Cheng, 1972;
Hudson, 1972; Brown, 1973a; Shmeleva and Syrovatskii, 1973). Empirical evidence
for this hypothesis in terms of the interrelationships of hard X-rays with thermal flare
characteristics has been considered by many authors (see, e.g., reviews by Hudson
(1973) and Brown (1973b)). Not only does this model result in some conceptual simplification of the overall flare problem, by linking the processes of particle acceleration
and flare heating, but it also permits detailed quantitative prediction of the heated
atmospheric structure (Brown, 1973a) and so testing of the model by calculation of
the expected optical line profiles (Canfield, 1974).

Clearly then it is essential to have the most reliable procedure possible for inferring
flare electron numbers and spectra. Uncertainties in the inference of these from X-ray
burst data stem from several sources:

(1) approximations used in the conversion from X-ray to effective electron spectra
in the source, both in the basic physics (e.g. bremsstrahlung cross section) and in the
mathematical procedure followed;

(2) model dependence of the flux and spectrum of electrons *at acceleration* as
against their effective values in the source;

(3) major gaps in our knowledge of the basic physical processes governing the
source electrons, and especially their energy losses.

After reviewing the basic observational features of bursts, I have aimed at discussing
the most important of each of these sources of uncertainty, posing the problems with
which the current state of the art of burst interpretation leaves us, and suggesting some
areas for investigation which might help us emerge from these.

II. The Mechanism of Burst Emission

It is now widely accepted that the source mechanism in the keV-MeV range is collisional bremsstrahlung of energetic flare electrons on protons (and on heavy ions) in
the flare plasma, synchrotron radiation and inverse Compton scattering having been
ruled out by arguments such as Korchak's (1967a, 1971). Virtually nothing is as yet
known of the role of *collective* interaction of the electron streams with the flare plasma
in hard X-ray sources (cf. Sections V(a) and VI(c)), but it would appear that these
interactions are incapable of direct generation of X-rays, due to the long wavelengths
of plasmons, though they may play a vital part in the overall energy balance of the
source (cf. Section VI(c)) – q.v. Tsytovich (1973) and references therein.

The source being optically thin, the total bremsstrahlung intensity may be written:

$$I(\varepsilon, t) = \frac{1}{4\pi R^2} \int_V n(\mathbf{r}, t) \int_\varepsilon^\infty F(E, \mathbf{r}, t) \, Q(\varepsilon, E) \, dE \, dV, \qquad (1)$$

where $I(\varepsilon, t)$ = mean instantaneous photon flux from whole source at $R = 1$ AU, differential in photon energy ε; $F(E, \mathbf{r}, t)$ = instantaneous electron number flux, differential

in electron energy E, at position $\mathbf{r}$ in the total (instantaneous) source volume V; $n(\mathbf{r}, t) =$ total proton number density at $\mathbf{r}$; $Q(\varepsilon, E) =$ bremsstrahlung emission cross-section at ε, E, differential in ε; Q should be written to include the substantial correction factor for heavy ions (Elwert and Haug, 1971); a correction may also be made for electron-electron bremsstrahlung though this is very minor except at relativistic photon energies (Takakura, 1969).

Source Equation (1) describes bursts observed with spectral and temporal resolution but without spatial resolution and averaged over all planes of polarisation and directions of observation. Generalisations to include these latter features will be implied where necessary in the text, details being available in the references.

In addition it must be emphasised that (1) refers only to the primary emission from the source and does not include the important contribution to the total observed burst from primary photons backscattered from the photosphere. The importance of this component, first pointed out by Tomblin (1972) and by Santangelo *et al.* (1973), has not yet been fully realized or investigated. From the theoretical viewpoint, it requires a quite distinct treatment from the primary component since it involves transfer of photons by multiple Compton scattering and hence modification of the intensity, spectral, directional, polarisation and spatial characteristics of the total emission. The influence of these changes on the considerations in this review are mentioned in the text in so far as they have currently been worked out (Tomblin, 1972; Santangelo *et al.*, 1973; Beigman and Vainstein, 1974; Brown *et al.*, 1974).

III. Resume of Observed Burst Characteristics

(a) TIME PROFILES

Kane (1969) and Kane and Anderson (1970) have described *small* events in terms of impulsive and gradual components, the former having a profile consisting of a spike (sometimes repeated) of *e*-folding rise and fall times $\lesssim 10$ s, and the latter a smooth rise and fall over several minutes (Figure 1). Dominance of the impulsive over the gradual component with increasing photon energy, and the relationship to radio bursts, probably identifies the former as non-thermal bremsstrahlung of fast electrons and the latter as gradual thermal flare emission.

Larger events with total durations of minutes to tens of minutes show a complex spiky time structure, sometimes of quasi-periodic form – Figure 2 (Frost, 1969; Frost and Dennis, 1971; Parks and Winckler, 1969; Van Beek *et al.*, 1974; Hoyng *et al.*, 1975). Additionally, Frost and Dennis (1971) and Frost (1974a) have reported a distinct second component in some large events, after the 'impulsive' phase and with a persistent hard spectrum (Figure 3), possibly associated with a second phase of particle acceleration (Frost and Dennis, 1971).

Comparitively little has yet been achieved in the quantitative time series analysis of burst time profiles, as against phenomenological modelling. Hoyng *et al.* (1975), in the first detailed study of this type, have, however, shown how much more information may be extracted from existing data than is generally done. In particular, they empha-

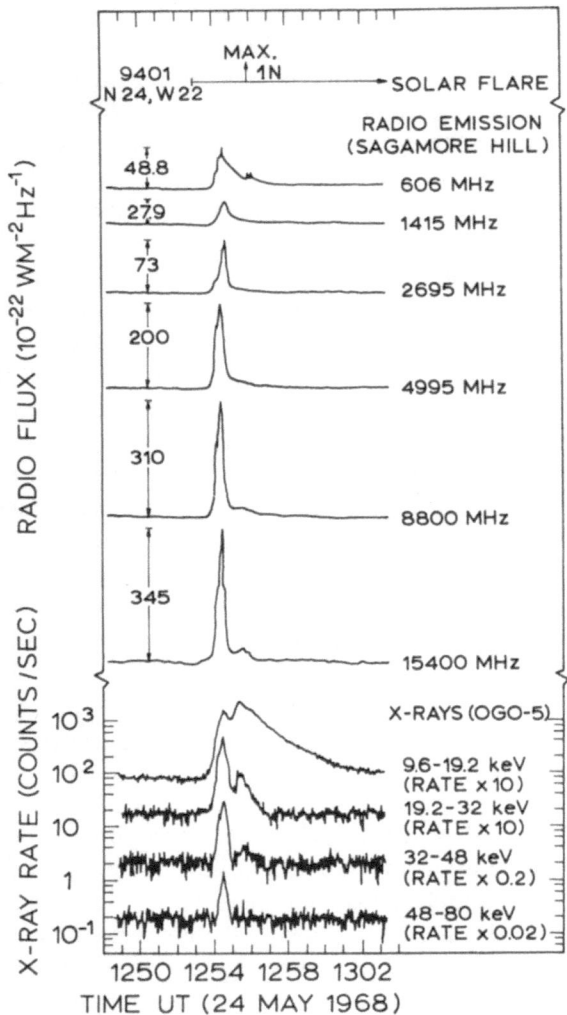

Fig. 1.   An example of the simple two component (impulsive and gradual) structure of some small
events reported by Kane and Anderson (1970).

sise the need for caution in the inference of time scales characterising the source from
time profiles and find e-folding times for the source as a whole, considerably longer
than their instrumental resolution (1.2 s).

(b) SPECTRAL CHARACTERISTICS

Flare X-ray spectra above about 10 keV are much harder than the exponential form
$(1/\varepsilon)\,e^{-\varepsilon/kT}$ from any *isothermal* plasma, a convenient and widely used representation
of such spectra in astrophysics being the negative power-law $\varepsilon^{-\gamma}$. Though the real

physical significance of this functional form is, as discussed later, very debatable, the representation is convenient for intercomparison of bursts, folding through instrumental responses, and for reasonable estimates of flare electron parameters in most events (cf. Sections IV and V).

For small events, a good spectral fit is often only obtained near the burst peak intensity. Kane (1971) has studied the statistics of this γ for many small events and finds a frequency distribution increasing from $\gamma \gtrsim 2$ up to $\gamma \gtrsim 6$ (above $\gamma = 6$, pulse

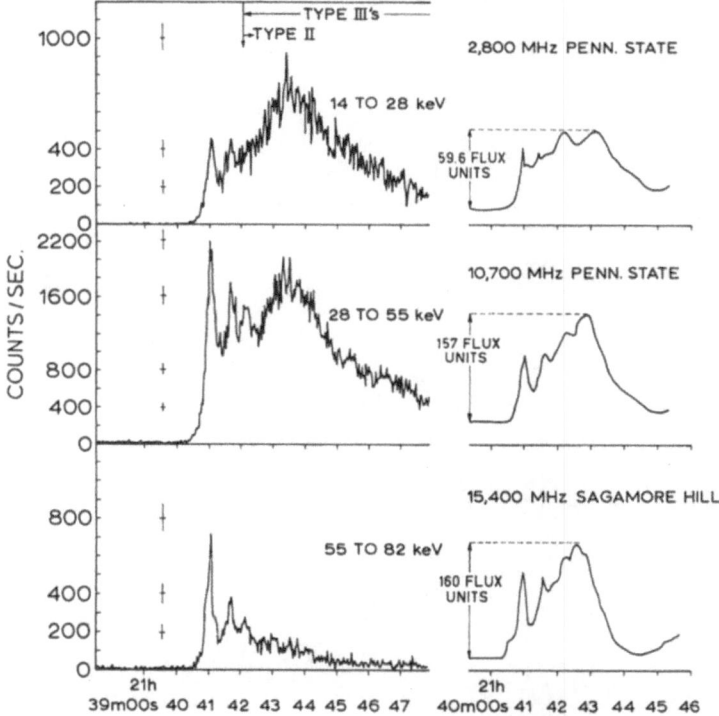

Fig. 2.   Complex time structure of a large event observed by Frost (1969).

pile-up prevents determination of the true γ). Similar studies have been reported by Datlowe (1975). Kane (e.g. 1974a) has also investigated the correlation of X-ray burst intensities and spectra with other flare characteristics. For larger 'impulsive' events a wide range of γ values up to over 7 has likewise been reported (e.g. Hoyng *et al.*, 1975) while Frost's second phase apparently always has a hard spectrum ($\gamma \simeq 3$).

Time development of spectra through events has also been studied (cf. Kane's review, 1974a). Kane and Anderson (1970) have reported that the impulsive spikes of some small events harden as they rise (γ decreasing) and soften as they fall (γ increasing) – Figure 4. Though similar behaviour has been reported for some other events (e.g. Parks and Winckler, 1969; McKenzie *et al.*, 1973), it does not in fact seem to be

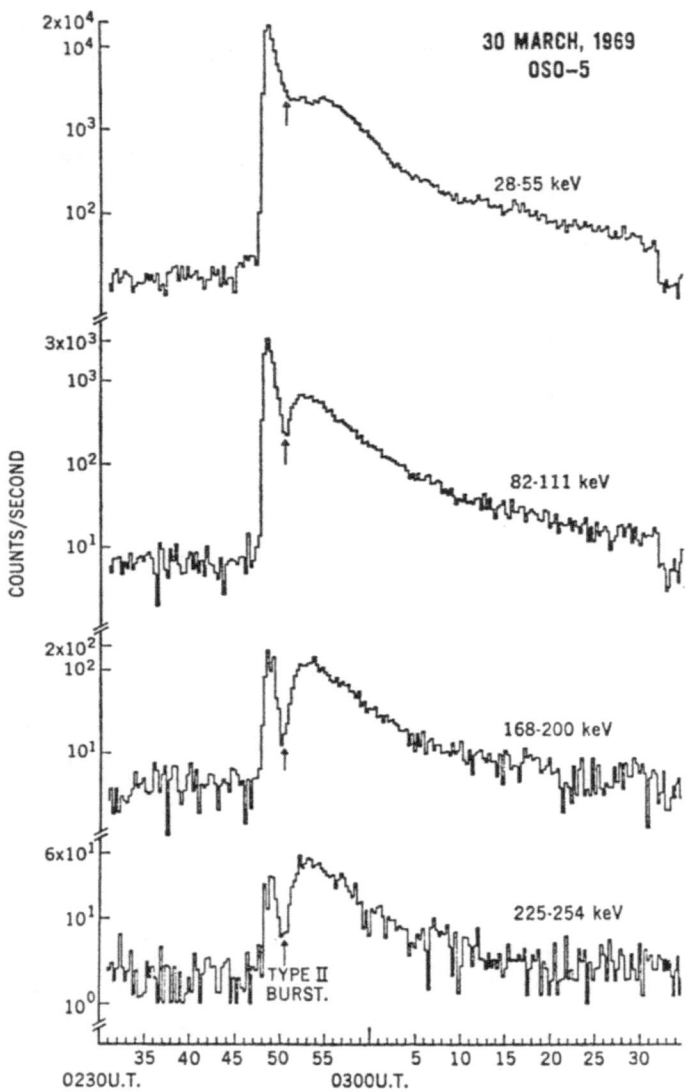

Fig. 3. Slow late component of a large burst, dominant at high energies, reported by Frost and Dennis (1971) as indicating a second stage of particle acceleration.

a general trend. In particular, Hoyng *et al.* (1975) have carried out the most detailed analysis available of the evolution of burst intensities and spectra, with 1.2 s time resolution. For none of the events studied was γ found to evolve similarly to the results of Kane and Anderson but rather, for example, showed an increase throughout one event (Figure 5). Hoyng *et al.* (1975) found, however, that in the long enduring event of 1972, August 4 γ followed a systematic variation related to the burst intensity,

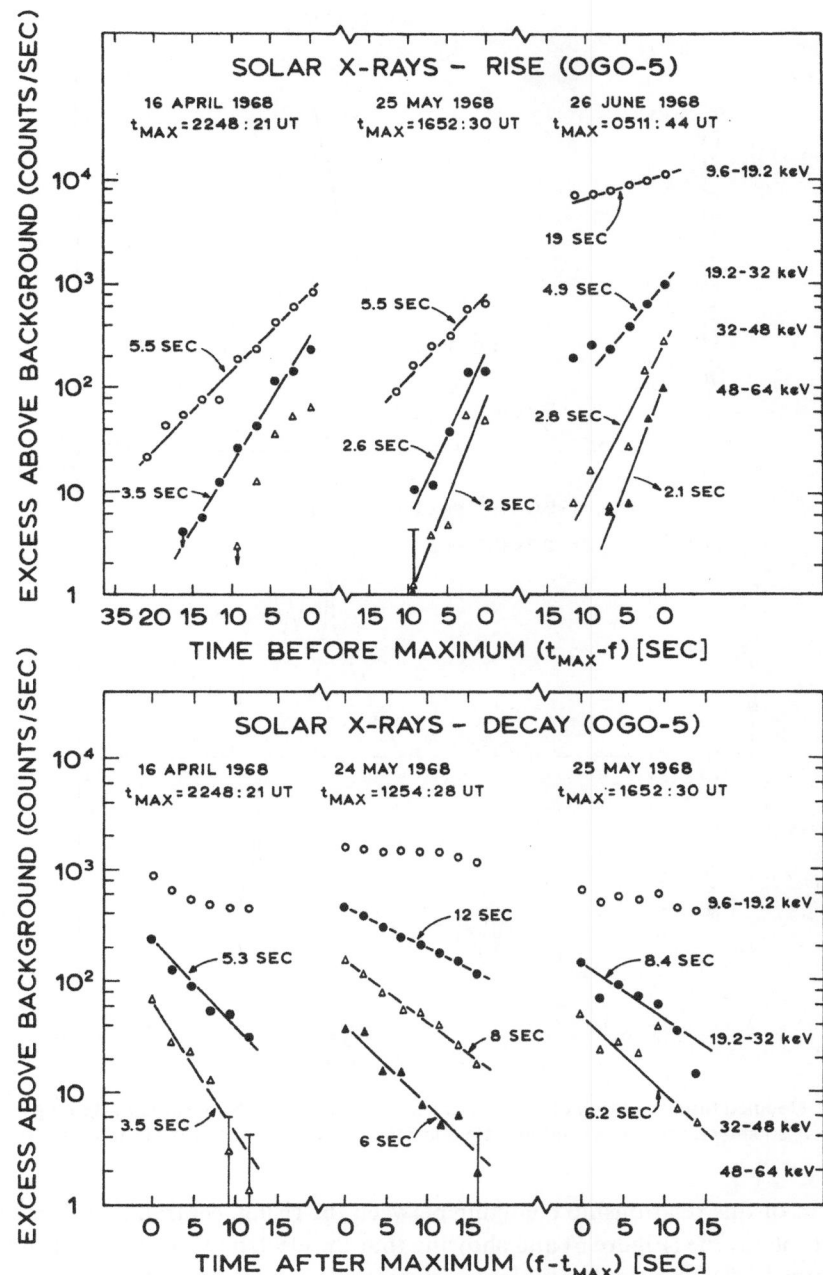

Fig. 4.   Rise and decay times of small impulsive bursts as a function of photon energy (from Kane and Anderson, 1970).

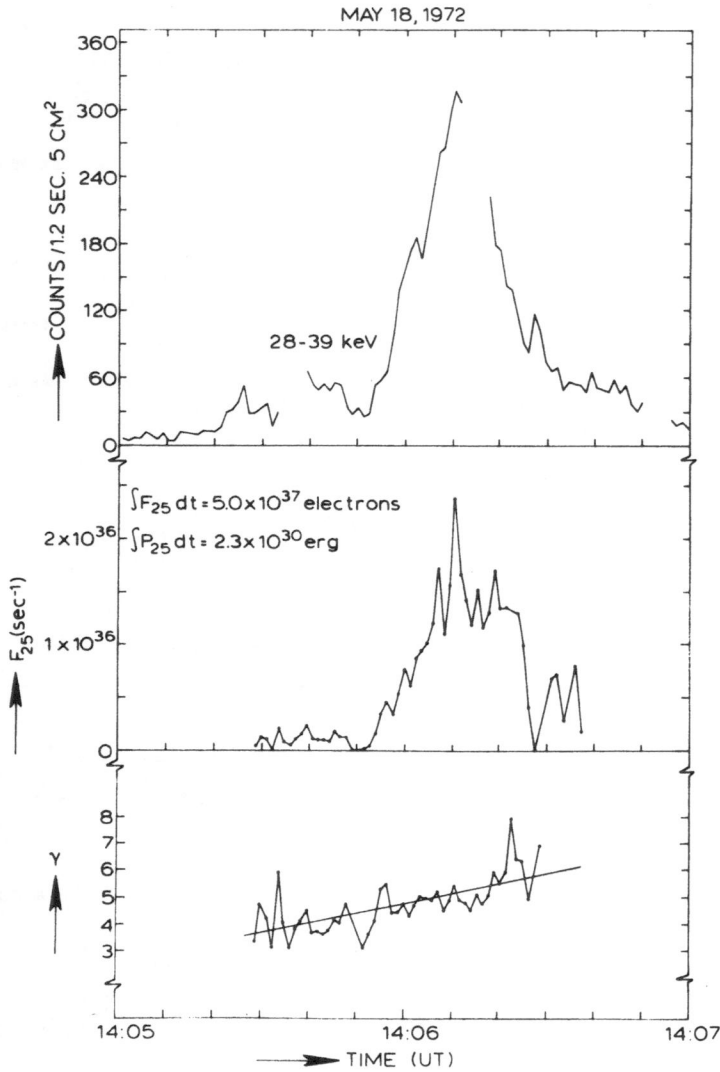

Fig. 5.   Detailed time evolution of spectral index γ in the event of May 18, 1972 (Hoyng *et al.*, 1975). Also shown are the raw count rate and inferred thick-target electron flux F_{25} (s^{-1}).

the form of this relationship changing between the rising, central, and decay portions of the time profile (Figure 6) and showing that the electron flux and spectral index are determined by some single physical parameter at each stage of this event.

It is important to resolve the discrepancy between the trend in γ reported by Kane and Anderson (1970) and that presented by other authors since much theoretical discussion of models has been based on the observation (e.g. Kane and Anderson, 1970; Brown, 1972a; Petrosian, 1973). In particular, it is not clear whether the discrepancy represents a true distinction between the spectral characteristics of small and large events or whether the temporal behaviour of γ in small events is not revealed at all by available time resolution, the apparent trend being barely extractable from the background and the rising gradual component (Takakura 1969) – cf. Figures 1 and 4.

Deviations of the burst spectrum from the power-law approximation are also important. In particular, the low energy end of the spectrum must break from a power-law at some point (otherwise the X-ray flux would be divergent at small ε). In practice

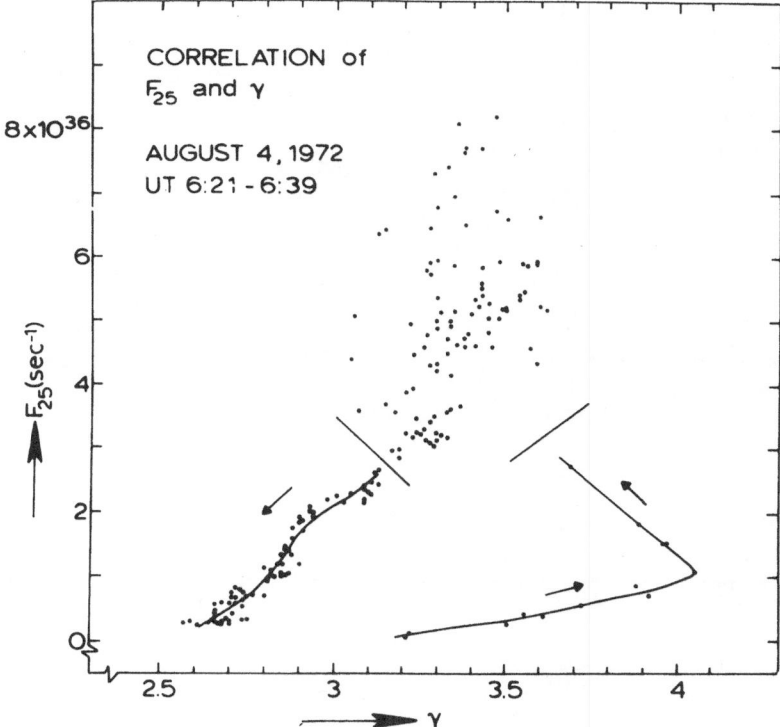

Fig. 6.   Relationship of spectral index γ to inferred electron flux F_{25} through the large event of 1972, August 4 (Hoyng *et al.*, 1975). Note the change in form of this relationship between rising and decaying phases.

the break is not directly observable since it occurs in the energy range where the gradual thermal component is dominant. Peterson *et al.* (1973) have however reported a power-law spectrum extending down as far as 6 keV (Figure 7) while Kahler and Kreplin (1971) have attempted to infer the downward extension of the power-law from the lowest photon energy at which the impulsive spike becomes visible above the gradual component (Figure 8), namely again a few keV in some bursts. The problem is particularly important since it is the low energy cut-off in the flare electron spectrum which governs the total electron energy inferred from X-ray bursts (Neupert, 1968; Brown,

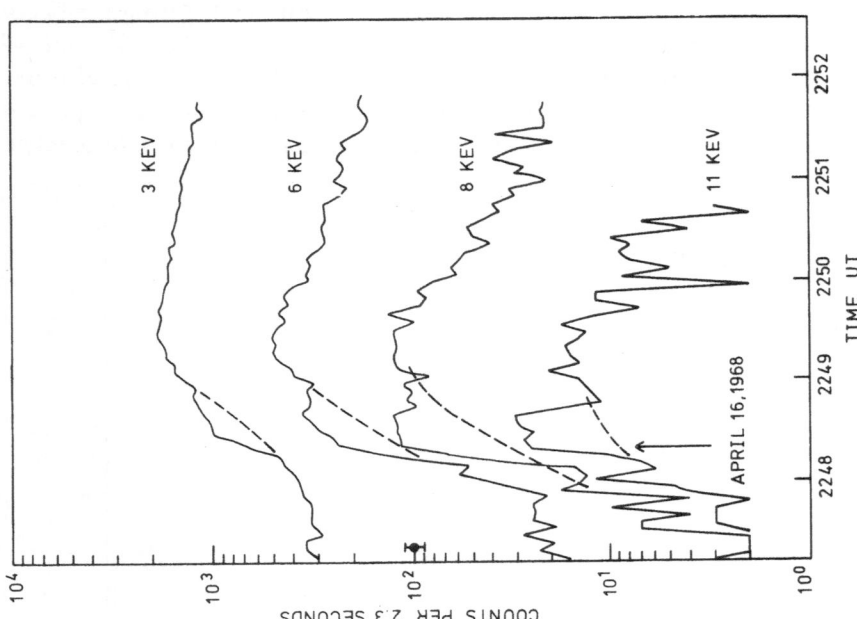

Fig. 8. 'Impulsive component' of emission visible in the time profile of a burst down to a few keV (Kahler and Kreplin, 1971).

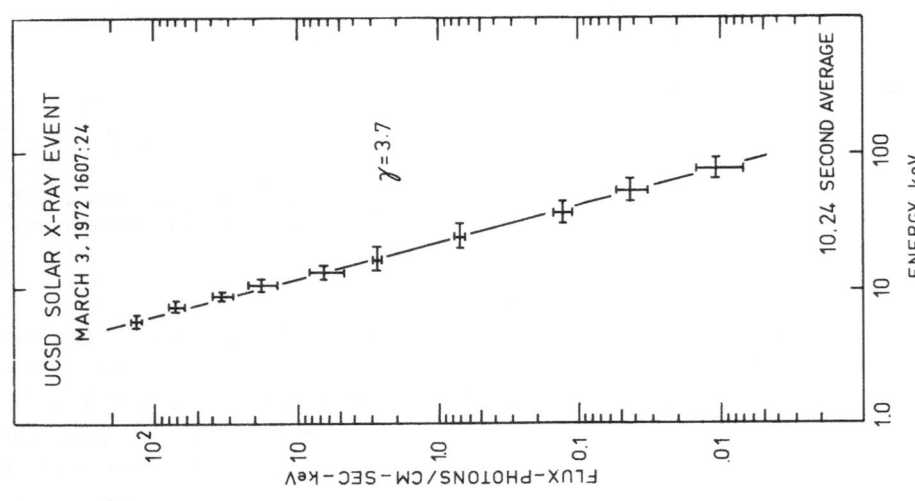

Fig. 7. Event exhibiting a power-law spectrum as far down as 5 keV (from Peterson et al., 1973).

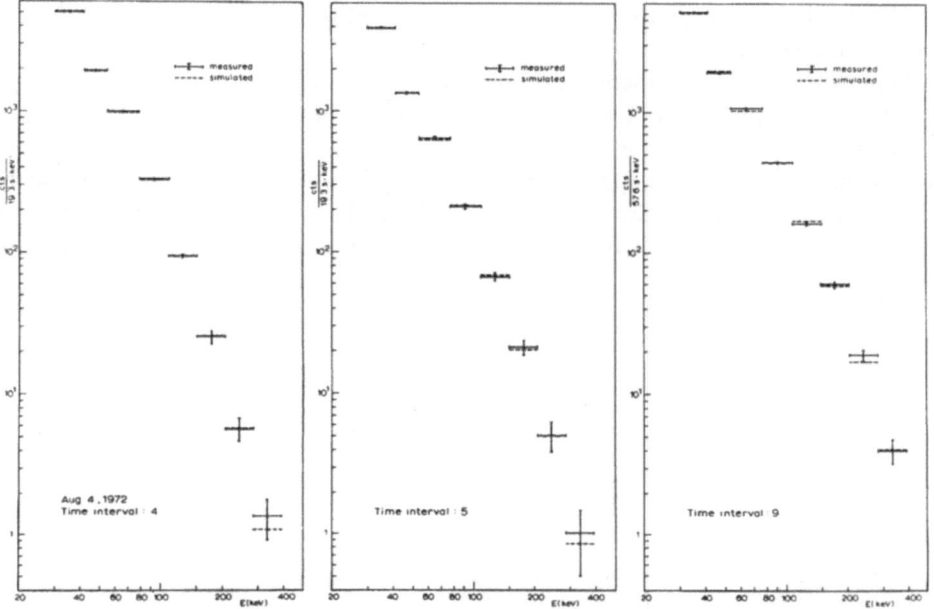

Fig. 9.   High energy bend ('knee') in the photon spectrum of the burst of 1972, August 4 (from van
Beek, 1973) in three different time intervals.

1971, and many subsequent authors) and so the importance attributed to the electrons
in the flare as a whole (Section VI).

Steepening of the spectrum at high energies has been widely reported in both small
and large events (Cline *et al.*, 1969; Frost, 1969; Kane and Anderson, 1970; Frost and
Dennis, 1971; Van Beek *et al.*, 1974), the bend occurring generally in the 60–100 keV
range (Figure 9) but sometimes as high as 500 keV (Cline *et al.*, 1969) and involving
a change of up to 2 in γ. The location and form of this spectral 'knee' and its evolution
have been considered by Van Beek *et al.* (1974). Its significance for the electron
spectrum is not yet known (or even fully proved – cf. Section IV(e)), though as Frost
(1969) has suggested it may indicate a real high energy cut-off in the acceleration
process.

(c) POLARISATION AND DIRECTIVITY

The potential importance of X-ray polarisation as a diagnostic of energetic electron
motions in flares has been expounded by Elwert (1968), Elwert and Haug (1970, 1971),
Haug (1972), Korchak (1967b, 1971, 1974) and Brown (1972b) while the basic methods
and problems of polarisation measurement in practice have been considered by Wolff
(1973) and by Thomas (1975). Only Tindo *et al.* (1970, 1972a, b, 1973) appear, however, to have so far successfully obtained actual data on flare X-ray polarisation. The
five bursts observed show polarisations between 20% and 40% (e.g. Figure 10) with
some evidence of increasing polarisation with distance from the disc centre, (cf. Fig-

JOHN C. BROWN

ure 12) while the polarisation plane has been found to lie along the solar disc radius through the source.

Some problems of the theoretical interpretation of these data and their time evolution have been pointed out by Frost (1974a) and Korchak (1974). In particular the energy range observed is really too low since much of the radiation is thermal. Frost (1974a) has combined the (0.6–1 Å) polarisation data for the event of 1970, November 5 at 0310 UT with OSO-5 hard X-rays (14–250 keV) and soft X-rays (2–8 Å) as shown in Figure 10. Frost points out that polarisation peaks correspond well with

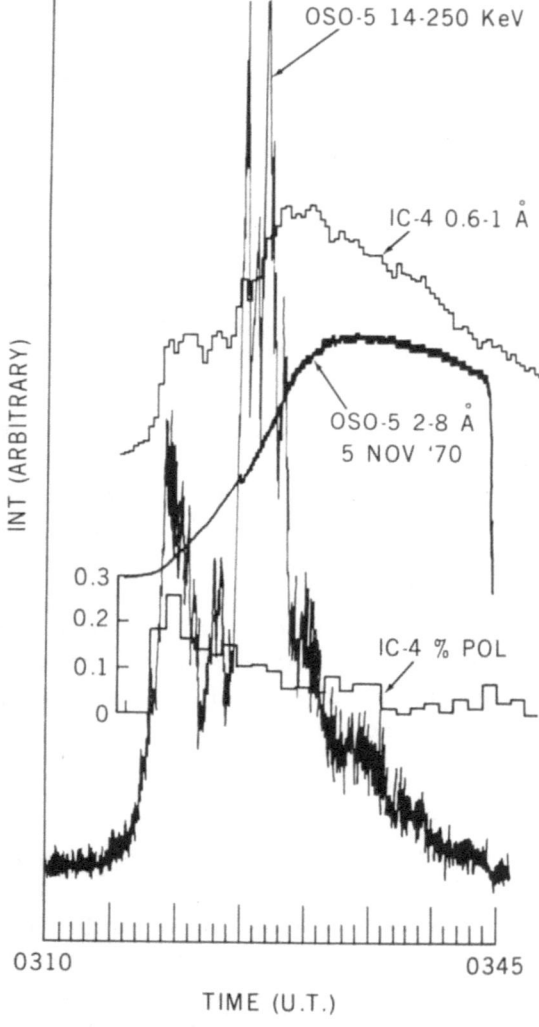

Fig. 10.   Evolution of the polarisation in the flare of 1970, November 5 from Tindo *et al.* (1972b) compared to OSO-5 hard X-ray and soft X-ray data (prepared and provided by K. J. Frost, 1974b, – see text for comments).

hard X-ray intensity spikes early in the event but that the monotonic decay of the polarisation after 03 15 UT may be due to the rising (unpolarised) thermal contribution and, therefore, that the 20% peak recorded must be regarded as a lower limit. Frost concludes that polarisation measurements must be made at higher energy and with high time resolution since both the degree and plane of polarisation may change between successive hard X-ray spikes (see also comments by Korchak (1974) on these problems).

In addition, Brown *et al.* (1974) have shown that some error may have been induced in the results of Tindo *et al.* due to their calibration procedure carried out by supposing the polarisation to be zero in the late (thermal) stages of the flare whereas even thermal flare X-rays may be somewhat polarized by photospheric backscattering. The extent of this error has not been evaluated satisfactorily (cf. Brown *et al.*, 1974; Beigman and Vainstein, 1974) but future experiments would yield more easily interpretable results if they operated at higher energies and were laboratory calibrated.

Directional characteristics of bremsstrahlung X-ray sources are closely related to their polarisation and can be used for the same purpose – i.e. the inference of the source electron velocity distribution. Various contradictory attempts have been made to deduce the mean directivity of bursts by examining their distribution across the solar disc (e.g. Ohki, 1969; Pinter, 1969; Drake, 1971; Kane, 1974b; Datlowe, 1975). However, such burst distribution studies seem unlikely ever to lead to meaningful directivity data for, firstly, the maximum directivity expected from any model is less than a factor of 10, even at high photon energies (Elwert and Haug, 1971; Brown, 1972b) while the dynamic range of intensities between different bursts exceeds a factor of 10^3. Secondly, this is much beyond the dynamic range of any single detector so that strong selection effects are present in data samples. A proper statistical analysis of the problem requires use of the actual distribution of burst intensities but the spread in mean intensity for N observed bursts will only decrease something like $1/\sqrt{N}$ so that a huge number of events must be observed before the spread in intrinsic intensities is averaged to much less than the directivity range sought (bearing in mind that the number of bursts is further diluted into a set of longitude intervals).

Distribution of spectral indices γ across the disc has also been analysed – i.e. the directional characteristics of burst spectra. Kane (1974b) finds no significant variation across the disc while Datlowe (1975) reports increase in mean spectral index towards and over the limb for OSO-7 bursts. The interpretation of this result both in terms of its statistical reality and of its physical significance, if real, is unclear at present.

(d) OTHER RELEVANT OBSERVATIONS

Information on the spatial characteristics of hard X-ray flares is exceedingly sparse. Takakura *et al.* (1971) have obtained the only direct resolution of a hard X-ray flare, showing the emission to be localized (horizontally) near the core of the optical flare, but only with a resolution of about 1′ and in one dimension. In addition, there are a

number of cases of occurrence of hard X-ray bursts from flares presumed behind the
limb which set a lower limit to the altitude of at least some of the emission in those
events ($\gtrsim 10^4$ km – Datlowe, 1975). In one commonly quoted instance of a behind-
the-limb burst, viz. 1969, March 30 (Kane, 1974a), however, this interpretation could
be incorrect since this vast event extended over much of a solar hemisphere at radio
frequencies (Wild and Smerd, 1972) and the X-ray source may have done likewise
(Frost 1974b – private communication). A basic problem is that there is no way to
distinguish purely coronal level emission from a small event and a small coronal component of a large event, when the flare itself is invisible.

Further limits are set by the interrelation of X-ray bursts to other flare flash-phase
characteristics (cf. reviews by Kane, 1973; Brown, 1973b). In particular the very close
synchronism of hard X-ray time profiles with (thermal) EUV bursts (Kane and Donnelly, 1971) which occur deep in the atmosphere ($n \gtrsim 10^{11}$ cm^{-3}) implies either that
the X-ray source itself is at low altitude or that the two emissions are closely linked to
a common energy supply, extended in altitude. Similar conclusions arise from associated optical flashes (Vorpahl, 1973, Zirin and Tanaka, 1973). On the other hand the
correspondence of time structure of hard X-rays with soft X-rays flare development
(Neupert, 1968) and with microwave bursts (e.g. Takakura, 1975) require a close
coupling of the hard X-ray source with events high in the atmosphere also. A high
altitude for the electron acceleration region itself seems to be well established by the
lack of collisional distortion of interplanetary electron spectra down to a few keV, as
shown by Lin (1974a, b). The discrepancy ($\gtrsim \times 10^3$) between the flux of electrons
needed for the X-ray and synchronous microwave burst is well known (Takakura and
Kai, 1966) and may be resolved by invoking the combined effects of microwave re-
absorption, magnetic field distribution, and the high energy electron spectral cut-off
(Takakura, 1973). Interplanetary electrons are also usually small in number (sometimes $\lesssim 0.5\%$) compared to X-ray source electrons (e.g. Datlowe and Lin, 1973; Lin,
1974a, b), apparently due to magnetic confinement of the bulk of accelerated particles in the flare region. Nevertheless it seems an odd coincidence that both microwaves and interplanetary electrons yield similar numbers, both much smaller than that
inferred from X-rays (cf. Section VI(c)).

IV. Inference of Mean Electron Spectra in Sources

(a) INTRODUCTION

In general the inference of the spectrum of flare electrons at acceleration or the variation of their spectrum through the X-ray source is quite strongly model dependent –
cf. Section V. However, burst observations without spatial resolution do permit determination of a 'mean source electron spectrum' (Brown, 1971), independent of models,
provided albedo and directional effects are neglected. Thus by defining

$$F(E, t) = \frac{1}{\bar{n}V} \int_V n(\mathbf{r}, t) F(E, \mathbf{r}, t) \, dV, \qquad (2)$$

where

$$\bar{n} = \frac{1}{V} \int_V n(\mathbf{r}, t)\, dV$$

then Equation (1) becomes

$$I(\varepsilon, t) = \frac{\bar{n}V}{4\pi R^2} \int_\varepsilon^\infty F(E, t)\, Q(\varepsilon, E)\, dE \qquad (3)$$

and $F(E, t)$ is the instantaneous space average of F weighted with respect to the ambient plasma density n.

Equation (3) is then an integral equation to be solved for $F(E, t)$ with $I(\varepsilon, t)$ given and $Q(\varepsilon, E)$ as kernel. In practice $I(\varepsilon, t)$ is not known except after folding through an instrumental response and, even then, only in a discrete set of photon energy bands set by discriminator levels. A common procedure is to assume a functional form for $I(\varepsilon)$ (usually $A\varepsilon^{-\gamma}$) and to adjust the function parameters so as to optimize the χ^2 fit of the calculated count rates to the data. This has the disadvantage that, although it optimises the fit for any chosen functional form, the functional form itself is somewhat arbitrary. (In the problem of distributed temperature fitting of soft X-ray spectra – e.g. Herring and Craig, 1973; Dere *et al.*, 1974; Craig, 1974; Craig and Brown, 1974 – this problem is a crucial one since there exists no *a priori* basis for adoption of any particular fitting function.) A more satisfactory approach is first to deconvolute $I(\varepsilon)$ from the count rates, knowing the instrumental response. Problems of and optimum procedures for this step have been considered in detail by Hoyng and Stevens (1974) and applied to the hard X-ray burst problem by Hoyng *et al.* (1975), or at least a discrete representation of it. Then the remaining problem of solution of Equation (3) (or its matrix equivalent in the discrete case) is a matter of the basic physics of the bremsstrahlung process, independent of the instrument (except in so far as this partially determines the observational and discretisation errors).

(b) THE INTEGRAL INVERSION PROBLEM

From the theoretical viewpoint, one is interested in how $F(E)$ is related to different functional forms of $I(\varepsilon)$ – i.e. in the analytic inversion of Equation (3). Though such a solution may never be achievable from practical data in discrete form (i.e. without use of fitting functions), it is nevertheless of fundamental importance since it determines how accurately and unambiguously electron spectra can *ever* be determined from the spectra of their bremsstrahlung emission, however good the X-ray spectral resolution may be. That is, fundamentally, how good are bremsstrahlung X-rays as a measuring device for electron spectra?

The possibility of analytic solution of (3) depends of course on the form of the kernel $Q(\varepsilon, E)$. For the full cross-section expression with relativistic, Coulomb and other corrections included (Koch and Motz, 1959), no such solution is obtainable in simple form and the equation must be converted to an equivalent discrete set of linear equations and inverted numerically. The most complex kernel for which analytic solution seems possible (provided directional effects are neglected – i.e. assuming isotropy of source electron velocities), is the non-relativistic Bethe-Heitler (BH) formula which is fortunately also a good approximation below 100 keV or so, namely (Koch and Motz, 1959)

$$Q_{BH}(\varepsilon, E) = \frac{K_{BH}}{\varepsilon E} \log \frac{1 + \sqrt{1 - \varepsilon/E}}{1 - \sqrt{1 - \varepsilon/E}},$$

(4)

where $K_{BH} = \zeta \times (8\alpha/3\pi) r_0^2 mc^2$, with α = fine structure constant, c = velocity of light, m and r_0 are the electron mass and radius, and ζ is the correction factor for the solar abundance of heavy ions ($\zeta \simeq 1.8$, Elwert and Haug, 1971).

The solution of (3), with (4) as kernel, has been shown by Brown (1971), via transformation to Abel's integral equation, to be

$$F(E, t) = \frac{4R^2}{K_{BH}nV} E^{1/2} \int_E^\infty \frac{I(\varepsilon) + 3\varepsilon I'(\varepsilon) + \varepsilon^2 I''(\varepsilon)}{\sqrt{\varepsilon - E}} \, d\varepsilon.$$

(5)

For the commonest case of a power-law, viz.

$$I(\varepsilon) = A\varepsilon^{-\gamma}$$

(6)

(with A, γ functions of t), solution (5) reduces to

$$F(E, t) = \frac{4R^2}{K_{BH}} \frac{(\gamma - 1)^2 \, B(\gamma - \frac{1}{2}, \frac{1}{2})}{\bar{n}V} AE^{-\gamma + 1},$$

(7)

where B is the beta function, or in numerical form* with ε, E in keV

$$F(E, t) \, (\text{electrons cm}^{-2} \, \text{s}^{-1} \, \text{keV}^{-1}) = \frac{6.7 \times 10^{50}}{\bar{n}V} \times$$

$$\times E^{1/2} \int_E^\infty \frac{\{I + 3\varepsilon I' + \varepsilon^2 I''\}}{\sqrt{\varepsilon - E}} \, d\varepsilon \quad (8)$$

in the general case and, for the power-law, in the same units

$$F(E, t) = \frac{6.7 \times 10^{50}}{\bar{n}V} (\gamma - 1)^2 \, B(\gamma - \frac{1}{2}, \frac{1}{2}) \, AE^{-\gamma + 1},$$

(9)

where I is in photons cm^{-2} s^{-1} keV^{-1}.

Other approximations have been used for $Q(\varepsilon, E)$ in the literature. In particular the approximation $Q \sim 1/\varepsilon^2$ and that used by Kawabata et al. (1973) ($\sim 1/\varepsilon E$) are readily shown to give the same electron/X-ray spectral index relationship as (9), namely $\delta = \gamma - 1$, for the power-law case, and to yield analytic inversions simpler than (5). However, both of these solutions involve a scale error in F of factors up to three or

---

* Equations (8) and (9), and also (17), (19), (22), and (23), incorporate heavy ion correction ζ and correct a numerical error in Brown's (1971) paper.

four, depending on γ, due to incorrect description of the behaviour of Q in the region $\varepsilon \lesssim E \lesssim 2\varepsilon$ which dominates the emission for the typically steep spectra involved. Though this factor is small in terms of the uncertainty of the *total* electron flux due to the unknown energy cut-off (cf. Sections III and VI), it is important in problems such as chromospheric heating by electrons where the low energy end of the spectrum is irrelevant (Brown, 1973a). Furthermore, in the case of spectra other than the power-law (6), these simpler approximations also yield the incorrect electron spectrum. For instance, Van Beek *et al.* (1974) have shown that a good fit to the X-ray spectral break, rather than a power-law with exponential cut-off, may be a pair of distinct power-laws joined at a sharply defined 'knee-energy'. Depending on just how 'sharp' this knee is, the contribution to solution (5) from the I'' term could dominate the others – a feature not present in the analytic solutions for simpler $Q(\varepsilon, E)$.

Thus, except for first approximations, it is wise to use the Bethe-Heitler result (5) for all analytic work.

(c) THE MATRIX SOLUTION

Present spectral resolution is not in fact capable of providing directly the data needed for use of the above solutions – e.g. of providing the second derivative of I – without use of some smooth fitting functions. It is therefore important to consider the problem of solution of Equation (3) in discrete form – this procedure being also required when using more general (e.g. relativistic) cross-section formulae for which no analytic solution exists (cf. Elwert and Haug, 1970). That is $I(\varepsilon)$ is replaced by a set of n discrete values $I_i = I(\varepsilon_i)$ $i = 1, n$ (in fact by $\int \hat{\Delta}\varepsilon_i I(\varepsilon) \, d\varepsilon$, but Hoyng and Stevens, 1974, have shown how the optimum discrete set $I(\varepsilon_i)$ can be extracted from finite band data). Then F is also replaced by a discrete representation $F_j (= F(E_j) \Delta E_j)$ where $j = 1, m$ with $m \leqslant n$, inequality corresponding to a least squares fitting problem and equality to an exact problem, only the latter case being considered here. Thus

$$\sum_{j=1}^{n} Q_{ij} F_j = I_i, \, i = 1, n$$

or

$$[Q] \{F\} = \{I\}, \tag{10}$$

where $\{I\}$ $\{F\}$ are the data and unknown $n \times 1$ vectors respectively and $[Q]$ is the (square) cross-section matrix, the formal solution being

$$\{F\} = [Q^{-1}] \{I\}. \tag{11}$$

The accuracy of matrix solution (11) is limited by the usual numerical instabilities in solution of linear equations (e.g. Fox and Mayers, 1968; Householder, 1964) i.e. by magnification of small errors in $\{I\}$ by large entries in $\{Q^{-1}]$ resulting in larger relative errors in $\{F\}$. In some cases these are so large as to render the solution obtained meaningless, the severity of the problem depending both on the form of the matrix $[Q]$ and of the data $\{I\}$. Considering first the effects of the matrix itself a convenient measure of the likelihood of such inherent instability is the matrix condition number N (e.g. Householder, 1964) given by $N = \|Q\| \times \|Q^{-1}\|$. $\|A\|$ is any convenient norm of a matrix A such as $[\sum_{i,j} a_{ij}^2]^{1/2}/n^{1/2}$ which implies $N = 1$ for the numerically ideal case of the unity matrix (or any scalar multiple of it). The actual case for the bremsstrahlung cross-section matrix Q is an upper triangular matrix with small diagonal elements – far from diagonal and so potentially ill-conditioned. As a simple illustration, if data on I are obtained at $\varepsilon_i = 20$, 40 and 80 keV and represented in terms of electron fluxes at $E_j = 25$, 50 and 100 keV, then with convenient scaling, the Bethe-Heitler cross section gives

$$Q = \begin{bmatrix} 19.2 & 20.6 & 14.4 \\ 0 & 4.82 & 5.08 \\ 0 & 0 & 1.21 \end{bmatrix} \quad \text{and so} \quad Q^{-1} = \begin{bmatrix} 0.052 & -0.224 & 0.310 \\ 0 & 0.207 & -0.870 \\ 0 & 0 & 0.828 \end{bmatrix} \qquad (12)$$

which implies $N \simeq 21.6$ (as against $N = 1$ for a unity 3×3) corresponding to a large inherent instability in the set of Equations (19) – i.e. in the derivation of electron from photon spectra. No improvement of spectral resolution (within a fixed ε range) can improve this situation since the introduction of more and more rows into Q merely increases their linear dependence and hence the ill-conditioning. The conclusion is that, for general data $\{I\}$, high accuracy is needed for solution $\{F\}$ to be guaranteed as meaningful, the necessary accuracy increasing with the dimension of $\{I\}$.

The actual severity of the consequences of an ill-conditioned matrix depends, however, on the problem itself – i.e. on the data (thus for example the errors introduced in F_j from I_i, $i \neq j$, by large off-diagonal elements of Q^{-1} may in fact be relatively small if I_i is itself small compared to I_j). In particular if, in (10), the F_j decrease rapidly with increasing j (i.e. E_j) then the effect is essentially to damp the magnification of errors inherent in the matrix form itself. This condition is probably satisfied by the power-law and other steeply decreasing functions so that *if one has grounds for expecting such a spectrum* (other than the X-ray data itself) then the instability problem need not to be so serious in practice. What is not true, however, is the usual supposition that, because substitution of (e.g.) a power-law for $\{F\}$ in (10) yields a 'good' fit to $\{I\}$ the power-law is necessarily near the true electron spectrum. It may merely show that the calculated X-ray spectrum is rather insensitive to the source electron spectrum as confirmed by the sensitivity of the inverse analytic expression (5), to $I(\varepsilon)$. One is therefore forced to the conclusion that, in general, bremsstrahlung X-rays are not a very accurate meter for electron spectra unless one has a pretty good idea beforehand what form these electron spectra take. Confidence in the ubiquitous use of power-laws for this purpose, other than as an act of faith in a model, then rests largely on consistency of this model with other related flare data. For this reason, in evaluating flare particle acceleration models, the constraint of reproducing a power-law spectrum should perhaps not be taken overseriously.

(d) THE PROBLEM OF THE THERMAL X-RAY SPECTRUM

It is apt to mention the analogous problem of inference of the temperature distribution in the hot flare plasma from soft X-ray spectral measurements. This is important for correct assessment of the total energy in the soft X-ray flare plasma (Craig 1974) and for its influence on the hard X-ray energy range (Brown 1974 and Section VI of this review). In this problem there is no definitive guide whatsoever, from theory or observation, as to what functional form the differential emission measure may take. Consequently various entirely arbitrary fitting functions have been used (e.g. Dere *et al.*, 1974) with parameters adjusted to 'optimize' the fit to observations. However, the equations involved are exceedingly ill-conditioned (Craig and Brown, 1974) and it is far from clear whether the fitting functions used are physically meaningful or merely suitable damping functions for the inherent instabilities of the problem. These difficulties arise by the nature of the thermal continuum emission function compounded by the substantial contribution from X-ray lines, and are as important for cosmic as for solar X-ray studies. The most soundly based attack on the problem and its consequences is that initiated by Herring and Craig (1973) and continued by Craig (1974' and currently being pursued further by Craig and Brown (1974).

(e) X-RAY SPECTRA FROM A POWER-LAW ELECTRON SPECTRUM

Conversely to the conclusions of IV(c), if one does favour the power-law as a likely form for the electron spectrum, it is important to know whether the X-ray observations are in fact consistent with it, even if not precise enough to establish it. That is, whether any observed deviations from an X-ray power-law – such as the high energy break – can be explained in a natural way other than in terms of deviation of the electron acceleration spectrum itself from a power-law. Though a number of factors relating to this question are discussed further in Sections 5 and 6, the chief causes of such deviations may be summarized here.

(i) *The Albedo Contribution*

Tomblin (1972) and Santangelo *et al.* (1973) have investigated the spectrum of photospheric albedo photons and their influence on the spectrum of total burst X-rays. At low energies ($\lesssim 15$ keV) the albedo contribution drops off due to photoelectric absorption. At high photon energies the albedo contribution is reduced by the relativistic anisotropy of Compton scattered photons and by the increasing energy loss of photons in the Compton scattering process. The net effect is a downward concavity (steepening) of the total X-ray spectrum around the 40 keV range.

(ii) *Directivity Effects*

Since the bremsstrahlung cross-section is anisotropic (Koch and Motz, 1959), if the source electron velocity distribution is not isotropic, the X-ray emission in general varies both in intensity and spectrum with direction (Elwert and Haug, 1970). These effects are particularly important at high photon energies due to relativistic beaming (Elwert and Haug, 1971). Petrosian (1973) has calculated that, for the case of a collimated electron stream moving vertically toward the photosphere, the emission spectrum should show a break above about 100 keV, and proposed that this cross-section effect may explain the observation of a high energy cut-off. This result has not been noted by previous authors on this topic (Elwert and Haug, 1971; Brown, 1972b) and it is not clear to what extent it is dependent on the different cross-sections used. Furthermore, Brown (1972b) has previously considered exactly the same problem but included collisional scattering of the electron stream (neglected by Petrosian) which reduces all anisotropic effects, and found no spectral break. Since the actual geometry of the source is quite unknown, however, the possibility of a spectral correction for directional effects should be borne in mind.

(iii) *Model Dependent Effects*

Examination of definition (2) shows that if the source plasma density n is homogeneous $(n \neq n(\mathbf{r}))$ or if the energetic electron spectrum is homogeneous $(F \neq F(\mathbf{r}))$ then the mean effective source spectrum (2) can be identified with the true spectral distribution of all electrons in the source. If, however, *both* n and F are *inhomogeneous* then in general the mean source spectrum (2) differs from the true spectrum in the sense that if electrons in some energy range move in part of the source where n is higher than elsewhere, then that part of the X-ray spectrum is enhanced (e.g. Brown, 1972a). Since no spatially resolved data are available, the form of such differences is dependent on the model adopted (see below).

Secondly, in models where the energy loss time of electrons is small, it is necessary to distinguish the spectrum of continuously accelerated electrons (before injection) and the instantaneous spectrum of electrons in the source due to the effects of the energy loss on the latter, as considered in V(a). This may cause deviations from a power-law (e.g. Brown, 1973d) as well as changing the power-law index (Brown, 1971).

In consequence of these three effects, it must be concluded that though there are doubts about the accuracy with which one can truly infer electron spectra, there is no conclusive observational evidence inconsistent with the power-law as a model of the entire electron spectrum at acceleration.

V. Burst Interpretation and Source Models

Discussions of the three main existing source models, viz. thick- and thin-target and electron trap, have been presented recently by Hudson (1973), Lin (1974), Brown (1973b), Takakura (1973) and Kane (1974a) while the possible importance of thermal emission in the hard X-ray range has been discussed by Brown (1974) and by Kahler (1975) in these Proceedings (cf. also VI(b) below). Here I will try to bring this controversy over models up to date, present the outstanding problems of these models, and speculate on how these might be resolved.

(a) THE THICK TARGET MODEL

This model postulates the injection of energetic electrons from a coronal source into the dense chromosphere where the bulk of bremsstrahlung X-rays are then generated (De Jager and Kundu, 1963; Arnoldy *et al.*, 1968; Acton, 1968; Brown, 1971, 1972b;

Hudson, 1972; Syrovatskii and Shmeleva, 1972; Petrosian, 1973; Brown and McClymont, 1974). The electrons are of course totally collisionally absorbed on very short time scales (Schatzman, 1966; Brown, 1973a) so that the bremsstrahlung target is thick and, furthermore, the time profile of the resulting burst is governed entirely by the injection rate of electrons from their source.

Thus, with reference to Equation (1), time variations in $I(\varepsilon, t)$ are entirely attributed to variations in $F(E, \mathbf{r}, t)$ due to electron source modulations, the minimum time scale of burst variation being set by the collisional loss term in dF/dt, electron escape being non-existent in this model (cf. Lin, 1974a). Since in fact the collisional losses, and hence the bremsstrahlung emission, occur almost entirely over about one chromospheric scale height (Brown, 1972b, 1973a; Brown and McClymont, 1974), this lower limit is in the millisecond range – well below available time resolution.

Satisfactory explanation of burst time profiles in the thick-target model is thus entirely a problem for the electron acceleration mechanism itself. This area of research, perhaps more than any other, is currently in need of intensive theoretical and laboratory investigation both in terms of the capabilities of any acceleration mechanism to produce the complex fine structure in bursts, and of quantitative analysis of burst time profiles to yield any basic characteristics of the electron flux and spectrum at acceleration (cf. Kane and Anderson, 1970; Hoyng et al., 1975) which could give any clue to the acceleration mechanism (cf. Vorpahl and Takakura, 1974). For the present it is generally merely assumed that the mechanism is capable of producing observed burst profiles.

(i) *Polarization and Directivity*

Following on the first descriptions of the possibility and importance of flare bremsstrahlung polarization (Korchak, 1967a, b, Elwert, 1968) Elwert and Haug (1970, 1971) and Haug (1972) have developed in detail the methodology for calculation of both the polarisation and directivity of bremsstrahlung from electron streams spiralling in a uniform magnetised plasma, emphasising the need to use fully relativistic cross-sections, though the electrons themselves are only semi-relativistic. Application of these methods to realistic source models are very laborious, involving, in general, four dimensional integration – along a curved guiding field with variable plasma density, as well as over the electron azimuth, pitch-angle, and energy distributions. Nevertheless, as recently emphasised by Korchak (1974), this is the only way in which satisfactory predictions are obtainable. The only published results for a specific model are those of Brown (1972b) for the case of a thick-target model with a purely vertical guiding field and an electron stream injected vertically but with collisional modification of the electron energies and pitch angles in the target. The plane of maximum intensity is that containing the source and the solar disc centre while the degree of polarisation increases from zero at disc centre to about 30% near the limb, with a slight energy dependence, as shown in Figure 11 (together with results neglecting scattering). These predictions are in agreement with the tentative observational results of Tindo *et al.* (1970, 1972a, b, 1973) both in terms of the polarisation plane and the dependence on solar central distance (Figure 12).

The directivity of this thick-target model is shown in Figure 13. When scattering of the electron beam is included, the directivity is seen to be quite small ($\simeq 3$) except at high photon energies but in all cases is in the sense of an increasing burst intensity toward the limb – a trend predicted by earlier qualitative descriptions (Ohki, 1969; Pinter, 1969). As discussed in III(c), however, no satisfactory directivity data are

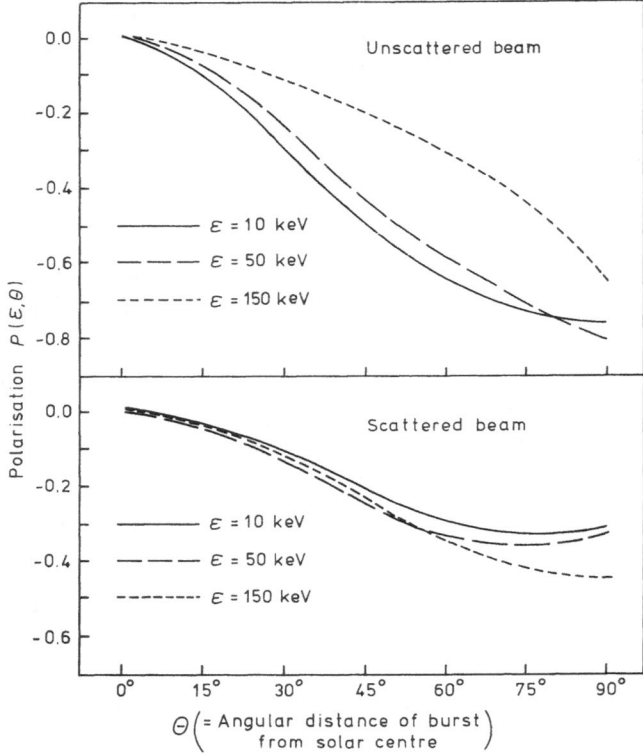

Fig. 11.   Polarisation of X-rays, for the thick-target source geometry analysed by Brown (1972b) (with and without collisional scattering of the electron beam) as a function of flare location and energy of observation.

available for comparison purposes and it is unlikely that the model can be tested on this basis until 'stereo' observations are made of individual bursts.

Brown's (1972b) analysis (cf. Petrosian, 1973) predicts a burst spectral index which decreases from center to limb, in conflict with the observational claims of Datlowe (1975). An important piece of analysis will be to consider the modification of the emission directivity by the photospheric albedo for this and other models with an anisotropic primary source. The effect will be greatest for the thick target model in which the bulk of X-rays is emitted downwards.

(ii) *Spatial Distribution of the Emission*

Brown and McClymont (1974) have recently made quantitative predictions of the height distribution of the thick-target emission and find, as expected, that it virtually all emanates from the narrow range of column depths ($\simeq$ one scale height) over which the electrons dump most of their energy collisionally. Association of this distribution with geometric height depends, however, on adoption of some model of the flaring atmosphere. When a quiet atmosphere model is used, the emission is highly localised

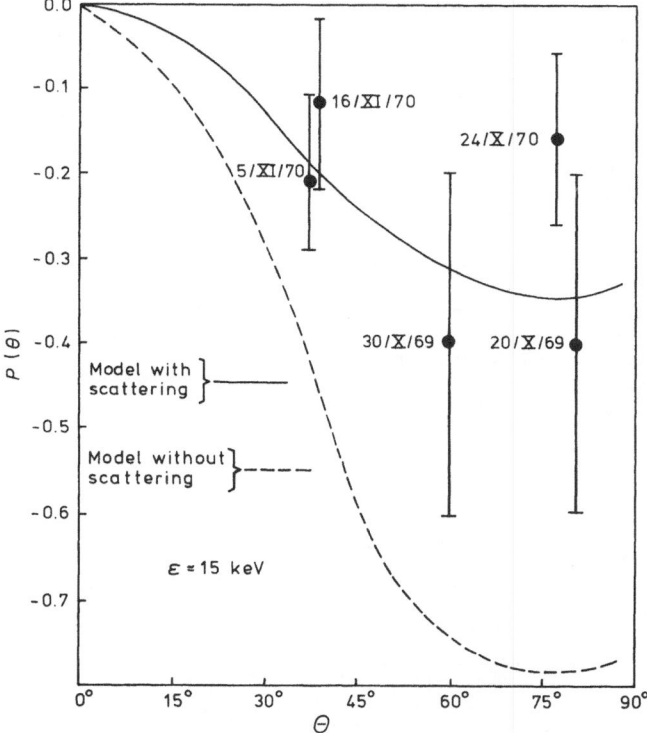

Fig. 12.   Comparison of the predictions of thick-target polarisation at 15 keV from Brown (1972b) with the observations of five bursts obtained by Tindo *et al.* (1970, 1972a, b).

at $\lesssim 1000$ km above the photosphere, in satisfactory agreement with the excellent correlation found between hard X-ray time profiles and those of the EUV and optical flares, known to be deep in the atmosphere. Though this correlation might only indicate a common energy source of the different emissions, rather than their spatial identity, it is doubtful whether any energy source, other than the fast electron streams themselves, is capable of producing the observed synchronism (Brown, 1973b). On the other hand it has been argued by Kane (1974a) and others that, if behind-the-limb events really do indicate a source altitude $\geqslant 10^4$ km, then they are incompatible with a thick-target description. Brown and McClymont (1974) have, however, pointed out that in some, and especially in large, flares the distribution of plasma in the atmosphere is radically altered by the flare itself and specifically that the 10^{16} gm of material ejected from an area of 10^{19} cm^2 at more than 10^3 km s^{-1} in the flash of a large flare (Sweet, 1969) is a thick plasma target to electrons in the X-ray energy range up to at least 70 keV. Hence the mass motion in an explosive flare is quite sufficient to raise

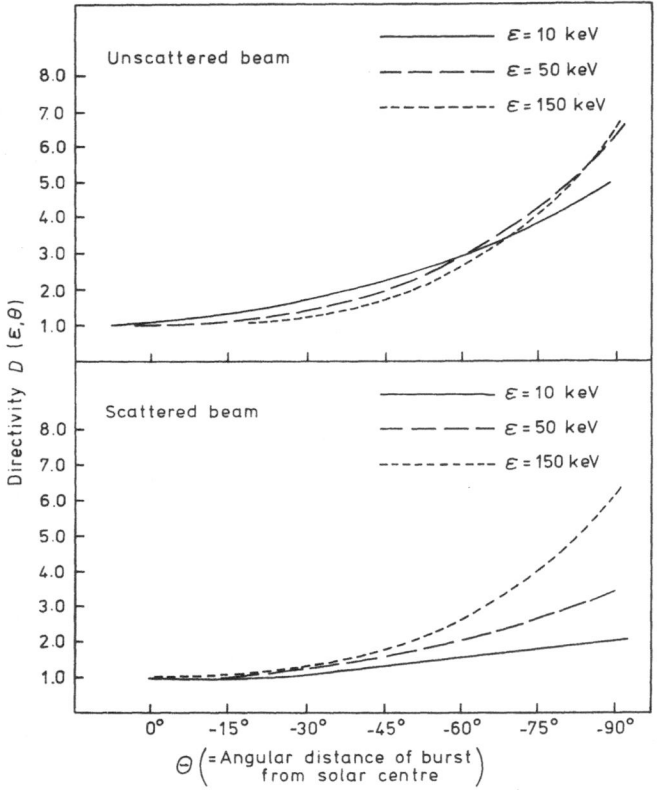

Fig. 13.   Directivity of X-ray emission from Brown's (1972b) thick-target model, as a function of photon energy and flare location, defined relative to a flare at disc centre.

the thick target plasma to visibility above the limb within time scales (10–100 s) shorter than the duraction of a large burst. Satisfactory tests in this direction may however, only be possible with the advent of a hard X-ray heliograph.

(iii) *The Electron Acceleration Spectrum*

An important feature of the thick-target model is the distinction between the electron injection spectrum and the mean electron spectrum in the source region (Brown, 1971 and subsequent authors). Since the electron lifetime is very short compared to the observational resolution, it proves possible to express (1) in the form

$$I(\varepsilon, t) = \frac{1}{4\pi R^2} \int\limits_{E_0 = \varepsilon}^{\infty} \mathfrak{F}(E_0, t) \int\limits_{E = \varepsilon}^{E_0} \frac{nQv \, dE}{(dE/dt)_{\text{TOT}}} \, dE_0, \tag{13}$$

where $\mathfrak{F}(E_0, t)$ is the acceleration spectrum – i.e. the number of electrons injected per s per unit E_0, $(dE/dt)_{\text{TOT}}$ is the total energy loss rate of an electron of injection energy E_0 when it has decayed to energy E and n is the local plasma density at that point. Due to the high density n involved, there is no doubt that collisional energy losses entirely dominate the synchrotron and inverse Compton contributions to $(dE/dt)_{\text{TOT}}$. It is further generally assumed (Brown, 1971; Kane, 1973; Lin, 1974a, b) that collective losses to plasma wave generation can also be neglected on the grounds that the electron beam is dilute. If this is valid, then

$$(dE/dt)_{\text{TOT}} \simeq \frac{K}{E} \, nv, \tag{14}$$

where $K = 2\pi e^4 \Lambda$ and Λ is the effective Coulomb logarithm, so that (13) reduces to

$$I(\varepsilon, t) = \frac{1}{4\pi R^2} \int\limits_{\varepsilon}^{\infty} \mathfrak{F}(E_0, t) \int\limits_{\varepsilon}^{E_0} \frac{EQ(\varepsilon, E)}{K} \, dE \, dE_0 \tag{15}$$

which is independent of the source geometry provided K is a function of E only, and in fact a slowly varying one (Brown, 1973d has considered the effect of decreasing hydrogen ionisation with increasing depth in a thick flare target and finds that, due to K decreasing with ionisation, the high energy end of the X-ray spectrum is somewhat enhanced). Again using the Bethe-Heitler formula for $Q(\varepsilon, E)$ it proves possible (Brown 1971) to solve (15) analytically for $\mathfrak{F}(E_0, t)$ to give

$$\mathfrak{F}(E_0) = -\frac{4KR^2}{K_{\text{BH}}} \frac{1}{E_0^{3/2}} \int\limits_{E_0}^{\infty} \frac{\varepsilon}{\sqrt{\varepsilon - E_0}} \{4I' + 5\varepsilon I'' + \varepsilon^2 I'''\} \, d\varepsilon \tag{16}$$

or

$$\mathfrak{F}(E_0) \, (\text{electrons s}^{-1} \, \text{keV}^{-1}) = -2.0 \times 10^{33} E_0^{-3/2} \, (\text{keV}) \int\limits_{E_0}^{\infty} \frac{\varepsilon}{\sqrt{\varepsilon - E_0}} \times$$
$$\times \{4I' + 5\varepsilon I'' + \varepsilon^2 I'''\} \, d\varepsilon \tag{17}$$

which, in the particular case of power law (6) reduces to

$$\mathfrak{F}(E_0) = \frac{4KR^2}{K_{\text{BH}}} (\gamma - 1)^2 \, B(\gamma - \tfrac{1}{2}, \tfrac{1}{2}) \, AE_0^{-\gamma - 1} \tag{18}$$

or, in numerical form,

$$\mathfrak{F}(E_0) \, (\text{electrons s}^{-1} \, \text{keV}^{-1}) = 2.0 \times 10^{33} \, (\gamma - 1)^2 \times$$
$$\times B(\gamma - \tfrac{1}{2}, \tfrac{1}{2}) \, AE_0^{-\gamma - 1}, \tag{19}$$

where E_0 is in keV and A such that $A\varepsilon^{-\gamma}$ is in $\text{cm}^{-2}\,\text{s}^{-1}\,\text{keV}^{-1}$ with ε in keV. Equation (19) shows that the thick-target electron injection spectrum is two powers steeper than the mean spectrum (7), a result with important consequences for the energy and number of electrons contained in the extrapolated low energy end of the electron spectrum. In addition, Datlowe and Lin (1973) have shown how it might be used as a test of the model, if the interplanetary electron spectrum can be taken as that of the electrons at acceleration. Their results for one small flare show an interplanetary electron index δ nearer to $\gamma - 1$ (as predicted by the thin-target model – cf. Section V(b)) than to $\gamma + 1$ (as predicted above for the thick-target). On the other hand, as pointed out by McClymont and Brown (1974), the fact that the interplanetary electrons in that event comprise $\lesssim 0.4\%$ of the number of electrons in the X-ray flare (Datlowe and Lin, 1973) puts the identification of their spectrum with that of the bulk of accelerated electrons in considerable doubt, since the few escaping particles, are, *a priori*, exceptional. Thus one cannot be sure at present whether energy dependent escape probability, or modification by plasma wave losses of the spectrum of electrons passing through the corona (cf. Section V(b)) may explain this discrepancy.

One major gap in our understanding of the thick-target and other electron stream models is whether and how the streams can in fact penetrate the plasma target against two stream instability losses to plasma waves. Though, as noted by Brown (1971), Lin (1974a, b) and Kane (1974a), the growth rate of such energy losses decreases as the stream becomes more dilute compared to the plasma, the growth time may still be small compared to the burst duration. Even without analysis of that problem, however, it is easy to see how radically the thick target model would be changed if collective losses were dominant. Firstly, any increase in $(\mathrm{d}E/\mathrm{d}t)_{\text{TOT}}$ in (13) would proportionally reduce the efficiency of the thick-target as an X-ray emitter (since electron lifetimes would be reduced), thus worsening the already considerable problem of electron number and energy requirements (Section VI). Secondly, the energy dependence of collective losses certainly differs from the collisional form (14) so that the relationship between the X-ray and the electron acceleration spectra (17) will no longer hold. Thirdly, the spatial distribution of the emission would be shifted higher in the atmosphere, and finally, since decollimation of streams is an even more important effect of the two stream instability, both the polarisation and directivity would be greatly reduced.

(b) THE THIN-TARGET MODEL

Motivated by the observations of behind-the-limb bursts and of the X-ray/interplanetary electron spectral index relationship, Datlowe and Lin (1973) have proposed a model in which the hard X-ray burst time profile is again produced by continuous modulation of an electron source but in which the electrons stream upward through a (thin-target) coronal region. Since, however, as argued in Section V(a) these observations may actually be compatible with the thick-target, it is necessary to examine the thin-target predictions of other observable quantities and to weigh the two models in terms of their compatibility with overall flare requirements.

As regards reproducing observed time profiles, the same remarks apply to the thin-target as already made for the thick-target except that the lower limit to fine time structure is set in the thin-target not by the energy loss time (which → ∞) but rather by the escape (or transit) time of electrons through the thin-target layer. This time may not be as small as in the thick-target since the target thickness cannot be less than the atmospheric scale height ($\simeq 10^{10}$ cm in the corona) but is certainly small enough to be compatible with present observations.

(i) *Polarisation and Directivity*

No-one has specifically considered these aspects of the thin-target model but results will clearly depend on the collimation of the upward moving electron stream and so on the form of the guiding field, of the initial acceleration and of collective scattering processes. Upper limits may, however, be set from the results for a purely radial stream (cf. Elwert and Haug, 1971; and Haug, 1972). By symmetry, the polarisation would evidently be zero at the disc centre and increase to the limb as shown in Figure 14, the plane of

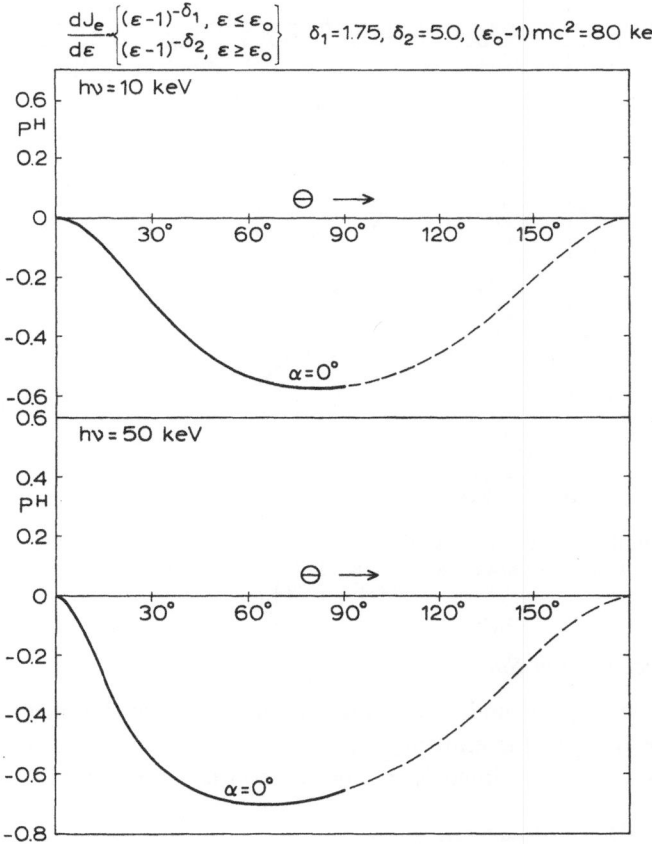

Fig. 14. Upper limit to the polarisation expected from the thin target model of Datlowe and Lin (1973) at 10 and 50 keV as a function of flare location, based on Haug (1972). θ is the distance from the solar centre, so dotted portions are for flares behind the limb.

polarisation being along the solar radius through the source. Figure 15 shows the maximum directivity of the model at 50 keV with a maximum at 30–60° from the disc centre then slight to severe darkening toward the limb depending on electron 'knee' energy E_0 (due to relativistic forward beaming of photons).

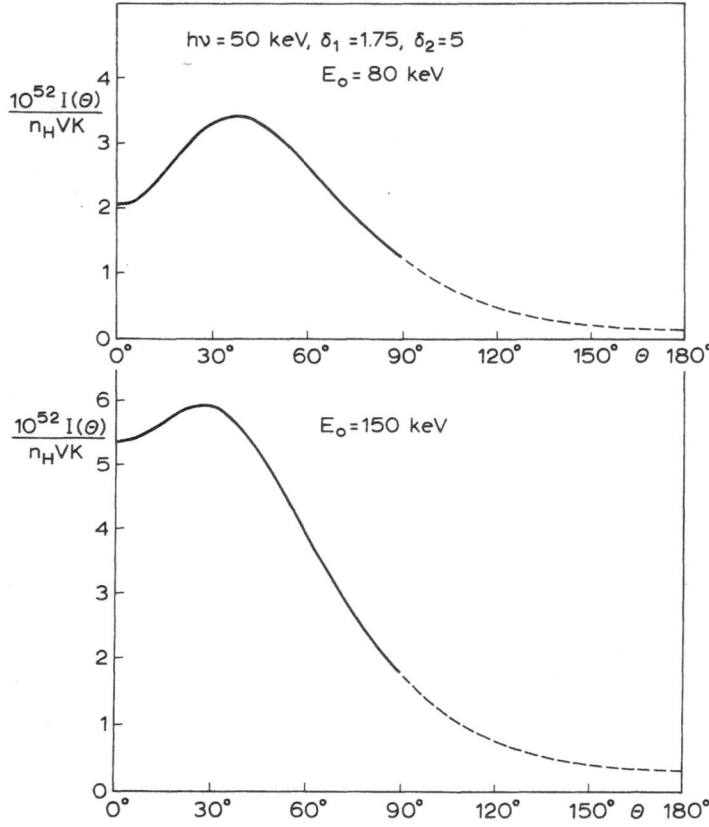

Fig. 15.   Upper limit to relative variation of total burst intensities at 50 keV across the disc in the thin-target model, (based on Elwert and Haug, 1971). E_0 is the electron spectrum 'knee-energy' (cf. Figure 14).

(ii) *Electron Acceleration Spectrum*

Since electrons passing through a *thin-target* do not undergo significant energy losses, the electron spectrum in the emitting layer is identical to the acceleration spectrum. Thus, in terms of a one-dimensional model, source Equation (1) simplifies (since $F \neq F(\mathbf{r})$) to

$$I(\varepsilon, t) = \frac{\Delta N}{4\pi R^2} \int_{\varepsilon}^{\infty} \mathfrak{F}(E_0, t) \, Q(\varepsilon, E) \, \mathrm{d}E_0, \tag{20}$$

where $\mathfrak{F}(E_0, t)$ is the total number of electrons per unit E_0 accelerated (and passing through the entire target area) per s and ΔN is the target 'thickness' in terms of ambient protons per cm² column. Again in the Bethe-Heitler approximation the solution is

$$\mathfrak{F}(E_0, t) = \frac{4R^2}{K_{BH}\Delta N} E_0^{1/2} \int_{E_0}^{\infty} \frac{\{I + 3\varepsilon I' + \varepsilon^2 I''\}}{\sqrt{\varepsilon - E_0}} \, d\varepsilon \tag{21}$$

or

$$\mathfrak{F}(E_0, t) \, (\text{s}^{-1} \, \text{keV}^{-1}) = \frac{6.7 \times 10^{50}}{\Delta N \, (\text{cm}^{-2})} E_0^{1/2}(\text{keV}) \int_{E_0}^{\infty} \frac{\{I + 3\varepsilon I' + \varepsilon^2 I''\}}{\sqrt{\varepsilon - E_0}} \, d\varepsilon \tag{22}$$

and for power law (6)

$$\mathfrak{F}(E_0, t) = \frac{6.7 \times 10^{50}}{\Delta N \, (\text{cm}^{-2})} (\gamma - 1)^2 \, B(\gamma - \tfrac{1}{2}, \tfrac{1}{2}) \, A E_0^{-\gamma + 1} \tag{23}$$

with ε, E_0 in keV throughout and I in cm⁻² s⁻¹ keV⁻¹.

The limit to ΔN for which a target becomes classed as 'thin' is to some extent arbitrary of course but $\lesssim 30\%$ '*collisionally*' thick' at 5 keV and above must be an upper limit, implying $\Delta N \lesssim 10^{19}$ cm⁻². Due to the low plasma density, however, collective energy losses are very much harder to circumvent in a thin than in a thick target. E.g. the large event of 1972, August 4 (Hoyng *et al.*, 1974) requires $\simeq 10^{37}$ electrons s⁻¹ even above 25 keV and so a stream density (over 10^{19} cm²) $\gtrsim 10^8$ cm⁻³, which is not small compared to that of the coronal plasma. Unless a means can be found of stabilising the resulting two stream instability (as required for type III radio bursts – e.g. Smith, 1974), the coronal thin-target plasma would have to be allocated a very small ΔN indeed to remain 'thin' for *total* energy losses and the thin-target efficiency problem (see below) would be made much worse. Even if such stabilisation can be established, it is again unclear, as for the thick-target, how the stabilisation process may modify the electron spectrum after acceleration and hence vitiate the assumption made by Datlowe and Lin (1973) in advocating the thin-target model on the basis of interplanetary electron spectra.

(iii) *Spatial Distribution*

Just how small ΔN is for the thin-target model depends on the height of the primary electron source. Brown and McClymont (1974) have considered the height distribution of thin target emission above this level, pointing out that it depends only on the density distribution of the target atmosphere. They conclude that for a thin-target X-ray source to be visible in behind-the-limb flares it must be high above the transition layer where $\Delta N \lesssim 10^{18}$ cm⁻² implying efficiencies for the model (compared to thick target emission) of only 3% at 5 keV, 0.1% at 25 keV and only 6×10^{-3}% at 100 keV. Hence the total energy and number of electrons needed for a burst are more than 30 times those required by thick target emission and so unacceptably high (cf. Section VI). If on the other hand the source lies below the transition region then ΔN may be

$\gtrsim 10^{19}$ cm^{-2} but with the source mostly lying only some 2000 km above the photo-sphere and not visible in behind-the-limb flares (just as for a chromospheric thick target model). Furthermore, not only does a thin-target require more electron energy in general than a thick-target (and perhaps more than available from a whole active region) but almost all of this energy is 'wasted' by being dumped in the low density corona and making no further contribution to the flare. Brown and McClymont (1974) have emphasised the problems of such dumping by quantitative consideration of the required trap parameters. In particular they conclude that, in a large event, more than 10^{40} electrons of $\geqslant 25$ keV would have to be trapped extremely high in the corona where $n \lesssim 10^7$ cm^{-3}. Such trapping would produce a vast microwave event unless the trapping field were exceptionally small, requiring in turn a large trapping volume – conservatively set at a cubic solar radius.

Finally, it has been widely recognized (e.g. Kane, 1973; Brown, 1973b; Brown and McClymont, 1974) that a purely thin-target model provides no explanation of the synchronism of hard X-rays and EUV bursts from the chromosphere.

(c) THE ELECTRON TRAP MODEL

Based on the original proposal by Takakura and Kai (1966), this model describes bursts in terms of the bremsstrahlung of electrons magnetically trapped in the low corona. Initially it was supposed that rapid ('impulsive') injection of electrons ex-plained the relatively short rising phase of bursts while the gradual collisional decay of the trapped electrons produced the burst decay profile. More recent observations, however, and particularly their complex temporal fine structure, have required con-siderable modification of the model.

(i) *Time Profiles*

Small bursts which apparently comprise a single spike might be direct examples of the simple impulsive injection/collisional decay hypothesis, but it has been pointed out that the observed softening spectral decay of these spikes conflicts with the harden-ing expected from collisions (Kane and Anderson (1970) and others). Brown (1972a) has, however, shown that this discrepancy can be resolved if the higher energy elec-trons encounter a higher mean plasma density along their paths. De Feiter (1974) has criticised this as 'artificial' but in fact the necessary decrease of electron pitch angle with increasing energy is a natural result of direct electric field acceleration (cf. Spei-ser, 1965; Petrosian, 1973) while some increase of plasma density down the arms of the trap is inevitable (cf. Benz and Gold, 1971). To that extent the explanation of impulsive spike characteristics in a trap model is more satisfactory than in continuous injection models since based on definitive physical features of the model rather than being an *ad hoc* requirement of the acceleration process.

The complex fine structure of larger events (Section III(a)) can be reconciled with an electron trap source in two ways. Firstly by a mixture of models (Kane, 1974a) in which the acceleration of electrons in the trap is repeated or continuous through the event but with a fraction of the electrons escaping downward (to a thick target) to produce the rapid burst fine structure and synchronised chromospheric EUV variations. Secondly, the rapid acceleration hypothesis may be maintained and the burst time structure attributed to MHD oscillations of the trap itself (Parks and Winckler, 1971; Brown, 1973c). This version of the model differs greatly, however, from the original impulsive acceleration concept since the magnetic field variations accompanying the required plasma density oscillations result in induced electric field acceleration of the trapped electrons throughout the event (Brown, 1973c; Brown and Hoyng, 1975).

(ii) *Polarisation and Directivity*

No adequate analysis of these features of a trapped electron source has yet been published. Earlier descriptions (Ohki, 1969; Pinter, 1969; Shaw, 1972; Elwert and Haug, 1970, 1971; Tindo *et al.*, 1972b) have considered either electrons moving along a horizontal field or circling horizontally near the trap ends (cf. Brown, 1972a; Benz and Gold, 1971; Haug, 1972), entirely neglecting the magnetic field curvature which is an essential feature of the model. Preliminary results of the detailed calculations necessary to include the field structure (McClymont and Brown, 1974) tend to suggest that both the polarisation and directivity are smaller than earlier predictions and than the raw observational results of Tindo *et al.* Results for both the degree and plane of polarisation are, however, much more complicated than previously supposed and their evaluation in terms of observations must await complete results of this analysis.

(iii) *Electron Acceleration Spectrum*

Since collisional distortion of the electron spectrum occurs only over the whole burst time scale, the instantaneous electron spectrum inferred from Equation (5) at the burst peak can be identified with the electron acceleration spectrum, provided correction is made for the higher plasma densities possibly encountered by higher energy electrons – Brown (1972a) (this correction may amount to over 2 powers steepening of the inferred electron spectrum and so greatly affect estimates of total electron numbers and energy in the model). When the entire evolution of the burst spectrum is considered, however, the time-integrated X-ray spectrum will correspond to that of a thick-target source instantaneously (with the same initial electron injection spectrum) since the collisional spectral evolution is just that within a thick target but on a long time scale.

(iv) *Spatial Distribution of Emission*

Quantitative predictions of the spatial distribution of emission from trapped electrons are not yet complete, though under way (McClymont and Brown, 1974). It is clear of course that at least some of the emission will emanate from high in the atmosphere, compatible with behind-the-limb events but there is in fact a considerable concentration of the emission deep in the two limbs of the trap due to the higher density there (Brown, 1972a; Haug, 1972; Benz and Gold, 1971; McClymont and Brown, 1974) and also to the greater path lengths of spiralling source electrons near their reflection points (McClymont and Brown, 1974).

TABLE I

Primary requirements of hard X-ray source models and their relation to existing models

Model description	Burst spectrum and time profile	Observed polarisation	Directivity (no data)	Interplanetary electron spectra	Synchronous EUV burst (chromospheric)	Behind-the-limb bursts	Electron energy & number requirements (above: 25 keV)
Thick target	Continuous modulation of acceleration mechanism (unspecified)	Good agreement with available data	Predicts limb brightening especially at high energy	Explained in terms of energy dependent escape or of collective energy losses	Heating by bombarding electrons themselves	Only explained by mass redistribution in the flaring atmosphere or a trapped electron component	2×10^{32} erg 3×10^{39} electrons in large event i.e. – severe
Thin-target	As thick-target	Not calculated in detail but compatible with data	Predicts limb darkening with maximum at 30°–60° from center	Same as X-ray source electron spectrum	Requires rapid energy transfer over large distances (unspecified)	High source region is feature of model	10^{33} erg 10^{40} electrons minimum in large event – impossible?
Electron trap	MHD oscillations of trap itself or escape of fraction of electrons thick-target	Results not yet available (McClymont and Brown, 1974)	Small (McClymont and Brown, 1974)	Same as X-ray source electron spectrum at burst peak	As thin target or by partial escape of electrons to thick-target	High source region is feature of model	As thick-target

(d) CONCLUSIONS

In Table I are summarised the key requirements of burst models and the abilities of the three above source models to meet these. The essence of the present situation is that the thick-target model may be capable with moderate economy of flare energy, of explaining all burst features including behind-the-limb occurrences, when mass motion of the target flare plasma is important, and interplanetary electron spectra if the electron escape is suitably energy dependent or if collective losses are involved. The thin target, on the other hand, provides an immediate interpretation of interplanetary electron spectra but only explains behind the limb events at the expense of demanding unreasonably large electron fluxes, and without explaining the EUV observations. Finally, coronally trapped electrons are readily compatible with behind-the limb emission and interplanetary electron spectra but may produce too small a polarisation and are hard to reconcile with chromospheric flare emissions unless a substantial (thick-target) downward escape of electrons also occurs.

VI. The Electron Energy and Number Problem

(a) INTRODUCTION

As discussed in previous sections, a thick-target is the most efficient of existing source models in terms of the electron flux required to produce any given X-ray burst intensity. As noted in the introduction, however, in some flares even the thick-target model needs a total energy of electrons comparable to or exceeding the total thermal energy. This conclusion was first reached by extrapolating the observed steep power-law spectra below the non-thermal energy range directly observable (Neupert, 1968; Brown, 1971; Kahler and Kreplin, 1971) but the result is now known sometimes to hold without any extrapolation of the X-ray power-law beyond the observed range (e.g. Syrovatskii and Shmeleva, 1972). Typical figures for a large (3B) event (1972, August 4) are those given by Hoyng $et\ al.$ (1975), namely 4×10^{39} electrons of $E_0 \geqslant 25$ keV with a total energy of 2×10^{32} erg (these requirements increasing tenfold if the spectrum is extended down to even 10 keV). Although flare heating models based on energetic electrons as the mode of energy transport have achieved some success (see Section I), it is a major theoretical problem to find a mechanism capable of accelerating such a number and energy of electrons on the required time scale. The severity of the energy release problem is well known (e.g. Sweet, 1969). The number problem is clearly seen on noting that the acceleration of even 4×10^{39} electrons from coronal material, where $n \lesssim 2 \times 10^9$ cm^{-3} (Lin, 1973), would require total involvement of 2×10^{30} cm^3 of plasma, or in practice involvement of a fraction of particles in an even greater volume, and this at a rate sufficiently fast to account for the burst time profile. It is therefore essential to consider any factor which might radically change our estimates of these figures, the following being two possible candidates.

(b) THE THERMAL HARD X-RAY CONTRIBUTION

Chubb $et\ al.$ (1966) (1971) has claimed that hard X-ray burst data have never satisfactorily excluded thermal interpretation. The subsequent development of this view and the debate over it have been reviewed elsewhere in this Symposium (Kahler, 1975). Brown (1974) has emphasised that, although thermal emission is hardly capable of explaining X-ray bursts throughout the hard X-ray energy range, it is essential to the problem of non-thermal electron energy in flares to determine the photon energies up to which thermal emission is important. This separation cannot be done on the basis of spectra alone since, as suggested by Chubb (1971) and proved quantitatively by Brown (1974), any hard X-ray spectrum can be produced by a suitable temperature distribution in a thermal plasma. In fact to produce a burst spectrum $I(\varepsilon)$ cm^{-2} s^{-1} keV^{-1} the emission measure distribution $\mu(T)$, differential in temperature $T(\mathrm{K})$ should be

$$\mu(T)\,(\mathrm{cm}^{-3}\,\mathrm{K}^{-1}) = \frac{1.4 \times 10^{45}}{T^{3/2}}\,\mathscr{L}^{-1}\left\{I(\varepsilon);\,\frac{1.16 \times 10^{7}}{T}\right\},\tag{24}$$

where $\mathscr{L}^{-1}$ is the inverse Laplace transform (Brown, 1974). In particular to produce power-law spectrum (8), the required distribution is

$$\mu(T)\,(\mathrm{cm}^{-3}\,\mathrm{K}^{-1}) = \frac{3.6 \times 10^{34}A}{\Gamma(\gamma-1)}\left[\frac{T}{1.16 \times 10^{7}}\right]^{-\gamma+1/2}\tag{25}$$

while to produce a power-law with high energy break – i.e. $A\varepsilon^{-\gamma}e^{-\varepsilon/\varepsilon_0}$

$$\mu(T) = \frac{1.4 \times 10^{45}A}{\Gamma(\gamma)\,T^{3/2}}\left[\frac{11.6 \times 10^{6}}{T} - \frac{1}{\varepsilon_0}\right]^{\gamma-1}\quad\text{if } T \geqslant 11.6 \times 10^{6}\,\varepsilon_0$$

$$\tag{26}$$

$$= 0 \text{ otherwise}.$$

(Contrary to De Feiter's (1974) claim, it should be noted that there is no more arbitrariness in this fitting procedure than in the adoption of a power-law electron spectrum for non-thermal models). Against the thermal mechanism Kahler (1971a, b) has claimed firstly that the thermal conduction intrinsic to a distributed temperature source would cool the source too fast to explain bursts and secondly that an electron temperature (i.e. Maxwell distribution) could not be established on the short time scale of rapid burst variations. Brown (1974) has countered that when Equation (24) is used to determine the temperature structure, Kahler's conduction figure is found to be an overestimate, and secondly that if the thermal flare plasma is highly turbulent, conduction is severely inhibited. Furthermore, energy exchange via the turbulent plasmons, greatly reduces the effective Maxwell relaxation time (M. Kuperus and J. Kuipers, 1974, private discussions.) Thus it is the reviewer's opinion that the thermal mechanism for hard X-ray emission still cannot be excluded in the energy range below about 100 keV. The important consequence in the present content, however, is that if thermal emission dominated the spectrum of the 1972, August 4 event up to 50 keV, the inferred non-thermal electron energy would be only 2×10^{31} erg and therefore a quite minor part of the flare.

Kahler and Kreplin (1971) and Kahler (1973) have attempted to separate thermal and non-thermal emissions by distinguishing the impulsive and gradual components of burst time profiles, and concluded that the non-thermal component can extend right down to a few keV in some flares. This method suffers from the problems that it does not take account of the dependence of the thermal contribution on the plasma density, the effects of a non-isothermal spectrum, or the possible impulsive profile of transient thermal emission itself. Though theoretical development of Kahler and Kreplin's work along these lines may help clarify the problem, no satisfactory answer may be forthcoming until spatially and spectrally resolved hard X-ray observations and also improved polarisation data over a wide energy range, are obtained. At present polarisation data support the view that thermal flare emission may dominate to well above 15 keV in energy (cf. Frost, 1974a; Brown *et al.*, 1974).

(c) ELECTRON REACCELERATION

The electron numbers quoted in Section VI(a) and previous sections, and throughout the literature, are based entirely on 'injection' models in which each electron is accelerated once only and subsequently injected into a source region where it (collisionally) emits bremsstrahlung and loses its energy once and for all. If, however, electrons could be accelerated and emit their bremsstrahlung during rapid deceleration, in one and the same region, then they would be available for *in situ* reacceleration within the source. Two mechanisms potentially capable of combining confinement and repeated acceleration are acceleration by reflection between moving magnetic mirrors (Fermi acceleration) and stochastic acceleration by collisions with plasmons in a turbulent plasma (e.g. Tsytovich, 1973). As shown in Section V(b) the generation of plasma turbulence by particle streams in a thick-target situation has the effect of reducing the electron lifetime and so the bremsstrahlung emission during *one* 'stopping'. If, however, an electron is contained in the acceleration region, then it may be reaccelerated repeatedly. In this case the total number of electrons obtained by integrating Equation (16) over the event duration is not of physical interest and the total number of electrons required for production of a burst depends instead on the frequency with which an average electron is reaccelerated. The much smaller number of electrons needed could then obviate the severe problem of total electron numbers stated in VI(a) and might reduce the total electron numbers required for a hard X-ray burst to the same order of magnitude as those needed to supply microwave bursts and interplanetary electrons.

The effect of such a reacceleration model on the total electron *energy* requirements for hard X-ray bursts is less clear. If all the energy lost by an electron in each acceleration cycle is irreversibly lost to heating the ambient plasma, the total energy which passes through the form of energetic electrons must turn out to be as before – i.e. $\gtrsim 2 \times 10^{32}$ erg in a large flare. Even this situation is quite distinct from that implied in 'injection' models, however, since a relatively small number of electrons play a continuous role in the transfer of stored flare energy into plasma heating, but no such total energy need be released into the final form of non-thermal particles. Secondly, it would be most important to establish whether the process of rapid energy exchange between plasmons and electrons in this situation may affect the *nett* energy loss rate of an electron, occurring in Equation (13) (with 'recycling' of energy between electrons and plasma) and so possibly even enhance the efficiency of the source relative to a thick-target injection model, in terms of energy.

Finally it must be recognized that though this line of thought may be worth pursuing from the point of view of efficiency, it is not clear how either an in situ reacceleration model, or a thermal model, could explain the relationship of hard X-ray bursts to chromospheric (EUV) flare emissions or the observed degree of burst polarisation, involving as they do relatively randomised electron velocities and a confined energy release region.

Acknowledgements

The author wishes to gratefully acknowledge the support of ESRO and of the Dutch National Committee for Geophysics and Space Research during preparation of this review and the provision of funds from the University of Utrecht and from the IAU for attendance of this Symposium.

Discussions of aspects of the review with P. Hoyng, I. Craig, A. McClymont and M. Kuperus and the provision of preprint material by K. Frost, S. Kane, A. Korchak, and A. McClymont were of invaluable assistance.

References

Acton, L. W.: 1968, *Astrophys. J.* **152**, 305.
Arnoldy, R. L., Kane, S. R., and Winckler, J. R.: 1968, *Astrophys. J.* **151**, 711.
Van Beek, H. F., De Feiter, L. D., and De Jager, C.: 1974, to appear in Proc. Eslab Symp. (Saulgau 1973).
Beigman, I. L. and Vainshtein, L. A.: 1974, *Soviet Astron. – A.J.* (in press).
Benz, A. O. and Gold, T.: 1971, *Solar Phys.* **21**, 157.
Brown, J. C.: 1971, *Solar Phys.* **18**, 489.
Brown, J. C.: 1972a, *Solar Phys.* **25**, 158.
Brown, J. C.: 1972b, *Solar Phys.* **26**, 441.
Brown, J. C.: 1973a, *Solar Phys.* **31**, 143.
Brown, J. C.: 1973b, Proc. Leningrad Symp. on *Solar Cosmic Rays*, June 1973.
Brown, J. C.: 1973c, *Solar Phys.* **32**, 227.
Brown, J. C.: 1973d, *Solar Phys.* **28**, 151.
Brown, J. C.: 1974, in G. Newkirk, Jr. (ed.), 'Coronal Disturbances', *IAU Symp.* **57**, 395.
Brown, J. C. and Hoyng, P.: 1975, submitted to *Astrophys. J.*
Brown, J. C., McClymont, A. N., and McLean, I. S.: 1974, *Nature* **247**, 448.
Brown, J. C. and McClymont, A. N.: 1974, *Solar Phys.* **41**, 135.
Canfield, R. C.: 1974, *Solar Phys.* **34**, 339.
Cheng, C. C.: 1972, *Solar Phys.* **22**, 178.
Chubb, T. A.: 1971, in E. R. Dyer (General Ed.), *Solar-Terrestrial Physics/1970*, Part 1, D. Reidel Publ. Co., Dordrecht, Holland, p. 99.
Chubb, T. A., Kreplin, R. W., and Friedman, H.: 1966, *J. Geophys. Res.* **71**, 3611.
Cline, T. L., Holt, S. S., and Hones, E. W.: 1969, *J. Geophys. Res.* **73**, 434.
Craig, I. J. D.: 1974, Ph.D. Thesis, University College, London.
Craig, I. J. D. and Brown, J. C.: 1974, unpublished work.
Datlowe, D. W.: 1975, This volume, p. 191.

Datlowe, D. W. and Lin, R. P.: 1973, *Solar Phys.* **32**, 459.

Dere, K. P., Horan, D. M., and Kreplin, R. W.: 1974, *Solar Phys.* in press.

Drake, J. F.: 1971, *Solar Phys.* **16**, 152.

Elwert, G.: 1968, in K. O. Kiepenheuer (ed.), 'Structure and Development of Solar Active Regions', *IAU Symp.* **35**, 144.

Elwert, G. and Haug, E.: 1970, *Solar Phys.* **15**, 234.

Elwert, G. and Haug, E.: 1971, *Solar Phys.* **20**, 413.

De Feiter, L. D.: 1974, *Space Sci. Rev.* **16**, 3.

Fox, L. and Mayers, D. F.: 1968, *Computing Methods for Scientists and Engineers*, Clarendon Press, Oxford.

Frost, K. J.: 1969, *Astrophys. J.* **158**, L159.

Frost, K. J.: 1974a, in G. Newkirk, Jr. (ed.), 'Coronal Disturbances', *IAU Symp.* **57**, 421.

Frost, K. J.: 1974b, Private communication.

Frost, K. J. and Dennis, B. R.: 1971, *Astrophys. J.* **165**, 655.

Haug, E.: 1972, *Solar Phys.* **25**, 425.

Herring, J. R. H. and Craig, I. J. D.: 1973, *Solar Phys.* **28**, 169.

Householder, A. S.: 1964, *The Theory of Matrices in Numerical Analysis*, Blaisdell Pub. Co.

Hoyng, P., Brown, J. C., Stevens, G. A., and Van Beek, H. F.: 1975, This volume p. 233.

Hoyng, P. and Stevens, G. A.: 1974, *Astrophys. Space Sci.* **27**, 307.

Hudson, H. S.: 1972, *Solar Phys.* **24**, 414.

Hudson, H. S.: 1973, in R. Ramaty and R. G. Stone (eds.), *High Energy Phenomena on the Sun*, NASA SP-342, p. 207.

De Jager, C. and Kundu, M. R.: 1963, *Space Res.* **3**, 836.

Kahler, S. W.: 1971a, *Astrophys. J.* **164**, 365.

Kahler, S. W.: 1971b, *Astrophys. J.* **168**, 319.

Kahler, S. W.: 1973, in R. Ramaty and R. G. Stone (eds.), *High Energy Phenomena on the Sun*, NASA SP-342, p. 124.

Kahler, S. W.: 1975, This volume, p. 211.

Kahler, S. W. and Kreplin, R. W.: 1971, *Astrophys. J.* **168**, 531.

Kane, S. R.: 1969, *Astrophys. J.* **157**, L139.

Kane, S. R.: 1971, *Astrophys. J.* **170**, 587.

Kane, S. R.: 1973, in R. Ramaty and R. G. Stone (eds.), *High Energy Phenomena on the Sun*, NASA SP-342, p. 55.

Kane, S. R.: 1974a, in G. Newkirk, Jr. (ed.), 'Coronal Disturbances', *IAU Symp.* **57**, 105.

Kane, S. R.: 1974b, paper presented at AAS meeting, University of Hawaii, Jan. 1974.

Kane, S. R. and Anderson, K. A.: 1970, *Astrophys. J.* **162**, 1003.

Kane, S. R. and Donnelly, R. F.: 1971, *Astrophys. J.* **164**, 151.

Kawabata, K., Sofue, Y., Ogawa, H., and Omodaka, T.: 1973, *Solar Phys.* **31**, 469.

Koch, H. W. and Motz, J. W.: 1959, *Rev. Mod. Phys.* **31**, 920.

Korchak, A. A.: 1967a, *Soviet Astron. – A.J.* **11**, 258.

Korchak, A. A.: 1967b, *Soviet Phys. Doklady* **12**, 192.

Korchak, A. A.: 1971, *Solar Phys.* **18**, 284.

Korchak, A. A.: 1974, preprint.

Kuperus, M. and Kuipers, J.: 1974, Private communication.

Lin, R. P.: 1974a, *Space Sci. Rev.* **16**, 184.

Lin, R. P.: 1974b, in G. Newkirk, Jr. (ed.), 'Coronal Disturbances', *IAU Symp.* **57**, 201.

McClymont, A. N. and Brown, J. C.: 1974, *Solar Phys.* in press.

McKenzie, D. L., Datlowe, D. W., and Peterson, L. E.: 1973, *Solar Phys.* **28**, 175.

Neupert, W. M.: 1968, *Astrophys. J.* **153**, L59.

Ohki, K.: 1969, *Solar Phys.* **7**, 260.

Parks, G. R. and Winckler, J. R.: 1969, *Astrophys. J.* **155**, L117.

Parks, G. R. and Winckler, J. R.: 1971, *Solar Phys.* **16**, 186.

Peterson, L. E., Datlowe, D. W., and McKenzie, D. L.: 1973, in R. Ramaty and R. G. Stone (eds.), *High Energy Phenomena on the Sun*, NASA SP-342, p. 132.

Petrosian, V.: 1973, *Astrophys. J.* **186**, 291.

Pinter, Š.: 1969, *Solar Phys.* **8**, 142.

Santangelo, N., Horstman, H., and Horstman-Moretti, E.: 1973, *Solar Phys.* **29**, 143.

Schatzmann, E.: 1966, in C. de Jager (ed.), *The Solar Spectrum*, D. Reidel Publ. Co., Dordrecht, Holland, p. 313.

Shaw, M. L.: 1972, *Solar Phys.* **27**, 436.

Shmeleva, O. P. and Syrovatskii, S. I.: 1973, *Solar Phys.* **33**, 341.

Smith, D. F.: 1974, in G. Newkirk, Jr. (ed.), 'Coronal Disturbances', *IAU Symp.* **57**, 253.

Speiser, T. W.: 1965, *J. Geophys. Res.* **20**, 4219.

Strauss, F. M. and Papaggianis, M. D.: 1971, *Astrophys. J.* **164**, 369.

Sturrock, P. A. and Coppi, B. A.: 1966, *Astrophys. J.* **143**, 3.

Sweet, P. A.: 1969, *Ann. Rev. Astron. Astrophys.* **7**, 149.

Sweet, P. A.: 1971, in R. Howard (ed.), 'Solar Magnetic Fields', *IAU Symp.* **43**, 457.

Syrovatskii, S. I. and Shmeleva, O. P.: 1972, *Soviet Astron. – A.J.* **16**, 273.

Takakura, T.: 1969, *Solar Phys.* **6**, 133.

Takakura, T.: 1973, in R. Ramaty and R. G. Stone (eds.), *High Energy Phenomena on the Sun*, NASA SP-342, p. 179.

Takakura, T.: 1975, This volume, p. 299.

Takakura, T. and Kai, K.: 1966, *Publ. Astron. Soc. Japan* **18**, 57.

Takakura, T., Ohki, K., Shibuya, N., Fujii, M., Matsuoka, M., Miyamoto, S., Nishimura, J., Oda, M., Ogawara, Y., and Ota, S.: 1971, *Solar Phys.* **16**, 454.

Thomas, R.: 1975, This volume, p. 25.

Tindo, I. P., Ivanov, V. D., Mandel'shtam, S. L., and Shuryghin, A. I.: 1970, *Solar Phys.* **14**, 204.

Tindo, I. P., Ivanov, V. D., Mandel'shtam, S. L., and Shuryghin, A. I.: 1972a, *Solar Phys.* **24**, 429.

Tindo, I. P., Ivanov, V. D., Valnicek, B., and Livshits, M. A.: 1972b, *Solar Phys.* **27**, 426.

Tindo, I. P., Mandel'shtam, S. L., and Shuryghin, A. I.: 1973, *Solar Phys.* **32**, 469.

Tomblin, F. F.: 1972, *Astrophys. J.* **171**, 377.

Tsytovich, V. N.: 1973, *Ann. Rev. Astron. Astrophys.* **11**, 363.

Vorpahl, J. A.: 1973, *Solar Phys.* **28**, 115.

Vorpahl, J. A. and Takakura, T.: 1974, *Astrophys. J.* (in press).

Wild, J. P. and Smerd, S. F.: 1972, *Ann. Rev. Astron. Astrophys.* **10**, 159.

Wolff, R. S.: 1973, in R. Ramaty and R. G. Stone (eds.), *High Energy Phenomena on the Sun*, NASA SP-342, p. 162.

Zirin, H. and Tanaka, T.: 1973, *Solar Phys.* **32**, 173.

SOLAR FLARE X-RAY MEASUREMENTS AND THEIR RELATION TO MICROWAVE BURSTS

L. D. DE FEITER

The Astronomical Institute, Utrecht, The Netherlands

Abstract. This review discusses the available observational material of solar hard X-ray bursts, their interpretation in terms of a model of the source region and their relation with other flash-phase phenomena, in particular the impulsive microwave bursts.

I. Introduction

This paper deals primarily with the flare-associated hard X-ray bursts and their relations to other simultaneously occurring flare phenomena, in particular the impulsive radio burst observed at microwaves and the metric type III bursts. Several examples can be found in the literature (e.g. Kane and Anderson, 1970; De Feiter, 1974). In terms of the solar sources of these radiations we will be dealing with high-energy electrons, in the 10–100 keV range, originating during the flash phase of the flare, and their interactions with the ambient magneto-active plasma.

One of the important problems in this part of solar-flare research has already been stated fifteen years ago by Peterson and Winckler (1959) in their discussion of the event of 1958, March 20, the first recorded observation of a solar X-ray burst at photon energies above 10 keV. High-energy electrons can produce the X-ray burst through Coulomb-bremsstrahlung and the microwave burst through magneto-bremsstrahlung, but the numbers of electrons involved in the two mechanisms differ by a factor of 10^4. Nevertheless the clear correlation in time of the two emissions points at a common source region, if not of the emissions themselves then of the electrons producing them.

A recent study by Kane (1972) has provided observational evidence that also the electrons that produce the metric type III bursts when moving outwards through the corona, are accelerated at the same time as those that give rise to the hard X-ray and the microwave bursts. Also, some impulsive EUV-bursts (e.g. Noyes, 1973; Donnelly, 1973) can be explained as enhanced emission of a region in the chromosphere-corona transition layer that is heated by penetrating electrons in the same energy range. When the emission mechanisms of all these emissions are properly understood observations in the various frequency domains can be used to determine more decisively the properties of the accelerated electrons, and the conditions under which the acceleration processes take place.

II. Hard X-Ray Bursts

(a) INSTRUMENTATION

Table I gives a summary of some of the characteristics of those satellite-born hard

Sharad R. Kane (ed.), Solar Gamma-, X-, and EUV Radiation, 283–297. All Rights Reserved.
Copyright © 1975 by the IAU.

X-ray spectrometers for which the results are most extensively quoted in the literature. For a good comparison of the instrumental performances important parameters, like instrumental background, dynamic range, measures taken against pulse pile-up, efficiency curves, spectral resolution, procedures applied for the correction for instrumental effects etc., are needed. As the reports on these matters often are not always published, I have refrained from adding this information to the table. It should be noted though, that some of the older data do have shortcomings in this respect (e.g. Kane and Hudson, 1970). In terms of the effective collecting area and time resolution

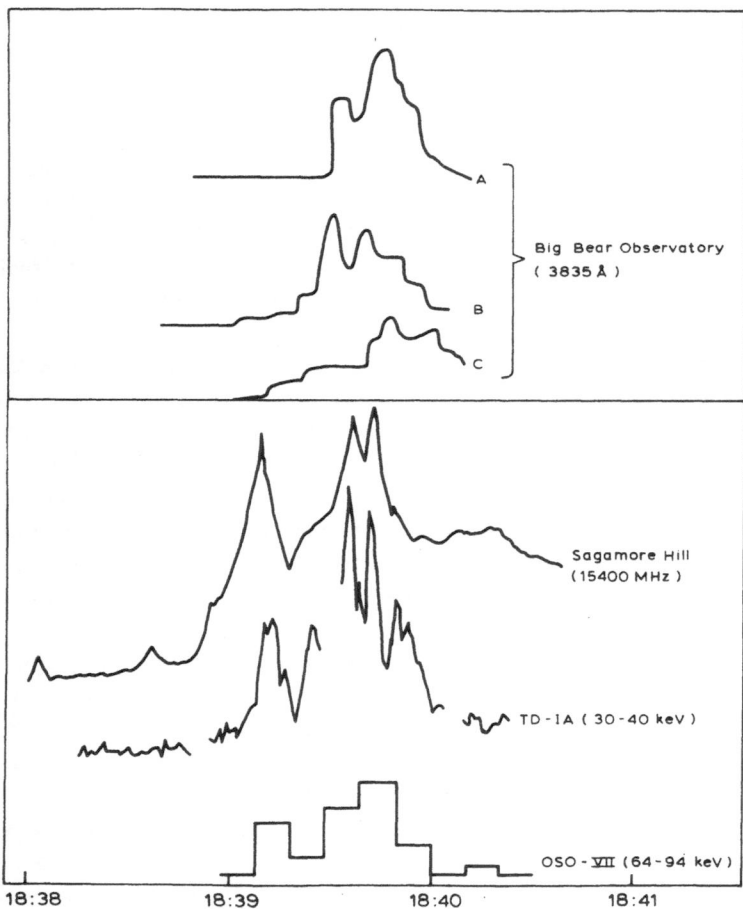

Fig. 1.   Time profiles for a small flare of 1972, Aug. 2 at 18 39 UT. From bottom to top the following profiles have been drawn: OSO 7 University of California, San Diego hard X-ray data (64–94 keV) taken at sampling intervals of 10 s; ESRO TD 1A Space Research Laboratory Utrecht hard X-ray data (30–40 keV) taken at sampling intervals of 1.2 s; Sagamore Hill microwave data (15400 MHz) and intensity profiles of specific bright spots identified in Figure 10 of Zirin and Tanaka (1973), observed with the 15 Å filter, centered on λ 3835 Å. The value of high-time resolution hard X-ray data is clearly brought out in this figure.

TABLE I

Satellite-borne solar hard X-ray spectrometers

Spacecraft	Operational period	Detector Type	Eff. area [a]	Energy range [b]	Time resol. [c]	References on the instrumentation
OGO 1	Sept. '64 –	Ioniz. chamber	~250 cm²	10–50 keV (1)	1–10 s	Kane et al. (1966)
OGO 3	June '66 –	Ioniz. chamber	~250 cm²	10–50 keV (1)	1–10 s	Kane et al. (1966)
OSO 1	March '62–	Na I (Tl)	4 cm²	20–100 keV	20 s	Frost (1964)
OSO 3	March 67–June 68	Na I (Tl)	0.6 cm²	7.7–210 keV (8)	15 s	Hicks et al. (1965)
OGO 5	March '68–	Na I (Tl)	9.5 cm²	9.6–128 keV (7)	2.3 s	Kane and Anderson (1970)
OSO 5	Jan. '69–	Cs I (Na)	5.4 cm²	15–250 keV (9)	1.8 s	Frost et al. (1970)
OSO 6	Aug. 69–March 72	Na I (Tl)	0.3 cm²	20–200 keV (4)	61 s	Brini et al. (1971); Brini et al. (1973)
OSO 7	Oct. '71–	Na I (Tl)	2.4 cm²	10.6–323 keV (9)	10.2 s	Datlowe et al. (1972)
TD 1A	March '72–	Cs I (Na)	5 cm²	24–912 keV (12)	1.2 s	Van Beek (1973)

[a] The effective area is derived from the sensitive area by multiplication with the fraction of the spin period during which the Sun is within the field of view of the instrument.

[b] The number of energy channels is given in brackets.

[c] In some cases the time resolution varies through the energy range; for instruments onboard spinning satellites the minimum time resolution has been set equal to the spin period.

the experiments on board OGO 5, OSO 5 and TD 1A are about equal; the spectro-meters on OSO 7 observe simultaneously the soft and the hard X-ray spectra which provides a possibility to separate clearly the thermal and the non-thermal components of solar X-ray bursts. Because of the highly eccentric orbit, the OGO 5 experiment operated for long stretches of time in a low-background environment outside the magnetosphere and therefore the data of this experiment are most suited for correlation studies with other flare phenomena; its dynamic range however is rather limited and restricts the observations to small flares. The importance of a resolution of time 1 s or less is clearly indicated in Figure 1 (see also Van Beek et al., 1974).

(b) THE OCCURRENCE OF HARD X-RAY BURSTS

Kane (1969) was able to show that many hard X-ray bursts, in the energy range around 20 keV, have a two-component structure, of which the first impulsive component has a hard non-thermal spectrum, whereas the second component varies slowly in time and has a soft thermal spectrum. The second component is absent above about 50 keV. Frost and Dennis (1971) discussed an event consisting of two hard non-thermal com-ponents. Peterson et al. (1973) find evidence for the existence of a slow hard X-ray phase usually associated with type IV radio emission. The relation between these three types of bursts is not completely clear at this moment; more information on this subject will be presented during this symposium. Here we will focus our attention primarily to the impulsive component. According to Kane and Anderson (1970) the impulsive component is essentially a flash-phase phenomena, and most flares with a well-pronounced flash-phase emit impulsive hard X-ray bursts.

(c) TIME VARIATIONS

When studied on a time scale of 1–2 s many hard X-ray bursts exhibit a clear fine structure. An example of this is given in Figure 1. Several other observations of rapid time variations have been reported; Beigman et al. (1971) have given indications of time scales as short as 0.05 s at wavelengths around 2 Å, Parks and Winckler (1971) and Frost (1969) also found rapid time variations. The high-time resolution data from the TD 1A experiment indicate that in some cases structures with characteristic times below 1 s do exist (Van Beek et al., 1974). Of particular interest in this connection is the event displayed in Figure 1. The rapid flashes found for this event are related to the fast and small flashes observed at λ 3835 by Zirin and Tanaka (1973) (see their Figure 10). The sizes of the sources of the individual UV flashes are as small as 1″ in diameter. A preliminary comparison with the radio spectrograph data from the Har-vard Radio Astronomy Station, Fort Davis (Maxwell, 1973) indicates that the X-ray peaks at 18 39 30, 18 39 35 and 18 39 43 may be associated with individual members of a group of type III bursts. Studies of this kind indicate strongly the necessity for a further reduction of the time resolution, also for ground-based observations. With modern data handling techniques it should be possible to overcome the problem of the tremendous data flow that will arise in that case.

(d) POLARIZATION

Polarization of X-rays indicates an anisotropy in the velocity distribution of the high-energy electrons that produce the bremsstrahlung X-rays (e.g. Elwert and Haug, 1970). On several satellites in the Intercosmos series the group at the Lebedev Physical Institute in Moscow has flown Thomson scattering polarimeters (see e.g. Tindo *et al.* (1973) for the wavelength range 0.6–1.2 Å. As the polarization measurements will be discussed more fully in another paper during this Symposium it suffices to state here that in most cases polarization has been observed for a rather large fraction of the duration of the X-ray bursts concerned. This should be taken as an indication of a prolonged injection of directed beams of energetic electrons in these events. A small amount of polarization ($\sim 2\%$) can be attributed to Thomson scattering of part of the radiation by the photospheric layers.

(e) POSITION AND STRUCTURE OF THE SOURCE REGION

All observations discussed thusfar refer to total flux measurements for which the association with a particular flare is made on the basis of time coincidence. For hard X-ray bursts only one observation has been reported for which the position of the source region has been determined directly using a balloon-born collimator (Takakura *et al.*, 1971). For this same event Tanaka and Enomé (1971) have determined the position line for the source of the associated microwave burst. The results of both measurements, drawn into a Hα filtergram of the optical flare, are presented in Figure 7 of Takakura *et al.*, 1971; the intersection of the two position lines falls near a bright nucleus of the Hα flare. The size of the source was smaller than the spatial resolution of the instrument, i.e. 1′.

For the wavelength region between 1 and 10 Å observations have been made with two crossed Soller-type collimators on board the cosmos 166 and 230, flown in 1967 (June 16–September 11) and 1968 (February 5–November 1) respectively (Beigman *et al.*, 1969, 1971). The resolution of these collimators was about 15″. The general structure, according to these observations of the X-ray flare is an elongated filament of 1′–2′ length and about 15″ width containing several small bright nuclei that may change rapidly in time.

The grazing incidence X-ray telescopes flown by the Solar Physics Group of A.S. & E. on rockets (Vaiana *et al.*, 1968; Vaiana and Giacconi, 1969) and on Skylab (Vaiana *et al.*, 1973) operate in essentially the same wavelength region, i.e. 3.5–13.5 Å. The spatial resolution of these instruments is better than 10″ throughout the field of view; on some of the best exposures details of a few arcseconds can be resolved. The rocket observations of the flare of 1968, June 8, indicate the existence of a large loop-like structure connecting two Hα emission centers on opposite sides of the magnetic neutral line.

The GSFC experiment on board OSO 7 contained among other experiments a multigrid collimator with a rectangular field of view of 20″ × 20″ FWHM for the wavelength region 1.74–7.95 Å. Observations of the flare of 1972, August 2 for which high-time resolution flash-phase phenomena are presented in Figure 1, confirm the earlier findings of an elongated shape with a length of about 27000 km (35″), but with a width that is smaller than the spatial resolution of the instrument.

Neither of the two last-mentioned reports refer to the bright points found by Beigman *et al.* (1969). However, comparison of hard X-ray time profiles with series of Hα filtergrams has led De Jager (1967) and Zirin and coworkers (Vorpahl and Zirin, 1970; Zirin *et al.*, 1971; Vorpahl, 1972) to conclude that the Hα features that occur simultaneously with the hard X-ray burst are bright knots, often referred to as 'flare kernels', with a diameter of 5″ to 10″ located near, to within 10″, the magnetic neutral line (Vorpahl, 1972).

One observation that to some extent is in conflict with the results discussed thus far has been made during the very rare occasion of the occurrence of a solar flare in an active region that was being occulted by the Moon's limb during the eclipse of 1970, March 7 (Kreplin and Taylor, 1971). The spatial resolution in this case was 20″, determined by the time resolution and the velocity with which the Moon's limb moves over the solar disk. The total area of the region at $\lambda < 16$ Å was found to be 136000 km, or 3′, but it contained a hot dense core of 54000 km diameter, or 1.′2. These figures are larger than the ones quoted above by a factor of 5 to 10. This discrepancy is probably due to the specific observing conditions, like e.g. the fact that the Moon's limb moved more or less along the flare filament. These observations seem to indicate that the hard X-ray emission takes place at the 'chromospheric footpoints' of a magnetic flux tube, whereas the hot and dense material along the whole flux tube emits X-rays at longer wavelengths. In magnetically complex regions several of these flux tubes may brighten up rather independently, thus leading to short flashes in hard X-rays at various places within the active region.

The relation between the hard and the soft X-ray bursts may be quite complex; in the initial phase the soft X-ray emission may indicate the presence of physical conditions that are favourable for the occurrence of impulsive phenomena and in the decay phase the soft X-ray source contains a component that is produced through collisional heating by the non-thermal electrons required for the production of the hard X-ray burst. It seems to me that total flux measurements are of limited use in extending the analysis beyond a mere comparison of the energies involved in both emissions. A study of the evolution in time of specific regions of a flare, using high-resolution X-ray images in various spectral regions, are required in order to resolve these questions.

The photospheric magnetic field data discussed by Rust (1972) for the 2B flare of 1969, October 24 (see also Zirin *et al.*, 1971) indicate the importance of small areas of one magnetic polarity embedded in a larger region of opposite polarity, the so-called 'satellite spots'. Important with respect to the occurrence of flares in these satellite spots is that they change with time; for this reason they were named 'Structures Magnétique Evolutives' by Martres *et al.* (1968).

(f) MODELS FOR THE HARD X-RAY EMISSION

The X-ray flux produced at the Earth's distance R by energetic electrons with energy distribution dn_e^*/dE interacting in a region with ion density n_i, which may be inhomogeneously distributed through the volume V, is computed from

$$\frac{\mathrm{d}F}{\mathrm{d}E_x} = \frac{1}{4\pi R^2} \int\limits_{E_x}^{\infty} \frac{\mathrm{d}\sigma}{\mathrm{d}E_x} \sqrt{\frac{2E}{m}} \left(\int\limits_V n_i \frac{\mathrm{d}n_e^*}{\mathrm{d}E} \mathrm{d}V \right) \mathrm{d}E. \tag{1}$$

For the differential cross section the Bethe-Heitler formula can be used in most cases:

$$\frac{\mathrm{d}\sigma}{\mathrm{d}E_x} = \frac{2\alpha\sigma_0}{\pi} \frac{mc^2}{EE_x} \ln\left(\sqrt{\frac{E}{E_x}} + \sqrt{\frac{E}{E_x} - 1} \right). \tag{2}$$

Isotropic emission has been assumed in (1); anisotropic emission due to relativistic beaming can be important at higher energies (e.g. Petrosian, 1973). The backscatter effects as studied by Tomblin (1972) and Santangelo *et al.* (1973) will not be discussed here; an interesting application of these effects which may allow a determination of the height above the photosphere of the emitting region will be reported during this symposium by Brown and Van Beek (1975).

Besides the spectrum of a hard X-ray burst its time evolution is an important observational datum; using the radio astronomical terminology we refer to the spectral development in time as the dynamic spectrum of the X-ray burst (Brown, 1971). This dynamic spectrum depends significantly upon the model of the source region, for which only a few limiting cases have been studied up till now. Brown (1972) analysed the so-called thin- and the thick-target models, where the differences are determined by the collisional losses of the energetic electrons, which are negligible in the former model and predominant in the latter. Also a distinction between models has been made on the bases of the time profile of the injection of energetic electrons into the source region; impulsive injection models (e.g. Takakura and Kai, 1966) and continuous injection models (e.g. Acton, 1968).

A comparison of these models should start with a discussion of the time-evolution equation of the non-thermal electron population (e.g. Blumenthal and Gould, 1970):

$$\frac{\partial n_e^*(E)}{\partial t} = \frac{\partial}{\partial E}\left(\frac{\mathrm{d}E}{\mathrm{d}t} n_e^*(E) \right) + q(E) - \frac{n_e^*(E)}{\tau_e}, \tag{3}$$

where the terms on the right-hand side indicate changes due to a systematic energy shift with time, e.g. collisional energy losses, the injection of energetic electrons from outside the emitting region and the escape from the emitting region respectively. If we indicate the characteristic times of these three processes by τ_c, τ_a and τ_e we can make the classification given in Table II.

As neither of these three characteristic times can be determined directly from the dynamic spectra the distinction between the four models listed in Table II cannot be made on the basis of the hard X-ray observations alone. However, assuming that the energy loss of the energetic electrons is entirely due to electron-electron collisions, a comparison of the observed decay times with the characteristic collisional loss time τ_c, and the variation of the decay time with energy allows a distinction to be made between the impulsive and continuous injection in the collision-dominated case. Energy considerations, and correlations with other flare phenomena provide means for a further discrimination.

TABLE II

Classification of hard X-ray emission models

Loss / Injection	Impulsive	Continuous
Collision-dominated $(\tau_e \gg \tau_c)$	$\tau_a \ll \tau_c \ll \tau_e$	$\tau_a \gg \tau_e \gg \tau_c$
Escape-dominated $(\tau_e \ll \tau_c)$	$\tau_a \ll \tau_e \ll \tau_c$	$\tau_a \gg \tau_c \gg \tau_e$

The energy loss rate through electron-electron collisions can be estimated from

$$\frac{\mathrm{d}E}{\mathrm{d}t} = \frac{2\pi r_e^2 mc^4}{v} \, n \ln \Lambda, \tag{4}$$

where r_e is the classical electron radius and $\ln \Lambda$ is the Coulomb logarithm (cf. Petrosian, 1973 and references therein). The collisional decay time derived from (4) is

$$\tau_c = \frac{E^{3/2}}{\sqrt{2}\,\pi r_e^2 m^{3/2} c^4 n \ln \Lambda} = 3.3 \times 10^8 \frac{E^{3/2}}{n} \text{ s}, \tag{5}$$

where in the numerical expression E is in keV.

If the observed decay times do not follow the $E^{3/2}$ behaviour, which is usually the case, we can conclude that the X-ray production takes place at high densities, or that the decay is determined by escape. In both cases a continuous injection of energetic electrons is required; in the former case, however, the total energy involved in the production of a given X-ray flux is smaller than in the second case.

Brown (1971), and several authors following him, has elaborated the thick-target case, which differs from the collision-dominated case introduced above. In that case the slowing down of the electrons is considered to happen in a time that is short as compared to the accumulation time of the instrument used for measuring the X-ray spectrum. Then we can follow, as Brown did, the X-ray production of a single electron until its energy is below the photon energy considered. Assuming $q(E)$ electrons with energy E are injected per unit of time, the X-ray flux produced at the Earth's distance is given by

$$\frac{\mathrm{d}F}{\mathrm{d}E_x} = \frac{1}{4\pi R^2} \frac{2\alpha\sigma_0}{\pi^2 r_e^2 mc^2 \ln \Lambda} \int_{E_x}^{\infty} Q(E) \ln\left(\sqrt{\frac{E}{E_x}} + \sqrt{\frac{E}{E_x} - 1}\right) \frac{\mathrm{d}E}{E_x}, \tag{6}$$

where $Q(E) = \int_E^{\infty} q(E)\,\mathrm{d}E$ is the integral injection spectrum. Equation (6) can also be derived from (1), using the appropriate solution of (3) for the instantaneous energy spectrum. The thick-target assumption means that there is at all times an equilibrium between the instantaneous energy spectrum of the energetic electron population and the spectrum of the injected electrons, i.e.

$$\frac{d}{dE}\left(-\frac{dE}{dt}\frac{dn_e^*(E)}{dE}\right) = q(E),$$ (7)

of which the solution is

$$\frac{dn_e^*(E)}{dE} = \frac{Q(E)}{-dE/dt}.$$ (8)

Substitution of (8) and (4) into (1) then leads to (6).

Similarly we can define the thin-target case, i.e. by requiring that there exists an equilibrium between the injected and the escaping electrons. In that case we have

$$\frac{dn_e^*(E)}{dE} = q(E)\tau_e.$$ (9)

Because of the lack of information on τ_e, in particular its energy dependence, the conclusion for the injection spectrum $q(E)$ derived from the observations of hard X-ray spectra are not as certain as in the thick-target case. The photon flux for the thin-target model then becomes

$$\frac{dF}{dE_x} = \frac{1}{4\pi R^2}\frac{2\alpha\sigma_0}{\pi}mc^2\sqrt{\frac{2}{m}}\,n_i V\tau_e \int_{E_x}^{\infty} q(E)\frac{\ln\left(\sqrt{\frac{E}{E_x}}+\sqrt{\frac{E}{E_x}-1}\right)}{E^{1/2}E_x}\,dE.$$ (10)

For a power-law injection spectrum, valid above a certain cut-off energy E_c, for which the total kinetic energy contents is P_K keV s^{-1} and the spectral index is δ, we find for the X-ray fluxes for the thick- and the thin-target case respectively:

$$\left(\frac{dF}{dE_x}\right)_{thick} = 3.46 \times 10^{-34}\frac{B(\delta-2,\tfrac{1}{2})}{2(\delta-1)} \times$$
$$\times E_c^{-1}(E_x/E_c)^{-\delta+1}P_K \text{ cm}^{-2}\text{ s}^{-1}\text{ keV}^{-1},$$ (11)

and

$$\left(\frac{dF}{dE_x}\right)_{thin} = 1.05 \times 10^{-42}\frac{(\delta-2)B(\delta-\tfrac{1}{2},\tfrac{1}{2})}{2(\delta-1)} \times$$
$$\times n_i\tau_e E_c^{-5/2}(E_x/E_c)^{-\delta-1/2}P_K \text{ cm}^{-2}\text{ s}^{-1}\text{ keV}^{-1}.$$ (12)

In these equations $B(x, y)$ is the Beta function (see also Brown, 1971) and all energies are expressed in keV. In order to produce a hard X-ray burst of the same size the ratio R of the kinetic energy of the energetic electrons required in the two limiting cases is

$$R = \frac{P_{K,thick}}{P_{K,thin}} = 3.03 \times 10^{-9}\frac{2(\delta-1)(\delta-2)B(\delta-\tfrac{1}{2},\tfrac{1}{2})}{(2\delta-1)B(\delta-2,\tfrac{1}{2})} \times$$
$$\times n_i\tau_e E_c^{-3/2}(E_x/E_c)^{-3/2}.$$ (13)

Without further specification of n_i and τ_e this ratio cannot be determined. From future observations of the structures of the hard and soft X-ray emitting regions we may find some indications as to these physical parameters; for the time being we can only give estimates. As τ_e should certainly be smaller than τ_c (Equation (5)), we find that in all practical cases $R < 1$, which means that the thick-target case is more efficient with respect to X-ray production than the thin-target case, which of course is more or less obvious.

With respect to the energy and mass balance for the flare flash phase phenomena, the thick-target case should be studied with preference, not only because of the fact that its analysis is rather straightforward, but especially because it requires the least amount of energy and mass in order to produce an X-ray burst of the observed size and thus it gives lower limits for these two important quantities. Converting (11) to an energy flux instead of a photon flux we find the following relation

$$P_{K,\,thick} = 5.78 \times 10^{33} \, \frac{\delta - 1}{B\left(\delta - 2, \frac{1}{2}\right)} \, (E_x/E_c)^{\delta - 2} \, \frac{dS_x}{dE_x} \text{ keV s}^{-1}, \qquad (14)$$

where dS_x/dE_x is expressed in keV cm^{-2} s^{-1} keV^{-1}. The energies involved in the energetic electrons range from 10^{28} erg, in small spikes, to 10^{30}–10^{31} erg and more, in long-enduring bursts as e.g. the one of 1972, August 4.

III. Microwave Bursts

(a) TYPES OF MICROWAVE BURSTS

The radio bursts as observed at microwave frequencies, i.e. above 1000 MHz, can be classified into three main categories: gradual bursts, impulsive bursts and type IV bursts (Boischot, 1972). In some cases the microwave burst can be decomposed into a gradual and an impulsive burst; the gradual burst may start before the impulsive burst (precursor) and it may attain its maximum after the impulsive burst (post-burst increase). The gradual bursts are generally fairly weak and they show a clear resemblance to the soft X-ray bursts. Following the original suggestion of Kawabata (1960) they are interpreted as due to thermal bremsstrahlung; recent analyses of Hudson and Ohki (1972) and of Shimabukuro (1972) have confirmed the common origin of the gradual microwave and the soft X-ray burst in a region with emission measure between 10^{49} and 10^{50} cm^{-3} and temperatures in the range of 2–6 × 10^6 K. According to Ramaty and Petrosian (1972) a thermal interpretation is only possible for the low-intensity gradual microwave bursts; the impulsive microwave as well as the microwave type IV bursts have a non-thermal nature. Of these non-thermal bursts the impulsive ones are closely related to the hard X-ray bursts.

The intensities of impulsive microwave bursts vary within wide limits, ranging from about 40 sfu (1 sfu = 10^{-22} watt m^{-2} Hz^{-1}) to more than 10^4 sfu; the largest burst ever recorded at microwavelengths reached a flux level of 47000 sfu at 71 GHz (Croom, 1970). All strong microwave bursts ($\gtrsim 100$ sfu) are associated with flares, whereas for weak bursts ($\lesssim 40$ sfu) the correlation is much weaker, about 65% (Kundu, 1965). This correlation in general is better for bursts at millimeter wavelengths (Shimabukuro, 1968; Croom and Powell, 1971). The association of microwave bursts and flares is different for different active regions; in active regions with magnetically complex configurations almost all flares are accompanied by microwave bursts (Fokker and Roosen, 1961; Matsuura and Nave, 1971).

(b) SPECTRAL CHARACTERISTICS

The spectrum of a microwave burst is a broad-band continuum with a maximum intensity around 3–10 GHz (Takakura, 1967). The frequency at which the maximum intensity occurs shifts to progressively higher values with increasing intensity, also the average intensity of the bursts with the same f_{max} is proportional to f_{max}^2 (Fürst, 1971). Hagen and Neidig (1971) have found a distinct difference in spectral characteristics for bursts that are associated with flares that occur very near or over sunspots and those that occur away from sunspots; the former have their maximum at frequencies above 10 GHz, whereas for the latter type of bursts the maximum emission occurs near 3 GHz. Also Guidice and Castelli (1973) noted a tendency for the higher values of f_{max} to occur in active regions in which the spots have a large magnetic field strength. These same authors found a linear relation between the low-frequency cut-off, f_{cut} and f_{max}, which for medium-size events (50–500 sfu) can be represented by $f_{max} = 3.4 f_{cut}$.

The relation between certain spectral characteristics of microwave bursts to solar proton acceleration has recently been discussed by Croom (1973, and references therein), who found that the association rate of microwave bursts and proton events increases with f_{max} and that the proton flux is related to the duration of the burst. Guidice and Castelli (1973) mention the necessity for the radio wave burst to have both strong microwave and meterwave components for it to be associated with a proton flare. These results are essentially an extension of earlier findings by e.g. Kundu and Haddock (1960) and others (see discussion by Kundu, 1965). To the present reviewer it seems that it is the microwave type IV burst rather than the impulsive microwave burst that is meant in these studies.

(c) POSITION AND STRUCTURE OF THE SOURCE REGION

The position of the source of the microwave burst corresponds roughly with that of the optical flare (Kundu, 1965), when determined with an interferometer having a resolution of a few arcminutes. Usually the source is more or less stationary, but in some cases a displacement can be observed. An interesting event in this connection is the flare of 1969, March 30 (Enomé and Tanaka, 1971), which was also exceptional in other respects (see e.g. Frost and Dennis, 1971; Smerd, 1970). The flare occurred 10° behind the west limb of the Sun, which means that the burst source was at 10 000 km or higher. After the burst maximum the source expanded to reach an altitude of 200 000 km (0.3 $R_\odot$). Enomé and Tanaka interpret this expansion in terms of diffusion of the energetic electrons in a turbulent magnetic field. It should be noted, however, that in between the two components of the burst a type II radio burst was observed. This prompted Frost and Dennis (1971) to suggest that the second component was due to electrons accelerated by repeated reflections at the shock front that produced the type II radio burst.

Recent observations with a 24″ fan beam show that the angular size of the source of the impulsive burst is about 0.5 or less (Enomé *et al.*, 1969), at 3.75 GHz. Also several cases of multicomponent bursts have been observed, which show independent intensity variations; to within the accuracy of the observations the sources of these components coincide with individual sunspots (Enomé and Tanaka, 1973). All these observations indicate that the microwave burst is produced near the feet of magnetic-field loops; in some cases emission takes place at both ends of such a loop with opposite senses of polarization.

(d) MODELS FOR HARD X-RAY AND MICROWAVE EMISSION

The theory of the radio emission and absorption of non-, or near relativistic electrons in a magnetic field has been elaborated by many authors, notably Ginzburg and Syrovatskii (1965, 1969), Takakura (1967), Ramaty (1969, 1973). In particular the emission- and absorption characteristics in a magneto-active plasma are still being debated (e.g. Ko, 1973). However, the theory may still not be in its final shape, the models to which the theory thus far has been applied are at best zero order approximations of the situation likely to exist in the solar atmosphere (e.g. Takakura and Kai, 1966; Holt and Ramaty, 1969; Takakura, 1972). The combination of these two circumstances makes the whole problem rather intractable, in particular when applied to total flux measurements in a frequency range of which a significant part corresponds to the plasma frequencies to be expected in the source region.

Briefly the problem can be stated as follows. The energy loss due to magnetic bremsstrahlung of a non-relativistic electron in a magnetic field with field strength B is

$$\left(\frac{\mathrm{d}E}{\mathrm{d}t}\right)_{\mathrm{mb}} = -3.89 \times 10^{-9} B^2 E, \tag{15}$$

where B is in Gauss. This energy is, depending on E/mc^2, emitted mainly in the frequencies near the local cyclotron frequency $f_{\mathrm{H}} \approx 2.8 \times 10^6\, B$ Hz, and a few low harmonics of this frequency. The total energy emitted by an ensemble of energetic electrons, $E > E_{\mathrm{c}}$, with total kinetic energy P_{K}, then is

$$P_{\mathrm{mb}} = 3.89 \times 10^{-9}\, B^2\, P_{\mathrm{K}}. \tag{16}$$

The radio flux at the Earth's distance, assuming isotropic emission, is

$$S_\nu = 1.38 \times 10^{-39} B^2 P_{\mathrm{K}}\ \mathrm{sf}^{-1}\ \mathrm{watt\ m^{-2}\ Hz^{-1}}. \tag{17}$$

For the maximum of the burst of 1966, July 7 we find from Takakura (1972): $S_{\nu,\mathrm{max}} \approx 10^{-18}\ \mathrm{watt\ m^{-2}\ Hz^{-1}}$, $f_{\mathrm{max}} \approx 10^{10}$ Hz, $\Delta f \approx 2 \times 10^{10}$ Hz. Hence the kinetic energy P_{K} of the electrons required to produce this radio flux is $P_{\mathrm{K,\,radio}} \approx 1.44 \times 10^{31} B^{-2}$ erg s^{-1}. The hard X-ray photon spectrum can be represented by $\mathrm{d}F/\mathrm{d}E_x = 6\,(E_x/100)^{-\gamma}$ photons cm^{-2} s^{-1} keV^{-1}, where γ is about 3. Using (11), we find for the kinetic energy $P_{K,x}$ of the electrons required to produce this X-ray flux 2×10^{28} erg s^{-1} for $E_c = 100$ keV. If we take $f_{max} \simeq 3f_H$, the magnetic field strength in the source region should be: $B \approx 1200$ G and therefore $P_{K, radio} \approx 10^{25}$ erg s^{-1}. Hence the kinetic energy of the electrons that produce the hard X-ray burst is 10^3–10^4 times that of the electrons that emit the microwave burst. This discrepancy, already noted by Peterson and Winckler (1959), has been extensively discussed over the past fifteen years; essential in all these discussions is the reduction of the radio emission of an energetic particle population due to the plasma effects; self absorption by the energetic electrons themselves, reduction of emissivity, absorption by the layers overlying the source region, non-isotropic pitch angle distribution and so on. These problems will be discussed during this symposium by Takakura (1975) and therefore I will not go into details of the computation of the radio spectrum. It suffices to state here that Takakura (1972) succeeded to make the two figures approach each other to within the uncertainties of the measurements, and of the assumptions of the model. It seems to me that this is about the maximum we can hope to obtain from total flux measurements. Real progress from here onwards seems to be possible only on the basis of spatially resolved observations for both the microwave and the hard X-ray region.

References

Acton, L. W.: 1968, *Astrophys. J.* **152**, 305.

Beigman, I. L., Grineva, Yu. I., Mandel'shtam, S. L., Vainshtein, L. A., and Zhitnik, I. A.: 1969, *Solar Phys.* **9**, 160.

Beigman, I. L., Vainshtein, L. A., Vasilyev, B. H., Zhitnik, B. N., Ivanov, V. D., Korneyev, V. V., Krutov, V. V., Mandel'shtam, S. L., Tindo, I. P., and Shurygin, A. I.: 1971, *Kosmich. Isled.* **9**, 123.

Blumenthal, G. R. and Gould, R. J.: 1970, *Rev. Mod. Phys.* **42**, 237.

Boischot, A.: 1972, in E. R. Dyer (General Ed.), *Solar-Terrestrial Physics*, Part I, D. Reidel Publ. Co., Dordrecht, Holland, p. 87.

Brini, D., Evangelisti, F., Fuligni Di Grande, M. T., Pizzichini, G., Spizzichino, A., and Vespigani, G. R.: 1973, *Astron. Astrophys.* **25**, 17.

Brini, D., Fuligni, F., Evangelisti, F., and Vespigani, G.: 1971, *Nuovo Cimento*, Series 11, 28.

Brown, J. C.: 1971, *Solar Phys.* **18**, 489.

Brown, J. C.: 1972, *Solar Phys.* **25**, 158.

Brown, J. C. and Van Beek, H. F.: 1975, This volume, p. 239.

Croom, D. L.: 1970, *Solar Phys.* **15**, 414.

Croom, D. L.: 1973, in R. Ramaty and R. G. Stone (eds.), *High-Energy Phenomena on the Sun*, NASA SP-342 p. 114.

Croom, D. L. and Powell, R. J.: 1971, *Solar Phys.* **20**, 136.

Datlowe, D. W., McKenzie, D. L. and Peterson, L. E.: 1972, Univ. California, San Diego SP-72-06.

De Feiter, L. D.: 1974, *Space Sci. Rev.* **16**, 3.

De Jager, C.: 1967, *Solar Phys.* **2**, 327.

Donnelly, R. F.: 1973, in R. Ramaty and R. G. Stone (eds.), *High-Energy Phenomena on the Sun*, NASA SP-342, p. 242.

Elwert, G. and Haug, E.: 1970, *Solar Phys.* **15**, 234.

Enomé, S. and Tanaka, H.: 1971, in R. Howard (ed.), 'Solar Magnetic Fields', *IAU Symp.* **43**, 413.

Enomé, S. and Tanaka, H.: 1973, in R. Ramaty and R. G. Stone (eds.), *High-Energy Phenomena on the Sun*, NASA SP-342, p. 78.

Enomé, S., Kakinuma, T., and Tanaka, H.: 1969, *Solar Phys.* **6**, 428.

Fokker, A. D. and Roosen, J.: 1961, *Bull. Astron. Inst. Neth.* **16**, 86.

Frost, K. J.: 1964, in W. N. Hess (ed.), *The Physics of Solar Flares*, NASA SP-50, p. 139.

Frost, K. J.: 1969, *Astrophys. J. Letters* **158**, L159.

Frost, K. J. and Dennis, B. R.: 1971, *Astrophys. J.* **165**, 655.

Frost, K. J., Dennis, B. R., and Lencho, R. J.: 1970, GSFC X-680-70-440.

Fürst, E.: 1971, *Solar Phys.* **18**, 84.

Ginzburg, V. L. and Syrovatskii, S. I.: 1965, *Ann. Rev. Astron. Astrophys.* **3**, 297.

Ginzburg, V. L. and Syrovatskii, S. I.: 1969, *Ann. Rev. Astron. Astrophys.* **7**, 375.

Guidice, D. A. and Castelli, J. P.: 1973, in R. Ramaty and R. G. Stone (eds.), *High-Energy Phenomena on the Sun*, NASA SP-342, p. 87.

Hagen, J. P. and Neidig, D. F.: 1971, *Solar Phys.* **18**, 305.

Hicks, D. B., Reid, L., and Peterson, L. E.: 1965, *IEEE Trans. Nucl. Sci.* **12**, 54.

Holt, S. S. and Ramaty, R.: 1969, *Solar Phys.* **8**, 119.

Hudson, H. S. and Ohki, K.: 1972, *Solar Phys.* **23**, 155.

Kane, S. R.: 1969, *Astrophys. J. Letters* **157**, L139.

Kane, S. R.: 1972, *Solar Phys.* **27**, 174.

Kane, S. R. and Anderson, K. A.: 1970, *Astrophys. J.* **162**, 1003.

Kane, S. R. and Hudson, H. S.: 1970, *Solar Phys.* **14**, 414.

Kane, S. R., Pfitzer, K. A., and Winckler, J. R.: 1966, Univ. of Minnesota Techn. Rept. CR-87.

Kawabata, K.: 1960, *Rept. Ion. Space Res. Japan* **14**, 405.

Ko, H. C.: 1973, in R. Ramaty and R. G. Stone (eds.), *High-Energy Phenomena on the Sun*, NASA SP-342, p. 198.

Kreplin, R. W. and Taylor, R. G.: 1971, *Solar Phys.* **21**, 452.

Kundu, M. R.: 1965, *Solar Radio Astronomy*, Interscience Publishers, New York.

Kundu, M. R. and Haddock, F. T.: 1960, *Nature* **186**, 610.

Martres, M.-J., Michard, R., Soru-Iscovici, I., and Tsap, T. T.: 1968, *Solar Phys.* **5**, 187.

Matsuura, O. T. and Nave, M. F. F.: 1971, *Solar Phys.* **16**, 417.

Maxwell, A.: 1973, in H. E. Coffey (ed.), *Collected Data Reports on August 1972 Solar-Terrestrial Events*, Report UAG-28 Part I, p. 255.

Neupert, W. M., Thomas, R. J., and Chapman, R. D.: 1974, *Solar Phys.* **34**, 349.

Noyes, R. W.: 1973, in R. Ramaty and R. G. Stone (eds.), *High-Energy Phenomena on the Sun*, NASA SP-342, p. 231.

Parks, G. K. and Winckler, J. R.: 1971, *Solar Phys.* **16**, 186.

Peterson, L. E. and Winckler, J. R.: 1959, *J. Geophys. Res.* **64**, 697.

Peterson, L. E., Datlowe, D. W., and McKenzie, D. L.: 1973, in R. Ramaty and R. G. Stone (eds.), *High-Energy Phenomena on the Sun*, NASA SP-342, p. 132.

Petrosian, V.: 1973, *Astrophys. J.* **186**, 291.

Ramaty, R.: 1969, *Astrophys. J.* **158**, 753.

Ramaty, R.: 1973, in R. Ramaty and R. G. Stone (eds.), *High-Energy Phenomena on the Sun*, NASA SP-342, p. 188.

Ramaty, R. and Petrosian, V.: 1972, *Astrophys. J.* **178**, 241.

Rust, D. M.: 1972, *Solar Phys.* **25**, 141.

Santangelo, N., Horstman, H., and Horstman-Moretti, E.: 1973, *Solar Phys.* **29**, 143.

Shimabukuro, F. I.: 1968, *Solar Phys.* **5**, 498.

Shimabukuro, F. I.: 1972, *Solar Phys.* **23**, 169.

Smerd, S. F.: 1970, *Proc. Astron. Soc. Australia*, **1**, 305.

Takakura, T.: 1967, *Solar Phys.* **1**, 304.

Takakura, T.: 1972, *Solar Phys.* **26**, 151.

Takakura, T.: 1975, This volume, p. 299.

Takakura, T. and Kai, K.: 1966, *Publ. Astron. Soc. Japan* **18**, 57.

Takakura, T., Ohki, K., Shibuya, N., Fujii, M., Matsuoka, M., Miyamoto, S., Nishimura, J., Oda, M., Ogawary, Y., and Ota, S.: 1971, *Solar Phys.* **16**, 454.

Tanaka, H. and Enomé, S.: 1971, *Solar Phys.* **17**, 408.

Tindo, I. P., Mandel'shtam, S. L., and Shuryghin, A. I.: 1973, *Solar Phys.* **32**, 469.

Tomblin, F. F.: 1972, *Astrophys. J.* **171**, 377.

Vaiana, G. S., Davis, J. M., Giacconi, R., Krieger, A., Silk, J. K., Timothy, A. F., and Zombeck, M.: 1973, *Astrophys. J. Letters*, **185**, L47.

Vaiana, G. S. and Giacconi, R.: 1969, in D. G. Wentzel and D. A. Tidman (eds.), *Plasma Instabilities in Astrophysics*, p. 91.

Vaiana, G. S., Reidy, W. P., Zehnpfennig, T., Van Speybroeck, L., and Giacconi, R.: 1968, *Science* **161**, 564.
Van Beek, H. F.: 1973, 'Development and Performance of a Solar Hard X-Ray Spectrometer', Utrecht University (Ph.D. Thesis).
Van Beek, H. F., De Feiter, L. D., and De Jager, C.: 1974, *Space Res.* **14**, 447.
Vorpahl, J.: 1972, *Solar Phys.* **26**, 397.
Vorpahl, J. and Zirin, H.: 1970, *Solar Phys.* **11**, 285.
Zirin, H. and Tanaka, K.: 1973, *Solar Phys.* **32**, 173.
Zirin, H., Pruss, G., and Vorpahl, J.: 1971, *Solar Phys.* **19**, 463.

RELATION OF MICROWAVE EMISSION TO X-RAY EMISSION FROM SOLAR FLARES

TATSUO TAKAKURA

Dept. of Astronomy, University of Tokyo, Japan

Abstract. Recent observations of impulsive microwave and hard X-ray emissions during the early phase of the flares are briefly reviewed in order to deduce the dynamics of energetic electrons consistently from two view points of the microwaves and X-rays. An emphasis is put on the necessity of distinction between temporal and spatial variations so far confused in the interpretation of the time histories of the X-ray and radio emissions. The role of plasma turbulence on the dynamics of the energetic electrons is shown to be important in deducing the model of X-ray and radio sources.

I. Introduction

The microwave emission and X-ray emission from the flares are intimately related to each other although the wavelengths are quite far apart, because both emissions are caused by energetic electrons – either thermal or nonthermal – created in the flares. The characteristic and dynamics of such energetic electrons in and around the flare region have been deduced from the microwave and X-ray observations. Recent reviews on such a subject are given by Kane (1973a, 1974), Hudson (1973), De Feiter (1972) and Takakura (1969, 1973), and a new text book on the microwave bursts is given by Krüger (1972).

In order to investigate the characteristics of the energetic electrons, both microwave observation and X-ray observation have disadvantages and advantages. For example, the radio spectrum of gyro-synchrotron emission depends strongly on the magnetic field whose distribution in the corona is hard to estimate, while the X-ray spectrum of bremsstrahlung depends on the controversial target model, thin or thick. The ambiguities can be reduced if we refer to both X-ray and radio observations.

The aim of the present review is to show recent progress on the relation of microwave emission to X-ray emission mainly at the impulsive phase of the solar flares. A brief remark on the gradual thermal emissions from the sporadic hot coronal condensation is also given.

II. Impulsive Emission

(a) OBSERVATIONAL

At the early phase of the flares, impulsive emissions occur on microwaves and hard X-rays which are also accompanied by EUV flashes, kernals of Hα (Vorpahl, 1972) and sometimes flashes of H_9 (λ 3835, Zirin and Tanaka, 1973). The microwave burst is ascribed to the gyro-synchrotron emission and the X-ray burst is ascribed to the collisional bremsstrahlung. Peterson and Winckler (1959) observed an impulsive hard X-ray burst for the first time and noticed its association with impulsive microwave burst. A good correlation between them was shown by Kundu (1961), and it has now

Sharad R. Kane (ed.), Solar Gamma-, X-, and EUV Radiation, 299–313. All Rights Reserved.

been confirmed by many observations (cf. review by Kane, 1974). Recently, Vorpahl (1972) has shown a better correlation between the peak radio flux at 8.8 GHz and the total number of electrons above 100 keV derived from the X-ray measurements though mainly below 80 keV.

The time difference between the maxima of hard X-ray and microwave bursts has been statistically given by McKenzie (1972) to be 0.3 ± 0.4 min using OSO-3 X-ray data and the reported time of microwave bursts. Kane (1972) has shown with OGO-5

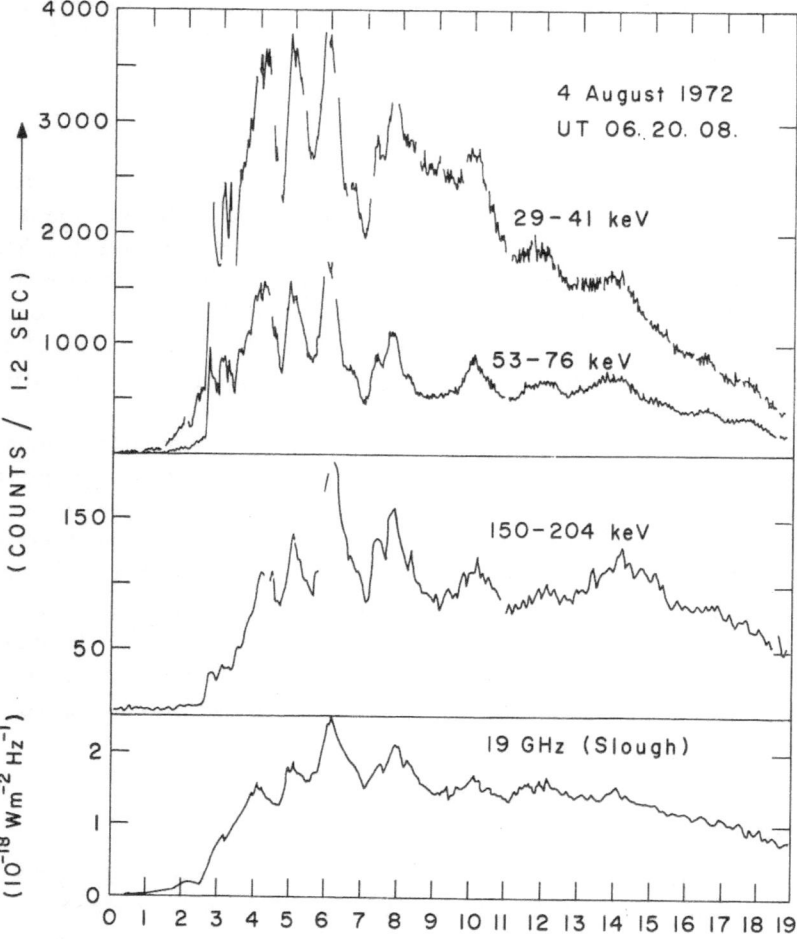

Fig. 1.   X-ray and microwave bursts of 1972, August 4 during the flare of importance 3B: 6^{h}20^{m}8^s UT at $t = 0$ on the time axis. The X-ray burst was observed with ESRO TD-1A satellite (Van Beek *et al.*, 1973). The microwave burst was observed at Slough, England (Croom and Harris, 1973). The timing of this microwave burst reported was in error by about 40 s. It has been corrected in the present figure in a comparison between the time profiles of 9.4 GHz observed at Slough and Toyakawa.

data that it is ± 6 s for small flares of importance 1 or less. Even for big flares, the time lag seems to be less than 10 s if we compare each correlated pulses as shown in Figure 1. The time difference, if any, is very important for the study of electron dynamics in the flare region. However the timing of microwave bursts so far recorded is generally not accurate to ± 6 s mainly due to low chart speed of the recorders.

The duration of microwave impulsive bursts ranges from 1 to 10 min in most cases when the time profile is simple and it is 2 to 30 min when the profile is complex. Great and prolonged bursts associated with intense flares have been designated as microwave type IV bursts, but we cannot find any distinctly different characteristics between the microwave impulsive bursts and the microwave type IV bursts from the microwave observations alone (Takakura, 1967). Their relation to hard X-ray bursts was thought to give a basis to divide the microwave bursts into subgroups, because the early measurements of hard X-rays showed that the hard X-ray bursts were only associated with the initial phase of the microwave bursts (Takakura, 1969). Therefore microwave bursts with longer duration are called 'the second phase'. Frost (1969) and Frost and Dennis (1971), however, showed two X-ray bursts having the second phase of longer duration. The other examples were obtained with the X-ray spectrometer on board the ESRO TD-1A (Hoyng et al., 1975, De Feiter, 1975). In August 1972, a series of large and small flares occurred in an active center and comprehensive data were collected at WDC, Boulder (Coffey, 1973 a, b). Time profiles of the hard X-ray bursts given in the report by van Beek et al. (1973) are reproduced in Figure 1 together with a microwave burst at 19 GHz given by Croom and Harris (1973). This radio burst is a typical great burst which may be called microwave type IV burst. It was associated with an importance 3B flare on August 4. A correlation between time profiles of X-rays and radio waves is remarkable even in such a great complex event of long duration. It is thus hard to distinguish between the impulsive phase and the 2nd phase, even if we refer to the time profile of hard X-rays. The burst seems to be composed of about 10 pulses. The individual peaks of X-rays and microwaves coincide with each other within about 10 s. The decay time of each radio pulse is slightly longer than that of X-rays. Generally, the correlation of time profiles of microwave burst with X-ray time profile seems best at about the frequency of maximum flux or at slightly higher frequencies, although the time profiles at the other frequencies are not shown in the figure. At the lower frequencies, the decay time becomes longer losing the similarity probably because of the optical thickness much greater than unity for the self absorption. The decay time of the microwave bursts at the frequency of maximum flux may correspond to the lifetime of electrons with effective energies of 100–300 keV, though the effective energy depends on the magnetic field, the size of the source and the number density of energetic electrons in the radio source. The effective energy increases with increasing frequency.

It is remarkable in Figure 1 that no appreciable difference in the duration of each pulse can be seen in the wide range of photon energies from 30 to 200 keV. Note that the duration of each pulse became longer accompanying the hardening of the spectrum at the later phase of the event.

Another great event was observed on August 7 starting at about 15^{h}13^m UT (Coffey, 1973a, b). This burst is also composed of several pulses and the correlation of time profiles of X-rays and microwaves (15.4 and 35 GHz, AFCRL) is also good without any time difference greater than 10 s. For a comparison, a typical impulsive event associated with a small flare with an importance − B on August 7 is shown in Figure 2.

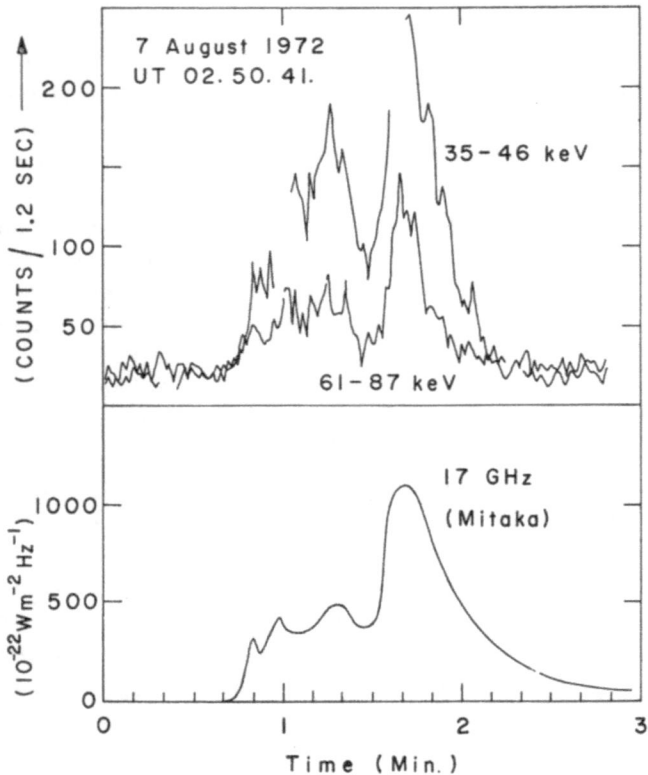

Fig. 2. X-ray and microwave bursts of 1972, August 7 during the subflare of importance −B. The X-ray burst was observed with ESRO TD-1A satellite (Van Beek *et al.*, 1973).

In a comparison between this impulsive burst and a great burst shown in Figure 1, there seems no distinct difference between them except for the duration of each pulse and the whole duration of the burst.

Intense short lived pulses of 2 s or less were superposed on the hard X-ray bursts observed by Anderson and Mahoney (1974). The associated microwave bursts (3.75 and 9.4 GHz, by courtesy of Toyokawa) are similar in time profile but such short and intense pulses are not superposed on the main burst (except for a small pulse) even though the time constant of the radiometer is 1 s.

Quasi-periodicity of the pulses has been reported (Parks and Winckler, 1969, 1971; Frost, 1969; Janssens and White, 1970; Janssens *et al.*, 1973).

In the analyses of the time history of microwave and X-ray bursts during a single flare, it is very important whether all pulses appearing in a single flare emitted from the same position or the individual pulses originate from different sources. The interpretations so far were implicitly based on the assumption that the X-ray source is only one during a single flare. Accordingly, the pulses on the time profile have been used to support the continuous injection model of electrons (e.g. Hudson, 1973), or they have been ascribed to the modulation of trapped electrons due to MH waves (Brown, 1973b) under the instantaneous injection model. The softening of hard X-ray spectrum during a single flare has also been used to support the continuous injection model by Kane and Anderson (1970), while Brown (1972) has suggested that it may also be ascribed to an energy-dependent pitch angle distribution of electrons and nonuniform target density inside the X-ray source. However, if the individual pulses originate from different *independent* sources, quite different interpretation is required for the time history of microwave and X-ray bursts. That is, a time variation of the various characteristic parameters during a burst, such as spectral index, measured without spatial resolutions gives a mixture of spatial and temporal variations, from one source to another, rather than the time variation of energetic electrons in a single source. Note the variation of photon spectral index may also be ascribed to the temporal or spatial changes of a ratio between coronal (thin target) and chromospheric (thick target) components even though the spectral index of the electrons is constant. For example, when the thick target component is higher the photon spectral index becomes apparently smaller.

Time variations of source structure of microwave bursts have been measured with radio interferometers mainly at Toyokawa with one-directional beam-width of 1'. The radio observations have given that the radio source of intense microwave bursts is composed of several smaller sources showing time variations independent of each other (Enome, 1972; Enome and Tanaka, 1973). One example is shown in Figure 3. Four sources were resolved in this case. Only one hard X-ray burst source has so far been observed with some spatial resolution (Takakura *et al.*, 1971). The spatial and time resolutions were not enough to resolve the source into smaller sources in that event. However, if we refer to the comparison made by Zirin and Tanaka (1973) between the optical flashes (λ 3835) and a hard X-ray burst (Figure 4) it is probable that the individual pulses in a single X-ray burst also originate from different sources. In this respect, hard X-ray observations with high spatial and time resolutions are highly desirable. The spatial resolution of the present microwave observations is also not high enough, although some high resolution measurements of one-directional source size were made (Hobbs *et al.*, 1973; Kundu *et al.*, 1974).

From the above consideration, it is highly probable that several independent sources are triggered successively during a single flare. The lifetime of individual source may be equal to the duration of each pulse appearing in the flux of X-ray and microwave bursts, e.g. about 1 min (full width at half maximum) in the case of Figure 1 and 15–20 s in case of Figure 2 for both X-ray and microwaves. In case of Figure 4, however, it is about 5 s for X-rays while 15–30 s for the microwaves. The difference of the

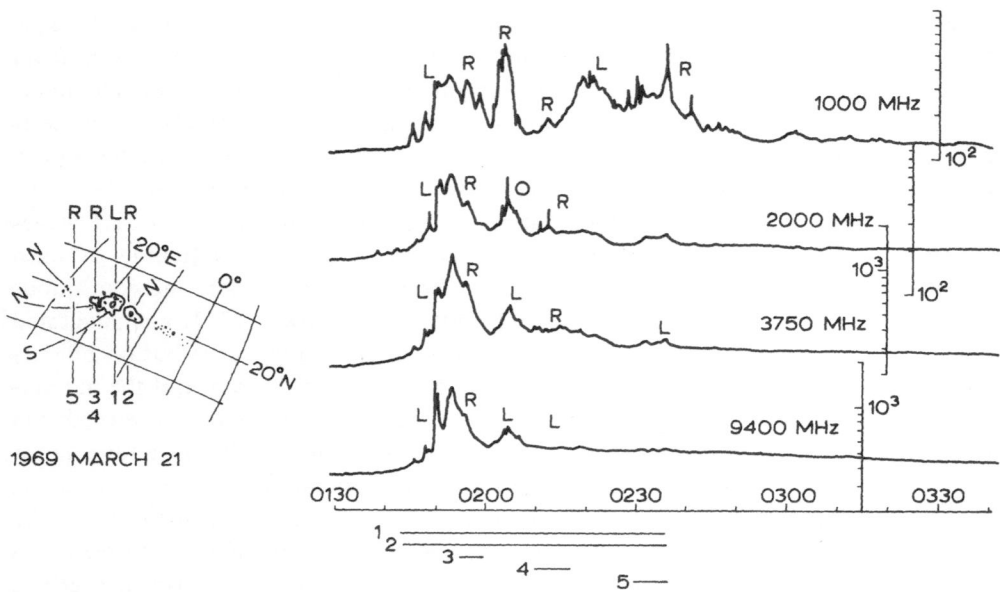

Fig 3.   Time history of the microwave burst of 1969, March 21 and one-directional locations of radio sources at 3750 MHz observed with the interferometer (Enomé, 1972; Enomé and Tanaka, 1973). The radio sources marked by 1 to 5 were detectable during the periods indicated by the same number on the time history. Individual sources showed independent time variation in intensity.

duration in this later case may indicate that their sources were not quite common and/or the lifetime of electrons with higher energies was longer.

(b) LIFETIME OF EACH PULSE

The problem to be discussed here is how to interpret such a short lifetime of each source as 1 min or less inferred from the observations.

As far as we can deduce from the August 2 event (Figure 4) the continuous injection model seems more reasonable as discussed by Zirin and Tanaka (1973). If we adopt such a thick target model, the duration of each pulse may merely correspond to the duration of a continuous acceleration of electrons in a magnetic flux tube. However, some hard X-ray bursts were associated with flares beyond the limb (cf. Takakura, 1973, Datlowe, 1975). It is a strong evidence that these X-ray bursts originated from the corona. There are also the other evidences favorable to as well as against the thin target model (Kane, 1974). Accordingly, at least in some events or during a certain period in a single event the coronal component of hard X-rays must be predominant. Then, if we adopt the thin target model, the decay of the burst is due to the energy loss of energetic electrons trapped in the magnetic flux tube and/or due to the escape of the electrons from the tube. Note that if the electrons escape mainly downwards into the chromosphere, the X-rays from the thick target cannot be ignored.

Two kinds of energy loss have been considered so far. One is collision with thermal electrons and another is gyro-synchrotron radiation loss. Note that the *e*-folding decay time of *number of electrons at a given energy* is shorter by a factor $(\gamma - 1)^{-1}$ than the *e*-folding decay time of *energy* of electrons if the power law electron spectrum with the index γ is nearly maintained during the decay, and that the former corresponds

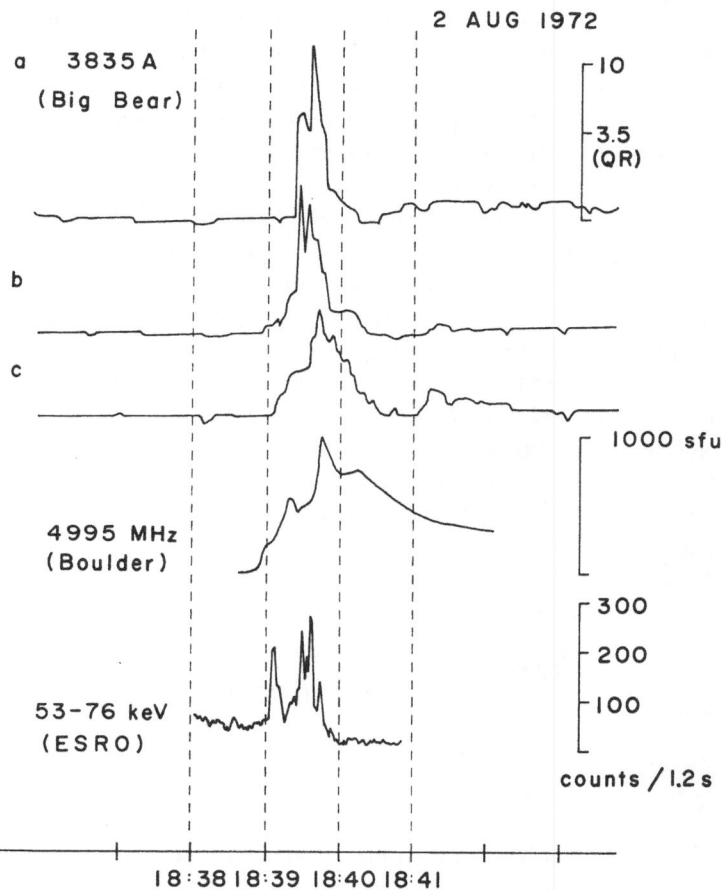

Fig. 4.   Light curves of H_9 (λ 3835) flashes measured at three positions, a, b and c at the flare of importance 1B of 1972, August 2 (Zirin and Tanaka, 1973). The lower two curves show microwave burst (Boulder) and X-rays (Van Beek *et al.*, 1973). Note that the lifetime of individual pulses of X-rays is about 5 s, while it is 15–30 s for microwaves.

to the decay time of fluxes of X-rays and microwave bursts. The *e*-folding decay time of microwave bursts at the frequency of maximum flux can be 30 s due to the collison loss if the effective number density of thermal electrons, n_e, is $(2-8) \times 10^9$ cm^{-3} and $\gamma = 4$ under the assumption that the effective energies of electrons emitting at the frequency of maximum flux are generally from 100 to 300 keV in small flares, though they could be higher in larger flares. With increasing frequencies, however, the required number density n_e increases roughly in proportion to the frequencies as the effective electron energy increases. Gyro-synchrotron loss requires a magnetic field of about 3000 G to give such a short decay time, even if $\gamma = 4$. This field strength is too strong in the corona.

In this problem the role of plasma turbulence has not been considered except for a brief remark by Brown (1971). Strong ion-acoustic waves or electron plasma waves are most probably existing in the acceleration and trapping regions of electrons at the flares when a current density exceeds a critical value (Friedman and Hamberger, 1969; Takakura, 1971; Coppi and Friedland, 1971). In the presence of the turbulent ion-acoustic waves, the mean free path of energetic electrons due to the scattering by the waves is estimated to be several order smaller in the corona than that of classical Coulomb collisions. Note that the energy loss due to the scattering is comparatively small but can be larger than that of classical Coulomb collisions. It would be worthwhile to consider this effect in more detail, although we don't know whether it is essential in the present problem or not, because solar electrons travel through the interplanetary plasma exciting plasma waves as observed as type III radio bursts but as if there were no efficient interaction with the plasma in the scatter free events (Lin *et al.*, 1973; Zaitsev *et al.*, 1972). The evidence of the plasma turbulence occurring in the acceleration region and/or in microwave burst source during the early phase of the flares would be an associated dm wave burst in the frequency range of about 1000–300 MHz ($10^{10} \lesssim n_e \lesssim 10^9$), although the association of such the burst is rather infrequent.

Smith and Priest (1972) have given a rate of change in electron velocity v due to an interaction with isotropic plasma waves.

$$\left\langle \frac{dv_r}{dt} \right\rangle = -\frac{2(2\pi)^3 e^2 c_s^2 \lambda_e^2}{m_e^2 v^4} \int k^3 W_k dk \simeq -\left(\frac{c_s}{v}\right)^2 \left\langle \frac{dv_\theta}{dt} \right\rangle, \tag{1}$$

$$\left\langle \frac{dv_\theta}{dt} \right\rangle = \frac{2(2\pi)^3 e^2 \lambda_e^2}{m_e^2 v^2} \left(1 + \frac{c_s^2}{v^2}\right) \int k^3 W_k \, dk, \tag{2}$$

where, $\langle dv_r/dt \rangle$ and $\langle dv_\theta/dt \rangle$ are respectively the average rate of change of velocity in the direction of motion and in the direction perpendicular to v,

$$c_s = (\kappa T_e/m_i)^{1/2} \qquad \text{is sound speed,}$$

$$\lambda_e = (\kappa T_e/m_e)^{1/2} \, 2\pi f_{p,e} \text{ is Debye length,}$$

the suffixes e and i indicate electron and ion respectively, T is temperature, f_p is plasma frequency and W_k is an energy density of waves for a wave number k. For the isotropic turbulent ion-sound waves, they have given after Kadomtsev (1965)

$$W_k \simeq 2 \times 10^{-4} \frac{v_{\text{eff}}}{v_e} \frac{T_e}{T_i} \frac{(\kappa T_e)^2}{e^2 k} \ln\left(\frac{1}{k\lambda_e}\right), \tag{3}$$

where v_{eff} is the critical velocity of electrons for the onset of turbulent ion- sound waves and it depends on T_i/T_e and $v_e = (\kappa T_e/m_e)^{1/2}$: we may set $v_{eff}/v_e \simeq T_i/T_e$. Equation (2) reduces to

$$\left\langle \frac{dv_\theta}{dt} \right\rangle \simeq 0.71 \times 10^{-3} \pi^4 f_{p,\,e} v_e^3 v^{-2} . \tag{4}$$

A deflection time τ_d^s due to the turbulent ion-sound waves may be given by

$$\left\langle \frac{dv_\theta}{dt} \right\rangle = v/\tau_d^s . \tag{5}$$

Thus we have

$$\tau_d^s \simeq \frac{14}{f_{p,\,e}} \left(\frac{v}{v_e} \right)^3 \simeq 1.8 \times 10^8 n_e^{-1/2} (E\,(\mathrm{keV})/T_e)^{3/2} \ \mathrm{s} , \tag{6}$$

n_e in cm^{-3}, T_e in degree and E (keV) is the kinetic energy of an electron in keV in non-relativistic range.

An energy exchange time τ_t^s due to the turbulent waves may be given by

$$\frac{dE}{dt} = m_e v \left\langle \frac{dv_r}{dt} \right\rangle = -\frac{E}{\tau_t^s} . \tag{7}$$

Accordingly, we have

$$\tau_t^s = -E \left\{ m_e v \left\langle \frac{dv_r}{dt} \right\rangle \right\}^{-1} \simeq \frac{m_i}{m_e} \left(\frac{E}{\kappa T_e} \right) \tau_d^s \simeq$$
$$\simeq 3.7 \times 10^{18} n_e^{-1/2} (E\,(\mathrm{keV})/T_e)^{5/2} \ \mathrm{s} \tag{8}$$

in the same units as used in Equation (6).

On the other hand, the deflection time τ_d^c and energy exchange time τ_t^c due to classical Coulomb collisions are

$$\tau_d^c \simeq \tau_t^c \simeq 2.1 \times 10^8 E\,(\mathrm{keV})^{3/2}/n_e \ \ \mathrm{s}$$

in the same units as used in Equation (6) for $m_e c^2 \gg E > 4\kappa T_e$. Therefore,

$$\alpha_d \equiv \tau_d^s/\tau_d^c \simeq 0.85 n_e^{1/2} T_e^{-3/2} \tag{9}$$

and

$$\alpha_t \equiv \tau_t^s/\tau_t^c = \frac{m_i}{m_e} \frac{E}{\kappa T_e} \alpha_d \simeq 1.8 \times 10^{10} n_e^{1/2} T_e^{-5/2} E\,(\mathrm{keV}) \tag{10}$$

in the same units as used in Equation (6). For example, if n_e is 10^9 to 10^{10} cm^{-3} and T_e is 10^7 to 10^6, the ratio α_d ranges from 10^{-6} to 10^{-4}, while incidentally α_t is 0.02 to 20 for 10 keV electrons. Accordingly, 10–100 keV electrons cannot stream freely into the chromosphere as has been thought in the thick target model if the plasma turbulence is occurring on the way of the electrons to the chromosphere.

A diffusion time $\tau_{D\parallel}$ for the electrons to move a distance l along the magnetic field may be given on the analogy of random walk, by

$$\tau_{D\parallel} \simeq \tau_d^s (l/v\tau_d^s)^2, \tag{11}$$

while the diffusion time $\tau_{D\perp}$ perpendicular to the magnetic field may be given by

$$\tau_{D\perp} \simeq \tau_d^s (l/\varrho_H)^2, \tag{12}$$

where $\varrho_H = v/\omega_{H,e}$ is the gyro-radius of the electron. Substituting (6), we have

$$\tau_{D\parallel} \simeq 1.6 \times 10^{-27} l^2 n_e^{1/2} T_e^{3/2} E \text{ (keV)}^{-5/2} \quad \text{s}, \tag{13}$$

$$\tau_{D\perp} \simeq 1.5 \times 10^4 \, (Hl)^2 \, n_e^{-1/2} T_e^{-3/2} E \text{ (keV)}^{1/2} \text{ s}, \tag{14}$$

n_e in cm^{-3}, l in cm, E (keV) in keV and H is magnetic field in Gauss. If we put $n_e = 10^9$ cm^{-3}, $T_e = 10^7$ and $l = 10^9$ cm, the diffusion time along the lines of force for 100 keV electrons is about 16 s and it decreases rapidly *with increasing energy* of electrons though the above equation is not valid in the relativistic energies. $\tau_{D\perp}$ is generally very long.

In conclusion, high energy electrons (>100 keV), which contribute to the radio emission as the gyro-synchrotron process, may have such a short escaping time as 20 s or less if turbulent plasma waves are excited in and around the acceleration region of electrons. On the other hand, lower energy electrons cannot escape from the acceleration region in their lifetime given by the energy transfer time (τ_t^s or τ_t^c). Furthermore, it seems difficult to have an unisotropic distribution of electron velocities required to have linear polarization and directivity of hard X-rays. The possible occurrence of such plasma turbulence in at least some events and/or during a limited period in a single flare may have caused many variations in the observed characteristics of hard X-ray and microwave impulsive bursts.

(c) MODEL OF RADIO AND X-RAY SOURCES FOR IMPULSIVE PHASE

In the first hard X-ray event, Peterson and Winckler (1959) made a rough comparison between the number of electrons producing hard X-rays in thick target and that producing microwave burst, finding a discrepancy of 10^4. This discrepancy was, however, mainly based on their misunderstanding as pointed out by Takakura (1962), that they compared the *time integrated* number of electrons which produced the whole X-rays in the thick target and the *instantaneous* number of electrons to emit the peak flux of microwaves. The discrepancy can be small if we adopt correctly the thick target model (Takakura, 1962; Hudson, 1972; Zirin and Tanaka, 1973).

Anderson and Winckler (1962) proposed a thin target model taking high target number density as 10^{12} cm^{-3}, i.e., the chromosphere, and supposed that the microwave burst is also emitted at the same height in the magnetic field of 10^3 G from common electrons producing X-rays. This number density is too high for the thin target model and also for the radio source. If we take more reasonable value as 10^{10}–10^9 cm^{-3} for the ion density under the thin target model and make more accurate estimate of electrons with power law distribution emitting microwaves in a uniform magnetic field of 500–1000 G, the discrepancy becomes again 10^3–10^4 at the same energy range. Therefore Takakura and Kai (1966) proposed a model in which the sources of X-rays and microwaves are not common.

Holt and Ramaty (1966), on the other hand, showed under the thin target model that the discrepancy can be reduced if we take into account the self absorption of gyro-synchrotron emission in rather weak magnetic field of 200–300 G together with a high energy cutoff of electrons. Further Takakura (1972, 1973) showed under the thin target model that the discrepancy can be negligible if we take into account two more factors, viz., the magnetic field in the radio source is nonuniform and the electron energy spectrum has steeper slope above 100 keV, though this latter factor is nearly equivalent to the high energy cutoff assumed by Holt and Ramaty. Following this model Anderson and Mahoney (1974) showed that X-ray and microwave bursts are consistently explained even for small flares.

Recently Kane (1973b, 1974) has proposed a combination model in which the hard X-ray source extends from the lower chromosphere to the lower corona. The acceleration of electrons is continuous and a part of the accelerated electrons is mirrored in the magnetic field and the rest is scattered into the loss cone and precipitates into the lower chromosphere as Model 3 shown in Figure 5. In his model the mirrored electrons are escaping into the corona. However, another possibility is that most of the mirrored electrons are mirrored again at the other end of the magnetic flux tube to be trapped in the tube, even though the injection of electrons is continuous.

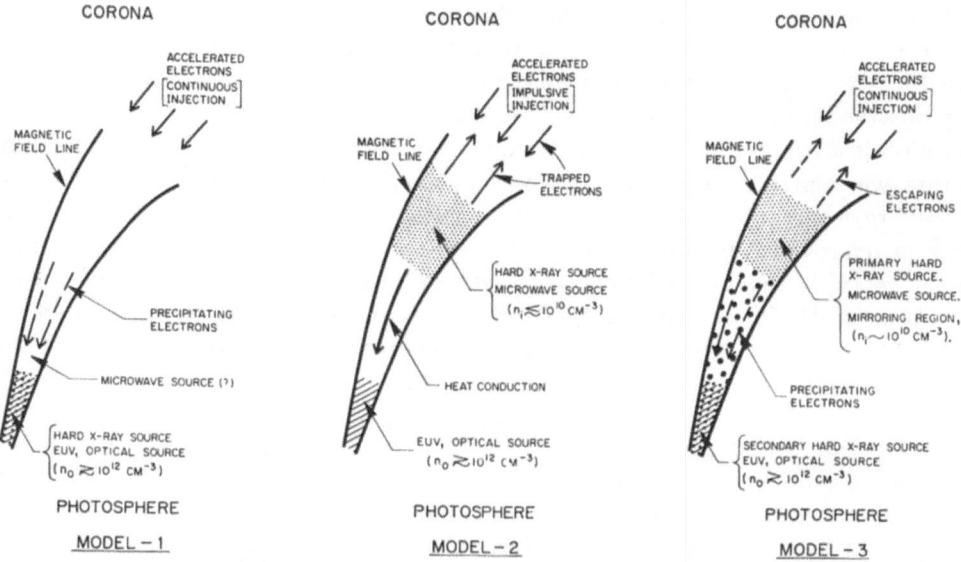

Fig. 5.   Models of impulsive phase (Kane, 1974). Model 1. Continuous injection, thick target model (Syrovatskii and Shmeleva, 1972 and Hudson, 1972). – Model 2. Impulsive injection, thin target model (Takakura, 1973). – Model 3. Continuous injection, thin and thick targets model (Kane, 1973b).

Recently Vorpahl and Takakura (1974) have suggested that the statistical result for the hard X-ray burst with shorter rise time to have harder photon spectrum could be ascribed to the existence of both chromospheric and coronal components; i.e., in cases where the chromospheric component predominates, the duration is shorter and the photon spectrum is harder and vice versa even though the electron spectrum index is almost constant from one event to another.

As already noted by Takakura (1973), it seems difficult in the present stage to say that the hard X-ray component from the chromosphere is less than that from the lower corona. The ratio between these two components probably varies with time and also spatially during a single flare and also from one flare to another. This would be one of the important problems to be solved by observations with high spatial and time resolutions. For example, very short and intense X-ray pulses as observed by Anderson and Mahoney (1974) and mentioned previously in subsection (a), would be emitted from the chromosphere since the corresponding microwave pulses were not observed.

Based on the interferometric observation at 35 GHz, Kawabata *et al.* (1973) tried to interpret the impulsive microwave burst by the gyro-radiation of 10–100 keV electrons under the continuous injection thick target model. They have found, however, that their model cannot account for the observed less steep spectral slope of the radio burst at high frequencies (Kawabata, 1974, private communication). They are trying to modify the model including the reflecting electrons of higher energies above 100 keV. This model is thus similar to that suggested by Kane as shown in Figure 5.

A thermal model for the microwave impulsive burst has been proposed by Fürst (1973) in order to account for the short duration of impulsive microwave bursts. The burst is ascribed to the gyro-emission from thermal electrons with high temperature ($8 \times 10^7 > T_e > 2 \times 10^7$). The radio source is optically thick for the gyro-emission at any observed frequencies up to 15.4 GHz due to the high temperature, and the short duration is attributed to the rapid cooling of the hot source due to heat conduction. Before we appreciate this model, however, comparison with associated soft- and hard-X-rays is required. A rough estimate shows that the expected soft X-rays at the peak of microwaves is more than one order higher than the observed flux. Another direct check is the measurement of brightness temperature of the radio source. Kundu *et al.* (1974) has given 1.2×10^9 K at 3.7 cm even for a very weak impulsive burst (18 sfu) observed with wide spacing interferometer. This result is against the thermal model.

III. Gradual Emission

There are small increases on microwaves showing gradual rise and fall (GRH burst). It may or may not be accompanied by an impulsive burst at its early phase. The correlation between the GRF burst and thermal soft X-rays is very good as shown by Hudson and Ohki (1972). Both are emitted due to a thermal process as shown by Kawabata (1960) from hot sporadic coronal condensation created at the flare. Based on a comparison between the X-rays and the microwaves, Hudson and Ohki (1972) have shown that the temperature of the condensation is effectively uniform. On the other hand Herring and Craig (1973) have proposed two temperature model, in which low temperature region has temperature of $(2–3) \times 10^6$ K and emission measure rises to 10^{51} cm^{-3} at the peak. If they have referred to the radio observation, however, this model could have been rejected as pointed out by Ohki (1974, private communication), since the low temperature region should emit the radio flux of 2 order higher than the observed value. Probably, the Skylab experiments have already given reliable informations about the structure of sporadic hot condensation.

In order to create coronal condensation excess gas should come from the chromosphere as pointed out by Hudson and Ohki (1972). The energy source to heat the chromosphere to cause expansion and supply the gas into the condensation is a controversial point. Hudson (1972) suggested that the chromospheric gas expands due to the heating by nonthermal electron streams emitting hard X-rays and Brown (1973a) has analysed this in more detail. However, the energy input into the hot condensation generally increases even after the decay of hard X-rays so that it is impossible for the nonthermal electrons to provide the heat input (Peterson *et al.*, 1973, Datlowe, 1975). Another evidence is given by Neupert *et al.* (1974) that chromospheric EUV brightening has already reached 18% of its maximum when hard X-rays are just beginning to be emitted. Accordingly, Hirayama (1974) and Ohki (1973) has proposed a model that the gas in the chromosphere cause rapid expansion due to the *heating by thermal conduction* from the very hot $(T \gtrsim 2 \times 10^7$ K) coronal region in which the heating of gas is assumed to continue during the increasing phase of soft X-rays, e.g., due to ohmic loss of current under the anomalous conductivity (Takakura, 1971). The front of heat flow propagates through the chromosphere as long as the energy flux exceeds the radiation loss thickening the transition region to emit excess EUV (Ohki, 1973). The speed of the front could be above 10 km s^{-1} at the layer with density of 10^{13} cm^{-3} if $T \gtrsim 2 \times 10^7$ K in the coronal heat source. Therefore, the flashes of EUV, Hα and H$_9$ could also be ascribed to the heat conduction when the coronal temperature exceeds $10^{7.5}$ K (Takakura, 1973). Accordingly stationary state as Shmeleva and Syrovatskii (1973) has treated could not be applied during the early phase of flares.

References

Anderson, K. A. and Mahoney, W. A.: 1974, *Solar Phys.* **35**, 419.

Anderson, K. A. and Winckler, J. R.: 1962, *J. Geophys. Res.* **67**, 4103.

Brown, J. C.: 1971, *Solar Phys.* **18**, 489.

Brown, J. C.: 1972, *Solar Phys.* **25**, 158.

Brown, J. C.: 1973a, *Solar Phys.* **31**, 143.

Brown, J. C.: 1973b, *Solar Phys.* **32**, 227.

Coffey, H. E.: 1973a, Collected Data Reports on August 1972 Solar-Terrestrial Events, World Data Center A for Solar-Terrestrial Physics, Report UAG-28, Part I.

Coffey, H. E.: 1973b, Collected Data Reports on August 1972 Solar-Terrestrial Events, World Data Center A for Solar-Terrestrial Physics, Report UAG-28, Part II.

Coppi, B. and Friedland, A. B.: 1971, *Astrophys. J.* **169**, 379.

Croom, D. L. and Harris, L. D. J.: 1973, see Coffey, 1973a, p. 210.

Datlowe, D. W.: 1975, This volume, p. 191.

De Feiter, L. D.: 1972, *Space Sci. Rev.* **13**, 827.
De Feiter, L. D.: 1975, This volume, p. 283.
Enomé, S.: 1972, Thesis.
Enomé, S. and Tanaka, H.: 1973, in R. Ramaty and R. G. Stone (eds.), *High-Energy Phenomena on the Sun*, NASA SP-342, p. 78.
Friedman, M. and Hamberger, S. A.: 1969, *Solar Phys.* **8**, 104.
Frost, K. J.: 1969, *Astrophys. J.* **158**, L159.
Frost, K. J. and Dennis, B. R.: 1971, *Astrophys. J.* **165**, 655.
Fürst, E.: 1973, *Solar Phys.* **28**, 159.
Herring, J. R. H. and Craig, I. J. D.: 1973, *Solar Phys.* **28**, 169.
Hirayama, T.: 1974, *Solar Phys.* **34**, 323.
Hobbs, R. W., Jordan, S. D., and Webster, W. J.: 1973, *Nature Phys. Sci.* **243**, 48.
Holt, S. S. and Ramaty, R.: 1969, *Solar Phys.* **8**, 119.
Hoyng, P., Brown, J. C., Stevens, G., and Van Beek, H. F.: 1975, This volume, p. 233.
Hudson, H. S.: 1972, *Solar Phys.* **24**, 414.
Hudson, H. S.: 1973, in R. Ramaty and R. G. Stone (eds.), *High Energy Phenomena on the Sun*, NASA SP-342, p. 207.
Hudson, H. S. and Ohki, K.: 1972, *Solar Phys.* **23**, 155.
Janssens, T. J. and White, III, K. P.: 1970, *Solar Phys.* **11**, 299.
Janssens, T. J., White, III, K. P., and Broussard, R. M.: 1973, *Solar Phys.* **31**, 207.
Kadomtsev, B. B.: 1965, *Plasma Turbulence*, Academic Press, New York.
Kane, S. R.: 1972, *Space Sci. Rev.* **13**, 822.
Kane, S. R.: 1973a, in R. Ramaty and R. G. Stone (eds.), *High Energy Phenomena on the Sun*, NASA SP-342, p. 55.
Kane, S. R.: 1973b, 3rd Meeting of the Solar Physics Division of AAS. Las Cruces, New Mexico.
Kane, S. R.: 1974, in G. Newkirk, Jr. (ed.), 'Coronal Disturbances' *IAU Symp.* **57**, 105.
Kane, S. R. and Anderson, K. A.: 1970, *Astrophys. J.* **162**, 1003.
Kawabata, K.: 1960, *Rept. Ionos. Space Res. Japan* **14**, 405.
Kawabata, K.: 1974, Private communication.
Kawabata, K., Sofue, Y., Ogawa, H., and Omodaka, T.: 1973, *Solar Phys.* **31**, 469.
Krüger, A.: 1972, *Physics of Solar Continuum Radio Bursts*, Akademie-Verlag, Berlin.
Kundu, M. R.: 1961, *J. Geophys. Res.* **66**, 4308.
Kundu, M. R., Velusamy, T., and Becker, R. H.: 1974, *Solar Phys.* **34**, 217.
Lin, R. P., Evans, L. G., and Fainberg, J.: 1973, *Astrophys. Letters* **14**, 191.
McKenzie, D. L.: 1972, *Astrophys. J.* **175**, 481.
Neupert, W. M., Thomas, R. J., and Chapman, R. D.: 1974, *Solar Phys.* **34**, 349.
Ohki, K.: 1973, Thesis.
Ohki, K.: 1974, Private communication.
Parks, G. K. and Winckler, J. R.: 1969, *Astrophys. J.* **155**, L117.
Parks, G. K. and Winckler, J. R.: 1971, *Solar Phys.* **16**, 186.
Peterson, L. E., Datlowe, D. W., and McKenzie, D. L.: 1973, in R. Ramaty and R. G. Stone (eds.), *High Energy Phenomena on the Sun*, NASA SP-342, p. 132.
Peterson, L. E. and Winckler, J. R.: 1959, *J. Geophys. Res.* **64**, 697.
Shmeleva, O. P. and Syrovatskii, S. I.: 1973, *Solar Phys.* **33**, 341.
Smith, D. F. and Priest, E. R.: 1972, *Astrophys. J.* **176**, 487.
Syrovatskii, S. I. and Shmeleva, O. P.: 1972, *Sov. Astron. – A.J.* **16**, 273.
Takakura, T.: 1962, *J. Phys. Soc. Japan* **17**, Suppl. A-II (International Conference on Cosmic Rays and the Earth Storm), p. 243.
Takakura, T.: 1967, *Solar Phys.* **1**, 304.
Takakura, T.: 1969, *Solar Flares and Space Research*, North-Holland, p. 165.
Takakura, T.: 1971, *Solar Phys.* **19**, 186.
Takakura, T.: 1972, *Solar Phys.* **26**, 151.
Takakura, T.: 1973, in R. Ramaty and R. G. Stone (eds.), *High Energy Phenomena on the Sun*, NASA SP-342, p. 179.
Takakura, T. and Kai, K.: 1966, *Publ. Astron. Soc. Japan* **18**, 57.
Takakura, T., Ohki, K., Shibuya, N., Fujii, M., Matsuoka, M., Miyamoto, S., Nishimura, J., Oda, M., Ogawara, Y., and Ota, S.: 1971, *Solar Phys.* **16**, 454.

Van Beek, H. F., Hoyng, P., and Stevens, G. A.: 1973, see Coffey, 1973b, p. 319.
Vorpahl, J. A.: 1972, *Solar Phys.* **26**, 397.
Vorpahl, J. A. and Takakura, T.: 1974, *Astrophys. J.* **191**, 563. Also in this volume, p. 237.
Zaitsev, V. V., Mityakov, N. A., and Rapoport, V. O.: 1972, *Solar Phys.* **24**, 444.
Zirin, H. and Tanaka, K.: 1973, *Solar Phys.* **32**, 173.

X- AND γ-RAY MEASUREMENTS DURING THE 1972, AUGUST 2 AND 7 LARGE SOLAR FLARES

R. TALON and G. VEDRENNE

Centre d'Etude Spatiale des Rayonnements – B.P. 4057–31029, Toulouse Cedex, France

and

A. S. MELIORANSKY, N. F. PISSARENKO, V. M. SHAMOLIN, and O. B. LIKIN

Institut de Physique Spatiale I.K.I., Académie des Sciences d'U.R.S.S., Moscou, U.R.S.S.

Abstract. The solar X- and gamma-ray experiment aboard Prognoz 2 recorded solar gamma-radiation bursts during the 1972, August 2, 4 and 7 events. This work analyses the general evolution of the August 2 and 7 events and the possible evidence for γ-ray lines during these two events.

I. The Experiment

The satellite Prognoz 2 was launched on 29 June 1972 for the purpose of studying the effects of solar activity. The satellite orbit is inclined at 65°, with an apogee of 200 000 km and a perigee of 1000 km.

At launch time the projection of the abside line on the ecliptic plane was close to the direction of the Sun.

The scientific equipment aboard the satellite consists of detectors for studying temporal changes in electromagnetic radiation and charged solar particles.

For the large events of 1972, August 2 and 7, data were analyzed from the RS 1 and SGL 1 X-ray spectrometers, a gamma radiation spectrometer from the Signe 1 experiment, and several charged particle detectors, all of which are described below.

The RS 1 spectrometer consists of a proportional counter shielded by passive (copper) and active (gas counter) anticoincidences. Its volume is 400 cm^3, and it is filled with a mixture of 90% xenon and 10% methane under a pressure of approximately 0.8 atm. The 0.5 cm^2 area window of the counter is made of Beryllium 60 μ thick.

The count rates of this detector were recorded in four energy ranges:

$$\Delta E_1 \simeq 4.1\text{–}9.7 \text{ keV}, \qquad \Delta E_2 \simeq 9.7\text{–}19.0 \text{ keV}$$
$$\Delta E_3 \simeq 19.0\text{–}33 \text{ keV}, \qquad \Delta E_4 \simeq 33\text{–}54 \text{ keV} \tag{1}$$

The detailed characteristics of the apparatus have been given elsewhere (Grigorov *et al.*, 1974).

The SGL 1 spectrometer is designed to measure the spectrum of X-rays in the range 38–340 keV. This energy interval is divided as follows:

$$\Delta E_5 \simeq 38\text{–}77 \text{ keV}, \qquad \Delta E_6 \simeq 77\text{–}130 \text{ keV}$$
$$\Delta E_7 \simeq 130\text{–}250 \text{ keV}, \qquad \Delta E_8 \simeq 250\text{–}340 \text{ keV} \tag{2}$$

This detector is a scintillation counter composed of a Cs I(T1) crystal, 39 mm in diameter and 8 mm thick, and an FEO 53 photomultiplier. The detector is shielded over 4π by a plastic scintillator 0.5 cm thick. The entry window consists of a sheet of 0.1 mm aluminum and 0.1 mm of reflective material. In addition to the X-ray data, this detector gives information on events when γ-rays (energy > 340 keV) and charged particles interact in the main detector. The detailed description of this SGL 1 apparatus is given elsewhere (Kudriatsev *et al.*, 1973).

In the Signe 1 apparatus designed to study solar gamma radiation and neutrons, a Stilbene scintillator, 38.1 mm in diameter and 38.1 mm high, shielded by a plastic scintillator, is used. The Compton electrons and recoil protons due respectively to photon and neutron interactions in the stilbene are detected by a Radiotechnique 416 F photomultiplier. The anticoincidence plastic scintillator coupled to a RCA 4441 A photomultiplier rejects charged particles. Pulse shape discrimination separates electrons from protons. The gamma radiation is measured in the following intervals:

$$\begin{aligned}
E_9 &\simeq 0.4\text{--}0.7 \text{ MeV}, & E_{10} &\simeq 0.7\text{--} 1 \quad \text{MeV} \\
E_{11} &\simeq 1 \text{ --}1.6 \text{ MeV}, & E_{12} &\simeq 1.6\text{--} 2.4 \text{ MeV} \\
E_{13} &\simeq 2.4\text{--}2.9 \text{ MeV}, & E_{14} &\simeq 2.9\text{--} 3.9 \text{ MeV} \\
E_{15} &\simeq 3.9\text{--}8.1 \text{ MeV}, & E_{16} &\simeq 8.1\text{--}11.8 \text{ MeV}
\end{aligned} \tag{3}$$

The detector works in two successive modes:
(1) Recording gamma photons between ΔE_9–ΔE_{14}
(2) Recording gamma photons between ΔE_{14}–ΔE_{16}
Each phase lasts 82 min; a more detailed description of this experiment has appeared in Vedrenne *et al.* (1973).

The count rate is recorded in the RS 1 and SGL 1 apparatus and in mode (1) of the Signe 1 apparatus once every 41 s and in mode (2) of the Signe 1 apparatus once every 160 s.

Data for proton fluxes were obtained above 500 MeV from a Cerenkov counter (Blioudov *et al.*, 1974) and above 15 MeV from an STS 5 discharge counter. The general characteristics of these detectors are given in Table I.

TABLE I

Detector characteristics

Detector	Energy range	Area (cm²)	Opening angle
(1) Be 60 μ window proportional counter	2.3–28 keV	0.5	20°
(2) CsI(T1) crystal scintillation counter, 8 mm thick, 39 mm diameter	38–340 keV	10	45°
(3) Stilbene scintillation counter 38.1 mm diameter, 38.1 mm thick	0.35–11.8 MeV	17.1	4π
(4) STS 5 gas discharge counter	Protons $E > 15$ MeV Electrons energy > 1 MeV	2.15	2π
(5) Cerenkov counter	Protons $E \geqslant 500$ MeV Protons $E \geqslant 100$ MeV	5 8	40° 30°

II. Experimental Results

A remarkable series of complex flares appeared in early August 1972, in particular on August 2, 4 and 7.

We observed these events in X-rays, but the time resolution (41 s) did not permit a fine time analysis. However the use of a satellite with a very eccentric orbit pointed at the Sun allowed us to follow the large scale evolution of the X- and gamma-radiation for all the events. Since the observations took place when the satellite was out of the magnetosphere, an increase in the background count rate was observed on all the detectors when the solar charged particles arrived near the Earth. One exception was the August 2 events, which will be analyzed first. In this case no contamination from these particles was observed.

(a) AUGUST 2 EVENTS

(i) *Description of The Events*

(α) *First event*. This event begins at 3 10 and shows a maximum at 3 30. This large, early event is associated with a type 1B burst from the Mac Math region 11976 located at 13°N and 35°–37°E.

The SGL 1 X-ray spectrometer recorded the beginning of the event at 3 11 20, with the first maximum (~ 1 photon cm^{-2} s^{-1} keV^{-1}) at 3 30 in the energy range 38–77 keV. For this energy range a less pronounced secondary maximum (~ 0.5 photons cm^{-2} s^{-1} keV^{-1}) appeared at 4 03 51.

The decreasing time of this event is about 55 min and the total duration is 143. (Figure 1).

The radio measurements we used were made at the Manila Observatory between 606 and 8800 MHz (Castelli *et al.*, 1973). High frequencies time profiles particularly at 8800, 4995 and 2695 MHz, coincide well with the X-ray measurements. The second radio maximum (~ 400) can be seen to be higher than the first at 4995 and 2695 MHz.

This radio burst has the usual characteristics of a proton flare burst (Castelli *et al.*, 1973):

– slow rise to maximum, long duration;
– values greater than 1000 fu at $\lambda = 3$ cm;
– intense meter wavelength emission;
– typical U-shape of the frequency spectrum at the time of the 2 maxima (Figure 2).

In spite of these characteristics, the Signe 1 experiment did not measure any gamma emission in the energy range 0.35–3.9 MeV.

An X-ray spectrum corrected for detector effects can be obtained from the experimental data using a Monte-Carlo program (Bui-Van *et al.*, 1973). This spectrum has the form $AE^{-\alpha}$ with $A = 6.3 \times 10^8$ photons cm^{-2} s^{-1} keV^{-1} and $\alpha = 4.5$.

Knowing this spectrum one can deduce the energy associated with the electron component using the method by Brown (1972). For electrons with energy $E_e > 10$ keV, an energy of $\sim 10^{30}$ ergs is obtained in the impulsive injection model and 5×10^{29} ergs s^{-1}

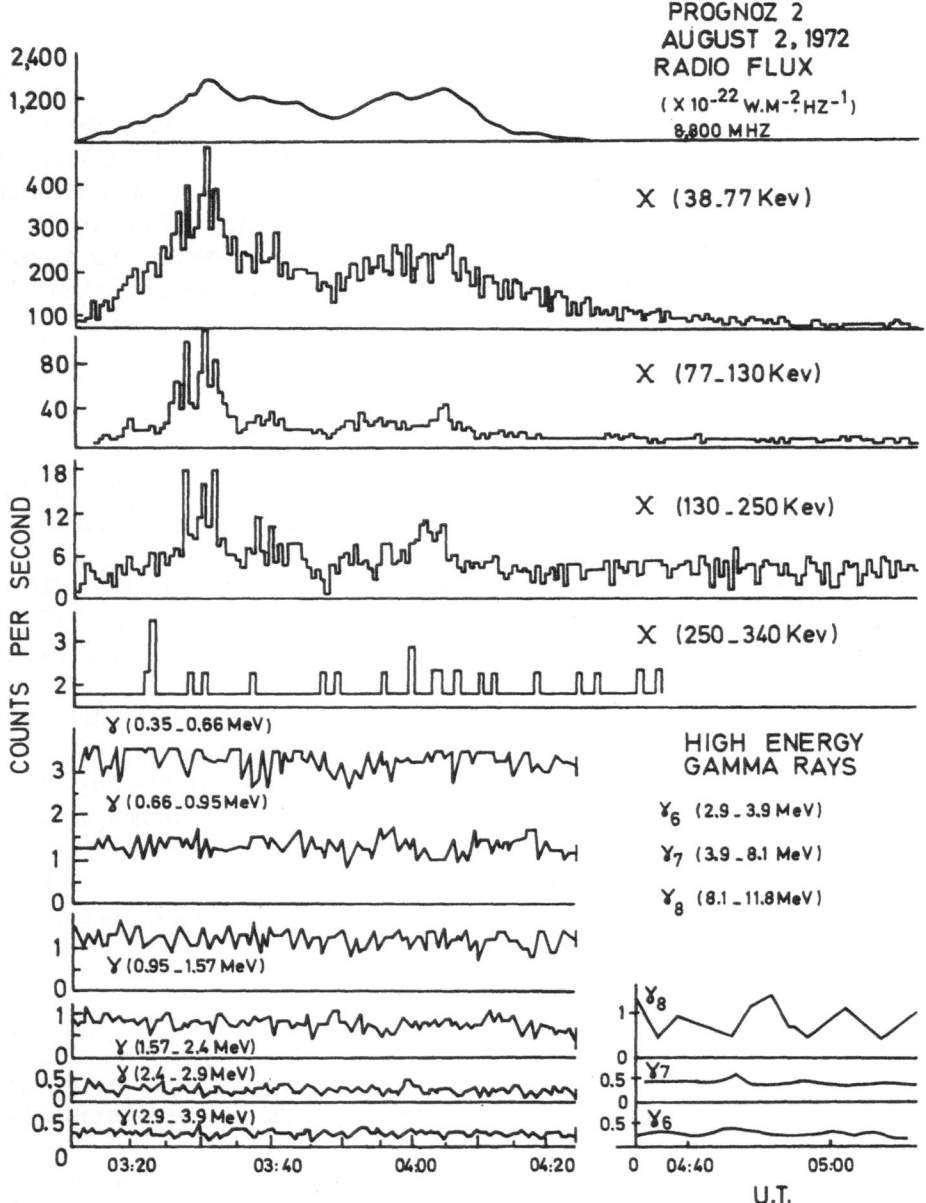

Fig. 1.   Time profile of the X- and γ-ray measurements of the SGL 1 and Signe I experiments at
0300 UT in 1972, August 2.

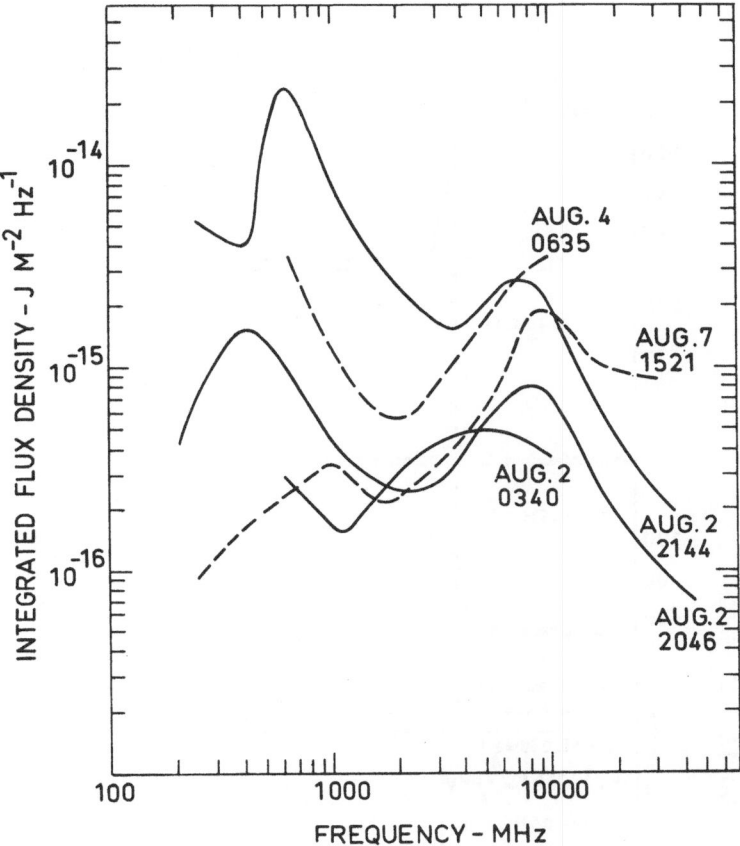

Fig. 2.   Integrated flux density for proton associated bursts of August 1972. Curves derived from Sagamore Hill and Manila Data.

in the continuous injection model. These values, as has already been observed several times, are well above the energy calculated from the radio fluxes (4.7×10^{24} erg in the range 600–8800 MHz).

(β) *Second event*. This very brilliant, impulsive Type 1B flare was located at 15° N and 25–26° E.

This event, observed in X-rays by our experiment and also by OSO 7 and TD 1 (Datlowe and Peterson, 1973; Van Beek *et al.*, 1973), was short. In the 33–77 keV band the X-ray emission appears at 18 38 18 and lasts about 8 min, the maximum, at 18 40, is 0.77 photons $cm^2\ s^{-1}\ keV^{-1}$. At higher energies, the X-ray emission is much shorter, about 41 s (Figure 3).

During the event the microwave spectrum was steeply rising to high frequencies which may indicate a low source height. At 18 40, about one minute after the maxi-

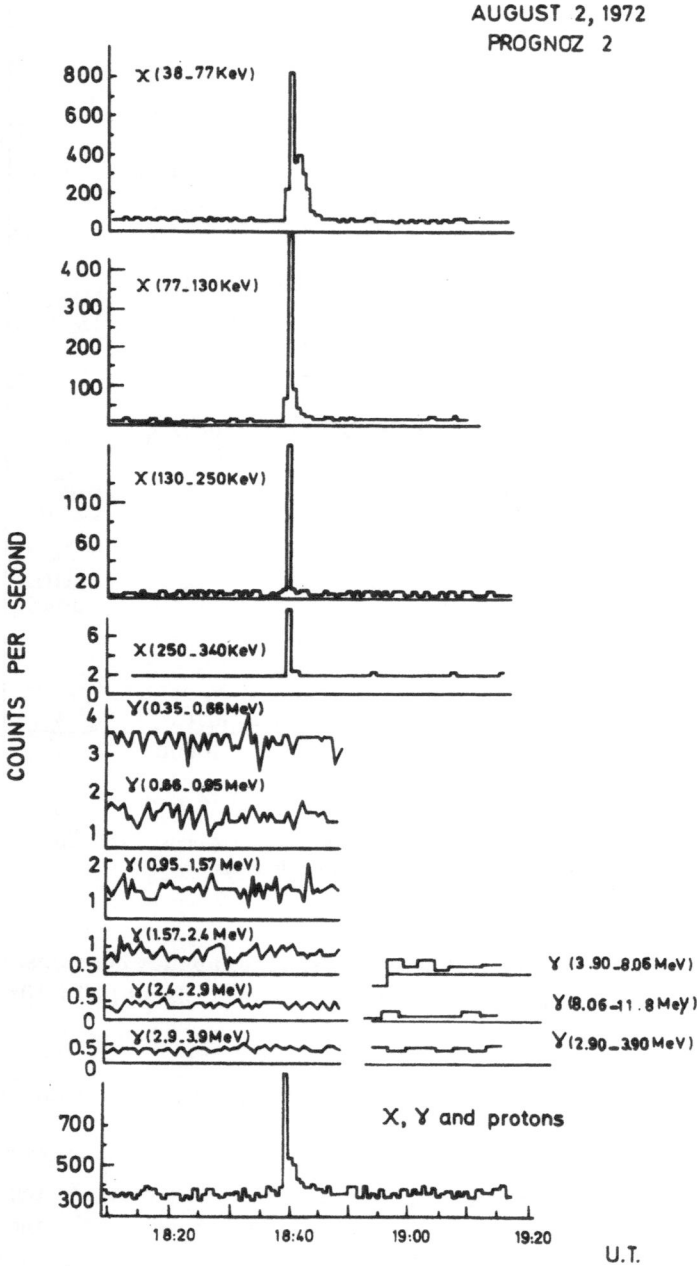

Fig. 3.   The event of 1972, August 2, 1840 UT.

mum, a rapid decrease of the flux around 35 GHz was observed. This decrease coincided with the end of the 3825 Å flash (Zirin and Tanaka, 1973).

Comparisons at optical (3835 Å), X- and radio frequencies are given by Zirin and Tanaka (1973) and their conclusion is that the thick target X-ray emission model is more consistent than the thin target model for this impulsive flare.

Using our Monte Carlo program, we have found the incident photon spectrum. At the time of maximum it has the form $AE^{-\alpha}$ with $\alpha = 3.7$ and $A = 1 \times 10^7$ photons cm^{-2} s^{-1} keV^{-1}. These values are in good agreement with those of OSO 7 (Datlowe and Peterson, 1973). The energy dissipated by the electrons in collisions has been computed using Brown's method (Brown, 1972). For the thick target model which applies best to this event, as Zirin and Tanaka (1973) have noted the energy involved at the time of the maximum is about 3×10^{29} erg s^{-1} for electrons above 10 keV.

During this event no gamma emission was observed.

(γ) *Third event.* This event is associated with a type 1B flare from the MacMath region 11976, 13° N, 26–28° E and is the most structured of all the August 2 X-ray events. It is, however, similar to the 03 10 flare is some ways: it is a long-lived, slowly rising flare. The characteristics of this flare are already given by Zirin and Tanaka (1973): reached maximum at about 2005 and gave a modest hard X-ray burst; at 2020 there was a new increase in area and brightness. The hard X-ray flux increased, and at 8800 MHz a large flux (6600 fu) was observed. The Hα front spread steadily to a maximum at 2050 UT. At 2140 the highest radio and X-ray flux peak can be noticed.

The entire event was observed by our X-ray detectors (Figure 4). For the first part of the event, OSO 7 could not give results because the satellite was in the Earth's shadow between about 2030 and 2110. However, during the period from 2110 to 2200 the second part of the event was observed by this satellite and a large continuum was reported up to 600 keV. Unfortunately at the time of this second large maximum in X-rays, our gamma ray detector was in a calibration mode during which the anti-coincidence was not working. Nevertheless it seems that an increase in the ΔE_9 and ΔE_{10} channels appeared at about 2140. But since these are the only data before the change in operating mode, no conclusive result is possible for this part of this event.

So we will concentrate on the first part of the event at 2040. Several bursts, particularly in the ranges 38–77 keV and 77–130 keV can be noted (Figure 5). This type of phenomenon can also be found in the August 7 events. The maximum for this event takes place at 204636 in the 38–77 keV channel. At the time of the maximum the X-ray spectrum has the following form: $8.10^6 \, E^{-3.3 \pm 0.3}$ photons cm^{-2} s^{-1} keV^{-1}. The calculations, again using the method developed by Brown (1972), give values for the electron energy of 2×10^{27} erg s^{-1} for $E_e > 40$ keV and 10^{33} electrons s^{-1} for $E_e > 100$ keV.

It is interesting to note that during this first part of the event, at about 2045 a very high increase in the gamma-ray flux above 350 keV (Figures 4 and 5) was observed. Figure 5 also shows the variation in radio flux at 8800 MHz. The good agreement

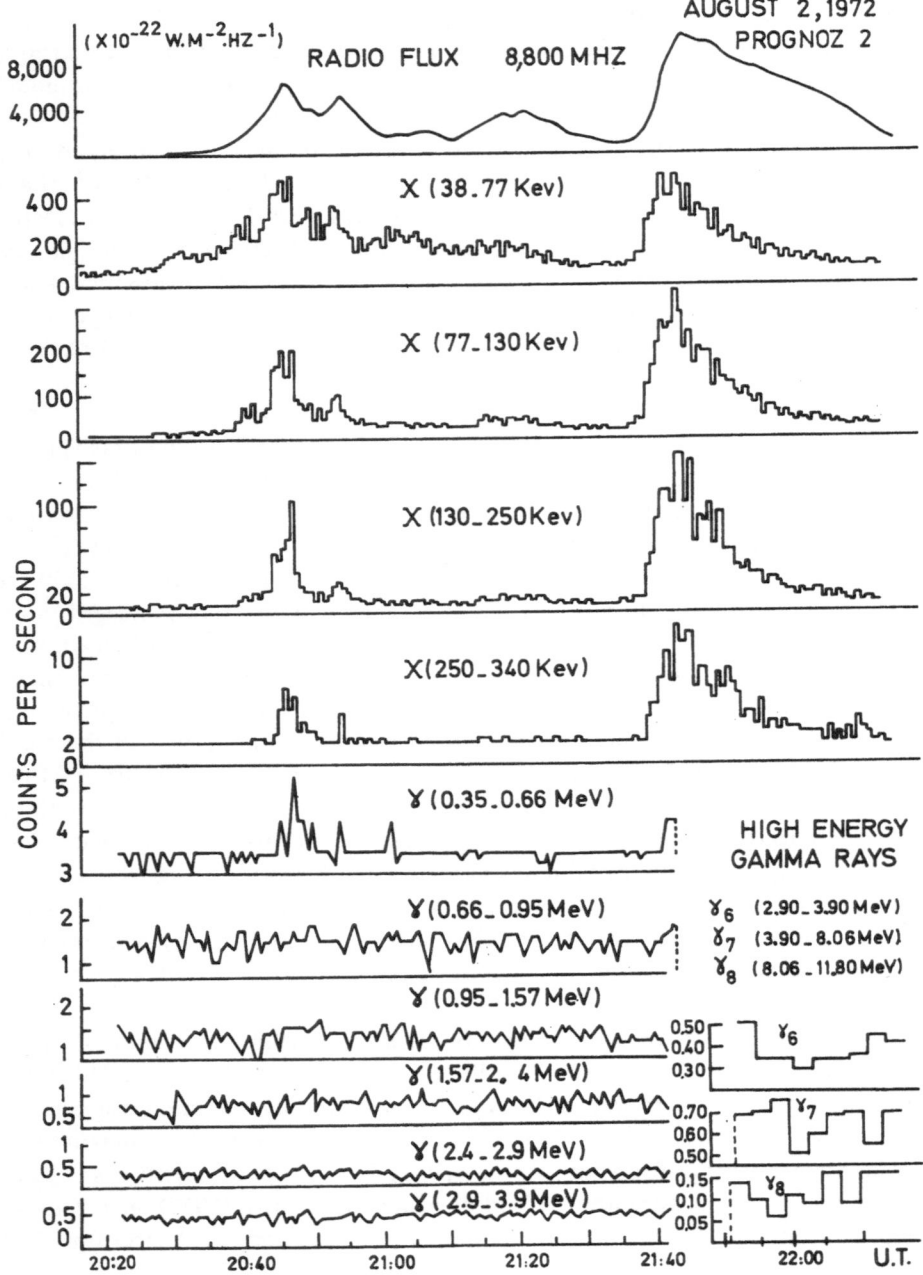

Fig. 4.   Time profiles of the August 2, events 2040 and 21 40 UT.

between the time variations of the count rate observed in the ΔE_9 channel and those of the radio and X-ray fluxes beyond 70 keV is evident. In addition, the radio spectrum can be seen to show the same characteristics as that of the 3 30 event (Figure 2). The U-shape of this spectrum is often taken to indicate the presence of charged particles.

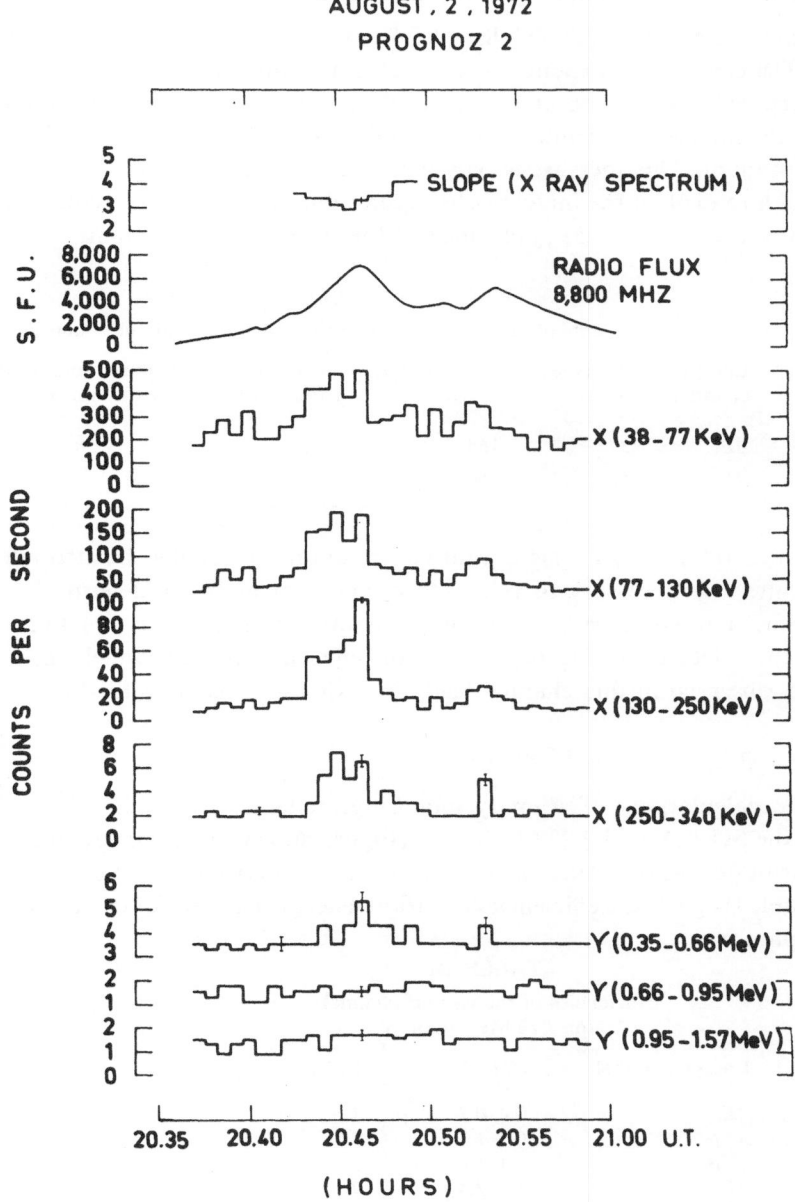

Fig. 5.   A detailed time profile of the 2040 August 2 event.

In order to interpret the flux increase in the ΔE_9 channel and to determine whether this flux increase could be due in part to the presence of a line a 0.511 MeV, it is necessary to calculate the contribution of the continuum in the ΔE_9 channel. We assume that the spectrum is a power law and that it can be extrapolated to higher energy.

(ii) POSSIBILITY OF A CONTRIBUTION OF THE 511 keV LINE

The X-ray spectrum calculated has the form $8 \times 10^6 \, E^{-3.3 \pm 0.3}$ photons cm^{-2} s^{-1} keV^{-1}. The error in the exponent is calculated assuming a possible deviation of 3σ in the experimental values obtained by the SGL 1 detector. Using this spectrum, we estimate the increase we would have had in the ΔE_9 channel if the event had included only continuum. This increase is only detectable in the first channel (ΔE_9), but it is not enough to explain the increase observed. Table II gives the 41 s count rate values obtained in the ΔE_9 and ΔE_{10} channels at the time of the maximum:

TABLE II

Counts observed during 41 s at the time of maximum intensity

Energy channel	Observed counts	Background	Continuum contribution (of solar origin)	Possible 0.5 MeV line contribution
ΔE_9	237	155	24	58
ΔE_{10}	56	56	5	–

While this table shows a significant contribution which could be attributed to the 511 keV line, it is necessary to take into account the errors in determining the continuum spectrum: specifically the continuum contribution in ΔE_9 is 24 ± 13 cts (41 s)$^{-1}$.

For a total detection efficiency of 33% the efficiency in the channel ΔE_9 is 14%, so the excess observed in this channel leads to a flux associated with the 511 keV line which is:

$$\phi_{0.511} = 0.6 \pm 0.5 \text{ photons cm}^{-2} \text{ s}^{-1}. \tag{4}$$

Using the calculations of Ramaty and Lingenfelter (1973) which give a relation between the 511 keV and 2.23 MeV fluxes (Figures 6, 7) we can estimate the 2.23 MeV line contribution in our detector if we know its sensitivity to this line.

The Table III gives the efficiencies of various energy channels for detecting 2.23 MeV gamma photons.

TABLE III

Efficiencies of the various channels
for 2.23 MeV photons

Energy channel	ΔE (keV)	Efficiency %
ΔE_9	0.4–0.7	1.6
ΔE_{10}	0.7–1	1.7
ΔE_{11}	1–1.6	3.4
ΔE_{12}	1.6–2.4	7.1
ΔE_{13}	2.4–2.9	0.6

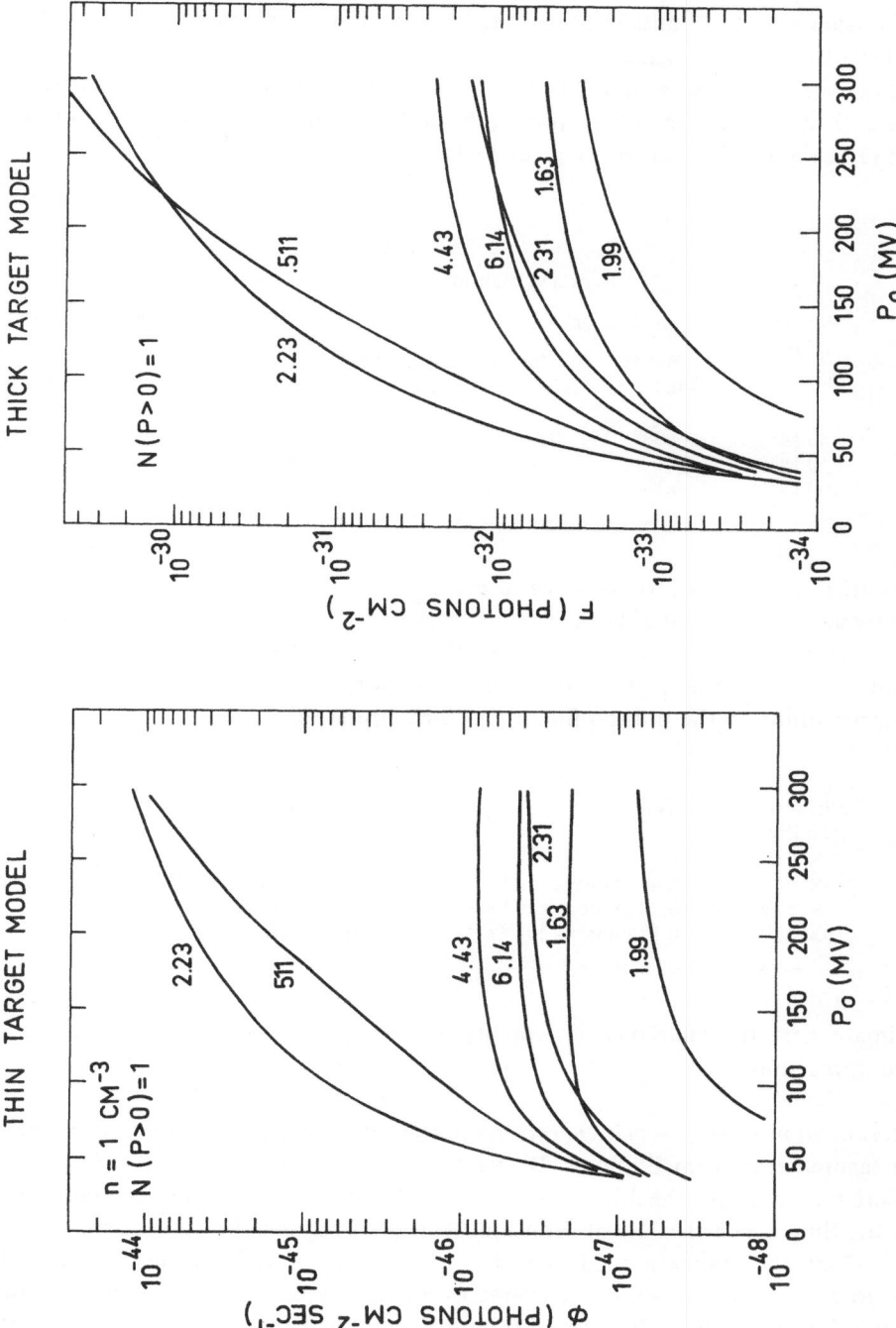

Fig. 7. γ-ray line flux normalized to one incident proton for the thick-target model.

Fig. 6. γ-ray line flux normalized to 1 cm⁻³ and for 1 incident proton for the thin-target model.

The sensitivity of our detector in the ΔE_{12} channel where the efficiency is greatest, is 0.15 photons cm^{-2} s^{-1} at 2.23 MeV at the 1 σ level and 0.33 photons cm^{-2} s^{-1} at the 2 σ level for 41 s.

The calculation of Wang and Ramaty (1974) enables us to specify the expected fluxes at 2.23 MeV for different rigidities in both thick and thin target models, knowing the 511 keV flux. The values are given in Table IV.

TABLE IV

Expected photons cm^{-2} s^{-1} at 2.23 MeV

Proton rigidity	Thick target model			Thin target model		
	ϕ 0.511 MeV (photons cm^{-2} s^{-1} keV^{-1})					
	0.1	0.6	1.1	0.1	0.6	1.1
40 MV	0.08	0.4	0.8	0.05	0.3	0.5
50 MV	0.1	0.6	1.2	0.1	0.6	1.1
100 MV	0.18	1	1.9	0.3	1.6	2.8

Under these conditions, the absence of a 2.23 MeV line contribution in the ΔE_{12} channel (where efficiency at 2.23 MeV is highest) makes it possible to define an upper limit for the gamma flux value at 0.5 MeV, assuming that Wang and Ramaty's model does indeed represent the gamma line emission processes.

The upper limits to the photon flux at 0.5 MeV are then:

Proton rigidity	Thick-target	Thin-target	
40 MV	0.44 photons cm^{-2} s^{-1}	0.66 photons cm^{-2} s^{-1}	(5)
50 MV	0.28 photons cm^{-2} s^{-1}	0.33 photons cm^{-2} s^{-1}	
100 MV	0.18 photons cm^{-2} s^{-1}	0.12 photons cm^{-2} s^{-1}	

We estimate next the characteristic rigidity of this event in order to give a more accurate upper limit.

(α) *Determination of the event's characteristic rigidity.* The rigidity can be estimated using measurements from Explorer 41 (Bostrom *et al.*, 1972) but for the August 2 event there is a problem: the flare responsible for the proton emission is located to the east on the Sun. Since the lateral diffusion of these protons is low compared to the diffusion along the magnetic field line, it is impossible to attribute the proton flux increase to any particular August 2 event; as we have seen, all these events can be responsible for proton emission. The Explorer 41 data however, indicate a plateau at 1200 UT August 3. This configuration being typical of the events occurring to the east on the Sun. We have therefore determined the characteristic rigidity using the Explorer 41 data, assuming that this rigidity was conserved in the scattering.

The flux values measured under these conditions will necessarily be lower than those at the Sun. The observed fluxes were as follows:

Proton energy E	Integral proton flux at the Earth $J(>E)$	
>10 MeV	500 protons cm^{-2} s^{-1}	(6)
>30 MeV	37 protons cm^{-2} s^{-1}	
>60 MeV	12.5 protons cm^{-2} s^{-1}	

From these values one can deduce the proton density in the interplanetary medium $U(P)$, since, if this density is put into the form $U(P) \simeq \exp(-P/P_0)$ then

$$U(>P) = J(>P) \cdot \frac{mc}{e} \cdot (P + P_0)^{-1} \text{ with } \frac{mc}{e} = 3.12 \times 10^{-8}. \qquad (7)$$

Under these conditions and with the above mentioned reservations the characteristic rigidity can be estimated at between 40 and 50 MV.

(β) *Number of protons emitted in the event assuming a gamma-ray line emission.* The estimate of the number of protons at the Sun in the thick target model are obtained for an average rigidity of 45 MV and for 511 keV fluxes of 0.1 photons cm^{-2} s^{-1} and 0.35 photons cm^{-2} s^{-1}. In this case as the event lasts 41 s, the total number of protons according to Figure 7 lies between 5×10^{33} and 2×10^{34} protons.

These values can be compared to that deduced from observations made at the Earth. But we have pointed out that the main problem lies in the low value of proton fluxes measured at the earth when emitted east of the Sun. However it is possible to give an approximation of these fluxes (>10 MeV) by using the curve of Straka (1970) (Figure 8). In order to do this, we have determined the energy from radio observations at 8800 MHz (8×10^{-16} J. m^{-2} Hz^{-1}); according to this curve, a proton flux of about 600 p cm^{-2} s^{-1} sr^{-1} results, or 15 times the value of the fluxes measured. Using this proton flux and a volume occupied by the particles in interplanetary space of 10^{39} cm^3 (Ramaty and Lingenfelter, 1973), the number of protons is equal to 2.6×10^{34}.

(γ) *Ambient density of the solar atmosphere.* In the thin target model taking the 0.511 MeV line flux between 0.1 and 0.45 photons cm^{-2} s^{-1} one obtains for nN (Figure 6):

$$7 \times 10^{45} < nN < 3 \times 10^{46}. \qquad (8)$$

Thus, taking the value derived from measurements at the Earth for the number of protons, the number density of the ambient solar material (cm^{-3}) becomes:

$$3 \times 10^{11} < n < 1 \times 10^{12}. \qquad (9)$$

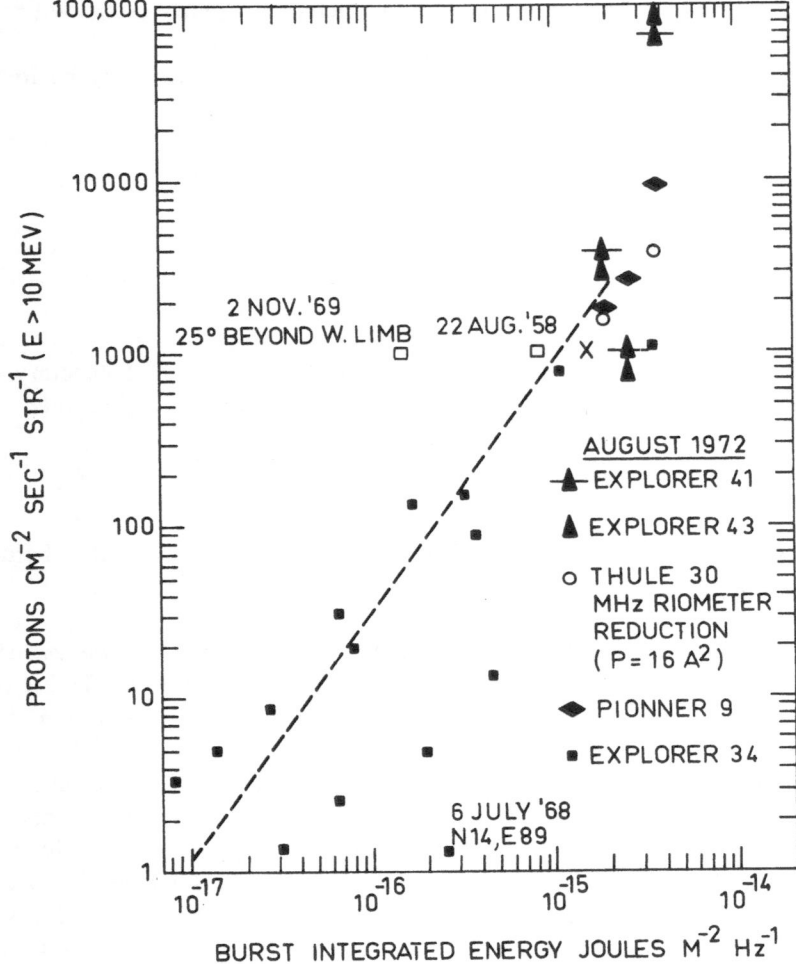

Fig. 8.   Burst integrated energy at 8800 MHz vs proton flux.

(δ) *Energy involved during the flare*. First we consider the thick-target model. Assuming that the proton population has a spectrum with the form $N(P) = P_0^{-1} \exp(-P/P_0)$, the average proton energy is P_0^2/Mc^2 or 3.4×10^{-6} erg. The total energy involved is then:

$$1.7 \times 10^{28} \text{ ergs} < W_T < 6.8 \times 10^{28} \text{ ergs} \tag{10}$$

In order to find the energy involved in the thin target model, we compute the energy W dissipated per unit time by a proton, in a medium having a density of 1 cm^{-3} (Ramaty and Lingenfelter, 1973).

The results are presented in Table V. Using the value 1.9×10^{-18} erg s^{-1} for a rigidity of 45 MV and the values of nN previously calculated, we obtain the energy

TABLE V

Energy dissipated by protons in the thin-target model

Rigidity P_0 (MV)	Energy dissipated $W \times 10^{-18}$ erg s^{-1}
20	2
30	2
40	1.9
60	1.8
80	1.6
100	1.5
120	1.4
200	1.2
300	1.1

involved, E, for an event lasting 41 s:

$$5 \times 10^{29} \text{ ergs} < E < 2 \times 10^{30} \text{ ergs} . \tag{11}$$

(b) AUGUST 7 EVENT

(i) *Description of The Event*

In the MacMath region 11976 with which the events of August 2 and 4 have already been associated, an event appeared on August 7. Figure 9 shows recording from RS 1, SGL 1, and Signe 1 detectors as well as the proton detector data for the burst period. For purpose of comparison, the same figure shows radio fluxes at a frequency of 15 400 MHz (Castelli *et al.*, 1973), as well as the periods when type III and type II bursts were observed (Lincoln *et al.*, 1972). From Figure 9 the low energy X-ray burst (curves 1, 2 and 3) can be seen to begin shortly before the burst appearing at higher energies (curves 6, 7 and 8); the highest energy component reaches a maximum towards 1521 UT, and then drops very rapidly. The X-ray time profile, particularly at high energies, coincides fairly well with the shape of the microwave radio burst. Figure 10 shows the time correlation between our data, radio flux measurements and TD1 and Intercosmos 7 experiments where the time resolution was better (Van Beek *et al.*, 1973, Valnicek *et al.*, 1973). The two types of radiation typically show a 'step' profile in time with intermediary maxima 1516, 151730, 1519 UT (Figures 9 and 10). This is probably related to the intermediary maxima indicated above in the X-ray and radio ranges.

The radio observation data in the frequency range 245 to 15 400 MHz (Castelli *et al.*, 1973) indicate that the beginning of the radio burst may be situated at 1422 and the observable increase in frequencies 930 to 2800 MHz begins at 1505 UT, simultaneously with an increase in Hα line intensity (Lincoln *et al.*, 1972).

Beginning at 1519 UT a white light burst appears lasting 7 to 8 minutes from the moment of maximum intensity in the radio spectrum (Castelli *et al.*, 1973). A type II radio burst also appears which seems to begin at about 1521 UT and which lasts

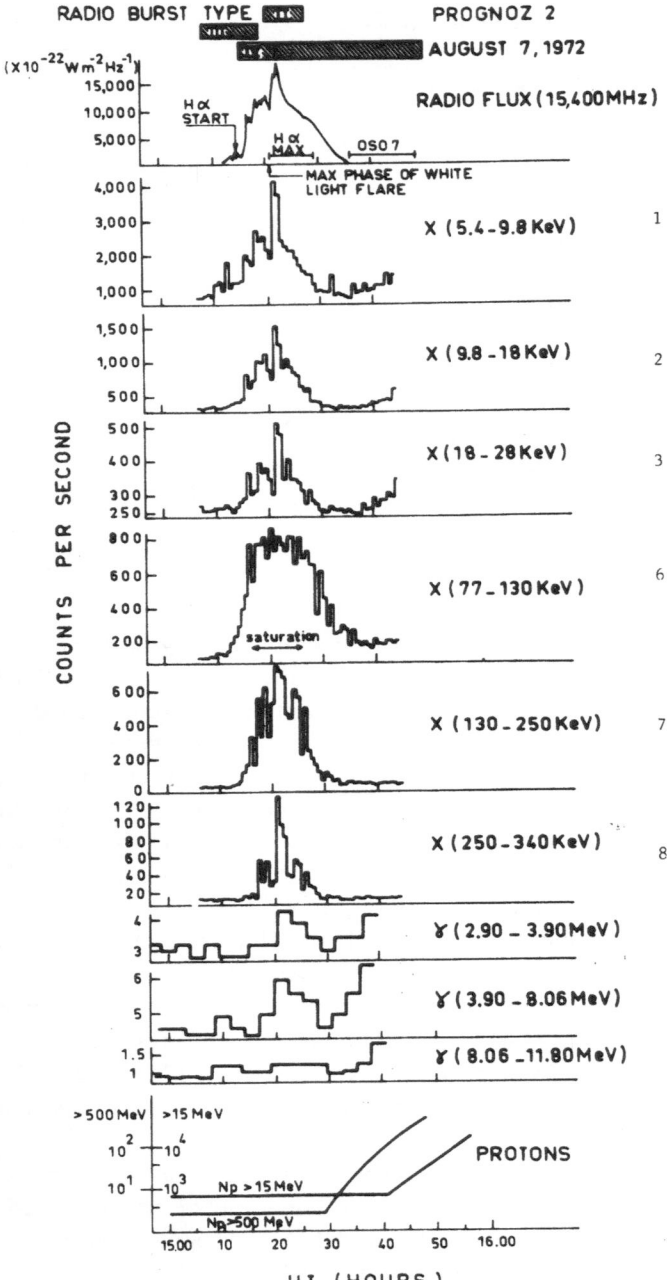

Fig. 9.   Time profiles of the X-ray, γ-ray and proton measurements in the August 7 event.

till 1535 UT (Dodge, 1973). At 152130 a flux maximum in the hard X-ray photon flux is noted almost simultaneous with an increase in the count rate in the range 2.9–7.8 MeV. The gamma burst lasts about 8 to 10 min.

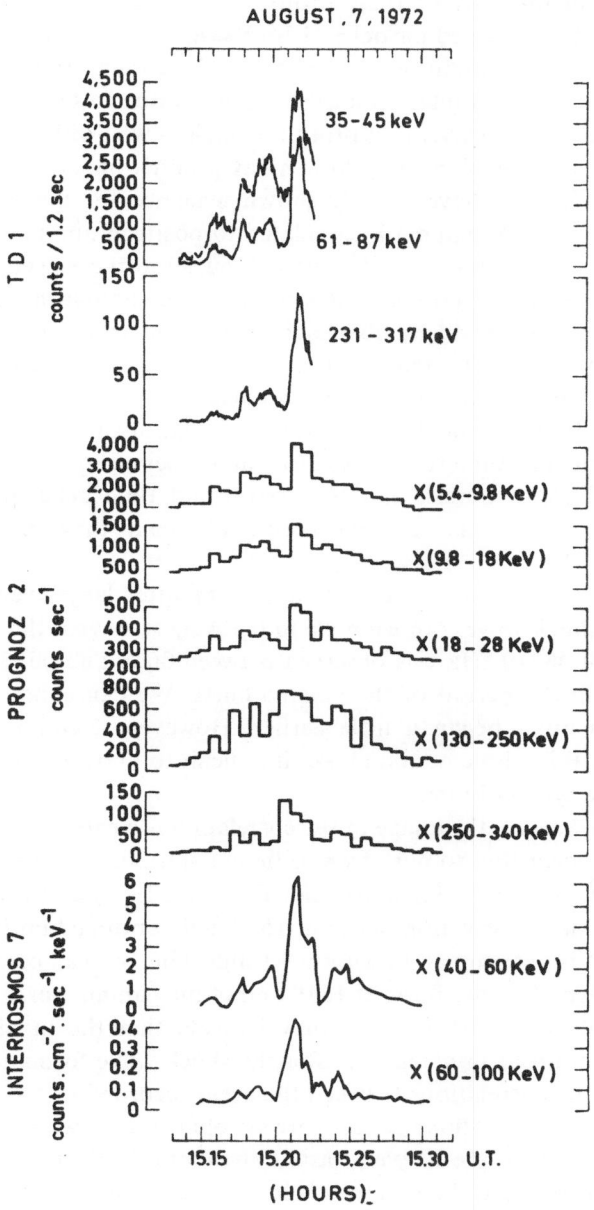

Fig. 10.   A detailed time profile of X-ray measurements in the August 7 event.

Since the X-ray measurements have a time resolution of 41 s it is not possible to show an accurate time correlation in the development of the photon fluxes for these two energy ranges. Nonetheless, the X- and gamma-events end at about 1527 UT, shortly before the high-energy protons appear.

The increase in intensity of these protons ($E_p \geqslant 500$ MeV) begins at 1529–1530, while the low energy charged particles (1 MeV $< E_p < 200$ MeV) appear only at about 1535 UT. Under these conditions the excess of the gamma ray fluxes between 1522 and 1530 cannot be interpreted as local production due to the arrival of protons. In addition we note that the arrival of protons coincides well with a very clear γ-ray flux increase which can be explained by local γ-ray production.

Chupp *et al.* (1973b) have already shown evidence for the 511 keV, 2.23 MeV, 4.43 MeV, and 6.1 MeV gamma lines related to positron annihilation, formation of deuterium, and excited states of C^{12} and O^{16} nuclei in these events. In our observations of this event, the spectral characteristics of the gamma-radiation were difficult to obtain due to the limited number of channels and to the measurement method. Nevertheless we shall try to analyze these data taking into account the contribution of the excited states of oxygen and carbon nuclei.

But first we shall comment on the time correlation in the different wavelength ranges. One of the most interesting results of the observational data is the correlation between type II radio bursts, and the arrival and time behavior of the energetic protons recorded over the gamma radiation which they produce in the flare region.

The radio bursts of this type usually appearing after large events result from the formation of a shock wave. For example in the August 4 event the type II radio burst at frequencies of 4824 MHz was observed between 0629 and 063305 (Castelli *et al.*, 1973), i.e. during the period of the gamma-burst. We can suppose at much higher frequencies this burst began a little earlier. However, a complex image of radio disturbances for this whole period makes it difficult to know the exact moment of the beginning of the type II burst.

In the August 7 event, the image of the correlation is clearer; the type II radio burst began at 1519 (according to data from different stations this moment can be determined to within $+1$ min, -0.3 min) (Lincoln *et al.*, 1972) and lasts until 1537. The gamma burst under observation began at 1522 and continued until 1530. The beginning of the burst is determined to within ± 1 min. Thus we can consider that between the start of the type II radio burst at 1519 and of the gamma burst at 1522, there was a certain time delay $t \simeq 10^2$ s which may indicate that the maximum proton flux density does not appear immediately after the shock wave formation.

Concerning time correlations between the event recorded in X- and γ-rays, we can note that the X- and γ-ray flux increases are simultaneous to within our time measurements errors. But when the X event lasts only about 2 min the γ-ray event extends over on period of about 8 to 10 min. This lack of correlation in the length of the 2 events is further evidence of a different origin for the X-rays (electron bremsstrahlung) and γ-ray fluxes (solar atmosphere proton interactions).

(ii) *Evaluation of a Possible Continuum Contribution above* 3 MeV

As we have just remarked time correlations between X- and γ-ray events seem to indicate a different origin for these emissions.

In spite of this argument, we have attempted to calculate the contribution of a continuum in channels ΔE_{14}, ΔE_{15}, ΔE_{16} taking into account the low energy measurements (<340 keV) given by SGL 1 detector. So this estimate is very approximate because we have no spectral information between 340 keV and 3 MeV.

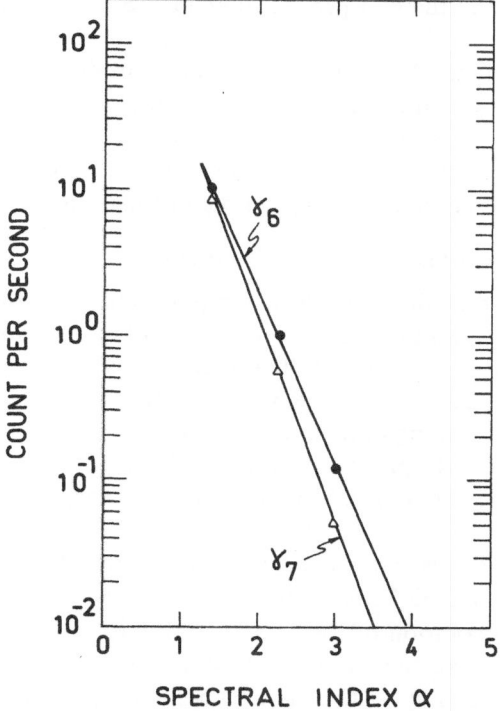

Fig. 11.   Counting rate in ΔE_{14} and ΔE_{15} vs. photon spectrum index α (normalized to channel 250–340 keV).

We have used the Monte Carlo program to calculate the contribution of the continuum in ΔE_{14}, ΔE_{15} for different slopes of this spectrum, the γ-ray spectrum being normalized to the value of the flux observed between 250 and 340 keV (Figure 11). If we retain a spectrum in E^{-3} which is rather close to that of the continuum for the August 4 event (Chupp *et al.*, 1974 – private communication), we notice that the expected gamma contribution is negligible compared with the counts observed. It must also be pointed out that whenever the exponent of the spectrum is greater than 1, it leads to contributions in ΔE_{14} and ΔE_{15} which are in ratio always greater than 1; for example, for $\alpha = 2.3$, $\Delta E_{14}/\Delta E_{15} = 2$, and this ratio increases at higher energy while $\Delta E_{14}/\Delta E_{15}$ observed during the event is 0.6 (Figure 11).

Let us further note that it is difficult to assume a spectrum of photons having a slope of less than 2 above 500 keV because the resulting gamma photon flux above 500 keV, would cause an increase in the anticoincidence count rate, which would be significant. However, no such systematic increase during the event was observed.

None of these arguments constitutes a definitive proof as to the absence of significant continuum contribution. It nevertheless seems clear that if we take a spectrum above 400 keV with a slope less than 2.5, the total count rate observed cannot be due to the continuum, the best proof still being the ratio of the fluxes observed in ΔE_{15} and ΔE_{16}.

Under these conditions, the presence of a contribution of monoenergetic gamma photons is reasonably possible. However, we can only give it as upper limit by supposing that the entire observed contribution is due to the presence of these lines. If a continuum exists, it can provide a contribution that is impossible to determine precisely, given the observational data at the time of the event.

(iii) *Evaluation of a Possible γ-Ray Line Contribution*

Calculations have been made under the assumption that only the 6.14 and 4.43 MeV lines are excited (Ramaty and Lingenfelter 1973). In fact perhaps the actual situation at the Sun is much more complicated than it is supposed here.

Due to poor time resolution the length of the event cannot be given exactly but lies between 8 and 11 min. Taking into account the detection efficiency for these two lines and using the Monte-Carlo program we have calculated the fluxes of 4.43 and 6.14 MeV γ-ray lines to be 1.8 and 0.7 photon cm^{-2} s^{-1} respectively for a γ-ray burst lasting 8 min.

We can try to check if these γ-ray line fluxes are in agreement with those measured by Chupp *et al.* (1973a) on OSO 7. But as these measurements are obtained later we have to consider the time evolution of the 2.23 MeV flux to estimate what the 2.23 MeV flux would have been at the moment of our measurements. For simplification we can choose a time evolution for the 2.23 MeV flux given by $\exp[-\lambda(t-t_0)]$ with $\lambda = \tau_c^{-1} + \tau_d^{-1}$ where τ_c is the capture time of neutrons on hydrogen and τ_d the mean neutron lifetime (Reppin *et al.*, 1973). In this case we neglect the dependence of ϕ on θ and E_n (Wang en Ramaty, 1974), (θ is the angle between the Earth-Sun line and the vertical to the plane stratified medium; E_n is the neutron energy). The capture time τ_c is ~ 100 s. for a density of the ambient solar atmosphere $n = 10^{17}$ cm^{-3} if we use the expression $\tau_c(s) \simeq (1.5 \times 10^{19}/n(cm^{-3}))$ (Wang and Ramaty, 1974).

The 2.2 MeV γ-ray line at the end of the neutron production is given by ϕ_{t0}:

$$\phi_{t0} = \phi_t \exp\left[-\lambda(t-t_0)\right]. \tag{12}$$

Since the radio and X-ray measurements seem to indicate that the event ends at 1530 we may suppose that the neutron production also ends at this time. So if we take for ϕ_t the 2.23 MeV γ-ray flux measured by OSO 7 at about 1538 ($\phi_t \simeq 4.8 \times 10^{-2}$ cm^{-2} s^{-1}) we obtain for $\phi_{t0} \sim 2$ photons cm^{-2} s^{-1}.

With these values of ϕ and a rigidity of ~ 50 MV for the event, the calculations of Ramaty *et al.* (1973) give a ratio of ~ 3 between the γ-ray lines at 2.2 and 4.4 MeV for the thick target model. If we take into account the revised γ-ray fluxes at 2.23 MeV (Wang and Ramaty, 1974), which are lower than the values already published by a factor 2–2.5, the ratio $\phi_{2.2}/\phi_{4.4} \simeq 1.2$ to 1.5 for 50 MV rigidity. In this case we have a reasonably good agreement between theoretical calculations and our measurements at the end of the event at about 1530.

We shall now compare Ramaty's theoretical estimates with our experimental results in order to evaluate some of the basic parameters of the solar flare. But before that we need to determine the rigidity of the protons spectrum.

(α) *Determination of the event's characteristic rigidity.* This rigidity is defined as for the August 2 event using the measurements made by Bostrom *et al.* (1972). For the August 7 event we have the following values:

$$J(> 10 \text{ MeV}) = 4.2 \times 10^4 \, p \text{ cm}^{-2} \text{ s}^{-1}$$
$$J(> 30 \text{ MeV}) = 4.56 \times 10^3 \, p \text{ cm}^{-2} \text{ s}^{-1} \qquad (13)$$
$$J(> 60 \text{ MeV}) = 8.4 \times 10^2 \, p \text{ cm}^{-2} \text{ s}^{-1}$$

From Equation (7) the characteristic proton rigidity can be seen to be between 40 and 50 MV, which corresponds to densities of $U(>10 \text{ MeV}) = 7.24 \times 10^{-6} \text{ cm}^{-3}$, $U(>30 \text{ MeV}) = 5 \times 10^{-7} \text{ cm}^{-3}$, and $U(>60 \text{ MeV}) = 6.8 \times 10^{-8} \text{ cm}^{-3}$, for an average characteristic rigidity of 45 MV. When curves $f(P)$ (Figure 12) are plotted, the values 40 and 50 MV corresponding to the slopes of these curves do reflect the average value of 45 MV. The particle rigidity derived from these observations is thus included between 40 and 50 MV. The results based on this and using of Ramaty's and Lingenfelter's calculations appear below.

We have to note, however, that the following estimates depend strongly on the characteristics of the proton spectra. In particular, the high values of the number of protons in the thick-target model, the density of the region where the γ-photons are produced in the thin-target model, and the energies involved would be reduced by an order of magnitude if the rigidity reached 100 MV.

(β) *Number of protons involved in the flare.* According to Figure 7 we obtain the following photons fluxes normalized to one proton:

Proton rigidity	4.43 MeV	6.14 MeV
40 MV	3.3×10^{-34}	2.4×10^{-34}
50 MV	8.7×10^{-34}	5.7×10^{-34}

(14)

Since the total fluxes at 4.43 MeV and 6.14 MeV are respectively 883 photons cm^{-2}

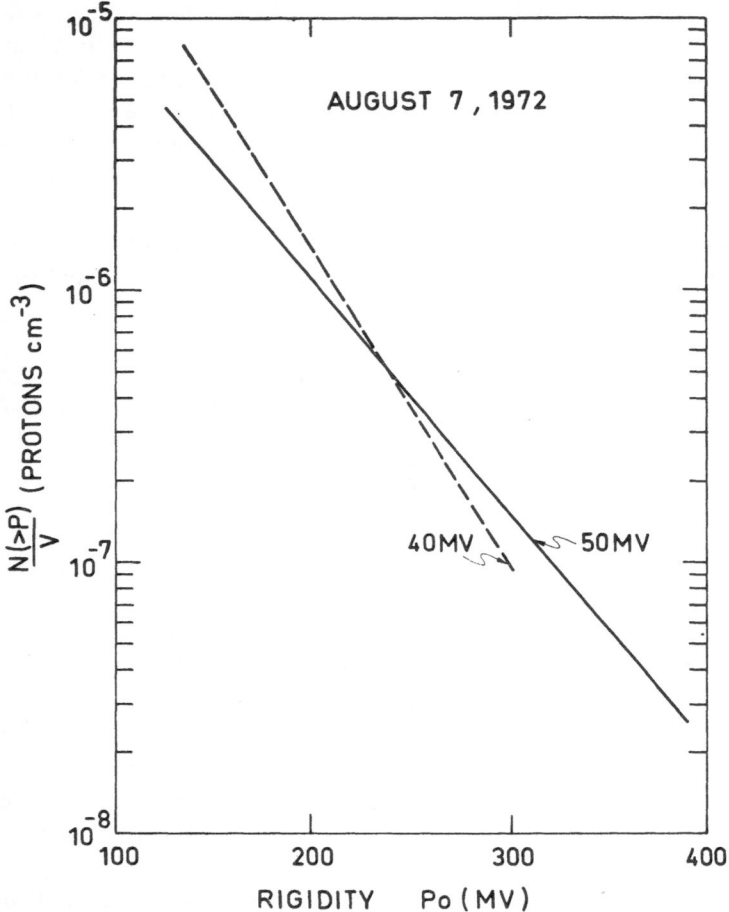

Fig. 12. Density of protons near the Earth vs rigidity for 2 characteristic rigidities in the 1972, August 7 event.

and 331 photons cm^{-2}, from Ramaty's curve it is possible to estimate the number of protons involved in the event:

Proton rigidity	4.43 MeV	6.14 MeV	Average value	
40 MV	2.7×10^{36}	1.4×10^{36}	2.0×10^{36}	(15)
50 MV	1.0×10^{36}	5.8×10^{35}	8.0×10^{35}	

From Figure 12 one can obtain the proton density in interplanetary space, and the proton flux at the Sun can be deduced:

$$2.5 \times 10^{35} < N(P > 0) < 5.6 \times 10^{34} \quad \text{for} \quad 40 \text{ MV} < P_0 < 50 \text{ MV} \quad (16)$$

This estimate is made on the assumption that the volume occupied by the particles in interplanetary space is a cone with a 30° opening angle whose vertex is on the Sun (Ramaty and Lingenfelter, 1973). This volume is estimated to be 10^{39} cm³. These two determinations of the proton flux at the Sun are not in good agreement.

(γ) *Density in the solar atmosphere deduced from the thin target model.* From Figure 6 we find the number of photons produced by one proton in a solar atmosphere having a density of 1 cm⁻³.

Proton rigidity	4.43 MeV	6.14 MeV	
40 MV	9.4×10^{-48}	6.9×10^{-48}	(17)
50 MV	1.7×10^{-47}	1.2×10^{-47}	

Since the photon fluxes measured are 1.8 photons cm⁻² s⁻¹ at 4.43 MeV and 0.7 photons cm⁻² s⁻¹ at 6.14, we can estimate the product nN expressed in cm⁻³, i.e.

Proton rigidity	4.43 MeV	6.14 MeV	
40 MV	1.9×10^{47}	1.0×10^{47}	(18)
50 MV	1.1×10^{47}	5.8×10^{46}	

The number of protons N is obtained by estimates given above: $N = 2.5 \times 10^{35}$ for $P_0 = 40$ MV and 5.6×10^{34} for $P_0 = 50$ MV. The density encountered by the protons at the Sun for the creation of measured photon fluxes is given in the following table:

Proton rigidity	4.43 MeV	6.14 MeV	Average value	
40 MV	7.9×10^{11}	4.0×10^{11}	6.0×10^{11}	(19)
50 MV	1.9×10^{12}	1.0×10^{12}	1.5×10^{12}	

(δ) *Energy involved in the flare.* First we consider the thick-target case. For a spectrum having the form $P_0^{-1} \exp(-P/P_0)$ and for non relativistic protons, the average proton energy is P_0^2/Mc^2; thus for the thick-target model the energies carried by 2×10^{36} protons and 8×10^{35} protons are 5.6×10^{30} erg and 3.5×10^{30} ergs for 40 and 50 MV respectively.

For the thin target model the rate of gamma photon production and the proton energy loss rate depend on the product of the ambient density and the number of protons emitted (Ramaty and Lingenfelter, 1973). The ratio of these two rates is independent of n and N depends on the accelerated proton spectrum, i.e. on P_0. Table V gives the energy lost by a proton with rigidity above 0 for $n = 1$ cm⁻³ and for various characteristic rigidities.

Between 40 and 50 MV, W is about 1.9×10^{-18} erg s^{-1} (Table V). Using the values for n and N given above, for the energy involved expressed in erg s^{-1} we have:

Proton rigidity	4.43 MeV	6.14 MeV	Average value
40 MV	3.7×10^{29}	1.9×10^{29}	2.8×10^{29}
50 MV	2.1×10^{29}	1.1×10^{29}	1.6×10^{29}

$$(20)$$

Since the event lasted 8 min, the total energy involved, expressed in ergs, is:

Proton rigidity	4.43 MeV	6.14 MeV	Average value
40 MV	1.8×10^{32}	9.1×10^{31}	1.3×10^{32}
50 MV	1.0×10^{32}	5.5×10^{31}	7.7×10^{31}

$$(21)$$

This corresponds to an average energy 1.2×10^{32} ergs involved in the entire γ-ray event.

III. Conclusion

After a general description of the 2 events on August 2 and 7, we have tried to present some evidence of the presence of γ-ray lines during these two events. It seems that for the August 2 event a line at 0.511 MeV is revealed but this does not exclude definitively the possibility of an X-ray continuum extending to about 600 keV. For the August 7 event, the evidence of high energy γ-rays up to 8 MeV is clearly shown, but as we have no spectral information between 500 keV and 3 MeV it is still difficult to estimate a continuum γ-ray contribution. We have nevertheless attempted to do it, but our results seem to be more consistent with the presence of lines at 4.43 and 6.14 MeV, if the Ramaty-Lingenfelter theory is used. Obviously without the support of this theory, it will be impossible to say anything about the origin of the observed γ-ray excess.

In any case it appears that the durations for the X- and γ-ray emission during the 7 August event are quite different, which seems to indicate a different origin for these emissions. In fact this is quite possible, even if there is only continuum, because, as Chupp *et al.* (1972) point out the origin of this continuum may not be entirely related to the presence of high energy electron bremsstrahlung and other lines, not resolved and associated with proton interactions, may contribute.

If as we think the excess of γ-ray flux recorded by our experiment can be attributed to proton interactions in solar atmosphere, the reasonably good time correlation at the beginning of the X- and γ-ray events seems to prove that the non-relativistic and mildly relativistic energy electrons are accelerated at about the same time as the relativistic electron and high energy protons (Švestka, 1973).

Acknowledgements

The authors wish to thank Mr. F. Cambou, I. A. Savienko, N. N. Volodichev, and A. A. Souslov, Mrs E. I. Morozova and Mr. N. I. Nazarov for their assistance in this work, as well as Mr F. Cotin for his great contribution to the preparation of the experiment.

References

Blioudov, V. A., Voloditchev, N. N., Grigorov, N. L., Kusine, Y. N., Likine, O. B., Netchaev, Y. Y., Podolsky, A. N., Savenko, I. A., Suslov, A. A., Ustinov, V. M., and Dritchikov, X.: 1974, *Geomagnetism i Aeronomiya* **13**, 1029.

Bostrom, C. O., Kohl, J. W., Mc Entire, R. W., and Williams, D. J.: 1972, 'The Solar Protons Flux – August 2–12, 1972', Preprint, The Johns Hopkins University, Silver Spring, Maryland.

Brown, T.: 1972, *Solar Phys.* **26**, 441.

Bui-Van, A., Giordano, G., Hurley, K., and Mandrou, P.: 1973, 'A Monte Carlo Program for Scintillation Counter Response, C.E.S.R. report 73–382 – August.

Castelli, T. P., Barron, W. R., and Aarons, T.: 1973, Solar Radio Activity in August 1972, AFCRL-TR-73-0086.

Chupp, E. L., Forrest, D. J., and Suri, A. N.: 1972, 'Solar Gamma Ray and Neutron Observations', Max Planck Institut für Physik and Astrophysik – M.P.I.-PAE/Extraterr. 77.

Chupp, E. L., Forrest, D. J., and Suri, A. N.: 1973a, Gamma Ray and Neutron Measurements and Their Relation to the Solar Flare Problem', 16th Meeting of Cospar, Konstanz, May–June.

Chupp, E. L., Forrest, D. J., Higbie, P. R., Suri, A. N., Tsai, C., and Dunphy, P. P.: 1973b, *Nature* **241**, 333.

Chupp, E. L., Forrest, D. J., and Suri, A. N.: 1974, Private communication.

Datlowe, D. W. and Peterson, L. E.: 1973, 'OSO-7 Observation of Solar X-Ray Bursts from 28 July to 9 August 1972', Report UAG-28, II, 291.

Dodge, J. C.: 1973, 'Interferometric Radio Spectrum of the Solar Corona, 1–11 August 1972 in Collected Data Reports on August 1972 – Solar Terrestrial Events', World Data Center, A, Report UAG-28, Part I, 242.

Grigorov, N. L., Melioransky, A. S., Nazarov, N. N., Pankov, V. M., Savenko, I. A., and Spirnov, S. P.: 1974, *Geomagnetism i Aeronomiya*, (in press).

Kudriavtsev, M. I., Likine, O. B., Melioransky, A. S., Savenko, I. A., Smirnov, V. V., and Chamoline, V. M.: 1973, *Geomagnetism i Aeronomiya* **13**, 406.

Lincoln, J. V. and Leighton, H. J.: 1972, 'Preliminary Compilation of Data for Retrospective World Interval July 26–August 14, 1972', World Data Center A., Report UAG-21, 1972a.

Ramaty, R. and Lingenfelter, R. E.: 1973, Nuclear Gamma Rays from Solar Flare, NASA X-660-73-14.

Reppin, C., Chupp, E. L., Forrest, D. J., and Suri, A. N.: 1973, 'Solar Neutron Production during the Events on 04 and 07 August 1972', *13th Int. Cosmic Ray Conf.*, Denver, August 1973.

Straka, R. M.: 1970, 'The Use of Solar Radio Bursts as Predictors of Proton Event Magnitude, AFCRL Space Forecasting Research Note, June 1970.

Švestka, Z.: 1973, 'Critical Problems of Solar Flare Research', 16th Meeting of Cospar, Konstanz, May–June.

Valnicek, B., Farnik, F., Horn, J., Letfus, V., Sudova, J., Komarek, B., Engelthaler, P., Ulrych, J., Moucka, L., Fronka, O., Vasek, T., Beranek, I., Pich, J., and Zderadicka, J.: 1973, *Bull. Astro. Inst. Czech.* **24**, No. 6.

Van Beek, H. F., Hoyng, P., and Stevens, G. A.: 1973, 'Solar Flares Observed by the Hard X-Ray Spectrometer on Board the ESRO TD1-A Satellite', Report UAG-28, II, 319.

Vedrenne, G., Talon, R., and Cotin, F.: 1973, 'Experience SIGNE 1', – C.E.S.R. Report 73–364, Avril, Toulouse, France.

Wang, H. T. and Ramaty, R.: 1974, *Solar Phys.* **36**, 129.

Zirin, H. and Tanaka, K.: 1973, *Solar Phys.* **32**, 173.

HIGH ENERGY GAMMA-RAY RADIATION ABOVE 300 keV
ASSOCIATED WITH SOLAR ACTIVITY*

E. L. CHUPP, D. J. FORREST, and A. N. SURI

Dept. of Physics, University of New Hampshire, Durham, N.H. 03824 U.S.A.

Abstract. The present status of our knowledge concerning the production of gamma-ray lines and continuum during the impulsive phase of solar flares is reviewed. Our data in this field is based solely on the OSO-7 observations made in 1972, August 4 and August 7. The experimental data will be reviewed. These observations along with theoretical work of Ramaty and Lingenfelter (1973a, b) and the charged secondary observations of the Chicago group (Anglin *et al.*, 1973) lead to the investigation of different hypothetical models to explain the production of neutral and charged secondaries in the solar atmosphere. At the present time it is not possible to rule out the preflare and postflare accumulation models if all the data is considered. We will discuss the outstanding experimental questions to be answered in future investigations.

1. Introduction

The experimental investigations to detect solar neutrons and gamma rays which were reported in the literature by 1970, were reviewed previously by Chupp (1971). Up to that time there was no conclusive evidence for either solar neutron or gamma-ray fluxes. On the other hand, there were at least three highly disputed claims of observations of both solar neutrons and gamma-rays, all in times of modest or low solar activity. None of these 'possible' events occurred in coincidence with the optical phase in any flare. Nonetheless, since they are published as positive fluxes, we should keep the reports in mind and the conditions of solar activity under which they were observed. The Tata result of Apparao *et al.* (1966) was obtained under very quiet solar conditions; that of Daniel *et al.* (1967) was made several hours before a subflare. This result was seriously questioned by Holt (1967) since no neutron decay protons were seen by the OGO-A satellite which was in orbit at the time and should have seen them if the neutron flux was 10^{-1} neutrons $cm^{-2} s^{-1}$ as reported. This criticism has now been countered by Daniel *et al.* (1971) who have revised their result downward nearly an order of magnitude to 1.5×10^{-2} neutrons $cm^{-2} s^{-1}$ based on a new measurement of the atmospheric neutron flux which allowed them to convert the measured solar neutron counting rates to an absolute flux. In the case of gamma rays, Kondo and Nagase (1969) reported an extremely large (800%) increase in the gamma-ray flux (3–10 MeV) 10 min after a 1N flare and associated radio burst. The last positive report of gamma-ray increase was given by Hirasima *et al.* (1970), who reported a gamma-ray line flux coincident with a 1000 MHz radio burst. As satellite experiments in the future continue to search for gamma-ray and neutron events, it will be interesting to see if any enhancements are found under similar activity conditions as in the cases just discussed, then we can decide if indeed these peculiar observations are most

---

* Dr R. Ramaty was kind enough to present this paper in absence of all the authors.

probably positive or spurious. A detailed discussion of recent work and several other experiments may be found in Chupp *et al.* (1973).

II. OSO-7 Gamma-Ray Observations in August 1972

The only evidence we found for gamma rays associated with solar flares was during the August 4 and 7 events. No description of the University of New Hampshire OSO-7 instrument will be given here since this has been described thoroughly elsewhere (e.g. Higbie *et al.*, 1973; Chupp *et al.*, 1973).

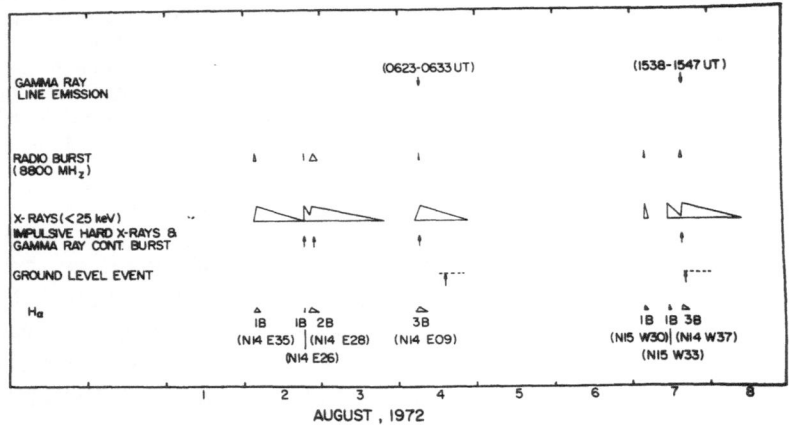

Fig. 1.   Event chronology – August 1972.

Figure 1 shows a chronological history of several associated phenomena during the first two weeks of August 1972. As noted, gamma rays were observed during the beginning phase of the 0621 UT August 4 flare in close time association with a radio burst, an X-ray burst and the *H*α flash. In this case a ground-level cosmic-ray event was observed, delayed by ~8 h from the flare. On August 7 gamma rays were observed after the maximum phase of the 1500 UT flare just after the satellite emerged into daylight.

(a) THE AUGUST 4 EVENT

The flare activity on August 4 started with a precursor flare in the X-ray band (0.5–3 Å) at 0507 UT (Dere *et al.*, 1973). The precursor activity continued for about an hour until 0610 UT when the main flare started in the X-ray band 7.5–15 keV as recorded by the UNH X-ray detector. The main optical flare started in Hα at ~0621 UT. Before the OSO-7 satellite was eclipsed by the Earth at 0633.8 UT, excess gamma-ray line and continuum emission was recorded by the University of New Hampshire Gamma Ray Detector on OSO-7. Strong radio emission accompanied this event (Castelli *et al.*, 1973; Croom and Harris, 1973). There is very good correlation between the gamma ray continuum observed on OSO-7 and the impulsive radio emission

(Suri *et al.*, 1975). The onset as well as the time of maximum are the same within a minute.

Figures 2 and 3 show plots of the intensity-time profiles of the event in different X-ray and gamma-ray energy regions on a long and short time scale, respectively. The observational data is incomplete because of:

(1) Saturation in the lower 3 X-ray channels.

(2) Discontinuity in the data near the peak of the impulsive phase when the satellite was eclipsed by the Earth.

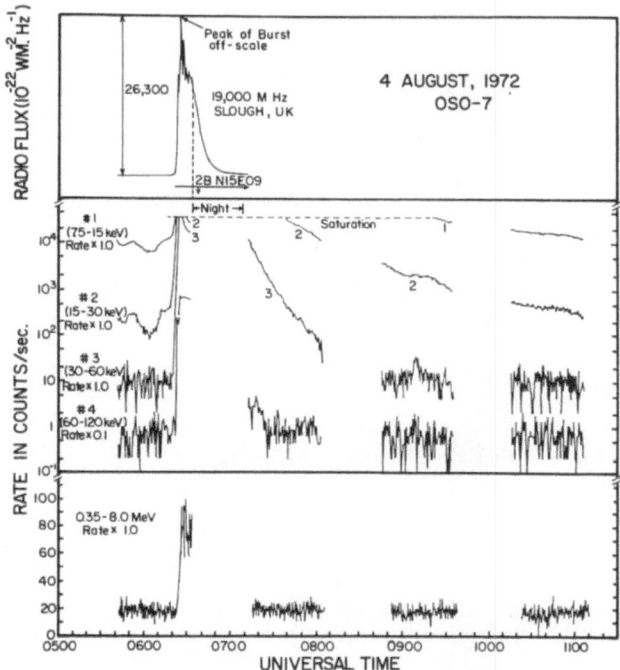

Fig. 2.   Coarse time history – 1972, August 4; X-rays, γ-rays and radio emission.

It can be seen from Figure 3, however, that the X-ray flux increases approximately exponentially with time with an *e*-folding rise time, which decreased with increase in the X-ray energy. Also, the onset of the impulsive phase occurs earlier at lower X-ray energies.

(i) *Gamma-Ray Line Emission*

Figure 4 shows the time integrated solar and background gamma-ray counting rate spectrum accumulated during the time interval ∼ 06 24–06 33 UT. The ordinate shows the total number of counts accumulated in each channel during the total live time of 91.4 s for the solar quadrant. The total number of counts in each channel up to channel 200 is shown and the sum of the counts in five consecutive channels there-

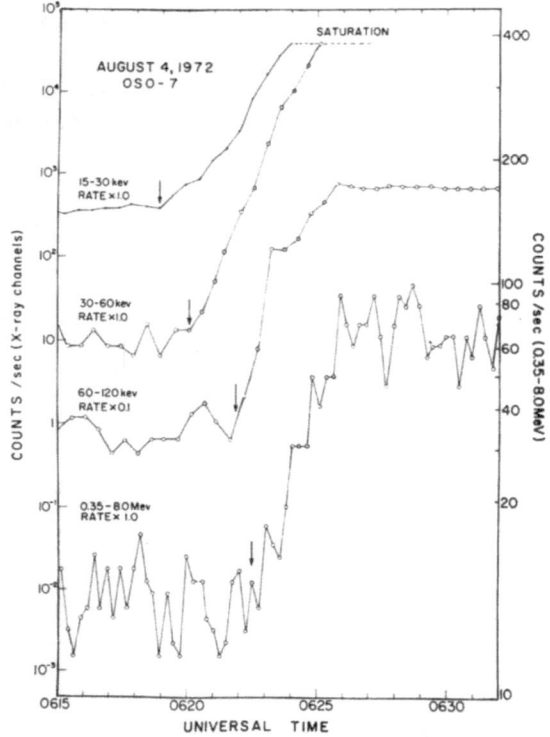

Fig. 3.   Fine time history – 1972, August 4, X-rays, γ-rays.

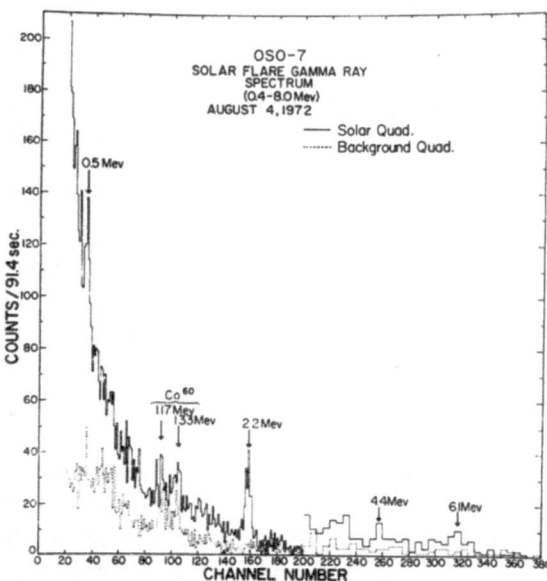

Fig. 4.   Complete gamma-ray spectrum – 1972, August 4; 0624–0633 UT.

after. The background spectrum has been normalized to the live time shown in the figure.

The flare spectrum shows a clear enhancement of the counting rate in both the 0.5 and 2.2 MeV spectral regions. The energy positions of these lines has been established from the calibration spectra and are at energies 510.7 ± 6.4 keV and at 2.24 ± 0.02 MeV. The 2.2 MeV line is about 15 σ above the continuum. The 0.5 MeV line is somewhat less significant but there is no question about its presence at about the 4 σ level. For the 0.5 MeV line a contribution in the background quadrant has been subtracted.

TABLE I

Flux Values for 1972, August 4 and 7

Time of flare observations	Gamma-ray flux at 1 AU (photons cm^{-2} s^{-1})			
3B (Hα) 1972, August 4, (0623:49–0633:02) UT Hα max – 0630 UT	510.7 ± 6.4 keV $(6.3 \pm 2.0) \times 10^{-2}$	2.24 ± 0.02 MeV $(2.80 \pm 0.22) \times 10^{-1}$	4.4 MeV $(3 \pm 1) \times 10^{-2}$	6.1 MeV $(3 \pm 1) \times 10^{-2}$
3B (Hα) 1972, August 7, (1538:20–1547:33) UT Hα max – 1530 UT	508.1 ± 5.8 keV $(3.0 \pm 1.5) \times 10^{-2}$	2.22 ± 0.02 MeV $(6.9 \pm 1.1) \times 10^{-2}$	4.4 MeV $<2 \times 10^{-2}$	6.1 MeV $<2 \times 10^{-2}$

The line features at 4.4 and 6.1 MeV are less significant ($\sim 3 \sigma$) and do not stand by themselves. Their presence is indicated in Figure 4 because these are the most intense deexcitation lines from ^{12}C (4.4 MeV) and ^{16}O (6.1 MeV) and are expected to be produced in solar flares. Table I summarizes the excess average flux above the gamma-ray continuum in three full spectral scans for the four peaks mentioned above. The excess counting rates in the peaks were obtained by *first subtracting the background* quadrant counting rates from the solar quadrant counting rates and then fitting a function of the form

$$N(n) = A_1 + A_2 n + A_3 n^2 + B \exp\left[(n - n_0)^2 / 2 \sigma^2\right] \tag{1}$$

to the spectral data using 20–30 channels around the gamma-ray peak position. This function represents a Gaussian peak superimposed on a quadratic continuum.

The normalizing peak number B, the line widths, σ, and the parameters, A_1, A_2, and A_3 were varied to find the best fit as determined by a minimum in chi-square.

Table I also gives the flux values at the Earth for the spectral features at 4.4 MeV (^{12}C) and 6.1 MeV (^{16}O).

(ii) *Time Profiles of the Positron Annihilation and Neutron Capture Lines*

Figure 5 shows the intensity-time profiles of the 0.5 and 2.2 MeV lines observed during the impulsive phase of the August 4 event. The time resolution of the instrument (3 min) and poor statistics (particularly for the 0.5 MeV line) do not allow us

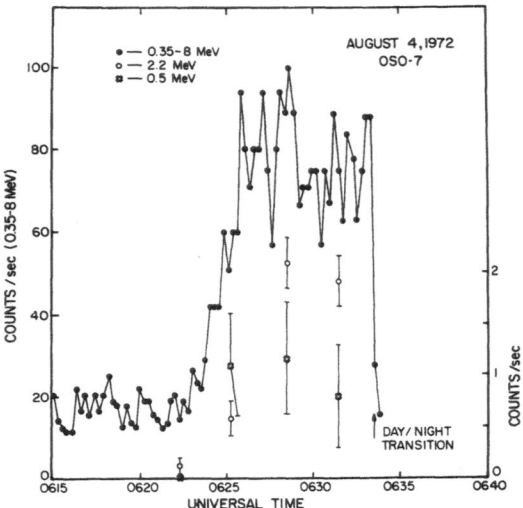

Fig. 5.   Fine time history – 1972, August 4; 0.5 and 2.2 lines.

to draw any final conclusions about the history of the production of these lines. However, we can say that

(1) The production of these lines takes place in coincidence with the impulsive hard X-rays and gamma-ray continuum.

(2) The 0.5 and 2.2 MeV line radiation rises to its maximum values in 3–6 min. The observed time history of the 2.2 MeV line shows that a reasonable value of the capture time of the neutrons in the photosphere is 100 ± 50 s (Reppin *et al.*, 1973).

(iii) *Preflare Upper Limits*

According to the flare model proposed by Elliot (1969), the flare energy is stored as energetic protons in the flare region. These energetic protons acquire their energy through a slow acceleration process which could be operating at the flare site for hours or days. If this is the case, then a weak emission of gamma rays could be taking place in the flare region prior to the onset of a flare.

We have searched the data for gamma-ray line radiation prior to the start of the August 4 event. No evidence was found for the emission of 0.5, 2.2, 4.4, and 6.1 MeV lines during the period 1437–2110 UT on August 3 and 0540–0618 UT on August 4. Data were rejected during the time the spacecraft repeatedly went through the South Atlantic anomaly.

The 2 σ upper limit fluxes are given in Table II. The upper limit fluxes were obtained using the relation

$$F_\gamma \leqslant \frac{2\,\sigma}{S} \leqslant \frac{2}{S}\left[\frac{R_S}{T_S} + \frac{R_B}{T_B}\right]^{1/2} \tag{2}$$

where R_S is the counting rate during the period T_S when the detector was pointing at the Sun. S is the sensitivity for a particular gamma-ray line and R_B is the counting rate when the Sun was not in the field of view of the detector for a time T_B.

Since the background quadrant counting rate is contaminated with the atmospheric radiation when the solar quadrant contains the Sun, we have taken R_B to be the background quadrant counting rate during the satellite night (looking away from the Earth).

TABLE II

Upper limits on gamma-ray line emission prior to
onset of the August 4 event
(photons cm^{-2} s^{-1})

0.5 MeV	2.2 MeV	4.4 MeV	6.1 MeV
$\leqslant 8.3 \times 10^{-3}$	$\leqslant 6.2 \times 10^{-3}$	$\leqslant 4.5 \times 10^{-3}$	$\leqslant 4.1 \times 10^{-3}$

(iv) *Shape of 0.5 MeV Line*

The possibility of observing thermal Doppler broadening in gamma ray lines produced during solar flares has been discussed by Kuzhevskii (1969) and Cheng (1972). The observation of these lines during the August 4 event allow us to put a limit on the temperature of the flare region in which these lines are produced.

Figure 6 shows the 0.5 MeV peak observed during the August 4 flare obtained by subtracting the background quadrant data from the solar quadrant data, and then subtracting a fit to the γ-ray continuum below the peak. The remaining peak was best

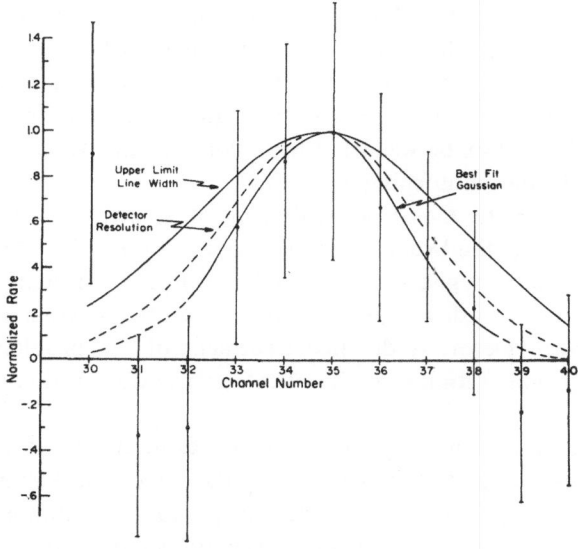

Fig. 6.   Fit of excess counts to 0.51 MeV line – 1972, August 4.

fitted by a Gaussian curve with a FWHM of 7.4%. The fact that the measured width (7.4%), within the uncertainty of the measurement, is the same as the expected width (8.8%), shows that there is no additional broadening of the positron annihilation line due to thermal effects. This gives an upper limit temperature in the flare region where annihilation takes place to be $<1 \times 10^7$ K at a 3σ confidence level.

(v) *Gamma-Ray Continuum*

In addition to the gamma-ray line emission, we also observed gamma-ray continuum emission extending up to 7 MeV as seen in Figure 4. The differential photon spectrum derived from Figure 4 along with that observed on TD-1A at lower energies (Van Beek, 1973) is shown in Figure 7. The spectrum shown was obtained by first subtracting the background quadrant counting rate spectrum from the solar quadrant counting rate spectrum. The counting rate contributions from gamma-ray lines at 0.5, 1.6, 2.2, 4.4, and 6.1 MeV were then subtracted including the Compton continuum at lower energies associated with the photopeaks. The photon spectrum incident on the detector was then found by transforming this counting rate spectrum, with the known line contributions removed, through the detector response by the 'strip-off' method (Burrus, 1960). The first step consisted of obtaining a reasonably accurate measure of the Compton continuum and of the first and second escape peaks for various energies of interest for the OSO-7 detector. Response function data collected by Higbie *et al.* (1973) at several energies were used for this purpose.

The Compton continuum correction was less than 20% of the observed counting rate at energies less than 0.7 MeV. However, the correction was about 30, 50, 44, 33, and 11 percent for energy bins 1–2, 2–4, 4–5, 5–6, and 6–6.5 MeV, respectively.

There are two effects which can give rise to the flattening of the observed gamma-ray continuum spectrum between 1–2 MeV. The first one is the Compton scattering of the 2.2 MeV upward moving photons in the photosphere which escape from the Sun. Wang and Ramaty (1974) have carried out Monte Carlo calculations on the transport of 2.2 MeV photons out of the photosphere and find that only a small fraction ($\sim 1\%$) of the observed flux between 1–2 MeV could be due to Compton scattering of 2.2 MeV photons in the photosphere.

The second effect is the Compton scattering of 2.2 MeV photons in the Earth's atmosphere. Because of the broad angular response ($\sim 100°$) of the instrument, it is likely that the detector registered scattered atmospheric radiation during a part of the event just prior to day/night transition when the Earth's atmosphere was in partial view of the detector. A detailed treatment of the problem is complicated; therefore, only a rough estimate of its effect on the observed continuum spectrum is made.

Considering the case of grazing incidence, we estimate that the flux of scattered photons in the energy range 1–2 MeV falling on the detector is 0.02 photons cm^{-2} s^{-1}. This is approximately 2% of the observed gamma-ray continuum flux between 1–2 MeV. It thus appears that the change in the spectral shape ~ 700 keV is real and not a local atmospheric effect.

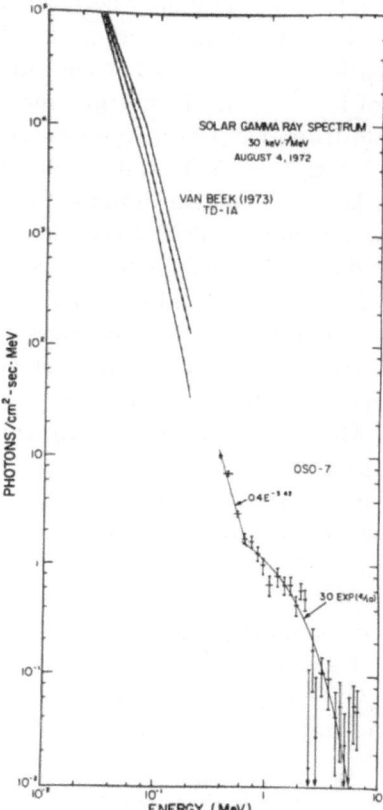

Fig. 7.   Solar γ-ray continuum in 1972, August 4; 30 keV to 7 MeV.

In the energy range 360–700 keV, the differential photon spectrum in Figure 7 was fit to a power law

$$dJ/dE = 0.4\, E^{-3.42 \pm 0.3} \text{ photons cm}^{-2}\text{ s}^{-1}\text{ MeV}^{-1} \qquad (3)$$

by the least square method. For gamma-ray energies in the range 0.7–7 MeV, however, a single power law is a poor fit. In this energy range the data was fit by the weighted least squares method to an exponential of the form

$$dJ/dE = k_1 \exp(-E/E_0) \text{ photons cm}^{-2}\text{ s}^{-1}\text{ MeV}^{-1}, \qquad (4)$$

where $E_0 = (1.0 \pm 0.07)$ MeV.

In the energy range 30–203 keV, hard X-rays from this event were observed by the University of Utrecht detector on the TD-1A satellite (Van Beek, 1973; Van Beek et al., 1973). The differential photon spectrum in the energy range 30–203 keV as observed on TD-1A is also shown in Figure 7.

The UNH X-ray data is not shown in Figure 7 because the X-ray detector saturated during the impulsive phase. The Utrecht detector on TD-1A did not saturate and also the pulse pile-up contamination is less than a few percent (Van Beek, 1973).

A power law spectrum of index 3.4 in the energy range 360–700 keV is consistent with the observations made from TD-1A at lower energies as shown in Figure 7. Over the energy range 29–203 keV, the TD-1A data (Van Beek, 1973) was fitted to a combination of two power law spectral distributions. Van Beek (1973) selected eight time intervals during the period 0623.5–0630.5 UT and determined parameters k and γ of the photon spectrum below and above the break. The spectral index varied from 2.7–3.5 below the break to 3.5–5.1 above the break. This is shown by the hatched area in Figure 7. The solid line in the hatched area represents the average spectral shape averaged over eight time intervals as given by Van Beek (1973).

The time averaged differential photon spectrum for the 1972, August 4 event over the energy range 30 keV–7 MeV shows two basic features:

(1) A change in the slope at 80–100 keV (Van Beek, 1973) when the power law spectral index changes from 3 to 3.9.

(2) A change in the spectral shape at ~ 700 keV.

(b) AUGUST 7 EVENT

The August 7 event was the second large flare of the August solar activity that gave evidence for the emission of gamma-ray lines in the solar flares. The flare began in Hα at 1455 UT when the OSO-7 spacecraft was behind the Earth. Approximately 40 min after the onset of the flare, the spacecraft emerged into sunlight and enhanced

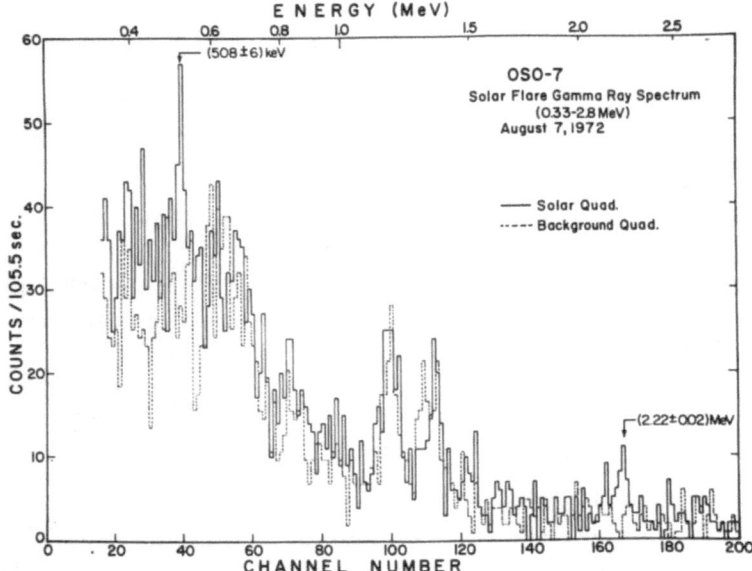

Fig. 8.   Complete gamma-ray spectrum – 1972, August 7; 1538–1547 UT.

counting rates in the spectral region around 0.5 and 2.2 MeV were observed. A final analysis of this data has not yet been completed so we will present only the preliminary results.

Figure 8 shows the solar and background spectra after the satellite emerged into daylight. Three full spectral scans are summed together covering the time interval from 1538.20–1547.33 UT. The lines at 0.5 and 2.2 MeV are the only lines clearly evident in the solar quadrant compared to the background quadrant. Table I gives a summary of the average gamma-ray line fluxes for the August 4 and 7 events.

III. Discussion and Interpretation of the August 1972 Solar Events

About seven years ago fairly detailed calculations carried out by Lingenfelter and Ramaty (1967) made predictions of the yield of the neutral secondaries at the Earth. In 1958, Severny predicted the production of neutrons associated with thermo-nuclear reactions occurring in shock fronts in the plasma associated with a solar flare, which could produce gamma rays from neutron-proton capture. Kuzhevskii (1969) has also made estimates of solar flare gamma-ray line fluxes.

Following the OSO-7 August 1972 observations, Ramaty and Lingenfelter (1973a, b) have extensively revised their calculations and this work provides the main basis on which interpretation can be made. It should be noted, however, that there is no complete geometrical flare model available at this time with which one can make more refined calculations.

The calculations of Lingenfelter and Ramaty (1967) and Ramaty and Lingenfelter (1973a, b) make the basic assumption that the differential spectrum of charged particles at the Sun is of the form $dN/dP \propto \exp(-P/P_0)$. The parameter P_0 determines whether the spectrum is a relatively hard or soft spectrum of charged particles. Ramaty and Lingenfelter have used the basic cross sections that are available in the literature and inferred ones when no experimental values are available.

Gamma ray yields have been calculated for a variety of production modes as shown in Table III. The yields of gamma rays have been calculated for two cases:

(1) The thick-target case, which assumes that the accelerated particles are undergoing nuclear interactions as they slow down and stop in sufficiently dense solar atmosphere. For example, in neutral hydrogen the amount of matter required to stop a 50 MeV proton is 1 gm cm^{-2}.

(2) The thin-target case, which assumes that the spectrum of charged particles is not modified as nuclear interactions are taking place. This means that the path length or amount of matter traversed by the particles is small compared with their nuclear interaction length or that ionization energy loss is just balanced by the gain in energy from a 'continuously' operating acceleration process.

These calculations also make the implicit assumption that in the thin target case the whole process is isotropic. This point is important when considering the production of high energy neutrons (>50 MeV) and π^0 mesons which preserve the momentum of the incident charged particles. Therefore, the neutrons in the first case may be emitted in a highly anisotropic manner from the Sun (see Chupp, 1971). Detection of these radiations at the Earth may be difficult, therefore, if the charged-particle acceleration process is highly anisotropic at the Sun.

The lower energy gamma rays listed in Table III should all be isotropically emitted independent of the angular distribution of the incident charged particles. The strong magnetic fields in flare regions might conceivably produce some effects on these lower energy gamma rays, but this refinement has not been made in the Ramaty and Lingenfelter calculations.

TABLE III

Gamma-ray line emission mechanisms from Ramaty and Lingenfelter (1973a)

Photon energy (MeV)	Origin	Production mode
0.511	Positron annihilation	$e^+ + e^- \rightarrow 2\gamma$
2.23	Deuterium de-excitation following neutron capture	$H' + n \rightarrow H^2 + \gamma$
1.63	$Ne^{20\,(1.63)}$ deexcitation	Ne^{20} (p, p') $Ne^{20(1.63)}$
	$N^{14(3.94)} = N^{14(2.31)}$ de-excitation	N^{14} (p, p') $N^{14(3.94)}$
2.31	$N^{14(2.31)}$ de-excitation	N^{14} (p, p') $N^{14(2.31)}$
		N^{14} (p, p') $N^{14(3.94)} \rightarrow N^{14(2.31)}$
		N^{14} (p, n) $O^{14} \rightarrow N^{14(2.31)}$
4.43	$C^{12(4.43)}$ de-excitation	C^{12} (p, p') $C^{12(4.43)}$
		O^{16} (p, pα) $C^{12(4.43)}$
5.2	$O^{15(5.20)}$ de-excitation	O^{16} (p, pn) $O^{15(5.20)}$
	$N^{15(5.28)}$ de-excitation	O^{16} (p, 2p) $N^{15(5.28)}$
6.14	$O^{16(6.14)}$ de-excitation	O^{16} (p, p') $O^{16(6.14)}$
7.12	$O^{16(7.12)}$ de-excitation	O^{16} (p, p') $O^{16(7.12)}$

We will not describe in further detail here the Ramaty and Lingenfelter calculations, but will show how these have been used to interpret the OSO-7 observations described above. Shown in Figure 9 is the theoretical ratio of a given gamma ray yield to the theoretical yield of 2.2 MeV gamma rays plotted versus the characteristic rigidity P_0 of the charged particles spectrum for a thick-target interaction model. The broad horizontal bands shown in the graph correspond to the experimentally measured OSO-7 gamma-ray flux ratios at the indicated energies with associated statistical errors in the flux measurements for the August 4 event.

The corresponding curve for the thin-target interaction model is shown in Figure 10. From either Figure 9 or Figure 10, it can be seen that a characteristic rigidity of the protons at the Sun of the order 70–100 MV is required to obtain a consistency for the observed ratio for all the measured gamma-ray yields. Recent slight modifications in the calculations and the experimental results do not change this conclusion*.

This range of P_0's agrees reasonably well with that deduced by Ramaty and Lingenfelter from the protons observed near the Earth by various spacecraft (cf. Kohl *et al.*, 1973). Thus, the gamma ray observations and the particle observations support the view that the August 4 flare was relatively soft in terms of accelerated particle energies.

* See notes added in proof.

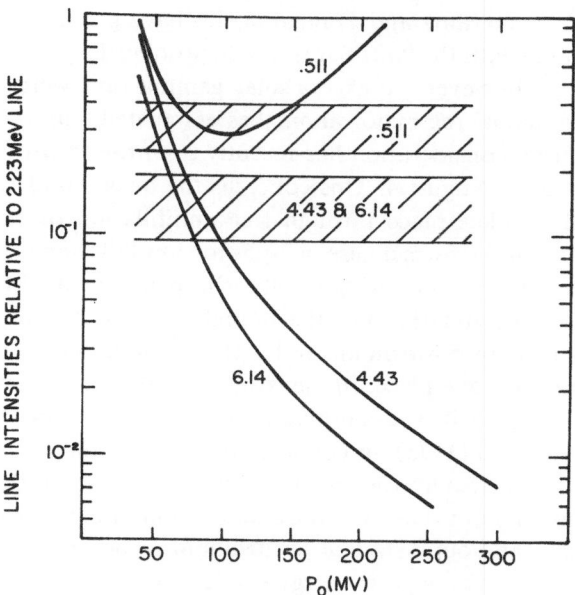

Fig. 9.   Theoretical relative intensity of γ-ray line flux relative to the 2.2 MeV theoretical flux in thick-target geometry and compared with experimental results.

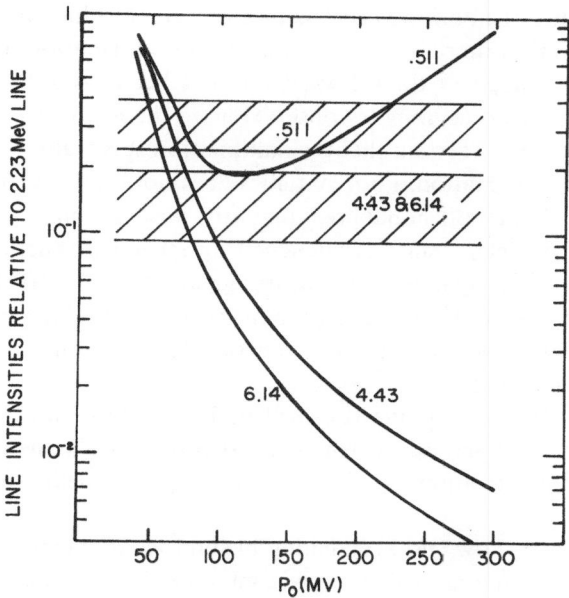

Fig. 10.   Theoretical relative intensity of γ-ray line flux relative to the 2.2 MeV theoretical flux in thin-target geometry and compared with experimental results.

On the other hand, Pomerantz and Duggal (1973) have given evidence for a harder component (associated with the 0621 flare), which produced a ground-level cosmic-ray event some 6 h later; however, no excess solar gamma rays were seen at that time. This observation is one of the major anomalies associated with the August 4 events.

A further interesting consideration has recently been raised with regard to the line at 0.5 MeV observed on August 4. This concerns the question of whether or not the positron annihilation takes place through free annihilation or through the bound state of positronium. In the former case, a 2-photon annihilation is the most predominant gamma-ray spectrum seen and gives the sharp line at 0.5 MeV. On the other hand, if annihilation occurs through the bound state, both 2-photon annihilation through the singlet state of positronium and 3-photon annihilation through the triplet state of positronium can take place. In the latter case, the gamma-ray spectrum is not a single line spectrum, but it is a continuous spectrum extending from the 0.5 MeV line downward. Leventhal (1973), in connection with the study of positron annihilation in a low density astrophysical medium where the atomic density is less than 10^{15} atoms cm^{-3}, points out that the continuum gamma-ray spectrum resulting from annihilation through the bound state of positronium could be predominant. If one is using a detector with relatively poor energy resolution, the apparent peak position of the 0.5 MeV line can be shifted to a lower energy as a result of the folding of the instrument resolution in with the triplet continuum spectrum and singlet positronium line spectrum.

As noted earlier, the energy of the 0.5 MeV line corresponds to 0.51 MeV within an error of ~ 5 keV. Annihilation through positronium formation could shift our photopeak to a lower limit of about 505 keV within our error. Thus one cannot conclude absolutely from this that there is not some positronium formation present in the spectrum observed. The shape of the 0.5 MeV line has been carefully studied and it is concluded that there is no asymmetry observable on the low energy side. This allows us to state that we are 99% certain that the annihilation spectrum we see is not a result of 100% positronium formation. Poor statistics does not allow putting a specific limit on the amount of P_s. Thus collisional breakup of the positronium at $n_{\text{atomic}} > 10^{15}$ cm^{-3} is possible, or the triplet P_s state is quenched by strong magnetic fields, or the P_s formation rate is reduced by flare temperatures greater than 7×10^5 K as suggested by Ramaty and Lingenfelter (1973b). Thus presumably we are looking here at a free annihilation spectrum; however, the limited statistics do not permit a definitive answer to this interesting question.

A definite resolution of this question will undoubtedly require observations with solid-state detectors. It is perhaps worth noting that there are many ways of quenching the triplet state of positronium, such as strong magnetic fields in the flare for the $m = 0$ state.

Since the line width shows no broadening beyond what is expected from the measured detector resolution, one is able to place an upper limit on the temperature of the annihilation region by considering a maximum width based on the error in our resolution measurements. This point was discussed above under the experimental observations. From the experimental observation that the 0.51 MeV line was evident within 200 s of the flare onset, the electron density in the annihilation region must be $\gtrsim 10^{12}$ electrons cm^{-3} since the mean capture time is $\alpha\ n_e^{-1}$.

Wang and Ramaty (1974) have recently carried out a detailed study on the time history of neutrons produced in nuclear reactions above the solar photosphere. Using Monte Carlo calculations they explore the fate of mono-energetic groups of neutrons produced isotropically, half of which go into the photosphere. This work takes into account several factors which include an ambient ^{3}He/H ratio of the solar photosphere, the radioactive decay of neutrons, the escape of neutrons directly to the Earth, and the escape of neutrons which scatter in the photosphere and then leave the Sun.

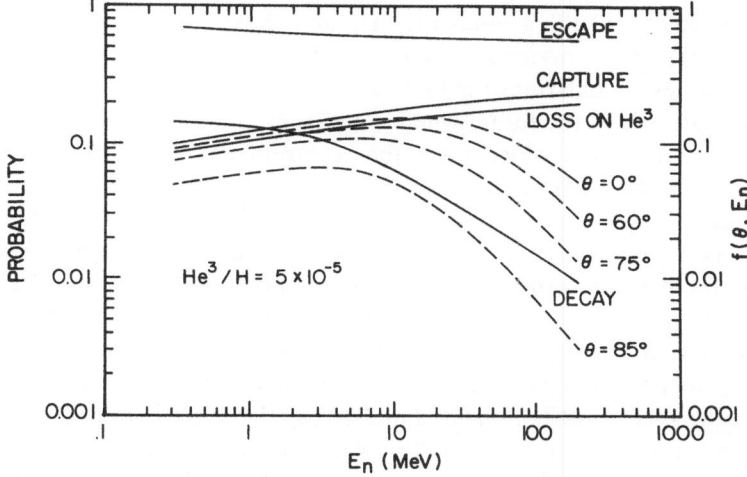

Fig. 11.   Neutron fate in the photosphere versus neutron energy on left ordinate. On right ordinate, the probability of escape of the 2.2 MeV gamma ray vs. neutron energy and different angles of emission.

They then calculate the capture of neutrons on protons to give the yield of 2.2 MeV gamma rays taking into account the Compton scattering of the photons as they leave the region of capture. The loss of neutrons in capture on ^{3}He could reduce the predicted 2.2 MeV line intensity. The resulting Mont Carlo probabilities for the neutrons and the relative photon yields for various initial neutron energies are shown in Figure 11, which shows the case when the ^{3}He/H abundance is 5×10^{-5}. At low neutron energies and for an emission angle of gamma rays relative to the solar vertical given by the angle θ, the relative photon yield per neutron is close to the neutron capture probability on protons. This means that gamma rays from low energy neutrons observed close to the vertical, escape essentially unattenuated from the Sun. At higher energies and at larger angles; however, there is a significant attenuation of the gamma rays. In the previous Lingenfelter and Ramaty (1967) calculations, it was assumed that all downward moving neutrons are captured and all upward moving photons escape from the Sun. In this case the relative gamma flux per neutron should be $\frac{1}{2}$.

However, from Figure 11 we see that depending on the energy of the neutrons, the location of the flare of the Sun, which determines the angle θ and the amount of the ^{3}He in the photosphere, one can overestimate the gamma yield by at least a factor of 2.5. Furthermore, if the flare occurs close to the limb of the Sun, the 2.2 MeV line could become essentially unobservable. For such limb flares the 0.5 MeV line and the nuclear de excitation lines would still be observable if these lines are produced above the photosphere. The August 4 flare was near the central meridian $\theta \sim 0$.

In case the ^{3}He/H ratio is zero, the relative gamma yields rise as also shown by Wang and Ramaty (1974). A basic result of these new detailed Monte Carlo calculations on the fate of neutrons produced in solar nuclear reactions is to reduce the calculated 2.2 MeV gamma-ray flux at the Earth from isotropically emitted neutrons from 50% to 20% depending upon the ^{3}He abundance as well as the angle or the amount of photospheric material. The time history of the 2.2 MeV line is also determined in these new calculations for each neutron energy. These Monte Carlo calculations will be valuable in the future when experiment and theory lead us to a more detailed acceleration model which gives the spectrum and time history of the solar cosmic rays at the Sun. These calculations also suggest that a limit may be placed on the ^{3}He/H abundance.

There are two other important considerations relating to these gamma-ray measurements. One point is concerned with the absolute number of protons required at the Sun in order to produce the observed yield of gamma rays at the Earth. The other is to take into account the observations on the charged secondary particles such as ^{3}He, deuterons, and triton nuclei, which are produced in similar nuclear reactions and observed at the space probes Pioneer 9 and 10 (by the University of New Hampshire and GSFC groups) and the IMP satellites 4, 5, and 6 (Anglin *et al.*, 1973). On the first point, there is a possibility of a discrepancy between the number of protons (> 30 MeV) required at the Sun to produce the August 4 gamma-ray lines and the number of protons seen at the Earth by various spacecraft. There may be 10–10^3 more protons observed near the Earth than required at the Sun as discussed by Ramaty and Lingenfelter (1973a, b) and Forrest *et al.* (1974). On the second point, Forrest *et al.* (1974) have argued that if one takes into account the ^{3}He secondary production observed by Pioneer 10 as well as the gamma-ray observations, then ~ 1 gm cm^{-2} of solar material must be traversed in order to explain both the charged secondary and the gamma-ray yields.

Another model that must be considered is the preflare acceleration model, such as envisioned by Elliot (1973) in which the solar cosmic-ray particles are accelerated over a long period of time of the order of days in a relatively thin solar atmosphere, and the flare phenomena is a manifestation of the release of these particles. The observed gamma rays could be produced in a thick target situation with the ^{3}He and other charged secondaries having been produced prior to the flare but released at the time of the impulsive flare. A density of 10^6–10^7 particles cm^{-3} in the solar atmosphere would be the medium in which such preflare acceleration could occur. This particle density is constrained by the total number of SCR for a thin-target situation and the absence of an observable preflare gamma-ray flux. The time required in order to integrate a path length of 1 g cm^{-2} amounts to something of the order of 10–100 days. This is too long a time because the drift of the charged particles across magnetic field lines would undoubtedly release them from any reasonable size trapping region. Clearly, the detailed understanding of the solar cosmic rays and the secondary yields of positrons and neutrons giving the gamma rays and the charged He, D, and T isotopes is in too primitive a state in order to completely determine a model at this time.

Most of the reactions producing gamma-ray lines that have been considered in the theoretical calculations have been due to direct proton reactions on ambient solar nuclei or reactions of protons or α particles producing neutrons, π^+ and π^0 mesons, and β^+ emitters which eventually give rise to the 2.2 MeV gamma ray, the 0.51 MeV gamma ray, and π^0 decay gammas of average energy 70 MeV. Recently Kozlovsky and Ramaty (1974) have considered production of gamma-ray lines by α–α reactions. In particular the following two reactions can produce lines at 431 keV and 478 keV.

$$
\begin{aligned}
&^4\text{He}\,(\alpha,\, n)\ ^7\text{Be*} && 431\ \text{keV} \\
&^4\text{He}\,(\alpha,\, n)\ ^7\text{Be} \xrightarrow[\varepsilon]{12\%} {}^7\text{Li*}\ 478\ \text{keV} && && (5) \\
&^4\text{He}\,(\alpha,\, p)\ ^7\text{Li*} && 478\ \text{keV}
\end{aligned}
$$

The first two reactions have thresholds at 9.7 MeV nucleon^{-1} and 8.5 MeV nucleon^{-1}, respectively, and the last reaction has a threshold at 8.5 MeV nucleon^{-1}. Kozlovsky and Ramaty argue that the cross-sections of all these reactions is ~ 100 mb at 10 MeV nucleon^{-1} and that the production cross-section of ^{7}Be in the ground state and first excited states are the same. They conclude that the intensities of these two lines from α–α reactions in solar flares should be as large or larger than the intensities of the 4.43 MeV line or 6.14 MeV line. Even though the OSO-7 flare spectrum on August 4 shows a suggestion of features at about the channels corresponding to 431 keV, they are not statistically significant. The question will have to be resolved by future gamma-ray experiments.

IV. Summary and Conclusions

We list in Table IV a summary of the principal conclusions that can be made from the gamma ray observations on August 4.

The OSO-7 observations give evidence for the emission of gamma-ray lines in only two of the largest flares of the August 1972 series. Therefore, the most critical need for future experiments is to make more frequent measurements and with higher sensitivity instrumentation in order to obtain the time history of the gamma-ray lines, especially those from de-excitation of excited nuclei in C, N, O, etc. In addition, gamma-ray detectors of much higher energy resolution are needed in order to fully investigate the line shapes of the gamma-ray lines that are produced so the positronium question can be studied and possible Doppler broadening and Doppler shifts determined.

TABLE IV

Principal conclusions from gamma-ray observations in 1972, August 4

Gamma-ray producing nuclear reactions begin in the first 200 s with the hard X-ray and before the optical maximum.

Nuclear reactions occur for $\geqslant 600$ s.

Density of annihilation region $\gtrsim 10^{12}$ (elec cm^{-3}).

Temperature in $e^+ + e^- \to 2 \to 3 \gamma$ region $< 10^7$ K.

Low-energy primary spectrum at Sun for gamma-ray production consistent with prompt low-energy spectrum seen at Earth and in space.

Total particle energy in thick target dump $< 10^{28}$ erg.

Acknowledgements

This work was supported by NASA under Contract Nas 5-11054 and NGR 30-002-021. We are thankful for the contributions of several persons in the fabrication and data analysis phase of the experiment, particularly Dr P. Higbie, A. A. Sarkady, I. U. Gleske, S. Foss, Sue Croteau, Margaret Simmons, Kishore and Dipika Patel, P. Ferguson, P. Dunphy and Dr C. Reppin. We also express our appreciation to Mary Miklos Chupp for editing the manuscript.

Notes added in proof: Some further revisions in the theoretical results shown in Figures 9 and 10 have been made by Ramaty and Lingenfelter (1975) and are discussed in another paper in this volume (page 363). These new calculations indicate that the characteristic rigidity P_0 for an exponential rigidity spectrum can nominally range from ~ 100–150 MV depending on thick or thin target assumptions, respectively. The experimental results can also be reconciled with a differential power law spectrum of form $E^{-\alpha}$ with the exponent ranging nominally from 2–3 depending on thin or thick target assumptions, respectively.

References

Anglin, J. D., Dietrich, W. F., and Simpson, J. A.: 1973, in R. Ramaty and R. G. Stone (eds.), *Symposium on High Energy Phenomena on the Sun*, NASA SP-342, p. 315.

Apparao, M. V. K., Daniel, R. R., Vijayalakshmi, B., and Bhatt, V. L.: 1966, *J. Geophys. Res.* **71**, 1781.

Burrus, W. R.: 1960, *IRE Trans. Nucl. Sci.* NS-7, 102.

Castelli, J. P., Barron, W. R., and Badillo, V. L.: 1973, World Data Center A, *Report UAG* **28**, Part 1, 183.

Cheng, C. C.: 1972, *Space Sci. Rev.* **13**, 3.

Chupp, E. L.: 1971, *Space Sci. Rev.* **12**, 486.

Chupp, E. L., Forrest, D. J., and Suri, A. N.: 1973, in R. Ramaty and R. G. Stone (eds.), *Symposium on High Energy Phenomena on the Sun*, NASA SP-342, p. 285.

Croom, D. L. and Harris, L. D. J.: 1973, World Data Center A, *Rept. UAG* **28**, Part 1, 210.

Daniel, R. R., Joseph, G., Lavakare, P. J., and Sunderrajan, R.: 1967, *Nature* **213**, 21.

Daniel, R. R., Gokhale, G. S., Joseph, G., and Lavakare, P. J.: 1971, *J. Geophys. Res.* **76**, 3152.

Dere, K. P., Horan, D. M., and Kreplin, R. W.: 1973, World Data Center A, *Rept. UAG* **28**, Part 2, 298.

Elliot, H.: 1969, in C. de Jager and Z. Švestka (eds.), *Solar Flares and Space Research*, North-Holland Publ. Co., Amsterdam, p. 356.

Elliot, H.: 1973, in R. Ramaty and R. G. Stone (eds.), *High Energy Phenomena on the Sun*, NASA SP-342, p. 12.

Forrest, D. J., Chupp, E. L., Suri, A. N., and Dunphy, P.: 1974, *Bull. Amer. Phys. Soc.* **19**, 458.

Higbie, P. R., Chupp, E. L., Forrest, D. J., and Gleske, I. U.: 1972, IEEE Trans. Nucl. Sci., *NS-19*, 606.

Higbie, P. R., Forrest, D. J., Gleske, I. U., Chupp, E. L., and Burtis, D. W.: 1973, *Nucl. Instrum. Methods* **108**, 167.

Hirasima, Y., Okudaira, K., and Yamagami, T.: 1970, *Acta Phys. Hung.* **29**, Suppl. 2, 683.

Holt, S. S.: 1967, *J. Geophys. Res.* **72**, 3507.

Kohl, J. W., Bostrom, C. O., and Williams, D. J.: 1973, World Data Center A, *Rept. UAG* **28**, Part 2, 330.

Kondo, I. and Nagase, F.: 1969, in C. de Jager and Z. Švestka (eds.), *Solar Flares and Space Research*, North-Holland Publ. Co., Amsterdam, p. 134.

Kozlovsky, B. and Ramaty, R.: 1974, *Astrophys. J. Letters* **191**, L43.

Kuzhevskii, B. M.: 1969, *Soviet Astron. − AJ* **12**, 595.

Lingenfelter, R. E. and Ramaty, R.: 1967, in B. S. P. Shen (ed.), *High Energy Nuclear Reactions in Astrophysics*, W. A. Benjamin Press, New York, p. 99.

Levanthal, M.: 1973, *Astrophys. J. Letters* **183**, L147.

Pomerantz, M. A. and Duggal, S. P.: 1973, World Data Center A, *Rept. UAG* **28**, Part 2, 430.

Ramaty, R. and Lingenfelter, R. E.: 1973a, in R. Ramaty and R. G. Stone (ed.), *High Energy Phenomena on the Sun*, NASA SP-342, p. 301.

Ramaty, R. and Lingenfelter, R. E.: 1973b, *Proc. 13th International Cosmic Ray Conf.*, Denver, Colo., **2**, 1590.

Ramaty, R. and Lingenfelter, R. E.: 1975, this volume, p. 363.

Reppin, C., Chupp, E. L., Forrest, D. J., and Suri, A. N.: 1973, *Proc. 13th International Cosmic Ray Conf.*, Denver, Colo., **2**, 1577.

Severny, A. B.: 1958, *Soviet Astron. − AJ* **2**, 310.

Suri, A. N., Chupp, E. L., Forrest, D. J., and Reppin, C.: 1975, submitted for publication.

Van Beek, H. F.: 1973, Ph.D. Thesis, Physics Dept., Univ. of Utrecht, Holland.

Van Beek, H. F., Hoyng, P., and Stevens, G. A.: 1973, World Data Center A, *Rept. UAG* **28**, Part 2, 319.

Wang, H. T. and Ramaty, R.: 1974, *Solar Phys.* **36**, 129.

MEASUREMENTS OF A GAMMA-RAY BURST ABOVE 1 MeV

R. KOGA, G. M. SIMNETT*, and R. S. WHITE

*Physics Dept. and Institute of Geophysics and Planetary Physics,
University of California, Riverside, Calif., U.S.A.*

Summary. Observations of a burst of gamma radiation, starting at 20 1247 UT, 1972, May 14 are reported. The measurements were made with a 0.5 m² actively shielded scintillator which was the front element of a double Compton telescope, during a balloon flight at an altitude of 5 mb from Palestine, Texas. The maximum intensity of the burst was 0.10 ± 0.02 cm^{-2} s^{-1} above 1 MeV, and the duration was 3.5 ± 0.4 min. The burst intensity during the first 2 min is constant to within 10%. No known electronic or detector malfunction could have produced this effect, and we believe it is a real event.

The origin of the event is still unexplained. The part of the detector used for this study has virtually an omnidirectional response. A search has been made for correlative evidence, but none has been found that is conclusive. The onsets of two subflares, at S13 E78 and N12, E08, occurred one minute after the rise in the gamma-ray data. A short lived soft X-ray burst was reported from Solrad 9 3 min after the gamma-ray burst. It cannot be ruled out that such associations are a chance coincidence. However, simultaneous subflares from separate plage regions are reported only every few days during this time period, so the probability of a chance coincidence is less than 0.001.

One possible explanation is that the simultaneous flares are caused by precipitating energetic protons and electrons from the high corona, with the bulk of the energy residing in the protons. Collisional bremsstrahlung from the electrons, leading to the gamma rays, might occur before the bulk of the protons lose their energy. In this way the delay in the Hα flares might be explained, but the delay in the soft X-ray emission is still puzzling.

We conclude that the origin of the gamma-ray burst is still unknown.

---

* Present adress: Department of Space Research, University of Birmingham, England.

GAMMA-RAY LINES FROM SOLAR FLARES

R. RAMATY

Laboratory for High Energy Astrophysics, NASA-Goddard Space Flight Center, Greenbelt, Md., U.S.A.

and

R. E. LINGENFELTER

Dept. of Planetary and Space Science, University of California, Los Angeles, Calif., U.S.A.

Abstract. We have treated in detail the theory of gamma-ray line production in solar flares. The strongest line, both predicted theoretically and detected observationally at 2.2 MeV, is due to neutron capture by protons in the photosphere. The neutrons are produced in nuclear reactions of flare accelerated particles which also produce positrons and prompt nuclear gamma rays. From the comparison of the observed and calculated intensities of the lines at 4.4 or 6.1 MeV to that of the 2.2 MeV line it is possible to deduce the spectrum of accelerated nuclei in the flare region; and from the absolute intensities of these lines it is possible to obtain the total number of accelerated nuclei at the Sun. The study of the 2.2 MeV line also gives information on the amount of He^3 in the photosphere. The study of the line at 0.51 MeV resulting from positron annihilation complements the data obtained from the other lines; in addition it gives information on the temperature and density in the annihilation region and on the anisotropy of the accelerated electron beam which produces continuum gamma rays at energies greater than about 1 MeV.

I. Introduction

Measurements of accelerated charged particles near the Earth clearly indicate that such particles are produced in great profusion in solar flares. These particles consist of both electrons and nuclei; but until the advent of solar gamma-ray astronomy, observations in the radio and X-ray bands had revealed only the existence of the electronic component in the flare region itself.

In a previous paper (Lingenfelter and Ramaty, 1967) we treated in considerable detail the nuclear reactions produced by accelerated charged particles in solar flares and we showed that in large flares these reactions produce detectable lines in the gamma-ray region. We found that the strongest lines should be at 0.5, 2.2, 4.4, and 6.1 MeV resulting from positron annihilation, neutron capture on hydrogen, and de-excitation of excited states in C^{12} and O^{16}, respectively.

The recent observations by Chupp *et al.* (1973) of the first gamma-ray lines from solar flares confirm these predictions. During the flash phase of the 1972, August 4 flare all of these lines were observed with relative intensities essentially consistent with our calculations.

Since these observations became available, several additional studies on gamma-ray line production in solar flares have been undertaken. Ramaty and Lingenfelter (1973a) have investigated the consistency of the observations with the theory of nuclear reactions in flares; these authors also considered the effects of positronium formation and neutron propagation in the solar atmosphere (Ramaty and Lingenfelter, 1973b); Reppin *et al.* (1973) treated the time dependence of the 2.2 MeV line; Wang and Ramaty (1974) have done a detailed calculation on neutron propagation and 2.2 MeV line formation and they pointed out the importance of photospheric He^3 as a non-radiative sink for the neutrons; Kozlovsky and Ramaty (1974a) have pointed out that $\alpha\alpha$ reactions produce the Li^7 and Be^7 lines at 478 keV and 431 keV which could be observable from flares; and Ramaty and Kozlovsky (1974) evaluated in detail the production of H^2, H^3 and He^3 in flares and they attempted to deduce the number of protons released from the flare of 1972, August 4 by combining the He^3 and gamma-ray observations.

In the present paper we wish to summarize the above material and to present updated calculations on the production of gamma-ray lines in solar flares. In Section II we define the interaction models that we use in our calculations; in Section III we consider neutron production and 2.2 MeV line formation; in Section IV we consider the production of prompt gamma-ray lines with special emphasis on the 4.4 MeV and 6.1 MeV lines for which observational data exists; in Section V we compare the results of Sections III and IV with data for the 1972, August 4 flare and we deduce the number and spectrum of accelerated particles at the Sun; in Section VI we treat problems concerning the formation of the 0.51 MeV line, and we summarize our results in Section VII.

II. Interaction Models

We consider two limiting interaction models (e.g. Ramaty and Lingenfelter, 1973a): A thin-target model in which the spectrum of accelerated particles is not modified during the time in which the nuclear interactions take place, and a thick-target model in which the accelerated particles move from the flare region downward into the Sun, undergoing nuclear interactions as they slow down in the solar atmosphere. In the thin-target model it is assumed that either the total path length traversed by the particles at the Sun is small in comparison with their interaction length, or that the particle energy loss from ionization and nuclear interactions is just balanced by energy gains from acceleration.

For the composition of the ambient solar atmosphere we use the abundances given by Cameron (1973). For the accelerated particle populations in both the thin- and thick-target models we consider power-law and exponential spectra. In the thin-target model these are

$$N_i(E) = k_i E^{-s}, \tag{1}$$

and

$$N_i(P) = k_i' \exp(-P/P_0), \tag{2}$$

respectively. Here $N_i(E)$ and $N_i(P)$ are the instantaneous numbers of accelerated particles of kind i in the interaction region per unit energy per nucleon, E, or unit rigidity, P; k_i and k_i' are constants determined by normalizing the N_i's to 1 proton of energy greater than 30 MeV and by using the composition of the ambient solar atmosphere; and s and P_0 are, respectively, the spectral index and characteristic rigidity assumed to be the same for all accelerated particle components. In the thick-target model we use expression similar to Equations (1) and (2), but we replace the instantaneous numbers N_i by total numbers, $\bar{N}_i$, such that $\bar{N}_i(N)$ and $\bar{N}_i(P)$ are the total number of accelerated charged particles per unit energy per nucleon or unit rigidity that are released from the flare region downward into the Sun. As with the instantaneous fluxes in the thin-target model, the normalizations of the $\bar{N}_i$'s are determined by using the composition of the ambient solar atmosphere and 1 proton of energy greater than 30 MeV.

In the thin-target model, the production rate of secondaries from a particular reaction is given by

$$q_i = n_i \int_0^\infty dE N_i(E)\, c\beta\, \sigma_i(E), \tag{3}$$

where n_i is the number density of target atoms in the solar atmosphere, $c\beta$ is particle velocity, and $\sigma_i(E)$ is the cross-section as a function of energy per nucleon. The units of q_i are secondary particles per second. In the thick-target model, the total production of secondary particles for a given reaction is

$$Q_i = \eta_i \int_0^\infty dE' \bar{N}_i(E') \int_0^{E'} dE\, \frac{dx}{dE}\, \sigma_i(E) \tag{4}$$

where η_i is the number of target nuclei per gram of solar material, and dE/dx is the stopping power of the primary particles in solar material due to both Coulomb and nuclear collisions. The quantities Q_i are the total number of secondaries. Expressions similar to Equations (3) and (4) can be written down for rigidity spectra.

By inverting the order of integration, Equation (4) can be written as

$$Q_i = \eta_i \int_0^\infty dE\, \frac{dx}{dE}\, \sigma_i(E) \int_0^\infty dE' \bar{N}_i(E'). \tag{5}$$

Because the inner integral is just the integral spectrum of $\bar{N}_i(E)$, Equation (5) can be further simplified,

$$Q_i = \eta_i \int_0^\infty dE\, \frac{dx}{dE}\, \sigma_i(E)\, \bar{N}_i(>E). \tag{6}$$

In our subsequent treatment of secondary particle production, we shall use Equations (3) and (6) or their equivalents for rigidity spectra.

III. Neutron and 2.23 MeV Gamma-Ray Production

Neutron production by accelerated charged particles was treated by Lingenfelter *et al.* (1965) and Lingenfelter and Ramaty (1967). Recently, Kozlovsky (1974, private communication) has updated the cross sections used by these authors and has added additional data at low energies including data on α-particle induced reactions. The neutron production cross-sections are shown in Figure 1. Here pp. pα, αα, pCNO and αCNO indicate neutron production in proton-hydrogen, proton-helium, α-particle-helium, proton-heavy nuclei, and α-particle-heavy nuclei reactions, respectively. The latter two cross-sections are the neutron production cross-sections for all nuclei with $A \geqslant 12$, normalized to one such nucleus by using the elemental and isotopic abundances of Cameron (1973).

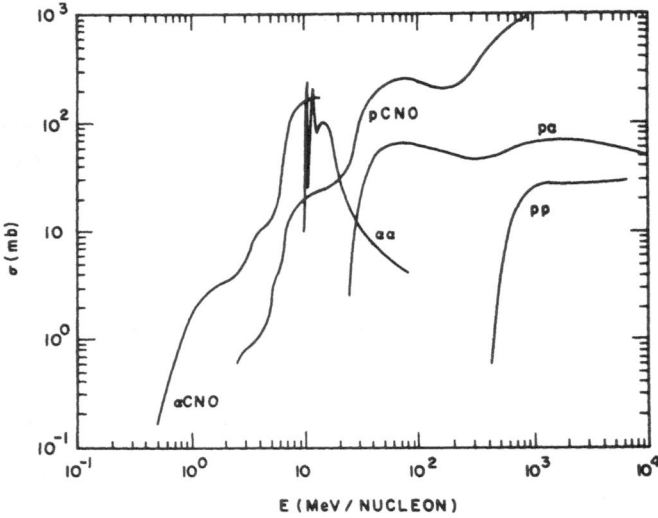

Fig. 1.   Neutron production cross-sections.

The instantaneous neutron production rates in the thin-target model for power-law and exponential spectra are shown in Figures 2 and 3, respectively. The various production modes are: pp (proton-hydrogen), pα (proton-helium), αp (α-particle-hydrogen), αα (α-particle-helium), pCNO (proton-heavy nuclei), CNOp (heavy nuclei-hydrogen), αCNO (α-particle-heavy nuclei), and CNOα (heavy nuclei-helium). As can be seen for flatter spectra (smaller values of s or larger values of P_0) the neutrons are produced mainly in pα, pp and αp reactions. For power-law spectra (Figure 2) neutron production at large values of s is mainly due to αCNO and CNOα reactions; the contribution of pCNO and CNOp reactions is small at all values of s; and αα reactions make a major contribution around $s=4$. For exponential spectra, (Figure 3) almost all of the neutrons are produced in pα reactions at most values of P_0. In this case the relative contributions of reactions induced by α-particles and heavy nuclei are lower than for power-law spectra. Because particles with $Z \geqslant 2$ have larger rigidities than protons, it follows from Equation (2) that their fluxes relative to the proton flux at the same energy per nucleon is lower than in the power-law case given by Equation (1). Since the nuclear cross-sections are the same for the direct and inverse reactions

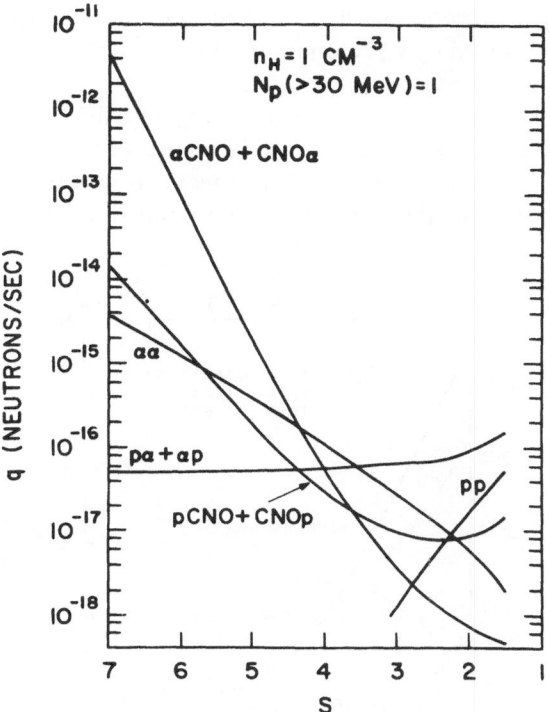

Fig. 2.   Partial neutron production modes in the thin-target model with power-law spectra.

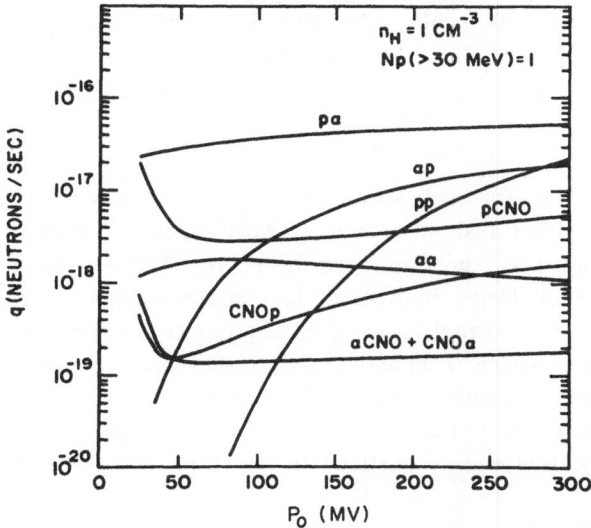

Fig. 3.   Partial neutron production modes in the thin-target model with exponential spectra.

at the same energy per nucleon, for particle spectra which are exponential in rigidity more neutrons are produced by a proton induced reaction than by the corresponding inverse process.

The total neutron production rates in the thin-target model, q, and the total neutron yields in the thick-target model, Q, are shown in Figure 4 for power-law and exponential spectra. In the thin-target model, for flat primary spectra the neutron production rates are about the same for the exponential and power-law cases. This result is due

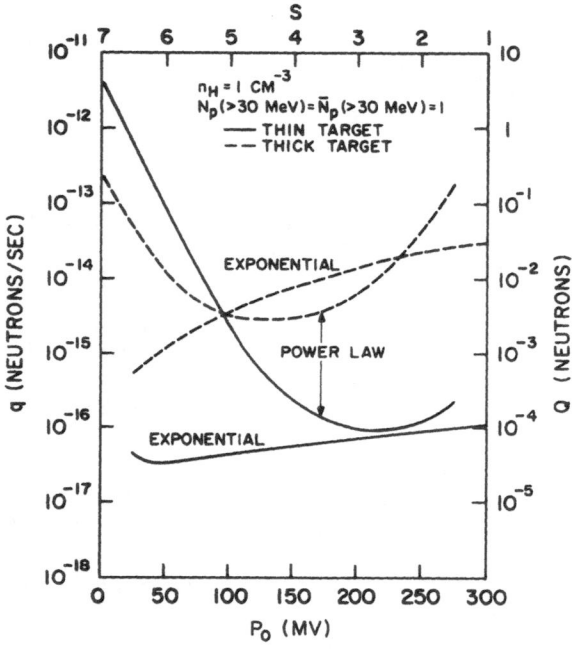

Fig. 4.   Total neutron production in the thin- and thick-target models with power-law and exponential spectra.

simply to the fact that for such spectra most of the neutrons are produced in pα reactions with effective threshold around 30 MeV nucleon^{-1} and all of the assumed spectra are normalized to 1 proton above this energy. But because a power law spectrum contains a much larger number of low-energy particles than an exponential spectrum with this same normalization, the steep power law spectrum can yield orders of magnitude more neutrons from αCNO and CNOα reactions which have thresholds more than an order of magnitude below 30 MeV nucleon^{-1}.

Similar effects are evident also in the thick-target model except that here the relative contribution of the low energy particles in general is diminished because of their shorter range.

Having considered the production of neutrons, let us now discuss their propagation and the ensuing gamma-ray line production. Wang and Ramaty (1974) considered in detail the effects of neutron propagation in the solar atmosphere on the production of gamma rays by the reaction

$$n + p \rightarrow d + \gamma. \tag{7}$$

In their treatment a distribution of neutrons was released in the chromosphere or corona, and the path of each neutron after its release was followed by a computer Monte-Carlo simulation. If the neutrons are released above the photosphere, any initially upward moving neutron escapes from the Sun. Some of the downward moving neutrons can also escape after being backscattered elastically by ambient protons, but most of these neutrons either are captured or decay at the Sun. Because the probability for elastic scattering is much larger than the capture probability, the majority of the neutrons are thermalized before they get captured. Since the thermal speed in the photosphere (where most of the captures take place) is much smaller than the speed of light, the gamma-rays from reaction (7) are essentially all at 2.2 MeV and the Doppler-broadened width of this line is negligible.

The bulk of neutrons at the Sun are captured either on H or on He^3. Whereas capture on H yields a 2.2 MeV photon, capture on He^3 proceeds via the radiationless transition

$$n + He^3 \rightarrow H^3 + p, \tag{8}$$

and hence produces no photons. The cross-sections for reactions (7) and (8) are $2.2 \times 10^{-30} \beta^{-1} \text{ cm}^2$ and $3.7 \times 10^{-26} \beta^{-1} \text{ cm}^2$, respectively, where β is the velocity of the neutron (for details see Wang and Ramaty, 1974). Thus if the He^3/H ratio in the photosphere is $\sim 5 \times 10^{-5}$ comparable to that observed in the solar wind, nearly equal numbers of neutrons are captured on He^3 as on H.

The results of the Monte-Carlo calculations of Wang and Ramaty (1974) are presented in Figures 5 and 6 for two assumptions on the photospheric He^3 abundance: $He^3/H = 0$ and $He^3/H = 5 \times 10^{-5}$. In these calculations an isotropic distribution of monoenergetic neutrons of energy E_n is released above the photosphere. The solid lines are the probabilities for the various indicated processes. As can be seen, the capture and loss probabilities increase with increasing energy, because higher energy neutrons penetrate deeper into the photosphere. This reduces their escape probability and leads to a shorter capture time, thereby reducing the decay probability. When $He^3/H = 5 \times 10^{-5}$, the probability for loss on He^3 almost equals the capture probability on protons. The escape probability is greater than 0.5, because all initially upward moving neutrons escape from the Sun. Note that the sum of all probabilities equals 1.

The dashed lines in Figures 5 and 6 are photon yields per neutron, $f(\theta, E_n)$, for various neutron energies, E_n, and angles, θ, between the Earth-Sun line and the vertical to the solar surface. The function f is defined such that for an average neutron production rate, q, the average 2.2 MeV photon flux at Earth is

$$\phi (2.2 \text{ MeV}) = qf/(4\pi R^2), \tag{9}$$

where $R = 1$ AU.

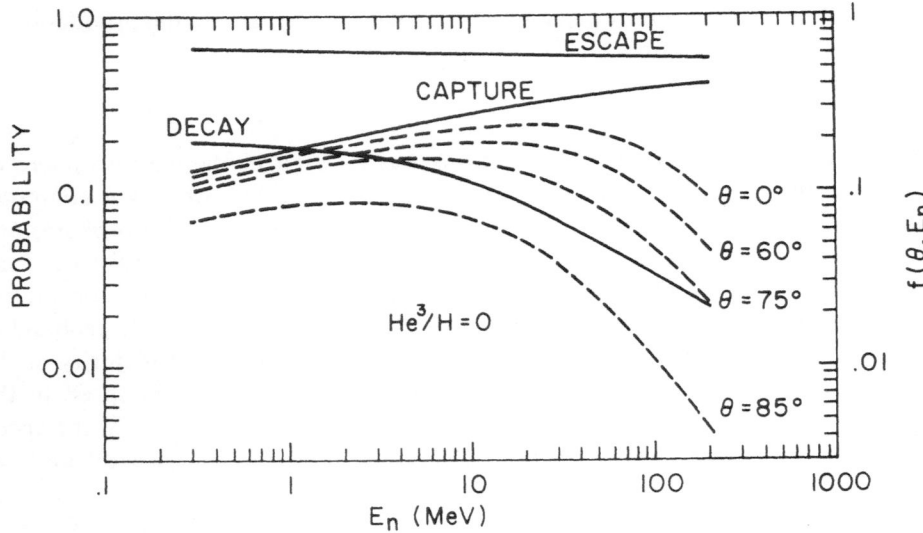

Fig. 5.   Probabilities for neutron escape, decay, and capture in the solar atmosphere (solid lines), and photon yields per neutron (dashed lines) for no He³ in the photosphere.

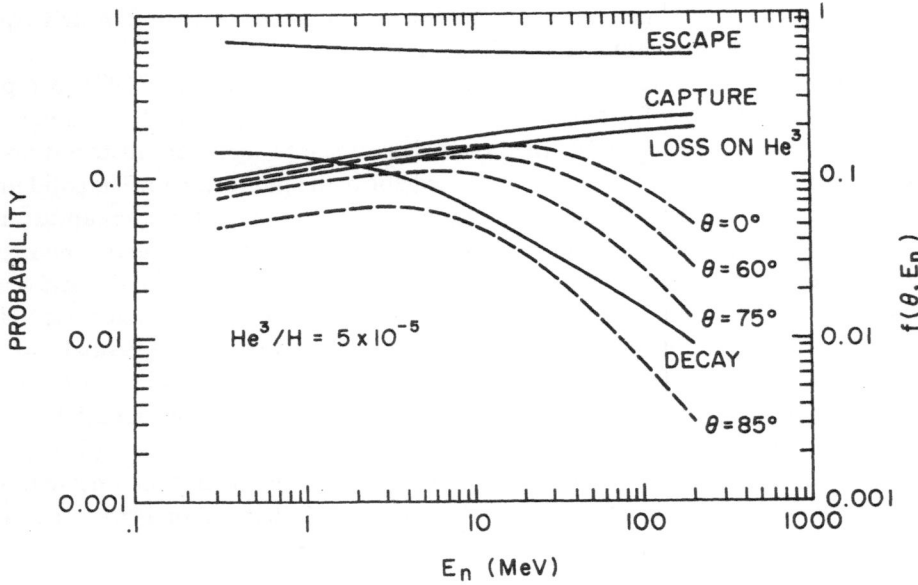

Fig. 6.   Probabilities for neutron escape, decay, and capture in the solar atmosphere (solid line), and photon yields per neutron (dashed lines) for He³/H = 5 × 10⁻⁵.

At low neutron energies and θ near zero, f is close to the capture probability on protons. This means that gamma rays from low-energy neutrons observed close to the vertical escape essentially unattenuated from the Sun. At higher energies and at larger angles, however, there is significant attenuation of the gamma rays due to Compton scattering in the photosphere. Even though f does depend on E_n, for flares sufficiently close to longitude and latitude zero on the Sun and neutron energies between about 1 and 100 MeV, we can approximate it by a constant. Most of the neutrons have energies in this range (Lingenfelter and Ramaty, 1967). Thus, for $He^3/H \simeq \simeq 5 \times 10^{-5}$ we use $f \simeq 0.12$, and for $He^3/H \simeq 0$, we take $f \simeq 0.2$. Note that these approximations are quite valid for the flare of 1972, August 4, since its solar longitude and latitude where E08 and N14.

It should be noted that Equation (9) is valid for the average neutron flux only, because the instantaneous 2.2 MeV flux lags behind the instantaneous neutron production rate. This lag is almost entirely due to the finite neutron capture time in the photosphere. Wang and Ramaty (1974) have investigated this effect, and some of their results are given in Table I. Here $<n>$ is the most probable density in the photo-

TABLE I

Most probable neutron capture densities, capture times, τ_c, and $\lambda = \tau_c^{-1} + \tau_d^{-1}$, where τ_d is the neutron mean life.

E_n (MeV)	$\langle n \rangle$ (cm^{-3})	τ_c (s)		λ^{-1} (s)	
		$He^3/H = 5 \times 10^{-5}$	$He^3/H = 0$	$He^3/H = 5 \times 10^{-5}$	$He^3/H = 0$
1	7×10^{16}	119	214	105	173
10	1.2×10^{17}	69	125	64	110
100	3×10^{17}	28	50	27	47

sphere where the captures take place, τ_c is the mean capture time, and τ_d is the neutron decay mean life. In terms of the parameter $\lambda = \tau_c^{-1} + \tau_d^{-1}$, the time profile of the 2.2 MeV photon flux from a monoenergetic burst of neutrons released at t_0 can be approximated (Wang and Ramaty, 1974) by

$$\phi \, (2.2 \text{ MeV}) \propto \exp[-\lambda(t - t_0)]. \tag{10}$$

IV. Prompt Gamma-Ray Lines Production

The various prompt gamma-ray lines that can be produced in solar flares together with their production mechanisms are listed in Table II. These lines have already been discussed by Ramaty and Lingenfelter (1973a) except for the Li^7 and Be^7 lines at 478 keV and 431 keV. The possibility of producing these lines in $\alpha\alpha$ reactions has been pointed out recently by Kozlovsky and Ramaty (1974a)

TABLE II
Prompt gamma-ray lines

Photon energy (MeV)	Origin	Production mode
0.431	Be^{7*} de-excitation	$He^4(\alpha, n) Be^{7*0.431}$
0.478	Li^{7*} de-excitation	$He^4(\alpha, p) Li^{7*0.478}$
1.63	$Ne^{20*1.63}$ de-excitation	$Ne^{20}(p, p') Ne^{20*1.63}$
	$N^{14*3.94} \rightarrow N^{14*2.31}$ de-excitation	$N^{14}(p, p') N^{14*3.94}$
2.31	$N^{14*2.31}$ de-excitation	$N^{14}(p, p') N^{14*2.31}$
		$N^{14}(p, p') N^{14*3.94} \rightarrow N^{14*2.31}$
		$N^{14}(p, n) O^{14} \rightarrow N^{14*2.31}$
4.43	$C^{12*4.43}$ de-excitation	$C^{12}(p, p') C^{12*4.43}$
		$C^{12}(\alpha, \alpha') C^{12*4.43}$
		$O^{16}(p, -) C^{12*4.42}$
5.2	$O^{15*5.26}$ de-excitation	$O^{16}(p, -) O^{15*5.20}$
	$N^{15*5.28}$ de-excitation	$O^{16}(p, -) N^{15*5.28}$
6.14	$O^{16*6.14}$ de-excitation	$O^{16}(p, p') O^{16*6.14}$
		$O^{16}(\alpha, \alpha') O^{16*6.14}$
7.12	$O^{16*7.12}$ de-excitation	$O^{16}(p, p') O^{16*7.12}$

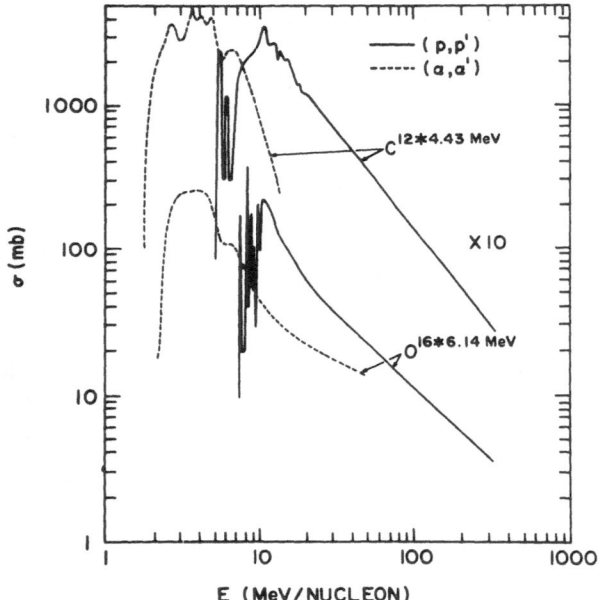

Fig. 7. C^{12*} and O^{16*} production cross-sections.

The cross-sections for the reactions $C^{12}(p, p') C^{12*4.43}$, $C^{12}(\alpha, \alpha') C^{12*4.43}$, $O^{16}(p, p') O^{16.*6.14}$, and $O^{16}(\alpha, \alpha') O^{16*6.14}$ are given in Figure 7. The cross-section for the proton induced reactions were summarized by Lingenfelter and Ramaty (1967), and the cross-sections for the α-particle induced reactions are from Kozlovsky (private communication). In addition, we also consider the reaction $p + O^{16} \rightarrow C^{12*4.43}$ $+ \cdots$ which was discussed previously (Ramaty and Lingenfelter, 1973a). By using these cross-sections and the interaction models discussed above, we can calculate the production rates of C^{12*} and O^{16*} and the resultant prompt photons at 4.43 MeV and 6.14 MeV. We must distinguish, however, between reactions induced by accelerated protons or α-particles, and reactions induced by accelerated heavy nuclei. For the former, the Doppler widths of the lines are small in comparison with available instrumental resolutions (about 150 keV). But for the latter the lines are significantly broadened by the motion of the excited fast nucleus which has lost little kinetic energy in the interaction. Because these lines are so broad that they cannot be resolved from the background with presently available instrumentation, in our treatment we consider the intensities of the 4.43 MeV and 6.14 MeV lines from proton and α-particle induced reactions only.

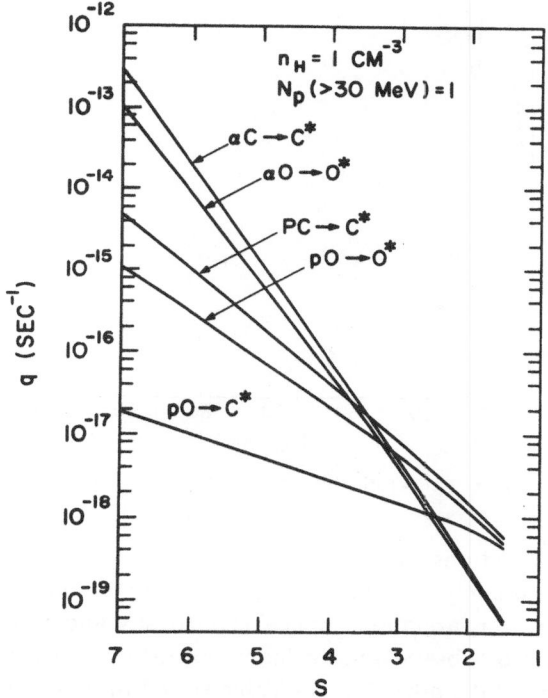

Fig. 8. C^{12*} and O^{16*} partial production modes in the thin-target model with power-law spectra.

The production rates of C^{12*} and O^{16*} in the thin-target model with power-law spectra are shown in Figure 8. We see that for flat spectra, the lines at 4.43 and 6.14 MeV are produced mainly by proton-induced reactions, whereas for steep spectra (large values of s), the contributions of the α-particles becomes important. For exponential spectra, proton-induced reactions are the principle source of the excited states at all values of P_0.

The cross-section for Li^7 production in $\alpha\alpha$ reactions is shown in Figure 9 (Kozlovsky and Ramaty, 1974b). The cross section for Li^{7*} production is about half of the total Li^7 production independent of energy. Similarly, the cross-section for Be^{7*} should also be about half of the total Be^7 production; and because Be^7 and Li^7 are produced by mirror reactions their production cross sections should be about equal, even though no data for Be^7 production in $\alpha\alpha$ reactions is available.

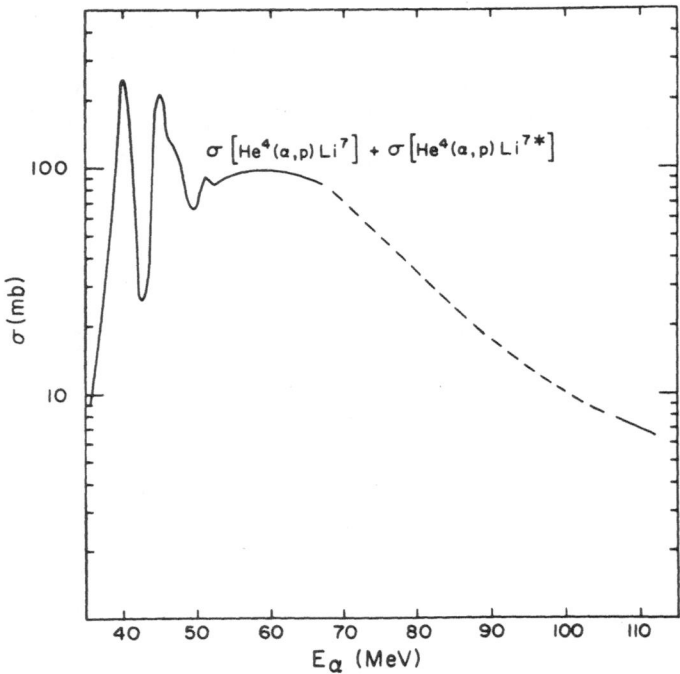

Fig. 9.   Li^7 production cross-section in $\alpha\alpha$ reactions.

Using these cross-sections we find that the intensities of the lines at 478 keV and 431 keV are approximately the same as the intensity of the 4.43 MeV line. Their Doppler width, however, are about 30 keV (Kozlovsky and Ramaty, 1974a); therefore, they should not be observed individually but rather as a broad spectral feature.

The cross section for the other lines of Table II and their intensities were discussed by Ramaty and Lingenfelter (1973a). The intensities of these lines are significantly lower than the intensity of the 4.43 MeV line for all interaction models and spectra.

V. Accelerated Particles at the Sun and in the Interplanetary Medium

The energy spectrum of the accelerated particles at the Sun can be deduced by comparing the calculated and observed ratios of the intensities of the strongest prompt line at 4.43 MeV to that of the neutron capture line at 2.23 MeV. This ratio for both the thin- and thick-target models is shown in Figure 10 as a function of s, for power-law spectra, and P_0, for exponential spectra. These were obtained from the calculations of Sections III and IV using a photon yield, f, of 0.2. This yield corresponds to a photospheric ratio $He^3/H=0$. For $He^3/H=5\times10^{-5}$ the curves in Figure 10 should be raised by about a factor of 2. The 4.43 MeV line is expected to be somewhat stronger than the 6.14 MeV line. Since the same observational limits were reported for

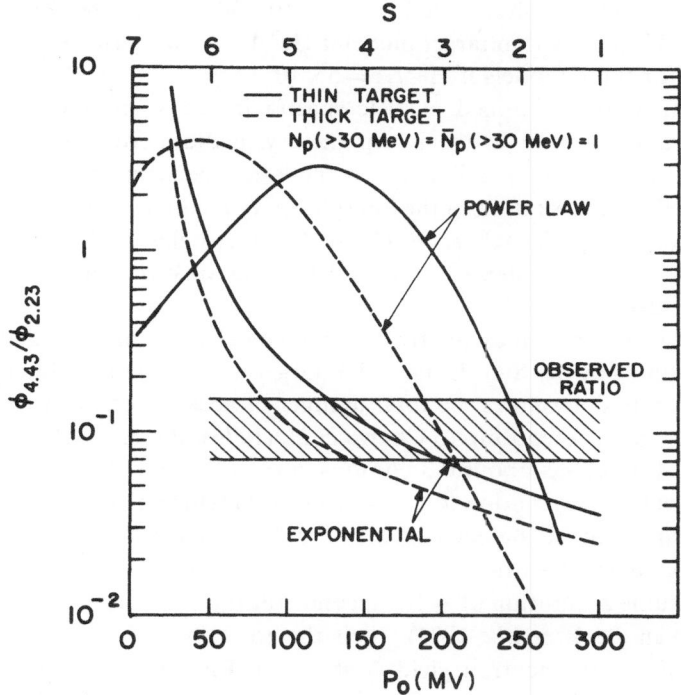

Fig. 10.   Ratios of the 4.43 MeV line intensity to the 2.23 MeV line intensity for thin- and thick-target models, and power-law and exponential spectra.

these two lines, if we normalize our calculations to the 4.43 MeV line, our results will also be consistent with the 6.14 MeV line within the uncertainties of the measurements.

As can be seen from Figure 10 for exponential spectra in both the thin and thick-target models $\phi_{4.43}/\phi_{2.23}$ decreases with increasing P_0. This results from the increase of the neutron production cross-section with increasing energy as opposed to the decrease of the excitation cross-sections (compare Figures 1 and 7). The same behavior can be seen for power-law spectra for s smaller than about 4.5 in the thin-target model and s less than about 6 in the thick-target model. For larger values of s, $\phi_{4.43}/\phi_{2.23}$ decreases with increasing s because these neutrons are produced mainly in αCNO and CNOα reactions which have lower thresholds than those for prompt gamma-ray production.

Let us compare now the calculations with the data. The observed (Chupp *et al.*, 1975) $\phi_{4.43}/\phi_{2.23}$ ratio of 0.11 ± 0.04 for the 1972, August 4 flare is also shown in Figure 10. As can be seen from this figure, assuming power-law spectra for the particles in the flare region, their spectral index s should lie between the values of 2 ± 0.2 deduced for the thin-target model and 3 ± 0.3 for the thick-target model if there is no He^3 in the photosphere; or between 1.8 ± 0.2 and 2.7 ± 0.2 for these models if the photospheric He^3/H ratio is 5×10^{-5}. Similarly, assuming exponential spectra, the implied P_0's should lie between 110 ± 30 MV for the thick-target model and 160 ± 35 MV for the thin target model if $He^3/H = 0$; or between 180 ± 50 MV and 230 ± 50 MV for these models if $He^3/H = 5 \times 10^{-5}$.

Comparison of these implied particle spectra in the solar flare region with the proton spectrum observed in the interplanetary medium from the 1972, August 4 flare is complicated by the possibility that the latter spectrum may have been significantly modified by acceleration in the interplanetary medium. Nonetheless, the proton spectrum obtained by Bertsch *et al.* (1974) from a rocket flight at 1916 UT 1972, August 4 had a spectral index $s \simeq 2$ in the 10 to 100 MeV region, in good agreement with our conclusions.

Knowing the spectral index for the various models, we can now deduce the total number of protons at the Sun. In the thick-target model, the time integrated photon flux from the flare determines the total number of accelerated particles that interact and stop at the Sun. According to Chupp *et al.* (1975), the 2.2 MeV intensity for the 1972, August 4 flare was about 0.3 photons $cm^{-1} s^{-2}$. This flux is the average over the time interval of observation (0623 to 0633 UT) which gives a total flux of about 180 photons cm^{-2}. Since the detector on OSO-7 was eclipsed by the Earth before the termination of the gamma-ray event, this is a lower limit to the total flux from the flare. If we assume a duration of $\sim 10^3$ seconds for the event, as indicated by the hard X-ray data (van Beek *et al.*, 1973), then the total flux of 2.2 MeV photons was ~ 300 cm^{-2}. From the neutron yield Q shown in Figure 4, we find that if $f = 0.2$, $\bar{N}_p(>30 \text{ MeV}) = 7 \times 10^{32}$ for $s = 3$, and $\bar{N}_p(>30 \text{ MeV}) = 1.0 \times 10^{33}$ for $P_0 = 110$ MV. The corresponding values for $He^3/H = 5 \times 10^{-5}$ are about the same, because the effect of a lower f is approximately cancelled by a higher neutron yield. Therefore, the total number of protons above 30 MeV released downward into the Sun in the thick-target model is about 10^{33} independent of the spectral form or the He^3/H ratio.

For the thin-target model, the instantaneous gamma-ray observations determine the product of the ambient proton or hydrogen density, n_H, and the instantaneous number of the accelerated particles in the interaction region. From the neutron yield q shown in Figure 4, the flux given above implies that if $f = 0.2$, $n_H N_p(>30 \text{ MeV}) \simeq 5 \times 10^{43}$ cm^{-3} for $s = 2$, and $n_H N_p(>30 \text{ MeV}) \simeq 10^{44}$ cm^{-3} for $P_0 = 160$ MV. As before, these values are not affected by the photospheric He^3/H ratio.

To obtain an estimate of the number of protons released into the interplanetary medium, the information obtained from the gamma rays in the thin-target model can be combined with data on the path length traversed by the nuclei before their escape from the interaction region. Such information can be obtained from studies of deuterons and helium-3 nuclei from flares. The H^2 and He^3 observations from the 1972, August events (Webber *et al.*, 1974), when compared with calculations of the production of these isotopes in nuclear reactions (Ramaty and Kozlovsky, 1974) imply that the amount of matter traversed by relativistic particles is about 1.5 g cm^{-2}. This means that the product $n_H t_1$ is about 3×10^{13} cm^{-3} s, where t_1 is the interaction time of the particles at the Sun. If t_1 is also interpreted as the escape time of the particles from the interaction region, then the protons were released into the interplanetary medium at an average rate $N_p (> 30 \text{ MeV})/t_1$ varying from about 1.6 to 3×10^{30} protons s^{-1}. The total number of protons released is the product of this rate and the acceleration time T. As indicated by the X-ray data (van Beek *et al.*, 1973), T is about 10^3 s; therefore the total number of protons released is between about 1.6 to 3×10^{33} protons. These numbers are larger by only about a factor of 2 to 3 than the number of protons released downward into the Sun in the thick-target model.

The number of protons released from the flare should be compared with estimates of the total number of protons in the interplanetary medium as obtained from charged particle observations. Ramaty and Lingenfelter (1973a) have estimated this quantity by taking the observed density of protons near Earth and by multiplying it with a storage volume in the interplanetary medium which they took to be $\sim 10^{39}$ cm^3. By using this volume and the peak proton flux greater than 30 MeV measured by Kohl *et al.* (1973), we get about 10^{34} protons. This number is larger by a factor of 3 to 6 than our estimate for the number of protons released in the thin-target model. But since the measured peak proton flux could consist to a large degree of particles accelerated by shocks in the interplanetary medium, it appears that there is no real discrepancy between the number of protons released from the Sun as deduced from the gamma rays and the number observed in interplanetary space.

VI. The Nature of the Positron Annihilation Radiation

Positrons can be produced in solar flares from the decay of π^+ mesons and radioactive nuclei which result from nuclear reactions of accelerated protons and nuclei with the ambient medium. The possibility that a significant fraction of solar positrons can also be produced by accelerated electrons in $e^+ - e^-$ pairs is being considered by Bai and Ramaty (1975). Annihilation of positrons can produce a gamma-ray line at 0.51 MeV. This line was observed by Chupp *et al.* (1973) for both the 1972, August 4 and August 7 flares.

The cross sections for the production of π^+ mesons and β^+-emitting radioactive nuclei were given by Lingenfelter and Ramaty (1967). Kozlovsky (1974, private communication) has recently updated the cross sections of the β^+-emitters. Using these cross-sections we calculate from equation (3) the instantaneous production rates of positron-emitters in the thin-target model for power-law spectra. The results are shown in Figure 11. As can be seen, for flatter spectra the principal positron source is π^+-mesons, but for steeper spectra, the pion contribution is negligible. This effect can also be seen in Figure 12, where the ratio of the pion yield to the total positron-emitter

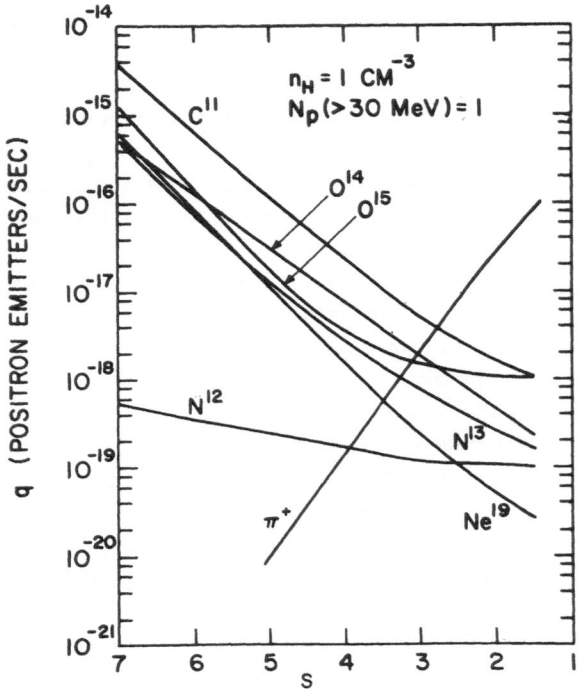

Fig. 11.   Partial positron production modes in the thin-target model with power-law spectra.

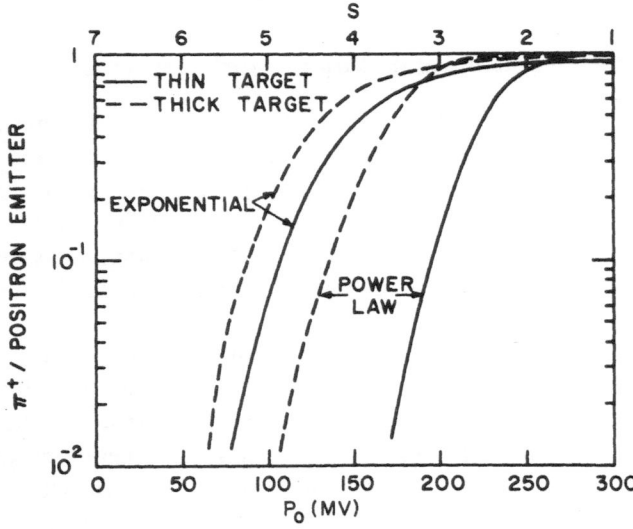

Fig. 12.   Ratios of the π^+-meson yields to the total positron yields for the thin- and thick-target models, and power-law and exponential spectra.

yield (including both pions and radioactive nuclei) is plotted as a function of s or P_0 for both interaction models. We see that, as for neutron production, the efficiency of π^+-production is larger in the thick-target model than in the thin-target model. The ratios of the total yield of positron-emitters to the neutron yield are shown in Figure 13 for the various interaction models.

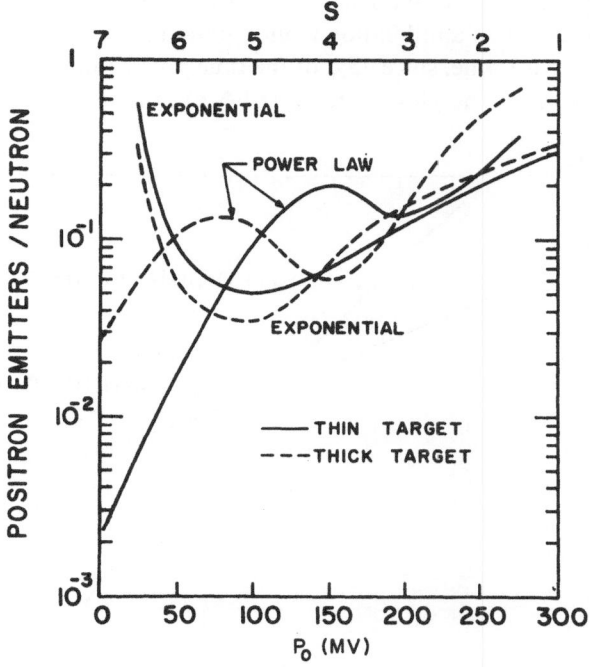

Fig. 13. Ratios of the total positron yield to the total neutron yield for the thin- and thick-target models, and power-law and exponential spectra.

TABLE III

	Positron emitter mean lives (s)
π^+	2.6×10^{-8}
N^{12}	0.016
C^{11}	1766
N^{13}	863
O^{14}	102
O^{15}	176

The intensity of the 0.51 MeV line, however, depends not only on the number of positron emitters produced, but also on the decay rate of the positron emitters and on the annihilation rate of the positrons. The mean lives against radioactive decay of the various positron emitters are shown in Table III.

The annihilation rate of the positrons depends on the density, temperature and state of ionization of the ambient medium. As discussed previously (Stecker, 1969; Ramaty and Lingenfelter, 1973b) in a low density and neutral medium, positrons nearly always annihilate via positronium formation; only if the ambient density exceeds about 10^{15} cm^{-3} do collisions destroy the ^{3}S state of positronium at such a rate that free annihilation becomes more important than positronium annihilation (Leventhal, 1973). Positron annihilation from a bound state of positronium results in an asymmetric 0.51 MeV line, since 75% of the time positronium annihilated from the ^{3}S state into 3 photons of energies less than 0.51 MeV, instead of 2 photons at precise-

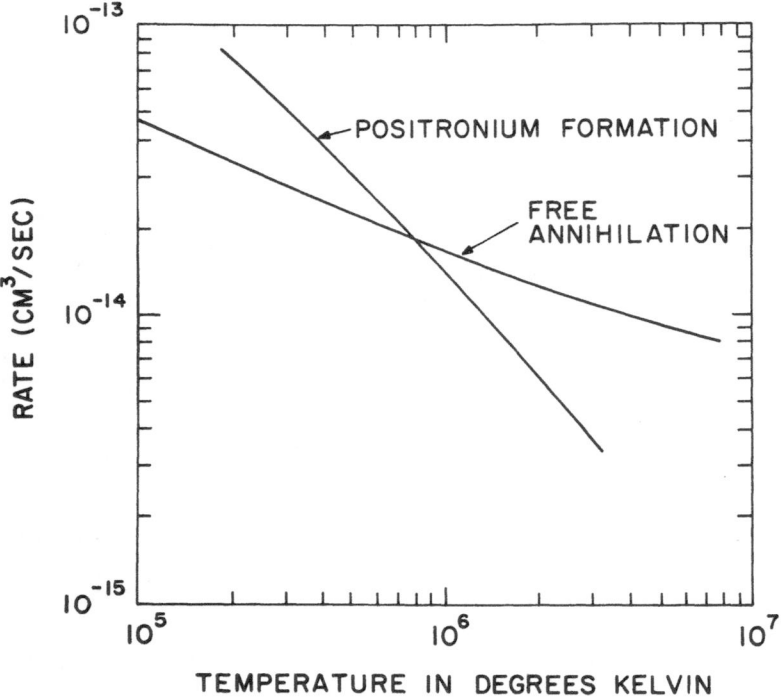

Fig. 14.   Positron free-annihilation rate and positronium formation rate in a hydrogen plasma.

ly this energy. The 0.51 MeV line from the Sun, however, appears to be symmetric, and the observational upper limit on the fraction of three-photon annihilations is about 20% of the total (D. Forrest, 1973, private communication).

As we noted, three-photon annihilation could be reduced by collisions, but the required density of $\gtrsim 10^{15}$ cm^{-3} is quite large. However, if the ambient medium is ionized, the rate of positronium formation is greatly reduced. In Figure 14 we show the rate of positronium formation and free annihilation per positron in a hydrogen plasma of unit density as a function of its temperature (C. Werntz and C. Crannell, 1973, private communication).

As can be seen, if the temperature is greater than about 10^6 K, most of the positrons annihilate without forming positronium. This is not an unreasonable temperature for the annihilation region; however, from the observed upper limit on the width of the 0.51 MeV line, the temperature in the annihilation region should be less than $\sim 10^7$ K (Chupp *et al.*, 1975).

According to the observations of the 1972, August 4 flare, the rise time of the 0.51 MeV line was within instrumental errors, similar to or perhaps even shorter than the rise time of the 2.23 MeV line (Chupp *et al.*, 1975). The latter was about 100 s, consistent with the expected lag between the production of the neutrons and the formation of the 2.2 MeV line (see Table I). Assuming that the time dependence of the production rate of the positron emitters is the same as that of the neutrons, the lag between the production of the positron emitters and the formation of the 0.51 MeV line should not be longer than about 100 s. From Figure 14, this result implies that the density of the ambient medium in the annihilation region is at least 10^{12} cm^{-3}. Furthermore if the positrons are of nuclear origin, then from Table III it follows that they should mainly result from π^+-mesons and short lived radioactive nuclei (O^{15}, O^{14} and N^{12}).

The observed average flux in the 0.51 MeV line for the flare of 1972, August 4 was about 0.06 photons cm^{-2} s^{-1} (Chupp, 1975). Hence the observed $\phi_{0.51}/\phi_{2.23}$ ratio was about 0.2.

Let f' be the 0.51 MeV photon yield per positron defined in the same way as the 2.23 MeV photon yield per neutron, f, in Equation (9). As discussed in Section III, f ranges from about 0.1 to 0.2. The maximum value of f' is 2. But because part of the positrons can escape from the Sun before they annihilate, and, furthermore, a fraction of the positrons can be trapped at the Sun in low density regions where the annihilation time is long, f' can be considerably less than 2.

According to our discussion in Section V, for power-law spectra $s \simeq 2$ in the thin-target model, and $s \simeq 3$ in the thick-target model. For these spectral parameters, from Figure 12 we see that the bulk of the positron emitters are π^+-mesons, and from Figure 13 we get that the positron emitter-to-neutron ratio is about 0.2. The observed ratio, $\phi_{0.51}/\phi_{2.23} \simeq 0.2$ then implies that $f' \simeq f$, i.e. the 0.51 MeV yield per positron is about the same as the 2.23 MeV yield per neutron. In view of the uncertainties involved in the deductions of f', this is not an unreasonable result. We can also deduce f' for the various values of P_0 obtained in Section V, and in all cases we find acceptable values $(0.1 < f' < 2)$. For these exponential spectra the contribution of π^+-mesons to the total positron production is greater than about 50% in all cases, except in the thick-target model with no He^3 in the photosphere where they contribute only about 25% of the positrons. A definite test for the possibility that the bulk of the positrons are due to π^+-mesons would be the observation of gamma rays from π^0 decay. As there is no published data on high energy solar gamma rays, we shall not discuss this possibility in the present paper.

As mentioned above, prompt 0.51 MeV photons could also result from positrons produced in $e^+ - e^-$ pairs. Such pairs would be produced mainly by $\gtrsim 1$ MeV electrons. Because these electrons also produce continuum emission by bremsstrahlung, it is possible to calculate the pair-to-bremsstrahlung yield (Bai and Ramaty, 1975). By comparing these calculations with the observed continuum emission from the 1972, August 4 flare (Chupp *et al.*, 1975), we find that if the bremsstrahlung is produced by an isotropic distribution of electrons at the Sun, pair production could account for less than 1% of the positrons required to produce the 0.51 MeV line. However, if the electron beam is directed downward into the Sun, the bremsstrahlung efficiency in the backward direction is greatly reduced for relativistic particles (Petrosian, 1973). In this case pair production could account for essentially all the prompt 0.51 MeV photons of the 1972, August 4 flare.

Finally, we wish to mention that delayed photons (by about 15 min) were observed for the 1972, August 7 flare (Chupp *et al.*, 1973). These photons are most likely due to the long lived radioactive nuclei such as C^{11} and N^{13}.

VII. Summary

The observed gamma-ray lines from the 1972, August 4 flare, at 0.5, 2.2, 4.4 and 6.1 MeV are due to positron annihilation, neutron capture on hydrogen, and deexcitation of excited states in C^{12} and O^{16}, respectively. The strongest line is at 2.23 MeV. It is due to fast neutrons produced by nuclear reactions of flare accelerated particles with the ambient solar atmosphere. These neutrons are thermalized and captured by ambient protons in the photosphere to produce deuterons and 2.23 MeV gamma rays. Photospheric He^3 competes with the protons in capturing neutrons. Because captures on He^3 do not lead to photon emission, the observation of 2.23 MeV line emission from the Sun implies that the He^3 abundance in the photosphere cannot be much larger than that observed in the solar wind ($He^3/H \sim 5 \times 10^{-5}$).

We have evaluated in detail the yield of neutrons and excited C^{12} and O^{16} nuclei from nuclear reactions of accelerated particles with the ambient solar atmosphere. For the 1972, August 4 flare the neutrons are produced mainly in pα and αp reactions by primary particles with energies greater than about 30 MeV nucleon^{-1}. The observed gamma rays at 4.43 MeV and 6.14 MeV are principally due to proton induced interactions. Reactions induced by fast carbon and oxygen nuclei lead to Doppler-broadened lines which cannot be distinguished from the continuum.

From the comparison of the calculated and observed ratios of the lines at 4.43 MeV or 6.14 MeV to the line at 2.23 MeV it is possible to deduce the spectrum of the accelerated particles in the flare region. The spectral index s defined in Equation (1) is 2 ± 0.2 for the thin-target model and 3 ± 0.3 for thick-target model if there is no He^3 in the photosphere; or 1.8 ± 0.2 and 2.7 ± 0.2 for these models if the photospheric He^3/H ratio is 5×10^{-5}. The characteristic rigidity P_0 defined in Equation (2) is 160 ± 35 MV for the thin-target model and 110 ± 30 MV for the thick-target model if $He^3/H = 0$; or 230 ± 50 MV and 180 ± 50 MV for these models if $He^3/H = 5 \times 10^{-5}$. The total number of protons above 30 MeV released downward into the Sun in the thick-target model is about 10^{33}. For the thin-target model about 2×10^{33} protons escape from the flare region if the thickness of the target for relativistic particles is 1.5 g cm^{-2} as deduced from deuteron and helium-3 observations.

The positrons which produce the 0.51 MeV line could be due to π^+-mesons, radioactive nuclei, and pair production. We have evaluated in detail the production of mesons and radioactive nuclei in nuclear reactions of accelerated charged particles with ambient nuclei. The relatively prompt nature of the 0.51 MeV line seems to favor positron production from π^+-decay or $e^+ - e^-$ pairs in the initial phase of the 1972, August 4 event. The delayed 0.51 MeV line emission observed for the 1972, August 7 event is very likely due to positrons from radioactive nuclei.

Acknowledgements

The research of REL was supported by the National Science Foundation under Grant GP 31620.

References

Bai, T. and Ramaty, R.: 1975 (to be published).
Bertsch, D. L., Biswas, S., and Reames, D. V.: 1974, *Solar Phys.* **39**, 479.
Cameron, A. G. W.: 1973, *Space Sci. Rev.* **15**, 121.
Chupp, E. L., Forrest, D. J., and Suri, A. N.: 1975, This volume, p. 341.
Chupp, E. L., Forrest, D. J., Higbie, P. R., Suri, A. N., Tsai, C., and Dunphy, P. P.: 1973, *Nature* **241**, 333.
Forrest, D.: 1973, Private communication.
Kohl, J. W., Bostrom, C. O., and Williams, D. J.: 1973, in H. E. Coffey (ed.), World Data Center *Rept. UAG* **28**, Part II, *Collected Data Reports on August 1972 Solar Terrestrial Events*, p. 330.
Kozlovsky, B.: 1974, Private communication.
Kozlovsky, B. and Ramaty, R.: 1974a, *Astrophys. J. Letters* **191**, L43.
Kozlovsky, B. and Ramaty, R.: 1974b, *Astron. Astrophys.* **34**, 477.
Leventhal, M.: 1973, *Astrophys. J. Letters* **183**, L147.
Lingenfelter, R. E. and Ramaty, R.: 1967, in B. S. P. Shen (ed.), *High Energy Nuclear Reactions in Astrophysics*, W. A. Benjamin, New York, p. 99.
Lingenfelter, R. E., Flamm, E. J., Canfield, E. H., and Kellman, S.: 1965, *J. Geophys. Res.* **70**, 4077.
Petrosian, V.: 1973, *Astrophys. J.* **186**, 291.
Ramaty, R. and Kozlovsky, B.: 1974, *Astrophys. J.* (in press).
Ramaty, R. and Lingenfelter, R. E.: 1973a, in R. Ramaty and R. G. Stone (eds.), *High Energy Phenomena on the Sun*, NASA SP-342, p. 301.
Ramaty, R. and Lingenfelter, R. E.: 1973b, *Conference Papers, 13th International Cosmic Ray Conference*, University of Denver, Colorado, p. 1590.
Reppin, C., Chupp, E. L., Forrest, D. J., and Suri, A. N.: 1973, *Conference Papers, 13th International Cosmic Ray Conference*, University of Denver, Denver, Colorado, p. 1577.
Stecker, F. W.: 1969, *Astrophys. Space Sci.* **3**, 479.
Van Beek, H. F., Hoyng, P. and Stevens, G. A.: 1973, in H. E. Coffey (ed.), World Data Center, *Rept. UAG* **28**, Part II, *Collected Data Reports on August 1972 Solar Terrestrial Events*, p. 319.
Wang, H. T. and Ramaty, R.: 1974, *Solar Phys.* **36**, 129.
Webber, W. R., Roelof, E. C., McDonald, F. B., Teegarden, B. J., and Trainor, J.: 1974 (preprint).
Werntz, C. and Crannell, C.: 1973, Private communication.

FAST ELECTRONS IN SMALL SOLAR FLARES

R. P. LIN

Space Sciences Laboratory, University of California, Berkeley, Calif. 94720, U.S.A.

Abstract. Because $\sim$ 5–100 keV electrons are frequently accelerated and emitted by the Sun in small flares, it is possible to define a detailed characteristic physical picture of these events. This review summarizes both the direct spacecraft observations of non-relativistic solar electrons, and observations of the X-ray and radio emission generated by these particles at the Sun and in the interplanetary medium. These observations bear on the basic astrophysical process of particle acceleration in tenuous plasmas. We find that in many small solar flares the $\sim$ 5–100 keV electrons accelerated during flash phase constitute the bulk of the total flare energy. Thus the basic flare mechanism in these flares essentially converts the available flare energy into fast electrons. These electrons may produce the other flare electromagnetic emissions through their interactions with the solar atmosphere. In large proton flares these electrons may provide the energy to eject material from the Sun and to create a shock wave which could then accelerate nuclei and electrons to much higher energies.

I. Introduction

Two distinct acceleration processes are observed to occur in solar flares. The first is the flash phase acceleration of electrons to energies of $\sim$ 5 to $\sim$ 100 keV. This flash phase acceleration occurs in many small solar flares. The second acceleration phase occurs only in a few large flares and accelerates protons and electrons to much higher energies. The bulk of this second phase acceleration occurs after the flash phase, and appears closely related to type II radio emission at the Sun and interplanetary shock waves observed at the orbit of the Earth. This review covers only the flash phase acceleration of electrons.

Non-relativistic electrons from the Sun were first directly observed near the beginning of the present solar cycle (Van Allen and Krimigis, 1965). Since then over 350 impulsive solar flare electron events have been observed, mostly from experiments on the IMP (Interplanetary Monitoring Platform) series of spacecraft (Anderson and Lin, 1966; Lin and Anderson, 1967; Lin, 1970). Some examples are shown in Figure 1. These electrons are the energetic particle species most frequently emitted by the Sun, and often originate in small flares or sub-flares.

Non-thermal X-ray and radio emission (Figure 2) indicate the presence of energetic electrons at the Sun. Low frequency radio observations (Figure 3) from spacecraft have provided a way of tracing these electrons from the vicinity of the Sun to and beyond the orbit of the Earth (see Fainberg *et al.*, 1972).

The $\sim$ 5–100 keV electrons constitute the bulk of the total flare energy in many small solar flares. Thus to a first approximation those flares can be thought of as a mechanism to convert the available energy, presumably contained in the magnetic field, into energetic electrons. These electrons may also produce the optical flare emissions – EUV (Figure 2) and Hα emissions (Figure 4) – by their interactions with the solar atmosphere.

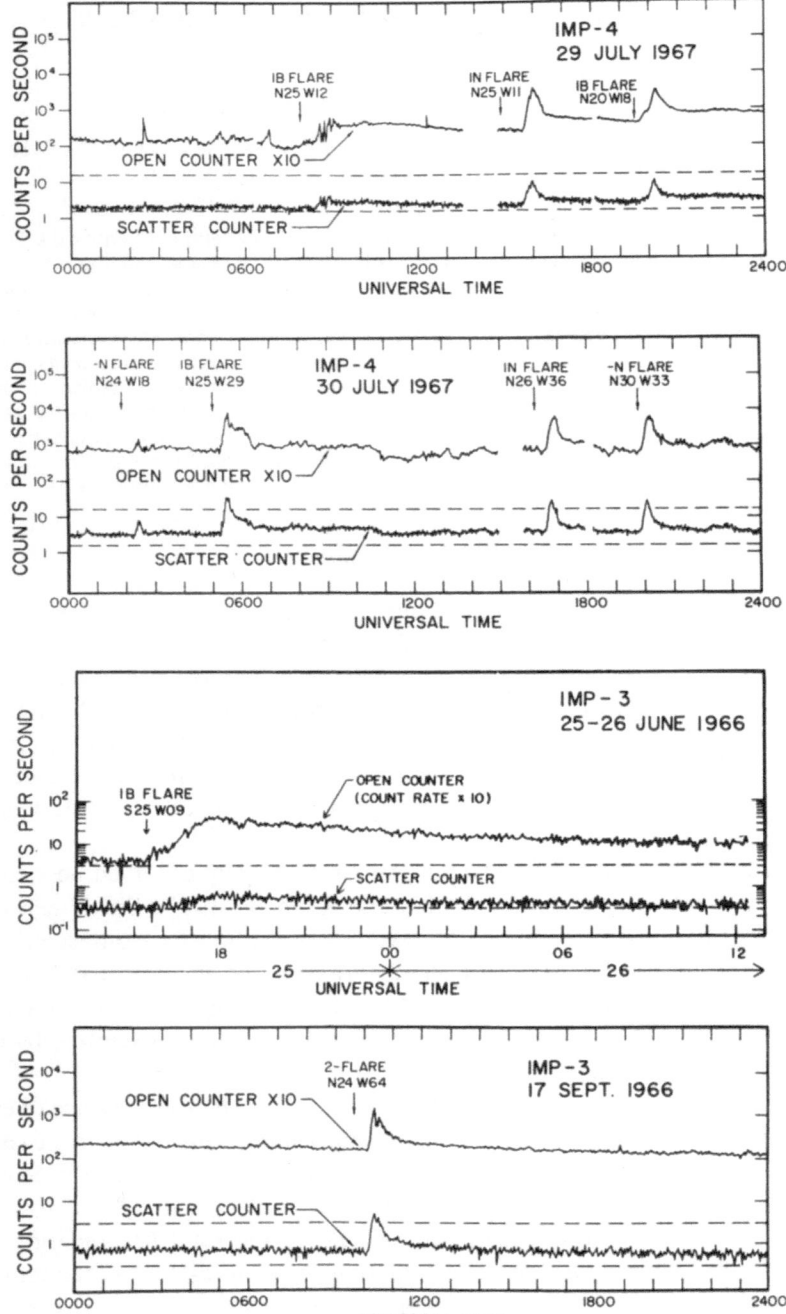

Fig. 1. Several examples of electron events observed by spacecraft at 1 AU are shown here. The upper two panels show a series of rapid rise-rapid fall events, all from the same active region. The scatter counter is sensitive only to electrons above 45 keV energy.

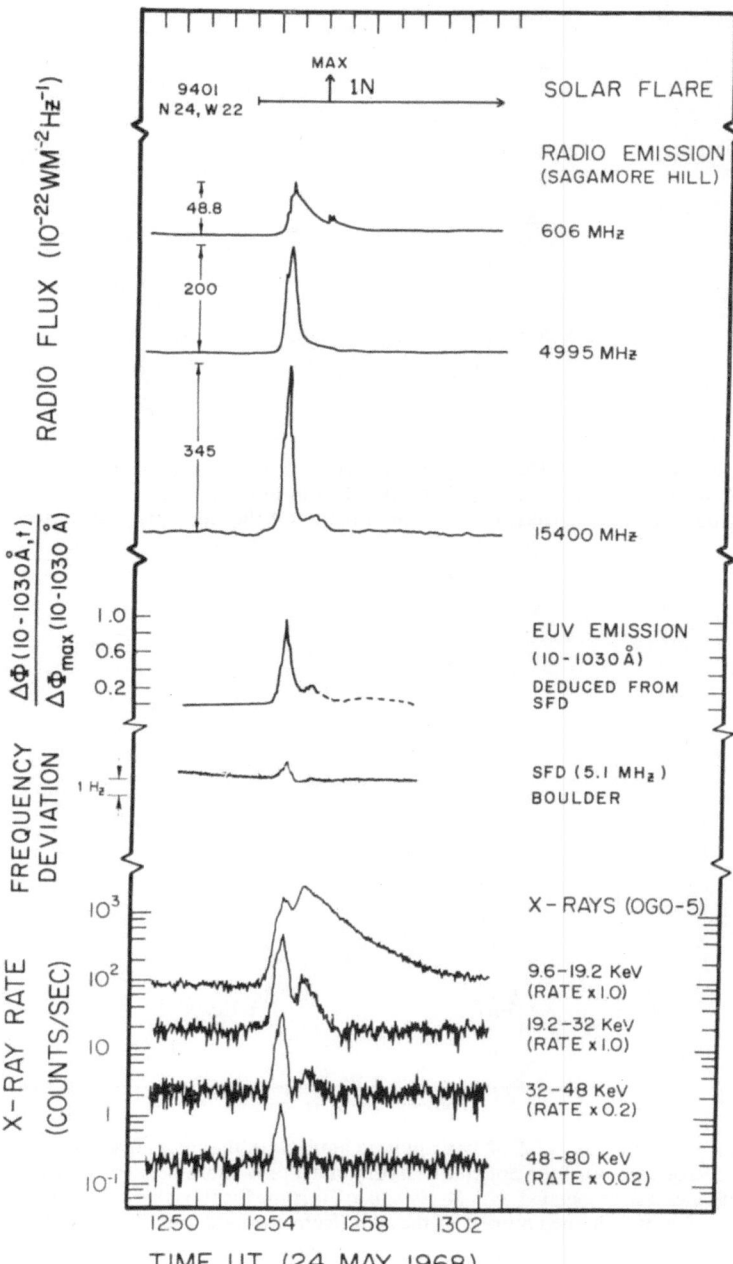

Fig. 2. An example of an impulsive non-thermal X-ray burst illustrating the close correspondence to the impulsive radio burst at cm wavelengths and EUV emission as derived from SFD measurements (Kane 1973b).

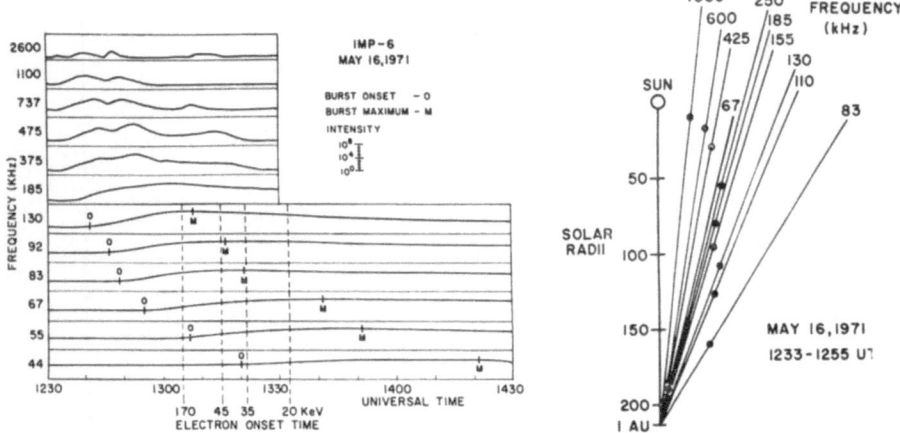

Fig. 3.   The type III burst of 1971, May 16 observed at frequencies from 2.6 MHz to 44 kHz. The 55 kHz emission originates closest to 1 AU. The location of the radio emission at different frequencies, as derived from the modulation of the signal, is also shown.

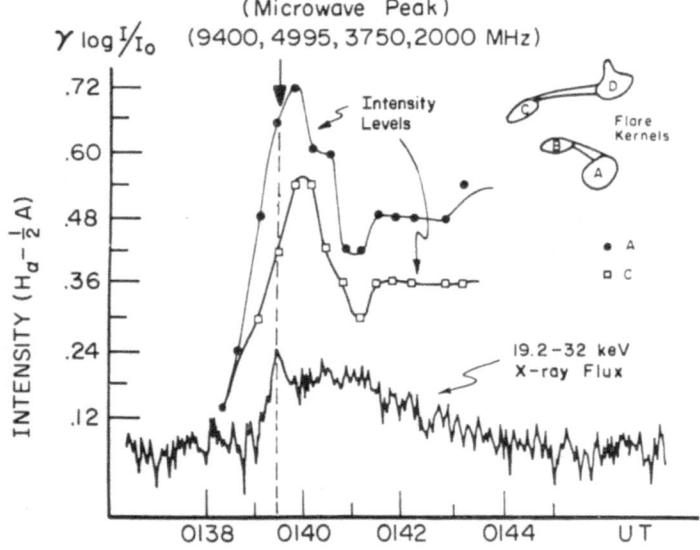

Fig. 4.   Impulsive X-ray flux (19.2–32 keV) plotted along with the light curve measures at Hα-1/2 Å during an impulsive flare. Of four potential bright points, only those labeled A and C exhibited the abrupt intensity increase associated with X-ray spikes. I_0 is the local background and γ, with a typical value of 3, is the contrast of the film (from Vorpahl and Zirin, 1970).

The electromagnetic emissions observed from small electron flares are those typical of the flash phase of solar flares. Figure 5 shows a typical but idealized picture of these emissions and their timing. Not all of these types of emissions accompany every flare which produces electrons, but when such emissions occur the timing is usually as

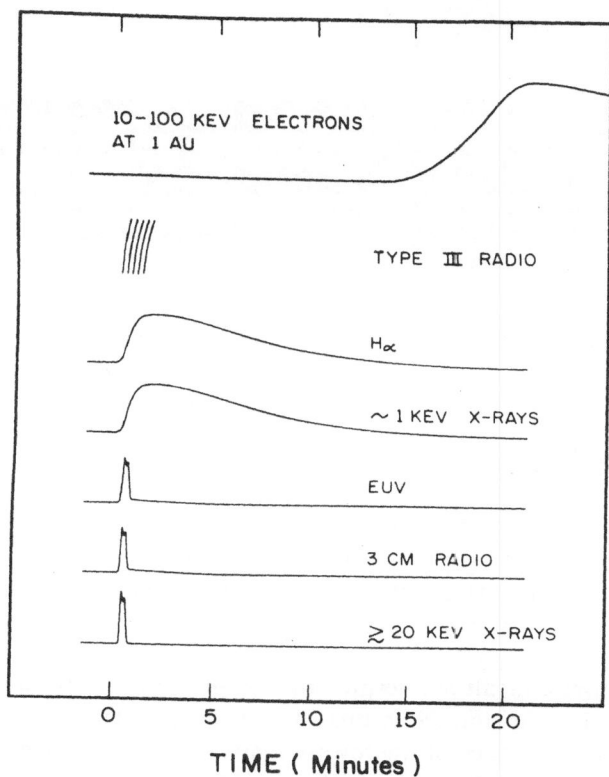

Fig. 5. An idealized picture of the flare flash phase event, showing the relative timing of each emission. Not all flare flash phases are accompanied by all these emissions but when they do occur, the timing is usually as shown.

shown. These emissions (and their appearance or non-appearance) can be interpreted as the direct consequence of the acceleration and subsequent motion of 10–100 keV electrons in the solar atmosphere.

II. Characteristics of Energetic Electrons at the Sun

The most direct information about the energetic electrons in a flare is obtained from the non-thermal X-rays emitted by the flare. Energetic electrons at the Sun produce non-thermal X-rays through bremsstrahlung in the solar atmosphere. Since the bremsstrahlung process is well understood and there is essentially no attenuation of these energy X-rays in the solar atmosphere, the observations of non-thermal X-ray bursts from the Sun can be directly and quantitatively interpreted in terms of energetic electrons at the Sun.

Microwave bursts and type III bursts are also observed in the radio range during the flash phase of solar flares (see review by Wild *et al.*, 1963). These emissions, how-

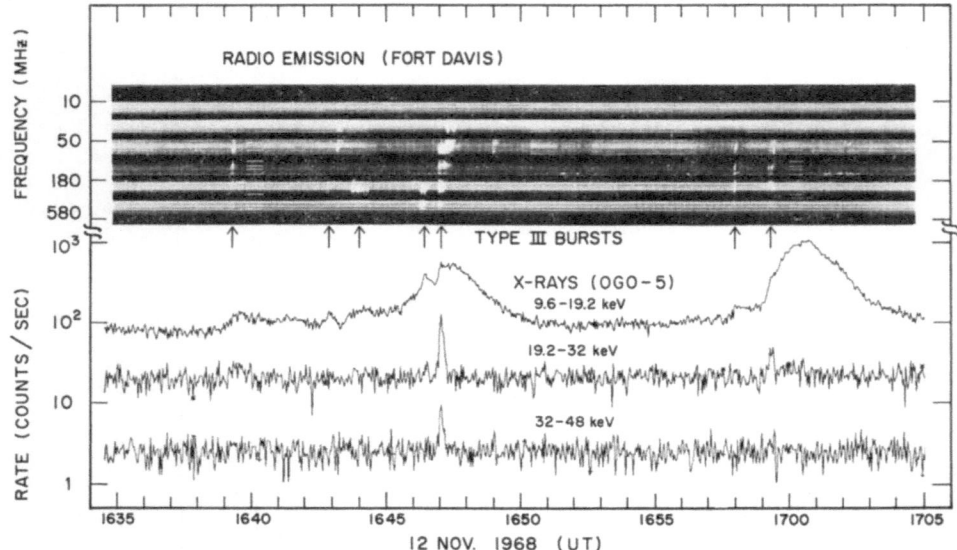

Fig. 6.   Simultaneous X-ray and type III burst records, illustrating the close temporal correlation
of the non-thermal X-ray bursts and type III bursts. (Kane, 1972).

ever, are much more difficult to interpret, due to substantial uncertainties in quantitatively evaluating the emission and propagation processes. However, the striking time
coincidence and resemblance of microwave emission to non-thermal X-ray emission
(Figure 2) (Anderson and Winckler, 1962; Arnoldy *et al.*, 1968), and the time coincidence of type III bursts and X-ray bursts (Figure 6) leave little doubt that those radio
emissions are due to the same electrons.

Several examples of > 10 keV non-thermal X-ray bursts during the flare flash phase
are shown in Figures 2 and 6. Their properties are (Kane and Anderson, 1970):

(1) The duration of the burst is ~ 10–100 s with *e*-folding rise and decay times of
2–10 s. There is evidence that many X-ray bursts are actually composed of many
spikes of ~ few seconds duration. In fact the close temporal coincidence between
type III bursts and X-ray bursts (see Figure 6) suggests that the acceleration time
scales may be faster than 1 s. The type III bursts typically occur in semi-periodic
groups with individual bursts lasting < 1 s and burst to burst separation of $\lesssim$ few s.
To the extent that can be observed with the available instrumental resolution, the
X-ray fine structure does correspond to type III radio burst group structure (S. R. Kane,
1974, private communication).

(2) The non-thermal burst typically occurs near the start of the soft X-ray burst
(see Figures 2 and 6) although on most occasions a rise in soft X-rays is evident prior
to the non-thermal burst. The non-thermal burst is usually coincident with the most
rapidly rising portion of the soft X-ray profile.

(3) The energy spectrum of the photons between ~ 10 to ~ 10^2 keV can be fitted to
a power law, $dJ(hv)/d(hv) = A(hv)^{-\gamma}$. A rapid falloff is often observed above ~ 10^2

keV when count rate statistics are sufficient at those energies. Such an X-ray spectrum would be produced by an electron population with power-law energy spectrum up to ~100 keV and cutoff beyond. Thermal spectra do not fit well to the data. Because the quasi-thermal component may mask observations below ~10 keV, most observations of the non-thermal component have been limited to above ~10 keV. However, the non-thermal component appears to extend to ≲5 keV on those few occasions when it can be observed above the thermal component (Peterson *et al.*, 1973; Kahler and Kreplin, 1971). The range of non-thermal X-ray spectra, illustrated in Figure 7, show

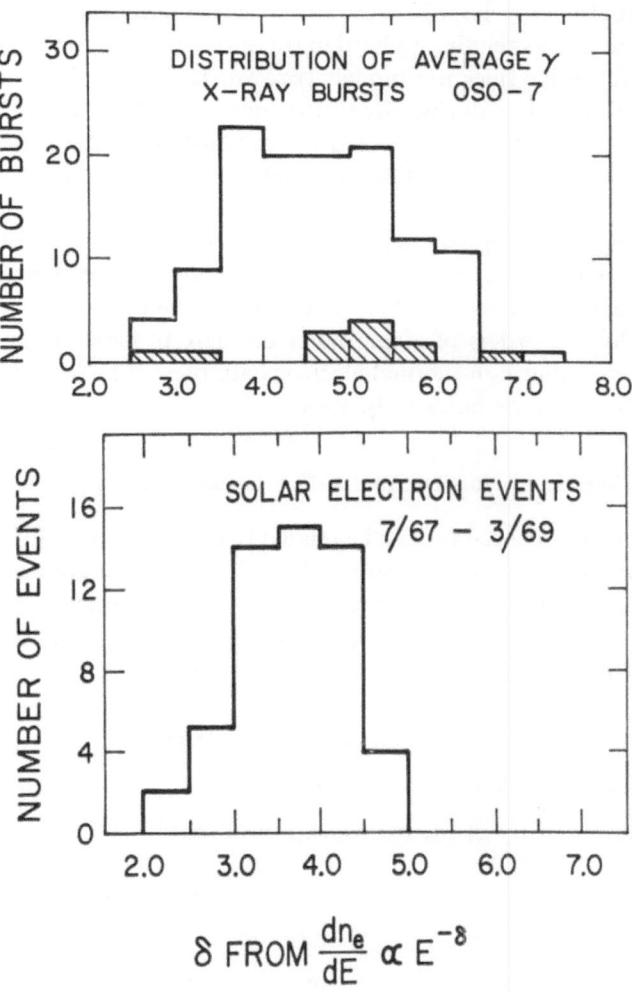

Fig. 7. The distribution of spectral exponents for non-thermal X-ray bursts (above) (from Datlowe, 1975) and electrons observed at 1 AU (below). The shaded events in the X-ray histogram are those from importance 1 flares; the rest are from subflares. For the thin target model the electron exponent
$$\delta = \gamma - \tfrac{1}{2}.$$

a lower limit at $\gamma = 3.0$ with very few bursts with harder spectra (Kane, 1971; Datlowe, 1975), and an upper limit of $\gamma \approx 6.5$. The limits $3.0 \lesssim \gamma \lesssim 6.5$ correspond to an instantaneous electron spectrum of $dn/dE \propto E^{-\delta}$ with $3.0 \lesssim \delta \lesssim 6.0$. Note the similarity to the range of electron spectra from interplanetary observations.

(4) The burst rise and decay times tend on many occasions to be smaller (more rapidly varying) for higher energies. Thus the X-ray spectrum and therefore the instantaneous electron spectrum varies from soft at burst onset ot hardest at burst maximum to soft again at burst end. This variation would be inconsistent with the simple model of impulsive injection of electrons, followed by decay of the electrons through collisional energy loss, because lower energy electrons would be lost most rapidly by collision, thus tending to harden the spectrum during the decay. Although more complex impulsive injection models (Brown, 1972) have been proposed to explain the energy variation during decay the most probable explanation is that the non-thermal electrons are being injected continuously over the period of the X-ray burst and the X-ray energy variations just reflect the variations of the acceleration source.

(5) Non-thermal X-ray bursts are commonly observed in flares. Datlowe (1975) finds that 2/3 of all flares observed by the OSO-7 soft X-ray detector were accompanied by non-thermal emission above 20 keV. The fact that the intensity of the non-thermal emission varies by 2–3 orders of magnitude from flare to flare within a given optical importance suggests that non-thermal electrons are present in every flare but may be too few in number in some flares to be observed.

III. Relationship of Non-Thermal X-Ray Emission to Energetic Electrons

The observed non-thermal X-rays can be related to the electrons at the Sun. Following Kane and Anderson (1970) we consider a volume V in the solar atmosphere containing a relatively cold ($T \approx 10^6$–10^7 K) hydrogen plasma and energetic ($\gtrsim 10$ keV) electrons. We assume the electron flux is isotropic; then the photon flux will also be isotropic. To obtain the X-ray flux at the Earth (distance $D = 1$ AU), let

E = kinetic energy of an energetic electron (keV)

$h\nu$ = energy of an X-ray photon (keV)

n_i = number density of the hydrogen nuclei (cm^{-3})

$\dfrac{dn(E)}{dE}$ = density of energetic electrons (electrons cm^{-3} s^{-1} keV^{-1})

$\dfrac{d\sigma(E, h\nu)}{d(h\nu)}$ = differential cross section for electron-proton bremsstrahlung

          (cm^2 ion^{-1} keV^{-1})

For non-relativistic electrons ($E \lesssim 100$ keV) the bremsstrahlung cross-section is approximately given by Bethe-Heitler formula (Jackson, 1962), written in our units $(E, hv$ in keV) as

$$\frac{d\sigma(E, hv)}{d(hv)} \approx 1.58 \times 10^{-24} \frac{1}{Ehv} \ln\left[\left(\frac{E}{hv}\right)^{1/2} + \right.$$
$$\left. + \left(\frac{E}{hv} - 1\right)^{1/2}\right] cm^2 \ ion^{-1} \ keV^{-1}. \tag{1}$$

Then at 1 AU the differential spectrum of the X-rays produced in V by bremsstrahlung is given by

$$\frac{dJ(hv)}{d(hv)} \approx 1.05 \times 10^{-42} \frac{n_i V}{hv} \int_{hv}^{100} \frac{1}{E^{1/2}} \frac{dn(E)}{dE} \ln\left[\left(\frac{E}{hv}\right)^{1/2} + \right.$$
$$\left. + \left(\frac{E}{hv} - 1\right)^{1/2}\right] dE \ photons \ cm^{-2} \ s^{-1} \ keV^{-1}. \tag{2}$$

Brown (1971) showed that this equation can be inverted for well-defined X-ray spectra to obtain the electron spectrum. For the special case of a power law-X-ray spectrum,

$$dJ/d(hv) = A(hv)^{-\gamma} \tag{3}$$

we obtain

$$\frac{dn}{dE} = 1.21 \times 10^{42} \gamma (\gamma - 1)^2 \ B\left(\gamma - \tfrac{1}{2}, \tfrac{3}{2}\right) \frac{AE^{-\gamma + 1/2}}{n_i V}, \tag{4}$$

where $B(x, y)$ is the beta function. The electron energy spectrum is of the form $dn/dE \propto E^{-\delta}$ where $\delta = \gamma - \tfrac{1}{2}$.[*]

The X-ray calculation also relates the X-ray intensity to the non-thermal 'emission measure', $n_e (> E_0) n_i V$, here defined with $n_e (> E_0) =$ density of non-thermal electrons above energy E_0, n_i the ambient ion density in the X-ray region and $V =$ volume of the X-ray region. The important thing to note here is that these relationships hold at a given *instant* of time. That is, the *instantaneous* number and spectrum of energetic electrons in the X-ray region is related to the instantaneous X-ray emission. These relationships hold for the electron population in the X-ray emitting region regardless of any assumptions about thick- or thin-target models. However, in order to obtain the total number and spectrum of electrons accelerated, the time each electron spends in the X-ray region (which may be a function of electron energy) and the ambient density, n_i, must be known.

Another important quantity which can be obtained independent of model is the minimum collisional energy loss of the electrons in the X-ray emitting region. This quantity can be obtained because both collisional and bremsstrahlung energy loss are

---

[*] This relationship has also been computed numerically with δ increasing sharply above 100 keV (Kane and Anderson, 1970; Lin and Hudson, 1971), and gives $\gamma \approx \delta + 0.7$ in the energy range 10 keV $< hv <$ 80 keV.

dependent on the ambient density so that the ratio of the two losses is independent of ambient density. For example, in fully ionized hydrogen, collision losses are (Trubnikov, 1965)

$$\frac{dE}{dt_{collision}} = -4.9 \times 10^{-9} n_i E^{-1/2} \text{ keV s}^{-1} \tag{5}$$

and from Berger and Seltzer (1964) we obtain for bremsstrahlung

$$\frac{dE}{dt_{brems}} = -6.2 \times 10^{-18} n_i E^{1/2} (E + 988) \text{ keV s}^{-1} \tag{6}$$

so that the ratio is

$$R(E) = \frac{dE}{dt_{brems}} \left(\frac{dE}{dt_{collision}}\right)^{-1} = 1.27 \times 10^{-9} E (E + 988) \tag{7}$$

Figure 8 shows these ratios averaged over power law spectra from $\sim$20–100 keV (Lin and Hudson, 1971) for both neutral and ionized hydrogen. By obtaining the *total* bremsstrahlung energy loss and multiplying by $R(E)$ one can obtain the minimum collisional energy loss in the X-ray emitting region.

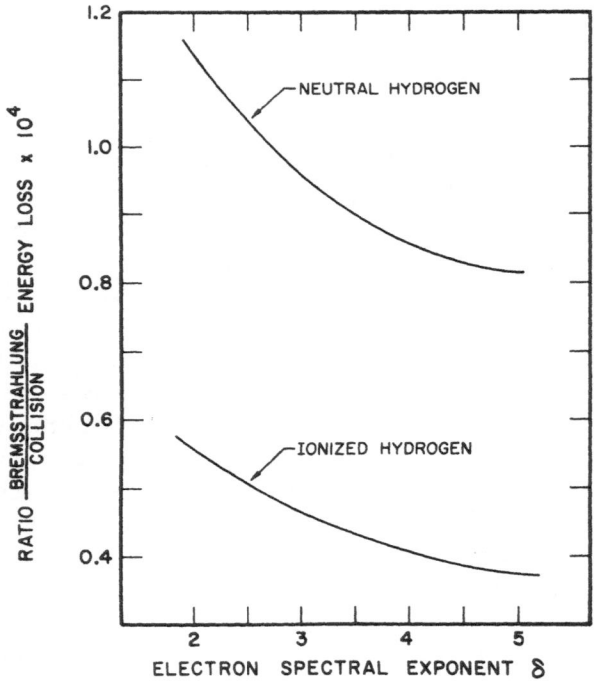

Fig. 8.   The ratio of electron energy loss by bremsstrahlung to energy loss by collisions for both neutral and ionized hydrogen is shown here as a function of the electron spectral exponent. The energy losses are averaged over the electron spectrum in the energy range 22–100 keV.

IV. The Evolution of the Electrons at the Sun

The relationship of the instantaneous X-ray producing electron spectrum to the *accelerated* electron spectrum depends on the evolution of the electrons subsequent to acceleration. Suppose the electrons are accelerated in one region and produce the bulk of the observed X-rays in another region (these two regions may be one and the same but for the sake of generality we will allow them to be different). The evolution of the electron distribution, $N(E, t) = V(dn/dE)$ where V = volume in the X-ray emitting region, can be described by the equation

$$\frac{\partial N(E, t)}{\partial t} = F(E, t) - \frac{N(E, t)}{\tau_e(E)} - \frac{\partial}{\partial E}\left[N(E, t)\frac{dE}{dt}\right], \tag{8}$$

where $F(E, t)$ is the input source of electrons $keV^{-1} s^{-1}$, $N(E, t)/\tau_e(E)$ is the number of electrons escaping the region per second, and the third term describes energy change processes for the electrons. These energy change processes could be loss by collisions with the ambient medium and/or loss by non-collisional processes such as wave-particle interactions. Although wave-particle interactions may be the dominant form of electron energy loss, the lack of relevant observations and the theoretical complexity of the interactions rule out the possibility of any quantitative estimates of their effects. Instead we shall treat just the collisional energy losses and bear in mind that these constitute a lower limit to the actual electron energy loss. Note that X-ray observations define $N(E, t)$ subject to a choice of ambient density n_i (Equation (4)). Thus this equation can be solved for $F(E, t)$, given n_i and given the form of $\tau_e(E)$, if only collisional energy losses are assumed to be important. This solution has been carried out numerically for several X-ray bursts for various n_i and $\tau_e(E)$ (Kane, 1973a).

To a good approximation we can consider $N(E, t)$ as constant over some time interval, Δt, and zero outside that interval. This removes the time dependence of the equation. Additionally we shall consider only the power-law case, $N(E, t) = BE^{-\delta}$, so that inserting for dE/dt the energy loss in ionized hydrogen (Equation (5)) we obtain

$$F(E) = N(E)\left[\frac{1}{\tau_e(E)} + \frac{4.9 \times 10^{-9}n_i(\delta + \frac{1}{2})}{E^{3/2}}\right]. \tag{9}$$

We have computed the anticipated energy dependence of $F(E, t)$ compared to $N(E, t)$ and $dJ(hv)/d(hv)$, and the energy dependence of the escaping electrons for two extremes:

(1) where the escape term is much larger than the collisional energy loss term. This situation is the *thin-target* approximation for X-ray emission;

(2) where the collisional energy loss term is much larger than the escape term. This situation is the *thick-target* approximation for X-ray emission.

We have used two obvious choices for the energy dependence of τ_e, although other forms might be appropriate. These two are: (1) τ_e = constant, and (2) $\tau_e \propto E^{-1/2}$, i.e., proportional to the scale size of the X-ray region divided by the particle velocity. The results are summarized in Table I.

The spectrum of electrons escaping from the X-ray region given in Table I is not necessarily the spectrum of the electrons escaping to the interplanetary medium. The electrons need not escape to the interplanetary medium to be lost from the X-ray region; they may also escape to the low density, $n_i \lesssim 10^9$ cm^{-3}, upper corona, where the flux of X-rays they produce will be below the threshold of current X-ray detectors. Also the acceleration region may be much higher in the solar atmosphere than the X-ray region, and the electrons observed in space may have come directly from the accelerated population.

TABLE I

Spectral dependence of electrons and X-rays

	Thick-target	Thin-target
Spectrum of X-rays	$\dfrac{\mathrm{d}J(h\nu)}{\mathrm{d}(h\nu)} = A(h\nu)^{-\gamma}$	$\dfrac{\mathrm{d}J(h\nu)}{\mathrm{d}(h\nu)} = A(h\nu)^{-\gamma}$
Spectrum of electrons in X-ray emitting region $N(E) \propto \dfrac{\mathrm{d}n}{\mathrm{d}E} \propto E^{-\delta}$	$\delta = \gamma - \frac{1}{2}$	$\delta = \gamma - \frac{1}{2}$
Spectrum of accelerated electrons $F(E) \propto E^{-\delta_a}$	$\delta_a = \gamma + 1$	$\delta_a = \gamma - \frac{1}{2}$ for $\tau_e = $ constant $\delta_a = \gamma - 1$ for $\tau_e \propto E^{-1/2}$
Spectrum of electrons escaping from the X-ray region, $S(E) \propto E^{-\delta_e}$	$\delta_e = \gamma - \frac{1}{2}$ for $\tau_e = $ constant $\delta_e = \gamma - 1$ for $\tau_e \propto E^{-1/2}$	$\delta_e = \delta_a = \begin{cases} \gamma - \frac{1}{2} & \text{for } \tau_e = \text{constant} \\ \gamma - 1 & \text{for } \tau_e \propto E^{-1/2} \end{cases}$

The time scale for loss of the electrons must be shorter than the most rapid decay observed for X-ray bursts. These decay time scales may be ~ 1 s (e-folding), faster than the resolution of current X-ray instrumentation. Similarly, the radio (type III) observations indicate the acceleration probably varies on a scale of $\lesssim 1$ s.

The collisional energy loss can be described in terms of a time constant

$$\tau_c = \frac{E^{3/2}}{4.9 \times 10^{-9}(\delta + \frac{1}{2})\, n_i}, \tag{10}$$

where E is in keV, n_i in cm^{-3} (Equation (5)). For typical values of δ, a collision time constant $\tau_e \approx 1$ s implies $n_i \approx 6 \times 10^{10}$ cm^{-3} for $E = 100$ keV; $n_i \approx 10^{10}$ cm^{-3} for $E \approx 50$ keV and $n_i \approx 2 \times 10^9$ cm^{-3} for $E = 10$ keV.

On the other hand τ_e may be as short as the rectilinear travel time to cross the X-ray emitting region. This region is likely to be a few thousand km thick to that $\tau_e \gtrsim 0.02$–0.1 s for ~ 5–100 keV electrons.

V. Thick- or Thin-Target?

Datlowe and Lin (1973) noted that it is possible to distinguish between thick- and thin-target cases under the assumption that the spectrum of electrons observed in the interplanetary medium is representative of the accelerated electron spectrum (i.e.,

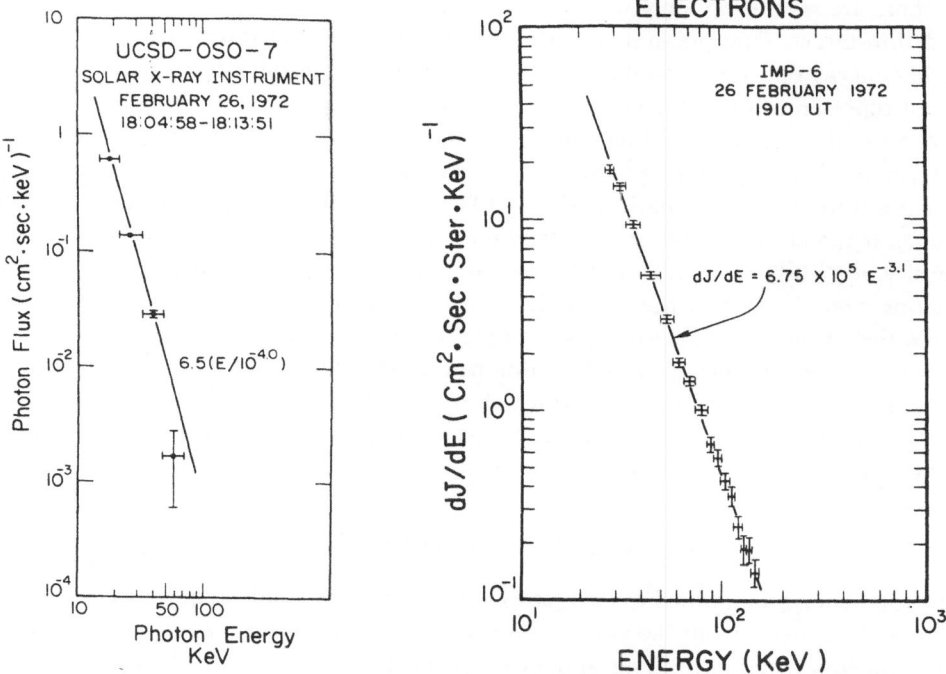

Fig. 9. The spectra of hard X-rays and electrons observed at 1 AU for the same flare event. The photons fit a power law spectrum $dJ(hv) = A(hv)^{-\gamma}$ where $\gamma = 4.0 \pm 0.3$, while the electrons fit a spectrum $dJ/dE = 6.75 \times 10^5 E^{-3.1}$. Since $dn/dE = v \, dJ/dE$ where v is the electron velocity, the electron fit a *density* spectrum $dn/dE \propto E^{-\delta}$ with $\delta = 3.6 \pm 0.1$. These two spectra are consistent with thin target emission under the assumption the escaping electrons have the same spectrum as the accelerated electrons.

$\delta_e = \delta_a$, see Table I). For one flare event where high energy resolution measurements were available for both the electrons and X-rays above 20 keV (Figure 9), the result was $\delta_a = \gamma - \frac{1}{2}$, favoring thin-target. Other X-ray electron events studied where only measurements with poor energy resolution were available are also generally consistent (see Lin and Hudson, 1971; and Kane and Lin, 1972) with a thin-target model. The thin-target case is also consistent with the location of the acceleration region $(n_i \lesssim 10^{10} \text{ cm}^{-3})$ derived from considerations of the low energy electron spectrum observed at 1 AU.

In favor of thick-target processes we note that if non-relativistic electrons penetrate to the dense $(n_i \gtrsim 10^{12} \text{ cm}^{-3})$ regions of the chromosphere-corona boundary and below, they could produce the observed EUV and peraps provide the energy for heating the Hα flare region through collisional loss (and possibly even heating the white light flare region) (Hudson, 1972). The close time coincidence between the hard X-ray spike and the EUV spike (Kane and Donnelly, 1971) is consistent with such an interpretation. At those densities the thick-target approximation would certainly be appropriate.

There are several possibly ways of reconciling the observations in support of thick-
and thin-target. One possibility is that the escape of the electrons from the flare is
highly energy dependent so that the spectrum observed in the interplanetary medium
is not representative of the accelerated electron spectrum.

A second possibility is that electrons of low energies, say below ~ 10 keV, are
described by the thick-target approximation while higher energy electrons are in
essentially thin-target situation (Kane, 1973a). This dichotomy could arise, for example, if the electrons are accelerated and contained by a magnetic 'bottle' in a low
density, $n_i \lesssim 10^{10}$ cm^{-3}, region. Electrons only appear in high density, $n_i \lesssim 10^{10}$ cm^{-3},
regions near the feet of the magnetic bottle if they are scattered into the loss cone.
Since the amount of scattering is a strongly decreasing function of energy, essentially
only the low energy electrons will be dumped into the loss cone. This interpretation is
consistent with the observations which show that the correspondence between rising
portion of the EUV emission and the rising portion of the non-thermal X-rays is best
for the lowest energy, ~ 10 keV, X-rays.

VI. Energy in Non-Relativistic Electrons

The total energy contained in energetic electrons in these small solar flares is obtained
by summing losses from the various processes which the electrons undergo, including
collision loss, bremsstrahlung and gyro-synchrotron emission, and escape into the
interplanetary medium. Almost a dozen events with both impulsive hard X-ray bursts
and electrons subsequently observed at 1 AU have been analyzed (Lin and Hudson,
1971; Kane and Lin, 1972) to obtain the energy in non-relativistic electrons. In addition, Datlowe (1975) has analyzed over 100 non-thermal X-ray burst events from subflares and importance 1 flares. Typical energy values are given in Table II.

(a) COLLISIONAL ENERGY LOSS

An important difference between thick- and thin-target for flare processes lies in the
amount of energy lost by collisions of the non-thermal electrons to the ambient
medium. This collision energy loss is the non-thermal electron energy input available
to support other flare processes such as heating of the quasi-thermal X-ray plasma and
production of EUV and Hα radiation. The thin-target approximation clearly will give
a lower limit to the non-thermal electron collisional energy loss, while the thick-target
approximation represents an upper limit. These limits on the collision energy loss
from electrons above energy E_0 can be computed from the observed X-ray emission.
Suppose

$$\frac{\mathrm{d}J\,(h\nu)}{\mathrm{d}\,(h\nu)} = A\,(h\nu)^{-\gamma}$$

then the rate of collisional energy loss is

$$P_{\mathrm{thin}}\,(> E_0) = 9.4 \times 10^{24} A\gamma\,(\gamma - 1)\,B\left(\gamma - \tfrac{1}{2}, \tfrac{3}{2}\right) E_0^{-(\gamma-1)} \text{ erg s}^{-1} \qquad (11)$$

TABLE II

Energy balance for a small electron flare

	Energy in erg	
	>5 keV	>20 keV
A. Non-thermal electron energy		
1. Electron energy loss through collisions		
(a) Thin-target	$\sim 10^{30}$	$\sim 5 \times 10^{28}$
(b) Thick-target	$\sim 4 \times 10^{30}$	$\sim 2 \times 10^{29}$
2. Other electron energy loss processes bremsstrahlung X-ray emission gyro-synchrotron radio emission escape to the interplanetary medium	$\lesssim 0.01$ of thin-target collision loss	
3. Total Energy in *Accelerated* Electrons		
(a) Thin-target	$\sim 5 \times 10^{30}$	$\sim 3 \times 10^{29}$
(b) Thick-target	$\sim 5 \times 10^{30}$	$\sim 2 \times 10^{29}$
B. Quasi-thermal flare energy		
1. Soft X-ray plasma	$\sim 5 \times 10^{29}$ erg	
2. EUV emission	$\sim 5 \times 10^{28}$ erg	
3. Hα and other optical emissions	$\sim 10^{28}$ erg	

for thin-target,* and

$$P_{\text{thick}} (> E_0) = 9.4 \times 10^{24} A\gamma (\gamma - 1) B (\gamma - \tfrac{1}{2}, \tfrac{3}{2}) E_0^{-(\gamma - 1)} \text{ erg s}^{-1} \qquad (12)$$

for thick-target,* so that the ratio

$$(P_{\text{thick}}/P_{\text{thin}}) = \gamma . \qquad (13)$$

Note that this ratio is independent of the low energy cut-off E_0.

The range of γ is observed to be from ~ 2 to ~ 7. Typical values of the collision energy loss for the small electron flares are given in Table II for both thick- and thin-target approximations.

(b) ENERGY IN ELECTRONS ESCAPING TO THE INTERPLANETARY MEDIUM

The energy lost in escaping electrons can be obtained by estimating the number of escaping electrons using simple propagation theory (Lin and Hudson, 1971) and multiplying by the average electron energy. Typical values for the energy lost in escaping electrons are 10^{-2} to 10^{-3} the collision energy loss, even in the thin-target case (Lin and Hudson, 1971), for flares where the energetic electrons contain the bulk of the energy. Thus, only about 0.1 to 1% of the electrons escape to the interplanetary medium in those cases.

* These expressions differ from what would be derived from Brown's (1971) expressions. The difference is due to a different expression for dE/dt and some numerical errors in Brown (1971).

(c) ENERGY LOST IN RADIO EMISSION

Another electron energy loss mechanism is gyro-synchrotron emission in the solar magnetic field. Gyro-synchrotron emission is the most likely explanation for the impulsive microwave burst. Takakura (1972) notes that careful consideration of a realistic model taking into account the non-uniformity of the magnetic field, and the self-absorption of the radio emission indicates that the electron population producing hard X-rays will also produce the microwave burst with the observed spectrum and intensity. Since these electrons are not relativistic, the Larmor formula for the power radiated (Jackson, 1962) per electron is appropriate and leads to

$$P \approx 3 \times 10^{-9} \ B_{\perp}^2 \ E \ \mathrm{s}^{-1}, \tag{14}$$

where B is the magnetic field in gauss and E is the electron energy in keV. For choices of B and n_i appropriate for the emission under this model we find that the collision loss is $\gtrsim 10^2$–10^3 times the loss by gyro-synchrotron emission for 10–100 keV electrons.

Type II radio bursts are produced through the excitation of plasma waves by a coherent Cerenkov mechanism (see next section). These non-collisional wave-particle interactions, although they are a negligible energy loss mechanism at 1 AU (Lin *et al.*, 1973), may be a substantial electron energy loss process nearer the Sun.

(d) TOTAL ENERGY IN ACCELERATED ELECTRONS

Under the thin-target assumption the fraction of electrons lost in collisions must be small. Therefore the collisional energy loss given by Table II for the thin-target case is only a small fraction of the total energy in *accelerated* electrons. The bulk of the accelerated electrons under the thin-target assumption must escape from the dense regions where they would be lost by collisions. Escape to the interplanetary medium is rather negligible. However, as mentioned earlier the electrons may also escape to the low density, $n_i \lesssim 10^9$ cm^{-3} upper corona, where the flux of X-rays they produce will be below the thresholds of current X-ray detectors.

We have, therefore, estimated the energy contained in accelerated electrons in the thin-target case by assuming that the electron lifetime (τ_e in this case) in the X-ray producing region is $\lesssim 1$ s. This time scale would be consistent with the observed variations of the X-ray bursts.

The thick-target collisional energy loss should correspond closely to the total accelerated electron energy since the thick-target assumption implies that the bulk of the electrons is lost through collisions.

Two low energy cutoffs (E_0) have been used in these computations: 20 keV because most of the hard X-ray and electron measurements have been made above that energy; and 5 keV since the few observations available of electrons and X-rays at low energies indicate that the non-thermal spectrum commonly extends to that energy. The energy in accelerated electrons above ~ 20 keV already constitutes a large fraction of the total flare energy.

(e) NUMBERS OF ELECTRONS

The total number of electrons accelerated can be obtained by dividing the total energy in accelerated electrons by the average electron energy. This method yields $\sim 10^{38}$ electrons above ~ 5 keV, or $\sim 10^{36}$ above ~ 20 keV. The number escaping to the interplanetary medium is $\sim 10^{33}$ above 20 keV$\sim$ or $\sim 0.1\%$ of the total number accelerated. This compares with estimates of Holt and Ramaty (1969) for the escape efficiency of relativistic electrons of between 0.6% to 23%.

If the > 20 keV X-ray production is assumed to occur at an ambient density of $\gtrsim 10^{10}$ cm^{-3} then the maximum instantaneous number of > 20 keV electrons in the X-ray region is estimated to be $\lesssim 10^{35}$, or at least one order of magnitude fewer than the total number accelerated. This confirms that the acceleration of electrons must have occurred continually over the duration of the X-ray burst.

VII. Total Energy in a Small Electron Flare

Sufficient observations are now available so that an essentially complete energy balance for a small solar electron flare can be constructed. For comparison this energy balance is also given in Table II.

(a) SOFT X-RAYS

Essentially all flares are observed to be accompanied by soft X-ray emission. The hard impulsive X-ray component appears to coincide with the most rapidly increasing portion of the soft X-rays' rise, as if the 10–100 keV electrons were giving up their energy to heating up the soft X-ray producing plasma (Kahler *et al.*, 1970; McKenzie *et al.* 1973). Although some heating of the soft X-ray plasma is probably provided by the 10–100 keV electrons it is unlikely that the entire soft X-ray burst is due to this process because (1) the soft X-ray burst is observed in many flares without any substantial hard impulsive burst; and (2) the soft X-ray emission appears to begin before the hard impulsive X-ray bursts and continues to rise after the end of the hard X-ray emission (Datlowe, 1975).

Measurements of the soft X-ray spectrum provide the temperature. T, and emission measure, $n_e n_i V$, of the soft X-ray plasma. Thus, the observations are able to define unambiguously the quantity (since $n_e \approx n_i$ for the 5 to 30×10^6 K flare plasmas)

$$n_e U = 3 \, kT \, n_e n_i V. \tag{15}$$

A typical emission measure for these small flares is $n_e n_i V \approx 10^{48}$ cm^3 and typical temperature is $T \sim 10^7$ K (Hudson *et al.*, 1969; McKenzie *et al.*, 1973), so that $n_e U \approx 5 \times 10^{39}$ erg cm^{-3}.

The volume of the soft X-ray region for a small flare is estimated to be $\sim 10^{28}$ cm^{-3} from measurements of the X-ray source size (Krieger *et al.*, 1972; Thomas and Neupert, 1971), so that the density $n_e \approx 10^{10}$. The energy in the soft X-ray plasma is then

$\sim 5 \times 10^{29}$ erg. In view of the approximations, this number is an order of magnitude estimate.

(b) EUV EMISSION

EUV emission in the wavelength range 10 to 1040 Å is commonly observed coinciding with and showing similar structure to the impulsive hard X-ray burst (Figure 2). The total energy emitted in EUV appears proportional to the energy in bremsstrahlung X-rays with a factor (Kane and Donnelly, 1971)

$$\varepsilon_{EUV} (10 - 1030 \text{ Å}) \approx 5 \times 10^4 \varepsilon_{X \text{ ray}} (10 - 50 \text{ keV}). \tag{16}$$

For small electron flares a typical ε_{EUV} is $\sim 5 \times 10^{28}$ erg. Kane and Donnelly (1971) noted that the best agreement is found between the EUV and X-rays of ~ 10 keV energy.

(c) Hα EMISSION

The total Hα emission energy can be estimated from the characteristics of the average importance 1 flare (Smith and Smith, 1963). The area is taken as ~ 100 millionths of the disk; the maximum bandwidth, ~ 3 Å, the peak intensity, ~ 0.8 of the continuum level, and the duration, $\sim 10^3$ s. The average emission in Hα is therefore $\sim 10^{28}$ erg.

Smith and Smith (1963) note that a significant fraction of the visible spectrum energy of a flare is contained in Hα, particularly for small flares. Energetically the optical emission constitutes a very minor part of an electron flare.

(d) TOTAL ENERGY IN FLARE

The energy in each part of the flare is listed in Table II. We have not included mechanical energy from mass motion in the flare or any dissipation/cooling which does not result in observable radiation. For the large flares which produce interplanetary shocks the energy contained in the shock is a major and sometimes dominant part of the total flare energy budget (Hundhausen, 1972). However, for these small flares no signs of shock phenomena are observed, either in the radio emission or in interplanetary space.

It is quite clear from Table II that the bulk of the energy in a small electron flare resides in the non-thermal electron population. In the thin-target cases the collisional energy loss of the > 5 keV electrons is of the same order as the energy needed to produce the other flare emission. Under the thick-target approximation too much energy, by about an order of magnitude, is available from the electrons.

VIII. Location of the Electron Acceleration Region

Measurements of the electron energy spectrum observed at 1 AU down to low energies can be interpreted in terms of an upper limit to the amount of material traversed by the accelerated particles (Lin, 1973). Suppose the electrons are accelerated at a height, h, in the solar atmosphere, and then pass through the overlying material (fully ionized hydrogen) to reach the vicinity of 1 AU. Using Trubnikov's (1965) expression for energy loss in ionized hydrogen (Equation (5)) we obtain for non-relativistic electrons

$$\frac{dE}{dx} = -2.6 \times 10^{-18} \frac{n_i}{E} \text{ keV cm}^{-1}.$$ (17)

Integrating from height h to 1 AU

$$\int_{E_1}^{E_2} E \, dE = -2.6 \times 10^{-18} \int_{h}^{1AU} n_i(x) \, dx \equiv K(h), \quad \text{and}$$

$$E_2 = (E_1^2 - 2K)^{1/2},$$ (18)

where E_1 is in initial accelerated energy of the electron and E_2 its energy at 1 AU. If the spectrum of the freshly accelerated electrons at the Sun is given by a power law in energy, as would be consistent with the X-ray observations,

$$\frac{dn}{dE_1} = AE_1^{-\delta}$$ (19)

then the spectrum of the electrons observed at 1 AU will be

$$\frac{dn}{dE_2} = \frac{AE_2}{(E_2^2 + 2K)^{(\delta+1)/2}}.$$ (20)

This spectrum has a maximum at

$$E_{2M} = \left(\frac{2K}{\delta}\right)^{1/2}$$ (21)

Actually for a given height h, the location of the maximum must be above this energy because:

(i) the direct radial distance outward through the solar atmosphere is used in the calculation without taking into account the helical paths followed by the particles;

(ii) scatterings which change the particle's direction are far more effective for low energy particles than for high energy particles and will subject the low energy particles to longer path lengths;

(iii) no other energy loss mechanisms such as generation of radio or plasma waves, etc., are taken into account.

None of the low energy solar electron spectra observed to date show a turnover above ~ 6 keV (Figure 10). Thus

$$\int_{h}^{1AU} n_i(x) \, dx \lesssim 3.5 \times 10^{19} \text{ cm}^{-2}.$$ (22)

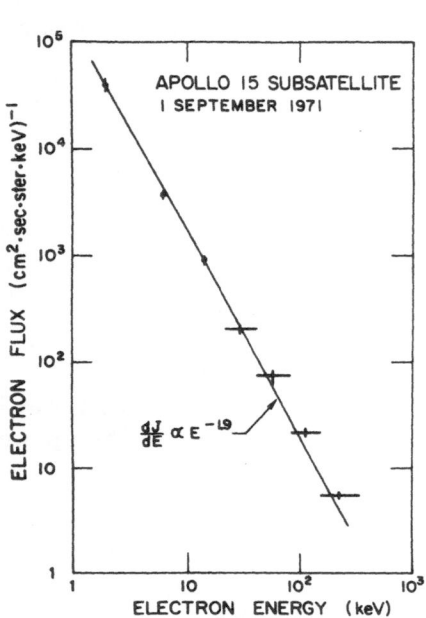
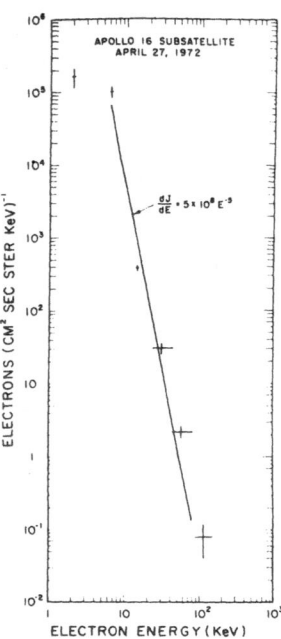

Fig. 10.   Two electron energy spectra extending to low energies. The September 1 event is accompanied by energetic protons while the 27 April event (from Lin *et al.*, 1973) is not. Both spectra extend
smoothly in a power law to below ∼6 keV.

We wish to re-emphasize the fact that this estimate is a *lower* limit to the actual
height of acceleration since the effects which were not taken into account would tend
to increase the minimum energy of the peak. Clearly the electron acceleration must
have occurred in the transition region or lower corona. Although only a few events
have been observed to energies below ∼20 keV, in no events has a turnover been observed at higher energies. Thus electron acceleration at the flash phase appears to be a
coronal phenomenon, at least for events observed to emit electrons into the interplanetary medium.

This location and ambient density is consistent with the observed starting frequencies (∼200–1000 MHz) of type III bursts, and is also consistent with the occasio-
nal observation of an electron event at 1 AU without detectable X-ray emission
(Kane and Lin, 1972). Presumably in those events the magnetic field structure in the
vicinity of the acceleration region is such as to prevent the electrons from entering
dense regions where a detectable flux of X-rays would be produced.

IX. Flash Phase Proton Acceleration

One question of importance in distinguishing possible flare particle acceleration
mechanisms during the flash phase is whether or not protons as well as electrons are

accelerated, and if so whether the acceleration mechanism accelerates particles to the same energy or the same velocity of rigidity, and what are the relative efficiencies for electrons and protons (Syrovatskii, 1969). Recently McDonald *et al.* (1972) noted that on occasion proton *micro-events* are observed. These events are characterized by extremely low fluxes, $\sim 10^{-2}$ (cm^2 s sr)$^{-1}$ above 10 MeV which, however, are observed up to $\gtrsim 50$ MeV, and no indications (such as type II and IV emission) of a second phase in the associated flare phenomena. We have plotted the maximum >45 keV

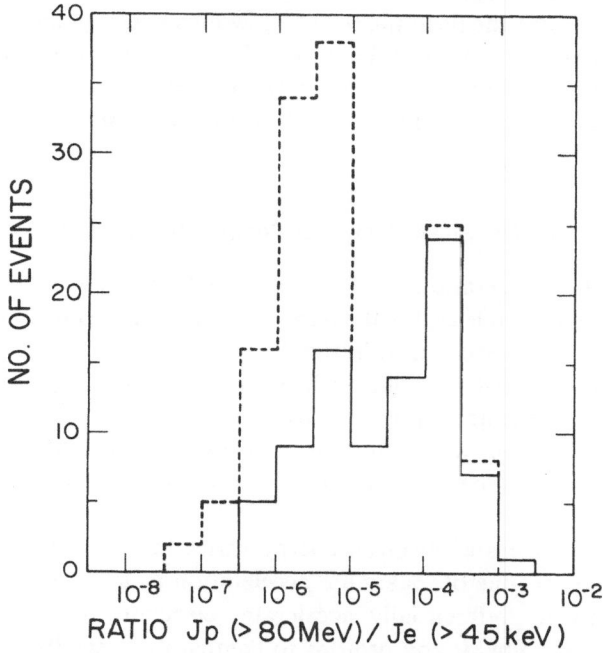

Fig. 11. The number of events plotted vs the ratio of protons to electrons of the same velocity. The solid curve is constructed from all the points for which a proton flux was actually observed. The dotted curve includes those points for which only an upper limit to the proton flux was obtained. Two peaks appear in the distribution, one at ratios of $\sim 10^{-6}$ to 10^{-5} and the other at ratios of $10^{-4.5}$ to 10^{-3}. The peak at ratios of $10^{-4.5}$ to 10^{-3} includes almost all energetic proton events as defined by the criterion given in the text.

electron flux vs the maximum > 10 MeV proton flux for all impulsive events which were observed by IMP or OGO spacecraft ionization chambers (S. R. Kane, 1974, private communication). These chambers were used because of their high sensitivity to ~ 10–15 MeV protons; events with maximum > 10 MeV proton flux of 3×10^{-3} (cm^2 s sr)$^{-1}$ could be picked out of the counting rates. No attempt was made to correct for any propagation effects so a great deal of scatter is present. Many events, including a majority of the pure electron events, were unaccompanied by a detectable proton increase. Figure 11 plots the number of events versus the ratio of proton flux above 80 MeV to electron flux, above 45 keV (i.e., above equal particle velocities), assuming an E^{-3} spectrum for the protons. Two peaks are evident in the distribution, one between ratios of 10^{-3} and $10^{-4.5}$ and the other beyond 10^{-5}. Almost all the large proton events – events with a second acceleration phase – fall in the 10^{-3} to $10^{-4.5}$ peak, while the micro-events fall into the $>10^{-5}$ portion. The events in the second group may be flash phase proton events. However, we note that most electron events (65 out of 95) are unaccompanied by *any* proton fluxes above the ionization chamber threshold. Confirmation of the micro-event proton origin in the flash phase could be obtained if an analysis of the velocity dispersion shows that the injection time is indeed coincident with the flash phase. The low fluxes may preclude sufficiently accurate analysis. Tentatively, we conclude that although some flash phase events may produce a small flux of energetic protons with velocities comparable to the electron observed from the same event, the efficiency of the acceleration is much lower than in usual proton events.

X. The Flash Phase Acceleration Mechanism

For many flares the acceleration of electrons to $\sim$5–100 keV energies must be the dominant form of energy release for the flare. We can place stringent requirements on the electron acceleration mechanism in the flare:

(1) The mechanism must be highly efficient in the sense that most of the flare energy is contained in the accelerated electrons.

(2) The time scale for the acceleration is $\lesssim$ few s, and more likely $\lesssim$ 1 s.

(3) Acceleration occurs over a period of 10 to 100 s, possibly in the form of a group of short pulses.

(4) The average accelerated electron energy varies through the event, going from low at onset to high at time of maximum acceleration to low again at the end.

(5) The mechanism preferentially accelerates electrons to $\sim$5–100 keV energy while accelerating very few, if any protons to comparable velocities.

(6) Approximately 10^{36} electrons are accelerated above $\sim$20 keV.

(7) The spectrum of the accelerated electrons is rarely harder than $dn/dE \propto E^{-2.5}$.

(8) The acceleration region is located in the lower corona at densities of $n_i \lesssim 10^{10}$ cm^{-3}.

(9) Only 0.1 to 1% of the accelerated electrons escape to the interplanetary medium.

Of these restrictions, clearly the first one is the most difficult to meet. Basically the flare mechanism must be such as to accelerate large numbers of electrons as the primary energy dissipation process. This high efficiency would rule out stochastic or resonant acceleration processes.

We note here that in most flare mechanisms (see Sweet, 1969, for review) there arise large currents and current densities. Laboratory studies of plasmas carrying substantial currents have shown that very efficient and rapid energization of the plasma electrons through collective wave-particle effects results for large enough values of the current density (Hamberger *et al.*, 1971).

This situation occurs for current densities $j = n_e v_d$ corresponding to drift velocities v_d greater than the electron thermal speed

$$v_d \gtrsim v_{th} = \left(\frac{kT_e}{m}\right)^{1/2}. \tag{23}$$

Under these conditions almost all of the input energy is transferred to the plasma electrons (see Figure 12) thus producing a population of energetic electrons.

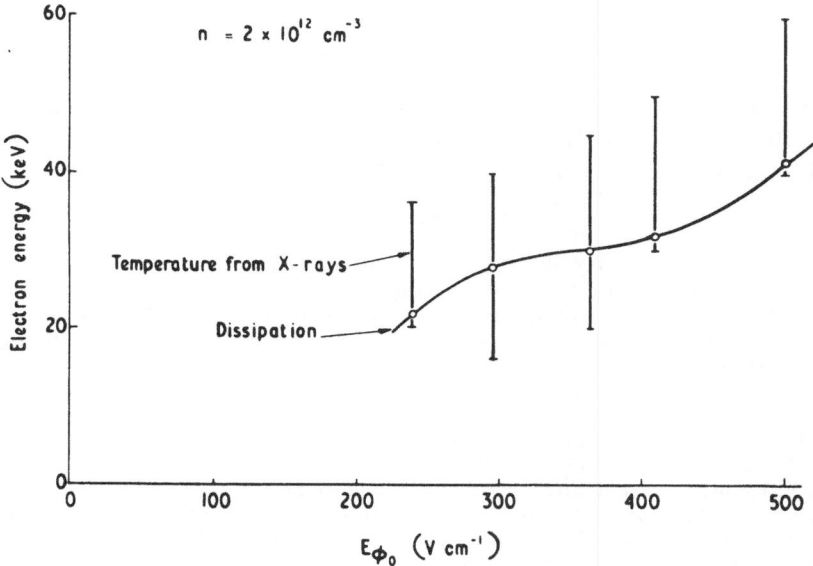

Fig. 12.   The results of laboratory experiments in turbulent heating of plasmas, showing that *most* of the energy fed into the plasma goes to energizing the electrons in the plasma in the Buneman anomalous resistivity region. The smooth curve gives the energy dissipation per plasma electron (obtained from the measured plasma current and applied electric field) while the bars indicate the average electron energy from X-ray measurements. Although this particular measurement is made at a density of 2×10^{12} cm³ the results are valid over a several order of magnitude range of densities (from Hamberger *et al.*, 1971).

In large proton flares the 5–100 keV electrons accelerated in the flash phase are still found to contain the bulk of the total flare energy (Hudson *et al.*, 1975; Hoyng *et al.*, 1975), only in these flares the total energy is $\sim 3 \times 10^{32}$ erg above 20 keV and $\gtrsim 10^{33}$ erg for any reasonable choice of the low energy cutoff to the electron spectrum. The rate of energy dissipation must be $\gtrsim 10^{30}$ erg s^{-1} and the bulk of the energy must appear as energetic electrons. These conditions are difficult to meet by any magnetic field merging model for flares because the rate of energy dissipation is very rapid, and because most of the energy in magnetic field merging comes out as heating and bulk flow of the plasma rather than as energetic electrons. An attractive alternative to magnetic merging is the following: currents are set up along magnetic field lines by the proper motion of the sunspots, and suddenly dissipated through some instability (for example, pinch or magnetic merging) which manages to increase the current density to the critical level for anomalous resistivity to set in. Evidence for this process is provided by the analysis of Tanaka and Nakagawa (1973) of the August 7 flare.

The second phase in large solar flares appears to be the result of a shock wave and ejected material, presumably caused by the explosive heating of the solar atmosphere by the flash phase electrons. When these electrons are few in number, as in small flares, no material is ejected and no second phase ensues.

Acknowledgements

I wish to acknowledge the support of Professor K. A. Anderson. Discussions with Drs J. C. Brown and S. R. Kane were helpful. This research was supported in part by NASA Grant NGL 05–003–017.

References

Anderson, K. A. and Lin, R. P.: 1966, *Phys. Rev. Letters* **16**, 1121.
Anderson, K. A. and Winckler, J. R.: 1962, *J. Geophys. Res.* **67**, 4103.
Arnoldy, R. L., Kane, S. R., and Winckler, J. R.: 1968, *Astrophys. J.* **151**, 711.
Berger, M. J. and Seltzer, S. M.: 1964, 'Tables of Energy Losses and Ranges of Electrons and Positrons', NASA SP-3012.
Brown, J. C.: 1971, *Solar Phys.* **18**, 450.
Brown, J. C.: 1972, *Solar Phys.* **25**, 158.
Datlowe, D.: 1975, This volume, p. 191.
Datlowe, D. and Lin, R. P.: 1973, *Solar Phys.* **32**, 459.
Fainberg, J., Evans, L. G., and Stone, R. G.: 1972, *Science* **178**, 743.
Hamberger, S. M., Jancarik, J., Sharp, L. E., Aldcroft, D. A., and Wetherall, A.: 1971, *Proc. International Conference Plasma Physics and Controlled Nuclear Fusion*, Vol. II, 37.
Holt, S. S. and Ramaty, R.: 1969, *Solar Phys.* **8**, 119.
Hoyng, P., Brown, J. C., Stevens, G., and Van Beek, H. E.: 1975, This volume, p. 233.
Hudson, H. S.: 1972, *Solar Phys.* **24**, 414.
Hudson, H. S., Jones, T. W., and Lin, R. P.: 1975, This volume, p. 425.
Hudson, H. S., Peterson, L. E., and Schwartz, D. A.: 1969, *Astrophys. J.* **157**, 389.
Hundhausen, A. J.: 1972, in C. P. Sonett, P. J. Coleman, Jr., and J. M. Wilcox (eds.), *Solar Wind*, NASA SP-308, p. 393.
Jackson, J. D.: 1962, *Classical Electrodynamics*, J. Wiley and Sons, New York.
Kahler, S. W. and Kreplin, R. W.: 1971, *Astrophys. J.* **168**, 531.
Kahler, S. W., Meekins, J. F., Kreplin, R. W., and Bowyer, C. S.: 1970, *Astrophys. J.* **162**, 293.
Kane, S. R.: 1971, *Astrophys. J.* **170**, 587.
Kane, S. R.: 1972, *Solar Phys.* **27**, 174.
Kane, S. R.: 1973a, *Bull. A.A.S.* **5**, 274.
Kane, S. R.: 1973b, in R. Ramaty and R. G. Stone (eds.), *High Energy Phenomena on the Sun*, NASA SP-342, p. 55.
Kane, S. R.: 1974, Private communication.
Kane, S. R. and Anderson, K. A.: 1970, *Astrophys. J.* **162**, 1003.
Kane, S. R. and Donnelly, R. F.: 1971, *Astrophys. J.* **164**, 151.
Kane, S. R. and Lin, R. P.: 1972, *Solar Phys.* **23**, 457.
Krieger, A., Paolini, F., Vaiana, G. S., and Webb, D.: 1972, *Solar Phys.* **22**, 150.
Lin, R. P.: 1970, *Solar Phys.* **12**, 209.
Lin, R. P.: 1973, in R. Ramaty and R. G. Stone (eds.), *High Energy Phenomena on the Sun*, NASA SP-342, p. 439.
Lin, R. P. and Hudson, H. S.: 1971, *Solar Phys.* **17**, 412.

Lin, R. P. and Anderson, K. A.: 1967, *Solar Phys.* **1**, 446.
Lin, R. P., Evans, L. G., and Fainberg, J.: 1973, *Astrophys. Letters* **14**, 191.
McDonald, F. B. and Van Hollenbecke, M.: 1973, in R. Ramaty and R. G. Stone (eds.), *High Energy Phenomena on the Sun*, NASA SP-342, p. 404.
McKenzie, D. L., Datlowe, D. W., and Peterson, L. E.: 1973, *Solar Phys.* **28**, 175.
Peterson, L. D., Datlowe, D. W., and McKenzie, D. L.: 1973, in R. Ramaty and R. G. Stone (eds.), *High Energy Phenomena on the Sun*, NASA SP-342, p. 132.
Smith, H. J. and Smith, E. P.: 1963, *Solar Flares*, MacMillan Co., New York.
Sweet, P. A.: 1969, *Ann. Rev. Astron. Astrophys.* **7**, 149.
Syrovatskii, S. I.: 1969, in C. de Jager and Z. Švestka (eds.), *Solar Flares and Space Research*, North-Holland Publ. Co., Amsterdam, p. 346.
Tanaka, K. and Nakagawa, Y.: 1973, *Solar Phys.* **33**, 187.
Takakura, T.: 1972, *Solar Phys.* **26**, 151.
Thomas, R. J. and Neupert, W. M.: 1971, *Bull. A.A.S.* **3**, 264.
Trubnikov, B. A.: 1965, *Rev. Plasma Phys.* **1**, 105.
Van Allen, J. A. and Krimigis, S. M.: 1965, *J. Geophys. Res.* **70**, 5737.
Vorpahl, J. and Zirin, H.: 1970, *Solar Phys.* **11**, 285.
Wild, J. P., Smerd, S. F., and Weiss, A. A.: 1963, *Ann. Rev. Astron. Astrophys.* **1**, 291.

NUCLEI OF HEAVY ELEMENTS FROM SOLAR FLARES

C. Y. FAN

University of Arizona Tucson, Ariz., U.S.A.

G. GLOECKLER

University of Maryland, College Park, Md., U.S.A.

and

D. HOVESTADT

Max Planck Institut für Extraterrestrische Physik, Garching B. Munchen, F.R.G.

Abstract. This paper presents a comprehensive review of the up to date knowledge on nuclear species of $Z > 2$ from solar flares. It covers the following five topics:
   (I) Solar flare particles of energies >15 MeV n^{-1} and solar composition;
   (II) solar flare particles of energies $\leqslant 15$ MeV n^{-1}, the enrichment of heavier elements;
   (III) theoretical interpretation of the enrichment;
   (IV) the charge states of solar particles; and
   (V) isotopic abundances of solar flare particles.

I. Solar Flare Particles of Energies >15 MeV n^{-1} and Solar Composition

The first evidence of the existence of nuclei heavier than helium among the energetic solar flare particles was detected in 1960, September 3 by Fichtel and Guss (1961), who used a sounding rocket to launch a recoverable nuclear emulsion stack above the Earth's atmosphere. Following this initial success, nine experiments were conducted by Fichtel and his associates over the period from 1960 to 1969 to study the composition of these heavy particles. In these measurements, the identification of the particles was accomplished by counting the number of δ-rays protruding from each primary track in the emulsion stack. The integral number of the δ-rays within a residual range R was then plotted as a function of R for each track. The result of such a plot has sufficient charge resolution to determine the composition of more abundant elements. Bertsch *et al.* (1972; 1973) summarized the results as follows:

(1) Within the experimental energy range from 12 to 100 MeV n^{-1}, the ratio of helium to medium nuclei is 58 ± 6 (Table I).

(2) The relative abundances, within experimental error, are almost invariant and are the same as the spectroscopic abundances of the Sun (shown in Table II).

In the energy ranges of these measurements, particles of nuclear charge $Z \leqslant 26$ are likely to be fully ionized, implying that all nuclear species from He to Fe have practically the same A/Z value. As a consequence, particles of the same velocity will not have their composition altered by the scatterings by the interplanetary magnetic fields, (including adiabatic deceleration and possible Fermi acceleration), as they propagate from the Sun to the Earth. Therefore, if (1) these elements were accelerated in the solar active regions *unbiased* and (2) particles of the same velocity were able to escape from

TABLE I
Ratio of helium nuclei to medium nuclei

Time of measurements	Energy interval		Reference
	(MeV n^{-1})	He/M	
1408 UT, 1960, Sept. 3	42.5–95	68±21	Fichtel and Guss (1961)
1840 UT, 1960 Nov. 12	42.5–95	63±14	Biswas et al. (1962)
1603 UT, 1960 Nov. 13	42.5–95	72±16	Biswas et al. (1962)
1951 UT, 1960 Nov. 16	42.5–95	61±13	Biswas et al. (1963)
0600 UT, 1960 Nov. 17	42.5–95	38±10	Biswas et al. (1963)
0339 UT, 1960 Nov. 18	42.5–95	53±14	Biswas et al. (1963)
1305–1918 UT, 1961 July 18	120–204	79±16	Biswas et al. (1966)
1443 UT, 1966 Sept. 2	12–35	48±8	Durgaprasad et al. (1968)
2233 UT, 1966 Sept. 2	14–35	53±14	Durgaprasad et al. (1968)
2319 UT, 1969 April 12	18–34	55±8	Bertsch et al. (1972)

Weighted average of above readings 58±5,

the regions with *equal probability*, then their relative abundances would represent the composition of the medium of the active regions. The results shown in Table II seem to indicate that the two key criteria are indeed met.

The composition of solar flare particles was also measured by a number of other investigators, using various types of detector systems on different vehicles. Mogro-Campero and Simpson (1972a, 1972b) used a dE/dx vs E solid state detector telescope on the OGO-5 satellite to study the particle composition of many flares observed from 1968, July 26 to 1971, February 1; Teegarden et al. (1973) measured the composition on the IMP-6 satellite of two flares occurring in 1971, April 6 and September 1, also with a dE/dx vs E telescope; Price et al. (1973), Sullivan et al. (1973), and Crawford et al. (1974) reported the composition of particles from He to Ni in several flares of widely varying intensities. All these results indicate that, in the energy range > 15 MeV n^{-1}, the spectra of different nuclear species are approximately parallel with each other. The relative compositions thus determined are listed in column 3, 4, and 5 respectively of Table II for comparison. It is seen that, except for the measurement of Mogro-Campero and Simpson which shows abnormally high abundances of elements above Si the abundances agree with each other within experimental error. In view of this general agreement, one could then use the relative abundances in solar cosmic rays as the solar abundances of the elements for which no spectroscopic estimates of the photospheric abundances are available, or modify the photospheric abundances for which the spectroscopic method is of doubt. Thus, Bertsch et al. (1972) proposed

$$\frac{Ne}{O} = 0.16 \pm 0.03, \qquad \frac{A}{O} \leqslant 0.017 \qquad \frac{He}{O} = 103 \pm 10, \tag{1}$$

and Crawford et al. (1974) suggested the following revision:

$$\frac{A}{Si} \leqslant 0.04 \qquad \frac{S}{Si} = 0.17. \tag{2}$$

TABLE II

Nuclear abundances relative to oxygen

Elements	Solar cosmic rays ($E>15$ MeV n^{-1})				Photosphere	Corona
	Bertsch et al. (1972)	Mogro-Campero and Simpson (1972)	Teegarden et al. (1973), 1971, Sept. 1 flare	Crawford et al. (1974)		
He	$(1.03\pm1.0)\times10^4$	–	$(4.2\pm0.25)\times10^3$	$\sim8.4\times10^3$	–	8.34×10^3
C	56 ± 6	57 ± 18	49 ± 3	50 ± 6	55.7	58.8
N	19^{+3}_{-7}	26 ± 19	11.6 ± 1.1	14.2 ± 2	17.2	19.2
O	100	100	100	100	100	100
F	<3	–	<0.6	–	–	–
Ne	16 ± 3	21 ± 8	12.7 ± 1.1	12.3 ± 2.0	–	10.3
Mg	5.6 ± 1.4	13 ± 6	18.2 ± 1.4	15.6 ± 2.5	5.16	7.05
Si	2.8 ± 1.0	36 ± 12	10.7 ± 1.1	9.1 ± 0.7	5.26	6.41
S	0.8 ± 0.6	–	2.5 ± 0.5	1.5 ± 0.3	2.47	1.54
A	<1.7	8 ∓ 5	<0.4	<0.4	–	0.96
Ca	<1.0	<2	<1.1	0.6 ± 0.2	0.32	0.26
Cr-Ni	1.1 ± 0.2	67 ± 16	2.8 ± 0.5	~8.5	4.1	5.8

II. Solar Flare Particles of Energies $\leqslant 15$ MeV n^{-1},
– The Enrichment of Heavies

The variability of He/M was first realized by Armstrong and Krimgisis (1971) and Krimigis *et al.* (1971). They made a statistical study of solar protons, α-particles and $Z \geqslant 3$ nuclei measured in 1967–1968 with a solid state detector aboard the Explorer-35 and the Mariner-5 spacecraft and found that the He/M ratio is 20 ± 10 instead of 60 as Bertsch *et al.* had indicated. The important difference between the earlier rocket experiments of Fichtel and his associates and the experiments of Krimigis *et al.* is the energy intervals of the measurements, the former covering the region of 12–204 MeV n^{-1} as opposed to the latter, which covers 0.5–2.5 MeV n^{-1}. It was then suggested that a mild energy dependence of the abundance ratio is one possible explanation of the observed difference (Armstrong *et al.*, 1972).

More conclusive evidence of the energy dependence of the abundances is shown by the results of Price *et al.* (1971), who studied the etched cosmic-ray particle tracks in a Surveyer-3 camera lens filter and an Apollo-12 spacecraft window brought back from the Moon. In both types of glass, tracks of heavily ionizing particles were revealed by chemical etching (Fleischer *et al.*, 1965; Price and Fleisher, 1971). Since these glasses are insensitive to particles of $Z < 16$ and the solar abundances of ions of $Z > 16$ is strongly peaked at Fe, the particle tracks must have been made by solar Fe nuclei. By comparing the energy spectrum of the Fe nuclei determined from the etched tracks with the solar He spectrum measured during the same time period by Lanzerotti (World Data Center A, *Rept UAG* **5**, 56; *UAG* **8**, 198; *UAG* **9**, 34) and by Hsieh and Simpson (1970), they found that Fe/He increases strongly with decreasing energies.

The enhancement of solar flare heavy nuclei in the low energy range is now established beyond any doubt. In Figure 1 we display the results of four solar flares, occurring 1971, January 25, and September 2 and 1972, April 18, and August 4, reported by Price *et al.* (1973). It is seen that above about 15 MeV n^{-1}, He, CNO, and Fe have similar spectra whereas towards lower energies their spectra diverge from each other with heaviers more abundant. These features were also observed by many others (Mogro-Campero and Simpson, 1972a; 1972b; Crawford *et al.*, 1972; Fleischer and Hart, 1973; Braddy *et al.*, 1973; Shirk and Price, 1972; Fleischer *et al.*, 1973; Biswas *et al.*, 1973; Nevatia *et al.*, 1973). The enhancement of solar heavy nuclei recorded in lunar rocks over the last half million years was reported by Bhandari *et al.* (1973).

The enrichment of heavy nuclei is usually expressed in terms of an enrichment factor of a species of nuclear charge Z at energy E, which is defined as

$$Q(Z, E) \equiv (Z/\mathrm{He})_{\mathrm{cr}} \ / \ (Z/\mathrm{He})_{\mathrm{s}}. \tag{3}$$

In this expression $(Z/\mathrm{He})_{\mathrm{cr}}$ and $(Z/\mathrm{He})_{\mathrm{s}}$ stand for the relative abundance of solar cosmic-ray particles and the solar photospheric composition respectively. Present experimental results show that Q increases with Z for a given E and decreased with

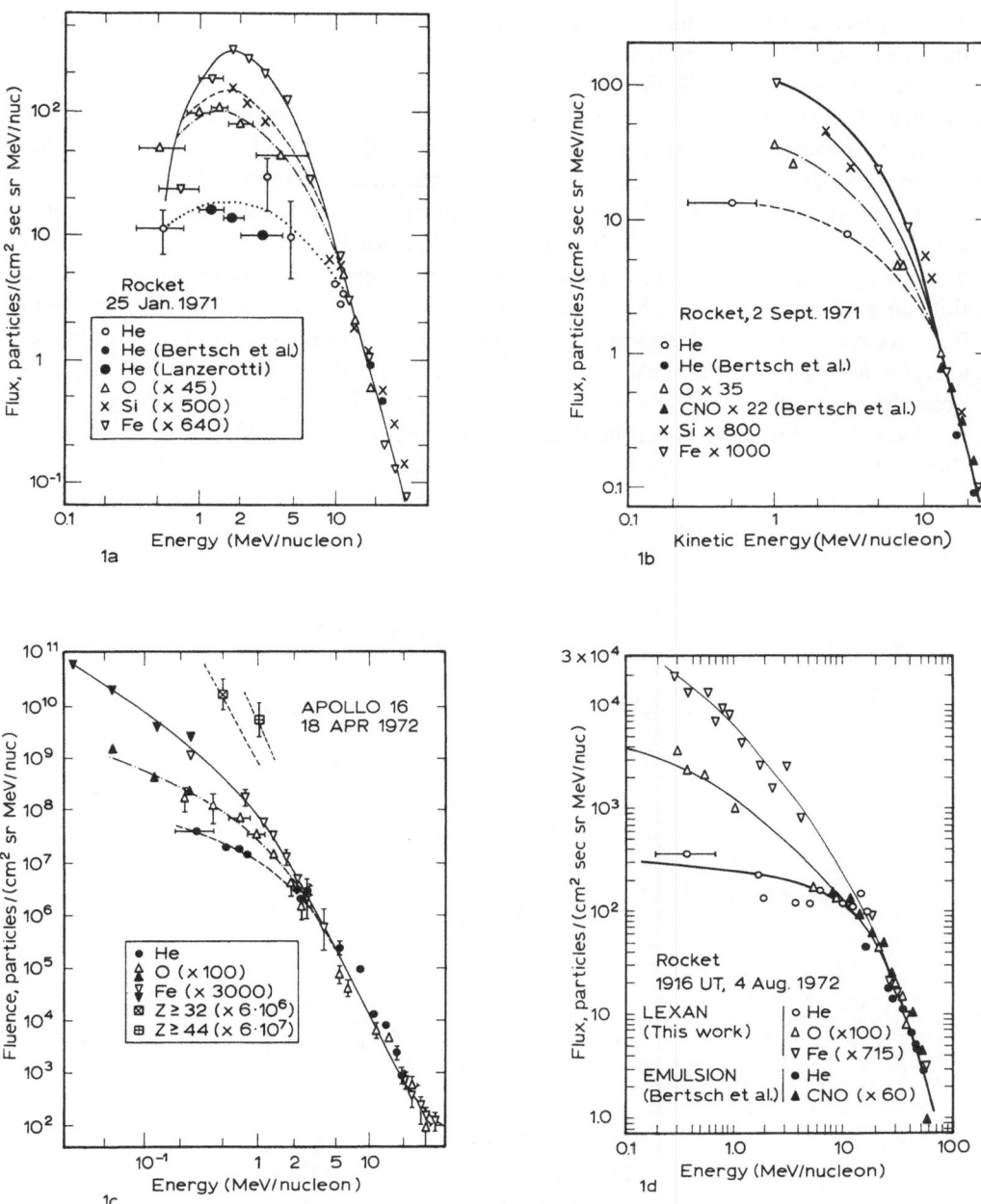

Fig. 1a–d. (a) Energy-dependent composition measured with Lexan stack during flare in 1971, January 25. – (b) Energy-dependent composition measured with Lexan stack during flare in 1971, September 2. – (c) Fe and heavier elements measured with SiO_2 glass, O and He measured with Lexan stack in 1972, April 18. – (d) Energy-dependent composition measured with Lexan stack during flare in 1972, August 4.

E for a given Z. One of the basic important questions is how does $Q(Z, E)$ vary from flare to flare and also within the same flare?

Mogro-Campero and Simpson (1972a; 1972b) first noted that Q varies from flare to flare. The variability is also apparent from the results of Price *et al.* as shown in Figure 1. A recent case which shows the variability is the 1972, October 29–November 4 event measured by Hovestadt *et al.* (1973) on the Imp-7 satellite. Their detector is a dE/dx vs. residual energy telescope, using a thin window isotubane filled proportional counter as dE/dx device and a surfcce barrier detector for the determination of residual energy. The lowest energy limit for each nuclear species is determined by the total thickness of material in front of the solid state detector, which amounts to about $0.328 \ \mu g \ cm^{-2}$ polyethylene equivalent, and the electronic thresholds. At entrance energies above about 400 keV n^{-1}, a clear separation of individual even Z nuclei is possible up to iron, a feature especially suitable for the study of low energy heavy particles. For this solar flare, the detector recorded a total of 24470 events. Figure 2 shows the differential energy spectra for C, O, and Fe. These spectra are practically

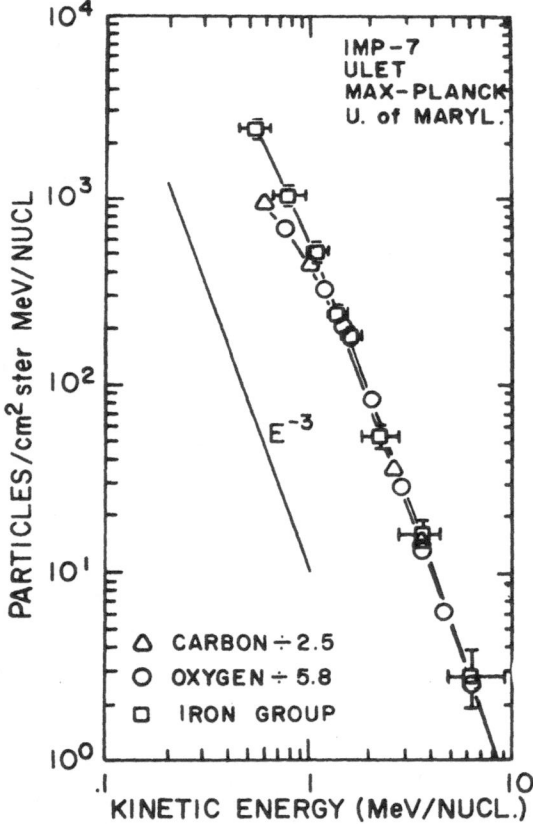

Fig. 2. Energy spectrum of C, O, and Fe- group measured on the IMP-7 satellite during the event in 1972, October 29–November 4. The E^{-3} spectrum is drawn for reference.

parallel down to ~ 1.5 MeV n^{-1} before they diverge. This feature is similar to that in Figure 1c but strikingly different from that in Figure 1a, b, and d, where the spectra diverge from each other at energy as high as 15 MeV n^{-1}. It is noted that the spectra in Figure 2 and those in Figure 1c were measured outside the magnetosphere and integrated over the entire event, whereas those in Figure 1a, b, and d were snapshots within the magnetosphere. Whether this is the cause for the difference in the spectral shape or the difference is merely a reflection of the variability of the enrichment factor Q, is not clear at this moment. For the same solar event, Armstrong and Krimigis (Armstrong and Krimigis, 1973; Krimigis and Armstrong, 1973) reported α/M in the energy range 1.6–3.2 MeV n^{-1} varied from 20 to 115 in six hour period and Fe/O in the energy range 3.3–7.6 MeV n^{-1} varied by a factor of 11.

III. Theoretical Interpretation of the Enrichment

Attempts have been made to explain qualitatively the enrichment of the heavier particles. Price *et al.* (1971) suggested that it may be due to preferential leakage of incompletely ionized heavy nuclei from the accelerating region. However, by using the Earth's magnetic field as a magnetic spectrometer, Sullivan and Price (1973) showed that the effective charge, Z^*, of 1.8 MeV n^{-1} Fe ions which they detected in 1971, January 25 (Figure 1a) is 22^{+4}_{-1} if the magnetic cut-off rigidity at Fort Churchill, where the rockets were launched, was at its normal daytime value, 150 MV. Therefore, if the explanation of Price *et al.* were correct, then the atomic electrons of the Fe ions must have been stripped off at a later stage of the escaping process (Mogro-Campero and Simpson, 1972a; Braddy *et al.*, 1973).

Cartwright and Mogro-Campero (1972) on the other hand suggested a three-stage model for the acceleration of solar particles to explain the enrichment. (1) Fully stripped ions are first accelerated to supra-thermal energies. (2) Subsequently a fraction of these ions are transported to a region for further acceleration. In this process, ions pick up electrons with their states determined by charge-exchange equilibrization process. (3) Finally ions are accelerated by hydromagnetic waves (Fermi-type acceleration) to the observed energies. The rigidity dependent efficiency of the final acceleration results in the observed enrichment of the heavier particles. This model has three functions which can be adjusted to explain the observed enrichment, but the requirement of three physically separated regions seems to be too artificial. Also this model fails to explain the energy dependence of the enrichment factor Q.

If one assumes that solar particles are accelerated in the lower chromosphere region where there are sufficient numbers of neutral hydrogen and helium atoms for the particles to establish their charge equilibrization by means of electron capture and loss in that medium, then the rate of the energy change of a particle can be expressed by the following general expression:

$$A \frac{d\varepsilon}{d\tau} = \alpha A \varepsilon - (Z^*)^2 f(\varepsilon). \tag{4}$$

In this equation, A, ε and Z^* stand respectively for the mass number, the energy per nucleon, and the effective charge of the particle; the first term on the right hand side is the rate of acceleration with α as the acceleration efficiency and the second term is the rate of energy loss to the ambient medium. By assuming functional forms for both α and Z^*, one can in principle integrate the equation to obtain the energy spectra for various nuclear species to be compared with experimental measurements. It may be entirely possible that the enrichment is merely the reflection of the ε and Z^* dependence of the energy loss in the medium (Cowsik *et al.*, 1973; Sullivan, 1974, private communication).

IV. The Charge States of Solar Particles

We have seen that to explain the increasing enhancement of heavy nuclei, all the proposed models require an establishment of charge state equilibrium for ions in the acceleration process (see also Ginzburg and Syrovatskii, 1964; Ramadurai, 1971). The measurement of Gloeckler *et al.* (1973) provides the first direct evidence for partially stripped carbon and oxygen at 100 keV n^{-1}.

The data have been obtained using the University of Maryland electrostatic deflection spectrometer on board the IMP-7 satellite which was launched in 1972, September 22. The method of particle identification is based on the fact that the amount of deflection, d, of an ion with an effective charge Z^* and kinetic energy T in a known electrostatic field is given by $d = gZ^*/T$. The constant g is determined by the geometry and the voltage of the deflection system. By measuring T with a solid state detector, the value of Z^* can thus be determined.

Solar particles were observed near Earth on October 17 and 18. Analysis of the data indicated the existence of Z^* equal to 5, 6, 7, and 8 particles. Assuming that very little Be, B, N, and Ne are present, one can then take all $Z^* = 5$ to be C^{5+} and all $Z^* = 8$ to be O^{8+} and estimate the abundance of C^{6+} and O^{7+}. The abundance ratios are

$$C^{5+}/C^{6+} = 1.8, \text{ and } O^{7+}/O^{8+} = 1.6 \tag{5}$$

at 0.1 MeV n^{-1}, and C/O = 0.8 which is in reasonable agreement with measurement at higher energies. The results are plotted in Figure 3.

The charge states of accelerated particles depend on their energies and the physical condition of the medium in which the particles are accelerated. The equilibrium charge states of carbon and oxygen at 0.1 MeV n^{-1} are found to be 2.5 and 2.9 respectively in a *neutral medium*. The measurements of Gloeckler *et al.* on the other hand, indicate much higher values: 5.4 for carbon and 7.4 for oxygen. To account for this discrepancy, they proposed two alternatives.

(1) Accelerated ions traverse the hot coronal region where the probability for electron pickup is negligible, and are stripped off most of their remaining electrons.

(2) After acceleration and escape from the Sun, the particles are adiabatically decelerated. In this case, the measured charge states are the equilibrium values at energies ~ 1 MeV n^{-1}. The limited amount of data they have does not allow them to decide which of the two alternatives corresponds to the actual condition.

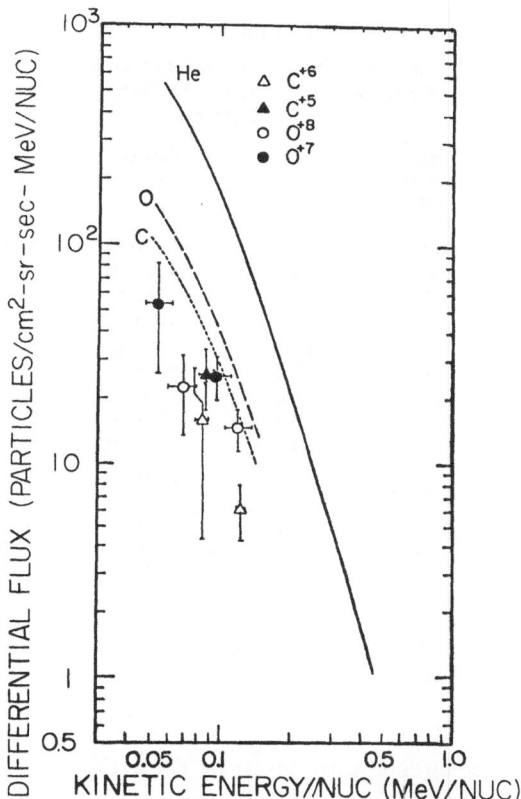

Fig. 3. The time averaged differential fluxes of C^{6+}, C^{5+}, O^{8+}, and O^{7+} ions measured on the Imp-7 satellite during 1972, October 17–18. Data have been corrected for energy defects in the solid state detectors and for background measured independently for each detector with deflection voltage commanded off. The C and O spectra are obtained by combining the measured fluxes of the partially stripped nuclei. The He spectrum is drawn for reference.

V. Isotopic Abundances of Solar Flare Particles

To conclude this paper, we would like to cite the isotopic composition of solar cosmic-ray particles measured during 1972, August 2–9 by Webber *et al.* (1973), to show that solar cosmic-ray measurement is a potentially powerful tool for solar physics research in other areas.

The measurement was done with a double dE/dx vs E telescope of the Goddard Space Flight Center – University of New Hampshire on the Pioneer-10 spacecraft. It consists of four solid state detectors, D_1, D_2, E, and F. F is used to reject penetrating particles while for stopping particles the energy losses in D_1 and D_2 and the residual energies in E are measured. The requirement of the D_1/D_2 pulse height ratios to follow the theoretical values within 5% provides a technique of background elimination, thereby improving impressively the mass resolution and making the isotopic composition measurement possible.

TABLE III

Isotopic abundance measurements
(energy approximately 8–15 MeV n^{-1}
for $Z \geqslant 2$ nuclei)

Isotope	Webber et al.		Natural Abundances[a]	Solar Abundances
	Events	%		
He^3	20	1.3	0.01	~0.03 %^{3}
He^4	1480	98.7	~100.0	99.9
C^{12}	173	98.8	98.9	98.9 [b]
C^{13}	2	1.2	1.1	1.1 [b]
N^{14}	42	97.6	99.6	
N^{15}	1	2.4	0.4	
O^{16}	312	96.6	99.7	99.65 [b]
O^{17}	8?	2.5?	0.04	0.05 [b]
O^{18}	2	0.6	0.26	0.30 [b]
Ne^{20}	37	86.2	90.9	92.3 [c]
Ne^{21}	2	4.6	0.25	0.3 [c]
Ne^{22}	4	9.2	8.82	7.4 [c]
Mg^{24}	30–33	81.1–89.2	78.7	
Mg^{25}	5	5.4–13.5	10.1	
Mg^{26}	2	5.4	11.2	

[a] From *Chart of the Nuclides* – 9th edn.
[b] Hall *et al*. (1972)
[c] Geiss (1972)

The results of the measurement are given in Table III. Assuming that He^3 nuclei are fragments of He^4 produced by the interaction with solar atmosphere, the amount of matter traversed by the He^4 nuclei would be about ~0.29 cm^{-2}. This path length would produce a negligible effect to alter the isotopic abundances of Mg, Ne, O, N, and C, that is, the following ratios listed in Table III may be tentatively regarded as the solar abundances:

$$\text{Ne}^{20}/\text{Ne}^{22} = 9 \pm 5, \quad \text{Mg}^{24}/\text{Mg}^{26} = 16 \pm 6, \quad \text{and}$$
$$\text{Mg}^{24}/\text{Mg}^{25} = 10 \pm 6, \tag{6}$$

for which there is no spectroscopic determination available.

References

Armstrong, T. P. and Krimigis, S. M.: 1971, *J. Geophys. Res.* **76**, 4230;
Armstrong, T. P. and Krimigis, S. M.: 1973, *13th International Cosmic Ray Conf.* (Conf. paper) **2**, 1504.
Armstrong, T. P., Krimigis, S. M., Reames, D. V., and Fichtel, C. E.: 1972, *J. Geophys. Res.* **77**, 3607.
Bertsch, D. L., Biswas, S., Fichtel, C. E., Pellerin, C. J., and Reames, D. V.: 1973, *Solar Phys.* **31**, 247.
Bertsch, D. L., Fichtel, C. E., and Reames, D. V.: 1972, *Astrophys. J.* **171**, 169.
Bhandari, N., Goswami, J. N., Lal, D., and Tamhane, A. S.: 1973, *13th International Cosmic Ray Conf.* (University of Denver, Colo., Conf. paper) **2**, 1464.
Biswas, S., Bertsch, D. L., Fichtel, C. E., Pellerin, C., and Reames, D. V.: 1973, *13th Int. Cosmic Ray Conf.* (Conf. paper) **2**, 1543.

Biswas, S., Fichtel, C. E., and Guss, D. E.: 1962, *Phys. Rev.* **128**, 2756;
Biswas, S., Fichtel, C. E., and Guss, D. E.: 1966, *J. Geophys. Res.* **71**, 4071.
Biswas, S., Fichtel, C. E., and Waddington, C. J.: 1963, *J. Geophys. Res.* **68**, 3109.
Braddy, D., Chan, J., and Price, P. B.: 1973, *Phys. Rev. Letters* **30**, 669.
Cartwright, B. G. and Mogro-Campero, A.: 1972, *Astrophys. J.* **177**, L43.
Cowsik, R., McClelland, J., and Sullivan, J. D.: 1973, *13th Int. Cosmic Ray Conf.* (Conf. paper) **2**, 1484.
Crawford, H. J., Price, P. B., Cartwright, B. G., and Sullivan, J. D.: 1974 (to be published).
Crawford, H. J., Price, P. B., and Sullivan, J. D.: 1972, *Astrophys. J.* **175**, L149.
Durgaprasad, N., Fichtel, C. E., Guss, D. E., and Reames, D. V.: 1968, *Astrophys. J.* **154**, 307.
Fichtel, S. and Guss, D. E.: 1961, *Phys. Rev. Letters* **6**, 495.
Fleischer, R. L. and Hart, H. R., Jr.: 1973, *Phys. Rev. Letters* **30**, 31.
Fleischer, R. L., Hart, H. R., Renshaw, A., and Wood, R. T.: 1973, *13th Int. Cosmic. Ray Conf.* (Conf. paper) **2**, 1486.
Fleischer, R. L., Price, P. B., and Walker, R. M.: 1965, *Rev. Nucl. Sci.* **15**, 1.
Geiss, J.: 1972, in C. P. Sonett, P. J. Coleman, and J. M. Wilcox (eds.), *Solar Wind*, NASA SP-308, p. 559.
Ginzburg, V. L. and Syrovatskii, S. I.: 1964, *The Origin of Cosmic Rays*, Pergamon Press, Oxford, p. 162.
Gloeckler, G., Fan, C. Y., and Hovestadt, D.: 1973, *13th Int. Cosmic Ray Conf.* (Conf. paper) **2**, 1492.
Hall, D. N. B., Noyes, R. W., and Ayres, T. R.: 1972, *Astrophys. J.* **171**, 615.
Hovestadt, D., Vollmer, O., Gloeckler, G., and Fan, C. Y.: 1973, *13th Int. Cosmic Ray Conf.* (Conf. paper) **2**, 1498.
Hsieh, K. C. and Simpson, J. A.: 1970, *Astrophys. J.* **162**, L191.
Krimigis, S. M. and Armstrong, S. M.: 1973, *13th Int. Cosmic Ray Conf.* (Conf. paper) **2**, 1510.
Krimigis, S. M., Roelof, E. C., Armstrong, T. P., and Van Allen, J. A.: 1971, *J. Geophys. Res.* **76**, 5921.
Mogro-Campero, A. and Simpson, J. A.: 1972a, *Astrophys. J.* **171**, L5;
Mogro-Campero, A. and Simpson, J. A.: 1972b, *Astrophys. J.* **177**, L37.
Nevatia, J., Durgaparasad, N., and Biswas, S.: 1973, *13th Int. Cosmic Ray Conf.* (Conf. paper) **2**, 1538.
Price, P. B., Chan, J. H., Crawford, H. J., and Sullivan, J. D.: 1973, *13th International Cosmic Ray Conf.* (Univ. of Denver, Colorado, Conf. paper) **2**, 1479.
Price, P. B. and Fleischer, R. L.: 1971, *Ann. Rev. Nucl. Sci.* **21**, 295.
Price, P. B., Hutcheon, I. D., Cowsik, R., and Barber, D. J.: 1971, *Phys. Rev. Letters* **26**, 916.
Ramadurai, S.: 1971, *12th Int. Cosmic Ray Conf.* (Univ. of Tasmania, Conf. paper) **1**, 385.
Shirk, E. K. and Price, P. B.: 1973, *13th Int. Cosmic Ray Conf.* (Conf. paper) **2**, 1474.
Sullivan, J. D.: 1974, Private communication.
Sullivan, J. D., Crawford, H. J., and Price, P. B.: 1973, *13th Int. Cosmic Ray Conf.* (Conf. paper) **2**, 1522.
Sullivan, J. D. and Price, P. B.: 1973, *13th Int. Cosmic Ray Conf.* (Conf. paper) **2**, 1470.
Teegarden, B. J., Von Rosenvinge, T. T., and McDonald, F. B.: 1973, *Astrophys. J.* **180**, 571.
Webber, W. R., Roelof, E. C., McDonald, F. B., Teegarden, B. J., and Trainor, J.: 1973, *13th Int. Cosmic Ray Conf.* (Conf. paper) **2**, 1516.

IMPLICATIONS OF NRL/ATM SOLAR FLARE OBSERVATIONS ON FLARE THEORIES

C. C. CHENG and D. S. SPICER

Naval Research Laboratory, Washington, D.C. U.S.A.

Summary. During the Skylab mission, many solar flares were observed with the NRL XUV spectroheliogram in the wavelength region from 150 to 650 Å. Because of its high spatial resolution ($\sim 2''$) the three-dimensional structures of the flare emission regions characterized by temperatures from 10^4 K to 20×10^6 K can be resolved. Thus the spatial relationship between the relatively cool plasma and the hot plasma components of a flare, and the associated magnetic field structure can be inferred. For example, the Fe XXIV plasma ($T \simeq 20 \times 10^6$ K) observed near the soft X-ray maximum during the $2B$ (M2) disk flare of 1973, July 15 is elongated perpendicular to the neutral line shown on the photospheric magnetogram, while lines of lower ionization temperatures such as He II–Fe XVI show the familiar double ribbon structures on either side of the neutral line (Widing and Cheng, 1974). The disk flare of 1973, September 5 shows the same spatial structures. The flare was observed at 18 31 UT at the X-ray maximum, and shows that the hot Fe XXIV cloud is located spatially between the ribbons of the He II, Fe XIV–Fe XVI emissions. As the flare cools at 18 37 UT, the Fe XXIV cloud disappears while the gap is filled with emissions from ions of lower ionization potentials, and exhibit loop structure. Many other flares also show similar spatial distributions in the XUV emissions. The linear dimensions of the Fe XXIV plasmas as measured from the photospheric plates range from 7000 km to 14 000 km, and the heights of associated loops range from 10 000 to 30 000 km. From the spatial distributions of the XUV emissions of the many flares, and the comparisons with the magnetograms, we concluded that the magnetic field configuration for the flares we observed are simple bipolar magnetic flux loops with the hot flare plasma located near the top and the cool plasma component on the footprints of the loop.

The implications of these observations are rather clear. Flare models requiring neutral points, neutral lines, or neutral sheets with the exception of Sturrock's inverted Y model, do not seem to agree with the observed magnetic field configuration. Sturrock's model, while rather appealing, appears in conflict with the observational fact that the loops associated with the flares are pre-existing and low-lying. It is hard to form a helmet type field at low altitude, and if this is to occur to a pre-existing loop, the opening of the loop would be the cause of the flare rather than its closing by tearing mode as in Sturrock's model. Elliott's model of high energy particle storage requires large volume and is incapable of explaining the spatial distributions and dimensions of the flare emissions we observed. The Alfvén-Carlqvist current interruption model, is phenomenologically in agreement with our data. However, as has been pointed out by Smith and Priest (1972) the ion-acoustic instability sets in before the space charge instability envisioned by Alfvén and Carlqvist. In addition, none of the above models can explain the periodic bursts in XUV, hard X-ray, and microwave often associated with flares. Recently, Cheng and Spicer (1974) proposed a Screw Pinch flare model which invokes the kink and/or the sausage mode instability associated with a simple flux loop. The model is based on the laboratory screw pinch effect and is capable of explaining the periodic bursts and spatial distributions of the hot and cool components of a flare emission. The screw pinch model seems in good agreement with the ATM data. A detailed account of the model will be published elsewhere.

References

Cheng, C. C. and Spicer, D. S.: 1974, 'A Screw Pinch Flare Model' (preprint, to be published).
Smith, D. F. and Priest, E. R.: 1972, *Astrophys. J.* **131**, 213.
Widing, K. G. and Cheng, C. C.: 1974, 'On the Fe xxiv Emission in Solar Flare of June 15, 1973', presented at COSPAR meeting (June 17–July 1), Sao Paulo, Brazil.

NONTHERMAL PROCESSES IN LARGE SOLAR FLARES

H. S. HUDSON and T. W. JONES

Physics Dept., University of California, San Diego, La Jolla, Calif., U.S.A.

and

R. P. LIN

Space Sciences Laboratory, University of California, Berkeley, Calif. U.S.A.

Summary. In many small solar flares the $\sim$10–100 keV electrons accelerated during the flash phase contain the bulk of the total flare energy output. In large flares, such as those in the period 1972, August 2–7, the flash phase electrons are present in substantially greater numbers. These electrons can explosively heat the chromosphere-lower corona and eject flare material. The ejected matter can produce a shock wave which will then accelerate nucleons and electrons to relativistic energies. We analyze energetic particle, radio, X-ray, gamma ray and interplanetary shock observations of the 1972 August flares to obtain quantitative estimates of the energy contained in each facet of these large flares. In general these observations are consistent with the above hypothesis. In particular:

(1) From the X-ray emission (van Beek *et al.*, 1973) the energy contained in >25 keV electrons is calculated to be $\gtrsim 2 \times 10^{32}$ erg for the 1972, August 4 event. Since the lower energy cutoff to the electron spectrum is known to be below 25 keV and possibly below 10 keV, the electrons contain enough energy to produce the following interplanetary shock wave, which has by far the bulk of the energy dissipated in the flare. Similar numbers are obtained for the large August 7 flare event.

(2) From the γ-ray emission (Chupp *et al.*, 1973) the energy in protons dumped at the same level of the atmosphere, assuming a thick target situation, is at least a factor of three smaller than the electrons. Moreover the γ-ray emission indicates that the bulk of the protons are accelerated at least several minutes after the electrons. Thus it is more likely that the electrons are responsible for the flare optical (Hα and white light) emissions which occur in the chromosphere.

(3) Approximately 5% of the electrons and $\gtrsim$99% of the protons escape into the interplanetary medium to be observed by spacecraft. This situation is consistent with the hypothesis of shock acceleration of the protons high in the solar corona.

(4) The four most intense X-ray bursts observed during the period July 31–August 11 are the only bursts followed by an interplanetary shock wave and a new injection of energetic protons into the interplanetary medium.

We conclude that the energy source for the interplanetary shock wave and ejected material is the collision loss of the flash phase electrons in the solar atmosphere, and that the energetic protons are likely to be accelerated by the shock wave. Since $\sim 10^{33}$ erg are likely to be contained in the electrons, that is, the bulk of the total flare energy and the equivalent of total conversion of 500 G in a volume of 10^{29} cm^3, it is clear

Sharad R. Kane (ed.), Solar Gamma-, X-, and EUV Radiation, 425–426. All Rights Reserved.
Copyright © 1975 by the IAU.

that the basic flare mechanism must be able to convert with high efficiency magnetic field energy into fast electrons during the flash phase.

Acknowledgements

This research was supported in part by NASA Grant NGL 05–003–017 at U.C., Berkeley, and NAS-5-11081 at U.C., San Diego.

References

Chupp, E. L., Forrest, D. J., and Suri, A. N.: 1973, in R. Ramaty and R. G. Stone (eds.), *High Energy Phenomena on the Sun*, NASA SP-342, p. 285.
Van Beek, H. F., Hoyng, P., and Stevens, G. A.: 1973, in H. E. Coffey (ed.), *Collected Data Reports on the August 1972 Solar-Terrestrial Events*, *UAG*-**28**, Part II, p. 319.

ON THE ACCELERATION PROCESSES IN SOLAR FLARES

Z. ŠVESTKA

*American Science and Engineering, Cambridge, Mass. 02139, U.S.A.**

Abstract. The paper summarizes what we know about the acceleration processes on the Sun. Four different instabilities are distinguished: (1) One with purely thermal consequences giving rise to the origin of any flare. (2) A non-thermal process at the flash phase of flares giving rise to ~100 keV electrons and protons, manifested through hard X-ray and impulsive microwave bursts (current interruption?). (3) An instability giving rise to streams of electrons, without accelerating protons, manifested by type III bursts (tearing-mode instability?). When (2) and (3) are linked, flare associated electron events in space are often recorded. (4) Finally an explosive instability produces a shock wave which manifests itself as a type II burst. This instability leads to a second-step acceleration of particles preaccelerated in (2) and gives origin to >10 MeV protons and relativistic electrons (probably stochastic acceleration).

I intend to summarize what we know about the acceleration processes in solar flares. When doing that we must distinguish very carefully what we know for sure, what we believe to know with some degree of certainty, and what we simply guess. Unfortunately there are only very few pieces of knowledge that we know for sure.

First of all we know that the majority of flares only emit kinds of radiation which can be completely interpreted as being thermal. Figure 1 shows an example of such a typical flare. There is no hard X-ray component present which would need an explanation through non-thermal processes. All the radiation we receive from such a flare can be interpreted as due to increased temperature in the solar corona to some 20 or 30×10^6 deg, admitting that there may be cores of higher temperature which we average when making a temperature estimate.

This leads to the first important conclusion, namely that the basic instability which gives rise to a flare must be a thermal instability**. If we make the reasonable assumption that energy losses are mainly due to losses through radiation, we can get some information on the height of the origin of this instability. As it is well known, the radiative losses increase with increasing temperature up to some 500 000 K but they become smaller if temperature further increases. This means that the instability which gives rise to a flare should take place higher than the layer with 500 000 K temperature, i.e., the flare must originate in the transition layer or higher in the corona. Only there, if some additional energy is put in and temperature increases, this leads to decreasing losses of energy and consequently further increase in temperature.

These conclusions are pretty sure but not yet definite. There is one observation which seems to contradict the thermal nature of the X-ray flare radiation and that is the observation of the polarization of X-rays. One would expect that X-rays are polarized if they are due to directed beams of electrons which happens during a non-thermal

---

* This lecture was prepared while working on a grant of Stifterverband für die Deutsche Wissenschaft at Fraunhofer Institute, Freiburg.
** By this term we mean any instability, the consequences of which are purely thermal.

Sharad R. Kane (ed.), Solar Gamma-, X-, and EUV Radiation, 427–439. All Rights Reserved.
Copyright © 1975 by the IAU.

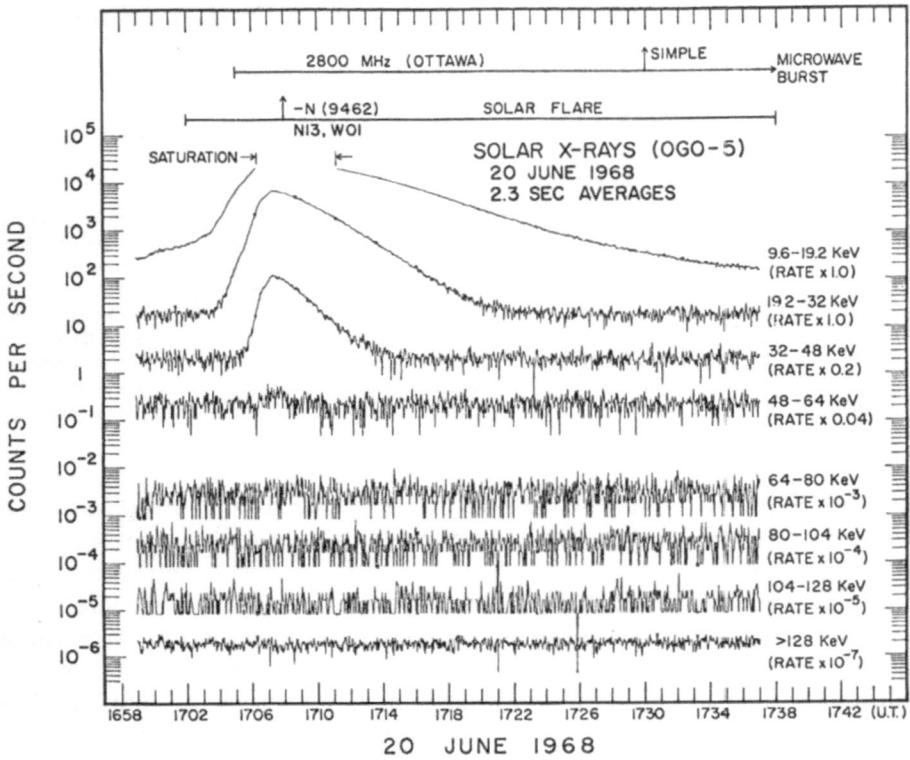

Fig. 1. A thermal X-ray flare burst, without any impulsive component. (After Kane, 1969; the increase at relatively high energies is due to pile-up and probably corresponds to X-rays below 10 keV.)

acceleration process. Such a process as we shall see is fairly short-lived and therefore also the polarization should last for only a very short time. The observations, however, show that polarization of X-rays in flares lasts for many, even tens of minutes (Tindo *et al.*, 1972; Thomas, 1975). This may be interpreted as due to continuous non-thermal acceleration of particles in the flare region so that the radiation only seems to be thermal but actually it is produced by a continuous nonthermal acceleration process. I mention this only to show that even this conclusion is open to doubt. Actually I believe that we are dealing here with thermal radiation. The observations of polarization are still fairly uncertain and even if one accepts these results of polarization measurements as real, there is still a possibility, e.g., to explain them as due to heat conduction in strong magnetic fields. In strong fields the heat conduction follows strictly the magnetic field lines and therefore we may encounter even in this case something like directed streams of particles.

Let us assume therefore that our first conclusion is correct. The flare originates through a thermal instability and in most flares we do not need any other process; only increase and subsequent decrease of temperature in the solar corona which affects

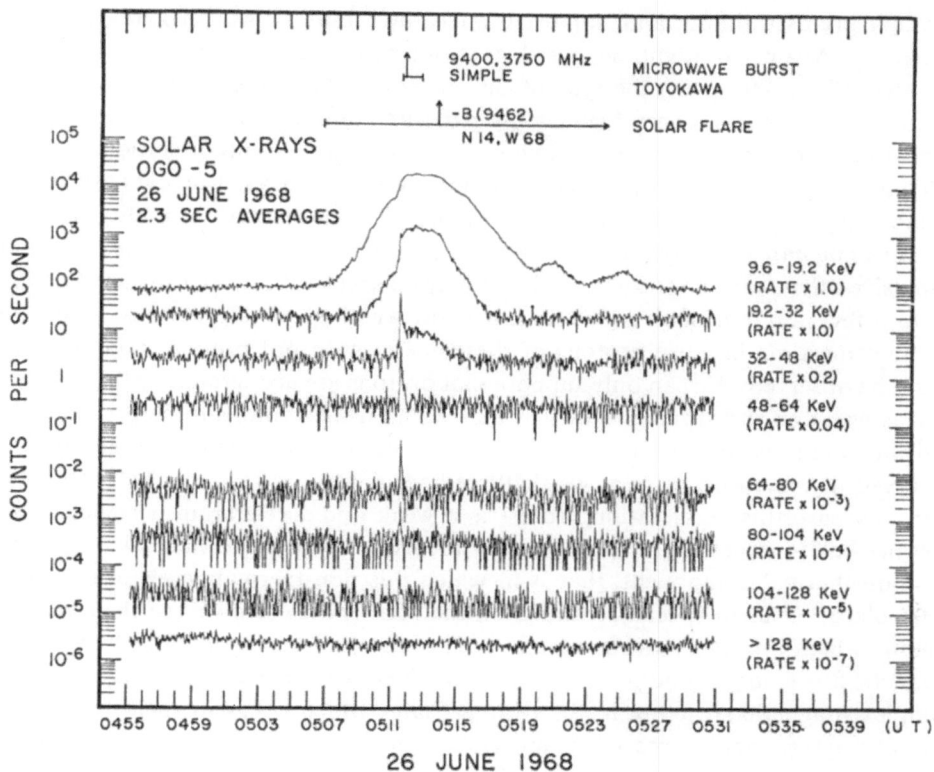

Fig. 2. An impulsive non-thermal hard X-ray burst occurring after the onset of the thermal flare. (After Kane, 1969.)

lower layers of the atmosphere through heat conduction. However, in many flares we observe something more: We observe, as Figure 2 shows, a short-lived emission in hard X-rays during the onset phase of the flare development. This hard X-ray phase can be very short in some flares, just tens of seconds, but in larger flares it may last for a few minutes. All this occurs before the flare maximum in the Hα line, that means during the flash phase of the flare, and it mostly starts after the onset of the soft X-ray emission, that means after the onset of the basic thermal flare phenomen. In some cases the hard X-ray origin and the soft X-ray origin coincide but there is no case when the hard-X-rays would start before the soft X-ray emission; therefore, we can conclude that the thermal instability we mentioned before is the basic instability which gives origin to all kinds of flares.

However, in the flares of the type demonstrated in Figure 2, we meet with an additional process which obviously is of non-thermal nature. It manifests itself through hard X-rays and impulsive microwave bursts. Both of these phenomena can be interpreted as due to streams of electrons with power-law spectrum in energy (Takakura and Kai, 1966; Kane and Anderson, 1970). These electrons produce hard X-rays through bremsstrahlung and impulsive microwave bursts through gyrosynchrotron radiation. Again, it might be possible to interpret these bursts as thermal ones (Kahler, 1975). However, the type of the spectrum, the impulsive behavior and the fact that in some cases very high energy particles coming from these sources are reported in space, are very much in favor of the non-thermal interpretation.

Thus we meet already with two different instabilities in flares. The first basic thermal instability giving rise to all kinds of flares, and the non-thermal instability which occurs in some flares shortly after the onset during their flash phase. The hard X-ray and impulsive microwave bursts give evidence only on acceleration of electrons to energies up to a few hundred of keV. We do not know whether protons are also accelerated at the same time because protons of these energies do not produce any effect that might be observed. We can only suppose that protons are accelerated by the same process as well and the argument for it is that we need the protons to be preaccelerated and thus prepared for another acceleration process to much higher energies which is observed in some rare cases as we shall see later on.

In any case this acceleration process is a weak one accelerating particles only to energies below or close to 1 MeV. It is of interest to see that the flares which contain this non-thermal component, that is in which this non-thermal acceleration process takes place, occur in special locations on the Sun. Most often we see them in active regions which are magnetically complex. That means in active regions where a new magnetic flux emerged and penetrated into an older existing one, thus producing a very complex magnetic structure in the region. Flares which occur in such magnetically complex active regions are much more likely to be associated with a non-thermal acceleration process than flares which occur in bipolar magnetic configurations. Another interesting feature is that flares which occur at about the same location produce very similar hard X-ray and impulsive microwave bursts; for example, radio bursts with maximum flux at about the same frequency (Švestka et al., 1974). This shows that the non-thermal instability is about the same in them and it can repeat several times during one or two days in flares which appear at about the same place in the active region.

Another manifestation of a non-thermal acceleration process on the Sun is the very well-known type III burst (cf. Lin, 1975). It is a fast drifting radio burst produced by a stream of electrons which propagate with a speed which is a significant fraction of the velocity of light. Many years ago De Jager and Kundu (1963) suggested that type III bursts and the microwave impulsive bursts or hard X-ray bursts are produced by the same instability which just accelerates electrons to high velocity. Those electrons which propagate downwards produce impulsive microwave bursts and hard X-ray bursts and those electrons which propagate upwards produce type III bursts. However, it appears that the situation is not that simple. First of all, when we try to correlate type III bursts and microwave bursts, we find that an exact time coincidence can be proved in less than 25% of cases (Švestka and Fritzová, 1974; Kane, 1972). In some other cases the coincidence may still exist since there is a series of type III bursts which lasts during the period when the microwave burst occurs. But there are at least 50% of cases when the type III burst occurs clearly at another time than the microwave burst or when only one of these kinds of bursts is observed. Apart from it the microwave bursts are closely and always associated with flares in contradistinction to the type III bursts of which only some 30% are associated with flares or sub-flares (Švestka, 1975).

The majority of type III bursts appear without any flare; on the other hand they are obviously restricted to particular active regions and to a particular phase of development of these active regions. There may be a large well-developed active region on the Sun which does not produce any type III bursts at all. Another region also inactive in type III bursts suddenly starts to produce type III bursts in large quantity, maybe several tens per day, and this production again suddenly stops after 20 h or one or two days. So far we are unable to find what makes particular active regions productive in type III bursts.

Figure 3a shows the dependence of the daily number of occurrences of microwave bursts on the daily flux value of radio emission of the Sun at 2800 MHz. This emission characterizes the importance of the solar activity and in particular it increases when magnetically complex active regions are present on the solar disk. We see that the occurrence of microwave bursts increases with the increasing radio flux as it is to be expected because microwave bursts predominantly occur in magnetically complex active regions. On the other hand, Figure 3b shows that no such dependence does exist for type III bursts and indeed type III bursts do not prefer magnetically complex active regions.

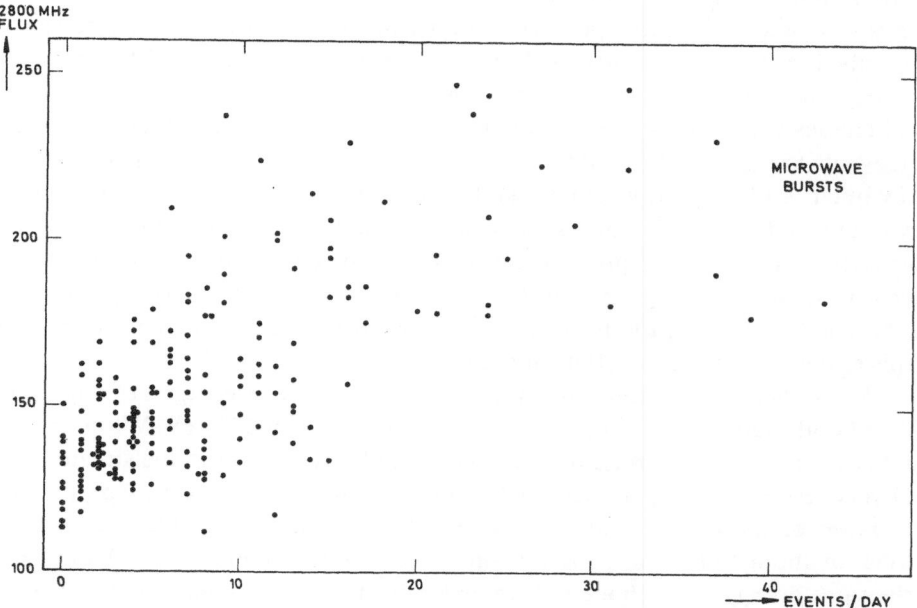

Fig. 3a.   Dependence of the daily number of occurrences of microwave bursts on the daily flux value
of radio emission of the Sun at 2800 MHz.

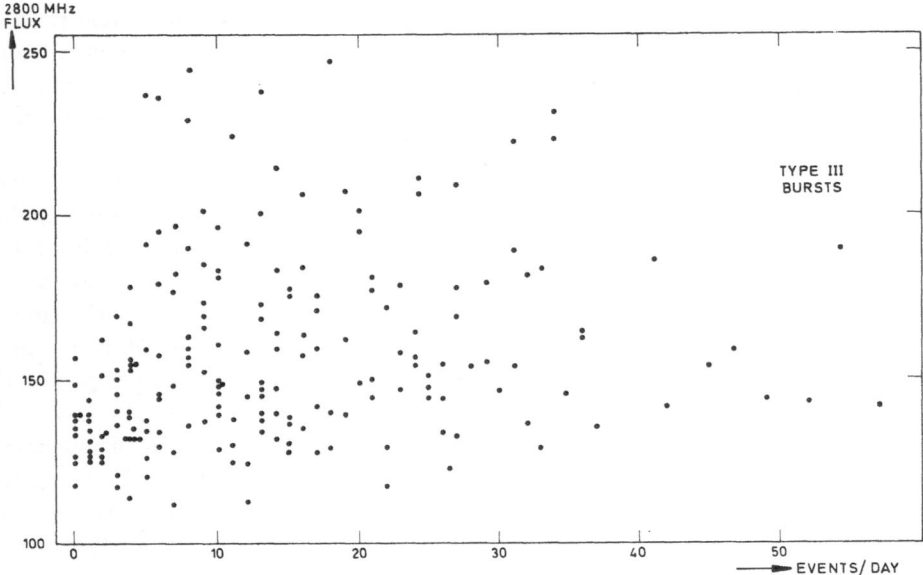

Fig. 3b.   The same dependence as in Figure 3a, plotted for the daily number of occurrences of type III bursts.

All these differences in the type III and microwave burst occurrence indicate that we meet actually with two different acceleration processes on the Sun. One of them, always associated with flares, gives rise to microwave and hard X-ray bursts; another one, only sometimes associated with flares, gives rise to the type III bursts. Only in relatively rare cases these two acceleration processes are linked. In that case we have simultaneously a hard X-ray burst, impulsive microwave burst and type III burst and such events are always flare associated. The type III bursts which coincide with hard X-ray bursts and microwave bursts and occur in the onset phase of a flare are usually very strong and they are accompanied with the type V radio continuum in most cases. The electron stream which produces such strong type III bursts can be followed quite often on radio waves very far from the Sun, deep into space up to the distance of the Earth. Satellites around the Earth then discover these electrons and electron streams in space, the so-called pure electron events.

We know that the electrons present in these streams are of energies of a few tens of keV and that there are no protons with energies in excess of 300 keV. We do not know whether protons with energies of a few tens of keV are present in the stream because until very recently these particles have never been measured in space. Apart from it a correlation of the electron and proton occurrence is quite difficult. Electrons with energies of about 40 keV need just 20 min to get from the Sun to the Earth but protons of the same energy need about 14 h to make this trip. Nevertheless, the fact that the 300 keV protons are essentially always missing in these streams is a strong argument pointing to the conclusion that protons are not accelerated in these cases. That means that the acceleration process which gives rise to the type III burst and electrons in space doesn't accelerate protons to energies comparable to the energies of electrons.

Hence we meet already with three different instabilities in flares. First with a thermal instability (instability No. 1) which gives origin to all flares. Second, with a non-thermal instability (instability No. 2) which accelerates electrons (and most probably also protons) to energies below or of the order of 1 MeV and which produces the hard X-rays and impulsive microwave bursts. This instability occurs in some flares. Finally, there is an instability No. 3 which is not necessarily associated with flares but which is related to particular configurations in particular active regions. It produces streams of electrons most probably without protons (e.g. particles are not accelerated to the same energy but to the same rigidity) and these streams of electrons manifest themselves as type III bursts on the radio waves. Although these bursts usually occur in active regions without any accompanying flare phenomenon, they tend to be associated with flares if flares occur in these particular active regions. In that case quite often the two acceleration processes giving origin to microwave bursts and type III bursts are linked and coincide exactly in time.

Still none of these instabilities can give rise to particles of extremely high energies which we observe from time to time in space. Cases when relativistic electrons and protons in excess of 10 MeV are observed in space are rather frequent and in infrequent but very important cases we observe protons up to energies of 1000 MeV. Naturally, they cannot be of thermal origin. The instability No. 2 gives rise to electrons of energies only up to 1 MeV and even if we assume that protons are also accelerated by this process, they hardly would be accelerated to higher energies. Finally, the instability No. 3 does not produce any protons with energies in excess of 300 keV. Therefore, we obviously need still another instability (No. 4) in order to produce energetic particles of these extremely high energies. What do we know about this strong acceleration process?

When protons of high energies and relativistic electrons are recorded in space we observe in the majority of cases a type IV burst on the radio waves, that is a continuum emission within the whole band of frequencies produced through gyrosynchrotron radiation of mildly relativistic electrons. Thus the type IV burst gives an evidence that mildly relativistic electrons have been accelerated in the region.

In essentially all such events we also observe the instability No. 2, that means we observe hard X-ray bursts and impulsive microwave bursts which usually form the first phase of the subsequent type IV burst. Therefore, several years ago De Jager (1969) suggested that relativistic electrons and high energy protons are accelerated in a second acceleration step following our instability No. 2. This supposition was a natural consequence of theoretical difficulties which one had with acceleration of protons to very high energies starting with the energies which protons have in a thermal flare. If the protons are first preaccelerated in the first acceleration step, it is much easier then to accelerate them further in a second acceleration step to higher energies.

This conception of two steps of acceleration is most probably correct. Since the metric type IV burst is sometimes delayed for minutes or tens or minutes, De Jager supposed that this second acceleration step would follow some ten minutes or maybe even later after the first acceleration process which gives rise to particles with energies of about 1 MeV. However, there are two observations which indicate very strongly that the second acceleration step must follow immediately the first one. One is the observation of white light flare emission which as we believe is produced by high energy particles impinging on the photosphere and which occurs essentially at the same time as the hard X-ray and microwave bursts (Švestka, 1970; McIntosh and Donnelly, 1972; Rust and Hegwer, 1975). We need protons with energies in excess of some 20 MeV to produce this emission so they must be produced in the second acceleration step. A similar result follows from gamma-ray observations. In the 1972, August 4 flare Chupp *et al.* (1973) and in the 1972, August 7 flare Vedrenne (1975) recorded the gamma-ray line produced by neutrons on the Sun and these neutrons need nuclei with energies in excess of 30 MeV to be produced. Therefore, the occurrence of this line again needs a second step of acceleration to be accomplished and similarly to the white light emission the gamma-rays occurred essentially simultaneously* with the hard X-ray burst. This leads to a definite conclusion that if the high energy protons and relativistic electrons are accelerated in the second step of acceleration, this second step must follow within one minute or less the first acceleration process.

We have seen that both the rather weak acceleration processes discussed before manifested themselves quite clearly on the radio waves as microwave bursts or type III bursts respectively. Therefore, one can suppose that the much stronger second-step acceleration process should manifest itself in some pretty obvious way as well. One of its manifestations of course is the type IV burst, but this is only a consequence of the process. Many high-energy electrons accelerated in the second step are trapped in the solar corona and they become visible through gyrosynchrotron radiation as the type IV burst. This emission can be delayed so that the type IV burst does not show us the actual time of occurrence of the acceleration. There is however one more type of radio bursts which may be directly associated with the second step process and this is the type II burst.

The idea that type II bursts which are produced by shock waves propagating through the corona may be closely associated with the acceleration of high energy protons is not new. It was expressed several times ago and demonstrated for a few individual events (e.g., Frost and Dennis, 1971). But statistics so far failed to prove any clear association between type II bursts and strong particle events on the Sun. We have recently repeated the statistical discussion of this kind having the great advantage of the complete list of all particle events of different sizes published in the newly prepared *Catalog of Solar Particle Events* compiled by the Working Group 2 of the former Inter-Union Commission on Solar-Terrestrial Physics (Švestka and Simon, 1975). This very complete set of solar particle data, combined with greatly improved records of the radio emission by dynamic spectrographs (that means, improved as to the

---

* A delay of about 2 min is to be expected due to the line formation process on the Sun.

sensitivity and as to the coverage in time) has made it possible to study the correlation between type II bursts and particle events on the Sun in full detail. This was not possible earlier when both particle data and type II burst observations were incomplete. This new statistical discussion (Švestka and Fritzová, 1974) has convinced us that indeed type II bursts are closely associated with the second acceleration step in flares.

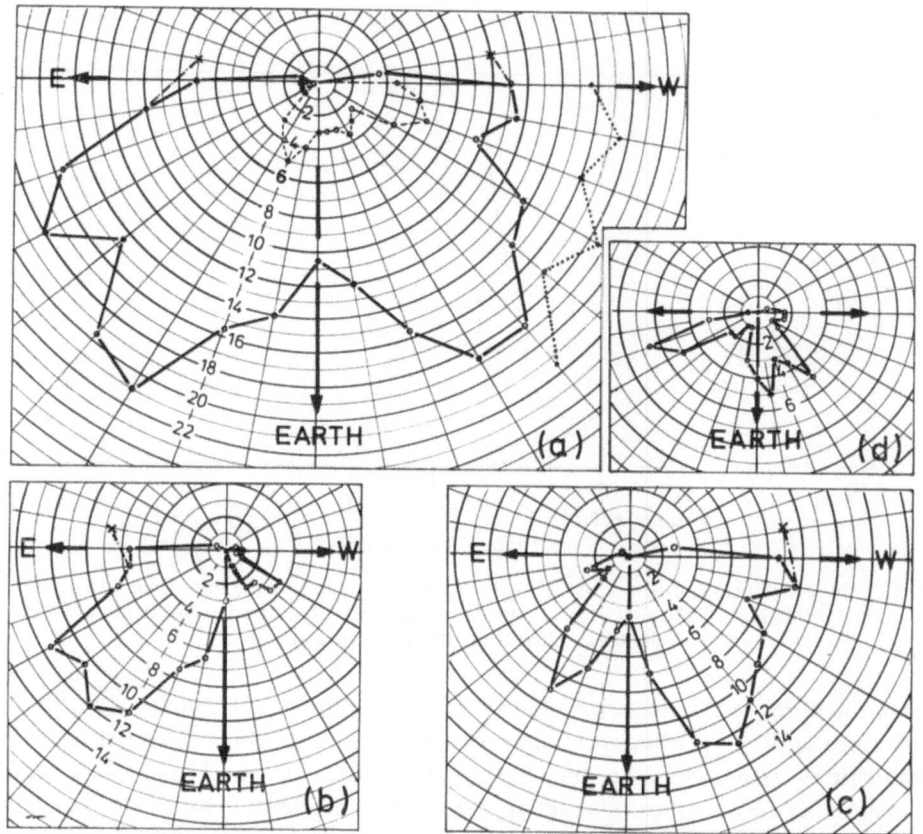

Fig. 4. The distribution in solar longitude of all type II bursts (heavy line in (a)), of those (c) that produced energetic particles and (b) that did not produce energetic particles near the Earth. (After Švestka and Fritzová-Švestková, 1974).

Figure 4a shows the distribution in solar longitude of all type II bursts during 30 months in the years 1966 to 1968. The position of each burst is identified with the position of the flare associated with it. Figure 4c shows the distribution in solar longitude of those type II bursts which were associated with particle events and Figure 4b shows the distribution of those type II bursts which were not associated with any particle event. The graphs show quite clearly that the proton producing flares prevailed on the western hemisphere while at least 84% of type II bursts without particle events were on the eastern hemisphere. This is exactly what we expect if essentially all type II

bursts are associated with acceleration of protons on the Sun and if particle propagation goes along the Archimedes spirals.

Figure 5 shows the distribution of time intervals between the maximum of the impulsive microwave burst and the type II burst onset, for metric type II bursts. The
metric type II bursts start on an average 2.6 min after the maximum of the microwave
burst. There are 20% of cases when the type II bursts originated prior to the microwave burst maximum, but there is *no case at all* when type II bursts would have started

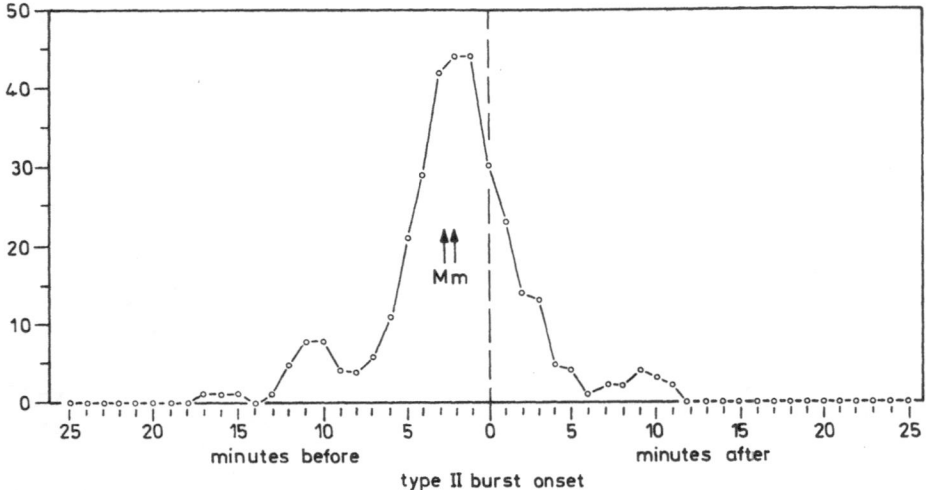

Fig. 5.   The distribution of time intervals between the maximum of the impulsive microwave burst
(i.e. maximum or end of the first step acceleration phase) and the metric type II burst onset. M is
the mean, and m the median time interval.

before the onset of the microwave burst. Therefore, the type II burst obviously originates close to the time of the primary acceleration process which gives rise to the non-
relativistic electrons which produce the impulsive microwave and hard X-ray bursts. If
we assume as the statistical average that the type II originates close to the time of the
microwave burst maximum, the 2.6 min difference means that the shock wave producing it propagates with the speed of about $1500\,\mathrm{km\,s^{-1}}$ which is an acceptable value.

We have found that the association of type II bursts with particles increases with the
importance of the particle event. For polar cap absorption events at least 73% and
potentially as many as 100% of the events were preceded by a type II burst. For particle events in which the proton flux in excess of 10 MeV exceeded 1 proton $\mathrm{cm^{-2}\,s^{-1}}$
$\mathrm{s^{-1}}$ as many as 91% of proton events still might have been associated with type II
bursts. On the other hand the association is sure for only 15% of cases if the flux of
protons with energies above 10 MeV is lower than 0.01 protons $\mathrm{cm^{-2}\,s^{-1}\,sr^{-1}}$, and
no association has been found for the pure electron events. This preference of strong
proton events indicates that the type II bursts indeed characterize the second acceleration step in which the energy of the primarily accelerated particles is further increased

It appears to be a reasonable assumption that this strong second step acceleration process is due to stochastic acceleration in turbulent plasma. Either the shock wave itself as it crosses the magnetic field lines or another effect produced by the same explosive instability gives rise to a powerful plasma turbulence in which stochastic acceleration of the particles preaccelerated in the first acceleration step can be accomplished. The short time differences between the maximum of the impulsive microwave burst and the onset of the metric type II burst, and in particular the fact that in some 20% of the cases the type II is observed even prior to the microwave burst maximum, indicates that the two steps of acceleration can be accomplished essentially simultaneously. While some particles are still being preaccelerated, other particles preaccelerated earlier already enter into the second acceleration step.

The first acceleration step is clearly characterized by the hard X-ray burst and impulsive microwave burst. These effects are essentially always observed (with the exception of limb events) when high energy particles are reported in space. On the other hand we know several strong and important particle events, like the cosmic-ray flare of 1966, July 7, or the flare of 1966, September 2, with 13 dB PCA effect, when no type III burst was observed at all. This fact, as well as the very frequent occurrence of type III bursts without any flare and the strong indication that protons are not accelerated in this kind of instability, gives an evidence that the type III burst type of acceleration is not identical with the primary acceleration step.

This leads us back to the problem whether protons are also accelerated in our acceleration processes 2 and 3. We have seen that there is a rather strong evidence that protons are not accelerated through the instability giving rise to the type III bursts. On the other hand nothing contradicts the supposition that protons are accelerated through the process which gives origin to the electrons which produce the hard X-rays and impulsive microwave bursts. We have no direct evidence for it. However, as soon as we suppose that this acceleration process, our No. 2, is the first step in the acceleration which gives rise to high energy protons, we must necessarily assume that protons are preaccelerated through this process as well.

While we can make a reasonable assumption that stochastic acceleration is associated with the shock wave, it is difficult to say with our present knowledge which type of instability accelerates the particles in the first acceleration step and in the type III bursts. We can only guess here. It seems that the most acceptable mechanisms for these two acceleration processes are current interruption in the case of the first acceleration step and tearing mode instability in the case of the type III bursts. There are no strong arguments for it. But, current interruption seems to be an attractive mode of particle acceleration and we know that it is difficult, maybe impossible, to accelerate particles through this mechanism to high energies. However, it still seems to be within reasonable possibilities to accelerate particles through current interruption to energies of the order of 100 keV as we observe. The tearing mode instability originally proposed by Sturrock (1968) for explanation of the whole flare phenomenon seems to be very appropriate for a type III burst because its duration is very short and in Sturrock's model it occurs high above the active region where type III bursts actually are observed.

Thus, summarizing, we have the following picture of the flare phenomenon: Somewhere in the transition layer or in the low corona an input of energy occurs which gives rise to a thermal instability. This thermal instability produces what we call a flare. In the majority of cases nothing more happens. Temperature increases, reaches a maximum and then the flare volume cools. Through heat conduction we observe the typical chromospheric flare. In some cases in addition to this thermal instability a current interruption occurs which accelerates electrons and protons to energies of a few hundred keV. In these cases we observe hard X-rays and impulsive microwave bursts. Apart from this kind of flare-associated instability there also exists another kind of instability, a tearing mode instability in coronal current sheets which gives rise, under conditions which we are unable to specify at the present time, to streams of electrons propagating up into the corona and giving rise to the type III bursts. In some particular cases these two instabilities are linked so that current interruption lower in the atmosphere and tearing mode instability in the corona occur at the same time. Then we have streams of electrons descending in the solar atmosphere and producing hard X-ray bursts and impulsive microwave bursts and another stream of electrons going up into the corona and producing type III bursts. This electron stream can propagate very deep into space and produce pure electron events observed aboard satellites near the Earth. Finally, in some relatively few flares still another kind of instability accompanies the current interruption and this instability gives rise to a shock wave and strong turbulent motions in the atmosphere. In that case the particles, both electrons and protons, preaccelerated through the current interruption, may be accelerated to substantially higher energies up to tens or hundreds of MeV and we observe then the so-called proton events in space.

One could try, of course, to simplify this picture. The instabilities No. 2 and 3 might be considered generally for one type of instability which behaves differently in different magnetic configurations, manifesting itself once through a microwave burst, once through a type III burst, and on rare occasion through a simultaneous occurrence of both these phenomena. However, the fact that we need protons to be accelerated at the time of the microwave bursts, while no protons are recorded in the type III streams, makes this simplification unlikely.

Similarly, one might suppose that the instability No. 4 which gives rise to the shock wave, is simply a strong fully developed instability No. 2. One cannot exclude this possibility; however, the energy involved in the shock is pretty large, and thus it seems that something qualitatively different happens in the flares that produce the shock waves.

I am fully aware of the highly speculative nature of some parts of my talk. The interpretations are difficult with our limited amount of knowledge. However, I have considered it useful to try to summarize the observations which give us some information on the acceleration processes in flares and to emphasize that we possibly need several different instabilities and several acceleration mechanisms in order to explain all the phenomena observed.

References

Chupp, E. L., Forrest, D. J., Higbie, P. R., Suri, A. N., Tsai, C., and Dunphy, P. P.: 1973, *Nature* **241**, 333.
De Jager, C.: 1969, in C. de Jager and Z. Švestka (eds.), *Solar Flares and Space Research*, p. 1.
De Jager, C. and Kundu, M. R.: 1963, *Space Res.* **3**, 836.
Frost, K. J. and Dennis, B. R.: 1971, *Astrophys. J.* **165**, 655.
Kahler, S.: 1975, This volume, p. 211.
Kane, S. R.: 1969, *Astrophys. J. Letters* **157**, L139.
Kane, S. R.: 1972, *Solar Phys.* **27**, 174.
Kane, S. R. and Anderson, K. A.: 1970, *Astrophys. J.* **162**, 1003.
Lin, R. P.: 1975, This volume, p. 385.
McIntosh, P. S. and Donnelly, R. F.: 1972, *Solar Phys.* **23**, 444.
Rust, D. M. and Hegwer, R.: 1975, *Solar Physics* **40**, 141.
Sturrock, P. A.: 1968, in K. O. Kiepenheuer (ed.), 'Structure and Development of Solar Active Regions', *IAU Symp.* **35**, 471.
Švestka, Z.: 1970, *Solar Phys.* **13**, 471.
Švestka, Z.: 1974, *Solar Flares*, D. Reidel Publ. Co., Dordrecht, Holland (in press).
Švestka, Z. and Fritzová-Švestková, L.: 1974, *Solar Phys.* **36**, 417.
Švestka, Z. and Simon, P. (eds.): 1975, *Catalog of Solar Particle Events, 1955–1969*, D. Reidel Publ. Co., Dordrecht, Holland.
Švestka, Z., Castelli, J. P., Dizer, M., Dodson, H. W., McIntosh, P. S., and Urbarz, H.: 1974, Flares of 8 June, 1972, at 13h19m and 13h52m UT, CINOF Report, presented at the STP Symposium, Sao Paulo, June.
Takakura, T. and Kai, K.: 1966, *Publ. Astron. Soc. Japan* **18**, 57.
Thomas, R.: 1975, This volume, p. 25.
Tindo, I. P., Ivanov, S. L., Mandelshtam, S. L., and Skurygin, A. I.: 1972, *Solar Phys.* **24**, 429.
Vedrenne, G.: 1975, This volume, p. 315.